SOCIETÀ ITALIANA DI FISICA

RENDICONTI
DELLA
SCUOLA INTERNAZIONALE DI FISICA
«ENRICO FERMI»

LXXXVIII Corso

a cura di M. GHIL
Direttore del Corso
e di R. BENZI e G. PARISI

VARENNA SUL LAGO DI COMO
VILLA MONASTERO
14 - 24 Giugno 1983

Turbolenza e predicibilità nella fluidodinamica geofisica e la dinamica del clima

1985

SOCIETÀ ITALIANA DI FISICA
BOLOGNA - ITALY

ITALIAN PHYSICAL SOCIETY

PROCEEDINGS
OF THE
INTERNATIONAL SCHOOL OF PHYSICS
«ENRICO FERMI»

Course LXXXVIII
edited by M. Ghil
Director of the Course
and by R. Benzi and G. Parisi
VARENNA ON LAKE COMO
VILLA MONASTERO
14 - 24 June 1983

Turbulence and Predictability in Geophysical Fluid Dynamics and Climate Dynamics

1985

NORTH-HOLLAND
AMSTERDAM · OXFORD · NEW YORK · TOKYO

Copyright © 1985, by Società Italiana di Fisica

All rights reserved. No part of this publication may be reproduced, stored in a retrieval system, or transmitted, in any form or by any means, electronic, mechanical, photocopying, recording or otherwise, without the prior permission of the copyright owner.

PUBLISHED BY:

North-Holland Physics Publishing
a division of
Elsevier Science Publishers B.V.
P.O. Box 103
1000 AC Amsterdam
The Netherlands

SOLE DISTRIBUTORS FOR THE USA AND CANADA:

Elsevier Science Publishing Company Inc.
52, Vanderbilt Avenue
New York, N.Y. 10017
U.S.A.

Technical Editor
P. PAPALI

Library of Congress Cataloging in Publication Data

International School of Physics "Enrico Fermi"
   (1983 : Varenna, Italy)
   Turbulence and predictability in geophysical fluid dynamics and climate dynamics.

   (Proceedings of the International School of Physics "Enrico Fermi" ; course 88)
   Title on added t.p. : Turbolenza e predicibilità nella fluidodinamica geofisica e la dinamica del clima.
   At head of title: Italian Physical Society.
   Bibliography: p.
   Includes index.
   1. Dynamic climatology--Congresses. 2. Fluid dynamics --Congresses. 3. Atmospheric turbulence--Congresses.
I. Ghil, Michael. II. Benzi, R. III. Parisi, G.
IV. Società italiana di fisica. V. Title. VI. Title: Turbolenza e predicibilità nella fluidodinamica geofisica e la dinàmica del clima. VII. Series.
QC981.7.D94I58 1983   551.5   84-27168
ISBN 0-444-86936-0 (U.S.)

Proprietà Letteraria Riservata
Printed in Italy

INDICE

M. Ghil, R. Benzi and G. Parisi – Introduction: turbulence, geophysical flows, predictability and climate dynamics . . . . pag. XIII

Part I – The onset of turbulence . . . . » 1

D. Ruelle – The onset of turbulence: a mathematical introduction.

   0. Introduction . . . . » 3
   1. Strange attractors and sensitive dependence on initial condition . . . . » 4
      1˙1. Differentiable dynamical systems . . . . » 4
      1˙2. Types of DDSs . . . . » 4
      1˙3. Sensitive dependence on initial condition . . . . . » 5
      1˙4. Attractors . . . . » 5
   2. Routes to chaos . . . . » 6
      2˙1. Some simple technical facts . . . . » 6
      2˙2. Generic bifurcations of an attracting fixed point . » 7
      2˙3. Frequency locking . . . . » 10
      2˙4. When does sensitive dependence on initial condition occur? . . . . » 10
      2˙5. Scenarios leading to chaos . . . . » 11
   3. Ergodic theory of differentiable dynamical systems . . . » 14
      3˙1. Theorem (multiplicative ergodic theorem) . . . . . » 14
      3˙2. Choice of an invariant measure . . . . » 15
      3˙3. Entropy (Kolmogorov-Sinai invariant) . . . . » 15
      3˙4. Dimension of attractors . . . . » 16
   4. Some recommended reading . . . . » 16

A. LIBCHABER – The onset of weak turbulence: an experimental introduction.

1. Basics of thermal convection . . . . pag. 18
2. Bifurcations from a limit cycle . . . . » 20
3. Frequency lock-in . . . . » 20
4. Saddle node bifurcation and intermittency . . . . » 22
5. Pitchfork bifurcations and the period-doubling cascade . » 24
6. Hopf bifurcation and the Ruelle-Takens scenario . . . . » 26

G. JONA-LASINIO – Small stochastic perturbations of dynamical systems.

1. Preliminary remarks . . . . » 29
2. Motion along the flow . . . . » 30
3. Motion against the flow . . . . » 31
4. The invariant measure . . . . » 34
5. Infinite-dimensional dynamical systems . . . . » 37

PART II – Fully developed turbulence . . . . » 43

H. TENNEKES – A comparative pathology of atmospheric turbulence in two and three dimensions.

1. Introduction . . . . » 45
2. Some properties of flow with friction . . . . » 47
3. Entropy and information . . . . » 49
4. Vorticity dynamics . . . . » 52
5. Energy and enstrophy in three-dimensional turbulence . » 54
6. The enstrophy cascade in two-dimensional turbulence . . » 57
7. Spectral characteristics of three-dimensional turbulence . » 59
8. Characteristic exponents . . . . » 63
9. Spectral characteristics of two-dimensional turbulence . . » 65
10. Epilogue . . . . » 68

U. FRISCH – Fully developed turbulence and intermittency.

1. Simple ideas and misconceptions about turbulence . . . » 71
   1‘1. Sharp transisions can occur when the Reynolds number is varied . . . . » 71
   1‘2. The flow is unstable and unpredictable . . . . » 71

1`3. Trajectories of marked particles are unstable and
       unpredictable . . . . pag. 72
1`4. Turbulent flow enhances transport . . . . »   73
1`5. High-Reynolds-number turbulence has a wide range
       of scales . . . . »   74

2. Fully developed turbulence: intermittency as a broken
    symmetry . . . . »   76

3. Turbulence with a spectral gap and predictability . . . »   80

Appendix. - On the singularity structure of fully developed
    turbulence (with G. PARISI) . . . . »   84

C. W. VAN ATTA – Stratified-turbulence experiments.

1. Introduction . . . . »   89
2. What should be measured? . . . . »   91
3. What can be measured . . . . »   92
4. Buoyancy effects in vertically homogeneous unsheared
    turbulent flow . . . . »   94
5. A laboratory experiment . . . . »   97
6. Some geophysical measurements . . . . »   103
7. Résumé . . . . »   105

S. A. ORSZAG – Lectures on spectral methods for turbulence computations.

1. Introduction to spectral methods . . . . »   107
2. Applications . . . . »   117
   2`1. Introduction . . . . »   117
   2`2. A transitional instability . . . . »   118
        2`2.1. Nonclassical character of transitional instabilities . . . . »   118
        2`2.2. Three dimensionality of transition . . . . . »   118
        2`2.3. Instability of two-dimensional nonlinear travelling waves . . . . »   118
        2`2.4. Competition between two-dimensional pairing
               and three-dimensional instabilities . . . . . »   121
        2`2.5. Spurious (numerical) turbulence . . . . »   122
   2`3. Computer simulations of turbulence . . . . »   122
        2`3.1. Turbulent channel flow . . . . »   122
        2`3.2. Turbulent spot . . . . »   124
        2`3.3. Taylor-Green vortex . . . . »   124
   2`4. Subgrid scale turbulence closures . . . . »   126
3. Conclusion . . . . »   127

PART III – Turbulence in geophysical flows . . . . pag. 131

R. SADOURNY – Quasi-geostrophic turbulence: an introduction.

1. Introduction . . . . » 133
2. Barotropic turbulence . . . . » 133
   2′1. Nonlinear transfers . . . . » 133
   2′2. Statistical equilibria . . . . » 136
   2′3. Stationary solutions: stability . . . . » 138
   2′4. Forced stationary turbulence: phenomenology . . . » 139
   2′5. Direct or semi-direct simulations . . . . » 142
   2′6. Spectral slopes and ZIR intermittency . . . . » 143
   2′7. Emergence of quasi-stationary vortices . . . . » 144
3. Baroclinic turbulence . . . . » 146
   3′1. Simplified formulations . . . . » 146
   3′2. Statistical equilibria . . . . » 148
   3′3. Turbulence with large-scale thermal forcing . . . . » 148
Appendix. - The quasi-geostrophic approximation . . . . . » 152

R. HIDE – Thermal convection in a rotating fluid subject to a horizontal temperature gradient.

1. Introduction . . . . » 159
2. Geostrophy . . . . » 160
3. Regimes of thermal convection in a rotating fluid annulus » 162
4. Patterns of regular nonaxisymmetric flow . . . . » 166
5. Atmospheric flows . . . . » 169

D. J. TRITTON – Experiments on turbulence in geophysical fluid dynamics. - I: Turbulence in rotating fluids.

1. Introduction . . . . » 172
2. Homogeneous turbulence . . . . » 173
3. Vibrating-grid turbulence . . . . » 173
4. Large vortex-generating mechanisms . . . . » 179
5. Turbulent shear flows . . . . » 181
6. An experiment on free shear layers . . . . » 182
7. Stewartson layers . . . . » 189
8. Concluding remarks . . . . » 191

D. J. TRITTON – Experiments on turbulence in geophysical fluid dynamics. - II: Convection of a very viscous fluid.

1. Introduction . . . . » 193
2. The experiment . . . . » 193

3. Observations . . . . pag. 194
4. Geophysical application . . . .   » 198

S. CHILDRESS – An introduction to dynamo theory.

Part I: Survey of the principal concerns of the theory . . . » 200

1. Introductory remarks . . . . » 200
2. Equations of the hydromagnetic dynamo . . . . » 202
3. Kinematic vs. hydromagnetic theory . . . . » 203
4. A disc model . . . . » 206
5. Some remarks on the underlying topology . . . . . . » 207
6. Cowling's theorem in a spherical core . . . . » 208
7. A necessary condition . . . . » 210

Part II: Projection methods for the kinematic problem . . » 211

8. General remarks . . . . » 211
9. Methods based on smoothing . . . . » 212
10. Braginsky's method . . . . » 215
11. Other results . . . . » 217

Part III: Hydromagnetic models . . . . » 218

12. Implications of linear and near-linear analysis . . . . » 218
13. Attempts to model the strong-field regime . . . . . . » 219
14. Concluding remarks . . . . » 223

JIM L. MITCHELL and T. MAXWORTHY – Large-scale turbulence in the Jovian atmosphere.

1. Introduction . . . . » 226
2. Turbulent planetary cascades . . . . » 226
3. Jovistrophic turbulence . . . . » 229
4. Global turbulent interactions and energetics . . . . . » 230
5. Rossby-wave energetics . . . . » 234
6. Vertical structure of the Great Red Spot . . . . » 235
7. Conclusion . . . . » 239

PART IV – Predictability of geophysical flows . . . . » 241

E. N. LORENZ – The growth of errors in prediction.

Part I: General aspects of error growth . . . . » 243

1. Introductory remarks . . . . » 243
2. The concept of error growth . . . . » 244

3. Simple numerical examples . . . . pag. 245
4. More general systems . . . . » 250
5. Approximate formulae . . . . » 252

Part II: Error growth in the atmosphere . . . . » 253

6. The general atmospheric problem . . . . » 253
7. A simple model . . . . » 255
8. Early experiments with global circulation models . . . » 256
9. Later studies with global circulation models . . . . . . » 258
10. An analogue study . . . . » 260
11. The influence of smaller scales . . . . » 261
12. Concluding remarks . . . . » 264

C. E. LEITH – Two-dimensional coherent structures.

1. Enstrophy cascade in 2D turbulence . . . . » 266
2. Selective decay . . . . » 267
3. Method of anticipated vorticity . . . . » 268
4. Minimum-enstrophy modon . . . . » 270
5. Minimum-enstrophy vortices . . . . » 271
6. MEV M . . . . » 272
7. MEV C . . . . » 277
8. Conclusion . . . . » 279

D. K. LILLY – Theoretical predictability of small-scale motions.

1. Summary . . . . » 281
2. Introduction and scope . . . . » 281
3. Conventional estimates of small-scale predictability . . » 281
4. Atmospheric intermittency and effects on predictability . » 285
5. Predictability of flow around widely spaced vortices . . » 287

G. D. ROBINSON – Scale and variance analysis of atmospheric motion.

1. Introduction . . . . » 290
2. The concept of velocity . . . . » 290
3. The concept of scale . . . . » 291
4. Scale analysis of atmospheric motion . . . . » 295
   4'1. Raw observations and manipulated data . . . . . » 295
   4'2. Examples of data analysis . . . . » 295
5. Analysis of the advective interactions . . . . » 305
6. Some aspects of the global kinetic-energy cycle . . . . » 307

PART V – Extended-range prediction and climate dynamics . pag. 309

E. KALNAY and R. LIVEZEY – Weather predictability beyond a week: an introductory review.

1. The problem of weather predictability: an example . . . » 311
2. Predictability and global forecast error growth . . . . . » 316
3. Predictability of time averages: internal dynamics *vs.* boundary forcing . . . . » 322
4. Predictability of long-lived atmospheric phenomena . . . » 329
5. Current status of long-range forecasting . . . . » 338
6. Current research and future outlook . . . . » 342

M. GHIL – Theoretical climate dynamics: an introduction.

1. Introduction . . . . » 347
2. Radiation budget of the Earth . . . . » 348
  2'1. Global radiation balance . . . . » 348
  2'2. Local imbalances and meridional fluxes . . . . . . » 350
3. Global energy-balance models . . . . » 352
  3'1. A model for global temperature . . . . » 352
  3'2. Stationary solutions and stability to perturbations » 353
  3'3. Structural stability . . . . » 355
4. Latitude-dependent models for surface temperature . . . » 356
  4'1. Horizontal heat transport . . . . » 356
  4'2. Model formulation . . . . » 358
  4'3. Stationary solutions . . . . » 360
  4'4. Internal stability . . . . » 362
  4'5. Structural stability . . . . » 364
  4'6. Concluding remarks on energy-balance models . . . » 365
5. Quaternary glaciations . . . . » 366
  5'1. Climatic variability . . . . » 366
  5'2. Paleoclimatic evidence of glaciations . . . . » 367
  5'3. Geochemical proxy data . . . . » 368
  5'4. The phenomenology of glaciation cycles . . . . . » 369
6. Physical mechanisms of glaciations . . . . » 371
  6'1. Internal mechanisms: model formulation . . . . . » 372
  6'2. Internal mechanisms: discussion . . . . » 375
  6'3. Orbital changes . . . . » 378
  6'4. Insolation changes and their climatic effect . . . . » 381
7. The forced behavior of a climatic oscillator . . . . . . » 385
  7'1. Free oscillations . . . . » 386
  7'2. Nonlinear resonance . . . . » 386
  7'3. Entrainment and combination tones . . . . » 388
  7'4. Multiple forcing: more combination tones . . . . . » 390
  7'5. Sharp peaks, aperiodicity and terminations . . . . » 392

8. Periodicity and predictability. . . . . pag. 394
   8˙1. Phase errors and frequency errors . . . . » 394
   8˙2. Measures of predictability. . . . . » 397
9. Concluding remarks . . . . » 399

R. Benzi and A. Sutera – The mechanism of stochastic resonance in climate theory.

1. Introduction . . . . » 403
2. Formulation of the model . . . . » 404
3. Some mathematical properties of Budyko-Sellers models » 408
4. Stochastic perturbations . . . . » 410
5. Effects of stochastic perturbations . . . . » 412
6. The orbital forcing . . . . » 415
7. The mechanisms of stochastic resonance. . . . . » 417
8. Conclusion . . . . » 420
Appendix . . . . » 420

Thomas L. Bell – Climatic sensitivity and fluctuation-dissipation relations.

1. Introduction . . . . » 424
2. The fluctuation-dissipation relation (FDR). . . . . . . » 426
   2˙1. Statement of the relation . . . . » 426
   2˙2. Markov processes and the FDR . . . . » 428
3. Will the FDR work? . . . . » 429
4. Sampling problems . . . . » 435
5. Additional remarks . . . . » 438
6. Conclusion . . . . » 439

Introduction: Turbulence, Geophysical Flows, Predictability and Climate Dynamics.

The LXXXVIII Course of the International School of Physics «Enrico Fermi» attempted to bring together two particularly active areas of fluid dynamics: turbulence theory and the dynamics of geophysical flows. *Turbulence theory* studies the spatial and temporal complexity of flow fields arising in various areas of continuum physics, from the molecular to the cosmic scale. *Geophysical fluid dynamics* studies all aspects of flows on the scale of our planet, with special regard to the effects of planetary rotation and gravity-induced stratification.

The two main topics of the course are supplemented by two closely related ones: predictability theory and climate dynamics. *Predictability theory* arises from applying turbulence theory to the practical questions of numerical weather prediction. The details of atmospheric behavior are sufficiently important to mankind to make the atmosphere rather unique among fluid systems. It is the only such system to be observed throughout much of its volume on a daily basis, and to have its local behavior predicted every day for a few days in advance. This daily prediction submits our knowledge of the atmosphere to a rather stringent test.

The successes and failures of weather prediction are the empirical basis of predictability theory. This theory in turn plays, as we shall see, an important role in our current understanding of turbulence in general.

The first and most important result of predictability theory is that the behavior of a fluid system cannot be predicted with prescribed spatial detail beyond a certain characteristic time. For large-scale atmospheric flow, this characteristic *predictability time* is of the order of ten days, or two-to-three weeks. Beyond this time, the best one can expect is to predict certain mean properties of the system.

A set of temporal and spatial mean properties of the daily local weather is usually defined as *climate*. The dynamics of climate are thus concerned with the low-frequency, small–wave-number portion of atmospheric variability. As such, climate dynamics also has to include the study of the oceans, snow and ice, the biosphere and, on even longer time scales, geodynamics.

Due to its economic and political importance, climate prediction on the time scale of years and decades would be highly desirable. On the time scale of months and seasons, it is currently performed on a routine basis by a few national weather bureaus. Hence, climate dynamics raises the same theoretical and practical questions about spatial and temporal complexity, and about its predictability, as does the dynamics of atmospheres and oceans in the first place.

To study these questions, 22 lecturers and approximately 50 other participants, mostly from Europe and North America, gathered at Villa Monastero on Lake Como during the second half of June 1983. They included fluid dynamicists, mathematicians, meteorologists, oceanographers and physicists. The superb view of the lake and the surrounding mountains, and the warm hospitality of the people of the small town of Varenna provided a most pleasant environment for lectures and discussions.

Lecturers were requested to arrive with a manuscript of their course in nearly final form. These manuscripts were duplicated by the local staff and circulated among the participants. As a result of the lively discussions which occurred, both in the lecture hall and in private, many of the lectures appearing here have been amended and extended. We hope they will continue to provide food for thought and discussion among the readers who were not as fortunate as the participants to share in the local food and beverages.

A book of finite size can only reflect a small part of active research in the four interrelated areas of the geophysical sciences outlined in this introduction. We hope to have selected some of the more lively and promising trends within the broader topics.

Part I deals with general aspects of the *onset of turbulent behavior* in fluids. D. RUELLE gives an introduction to the mathematical theory of dynamical systems. The temporal complexity of solutions to such systems is shown to increase as a parameter changes. This increase in complexity provides a useful analog to the transition from stationary to periodic and finally aperiodic, irregular motion in fluids.

A. LIBCHABER provides an experimental illustration of various types of transition to chaotic, irregular motion. The experiments are carried out in a Rayleigh-Bénard convective cell in which severe constraints reduce the number of degrees of freedom active in the fluid. As a result, simple « routes to chaos » can be observed in great detail.

The first two chapters of Part I deal exclusively with « deterministic chaos », in which irregular behavior obtains due to the nonlinear, purely deterministic interaction of a few degrees of freedom. G. JONA-LASINIO introduces some of the effects due to a large number of additional degrees of freedom, modeled as a stochastic process with simple statistics, such as white noise. These effects include sharp transitions between coexisting, deterministically stable types of behavior.

Part II treats certain aspects of *fully developed turbulence*. H. TENNEKES compares various approaches to the description of atmospheric turbulence. He discusses the equations of motion in the absence of rotation, namely the Navier-Stokes equations, and their inviscid limit, the Euler equations. Vorticity is shown to play a central role in a description of the spatial structure of turbulence in a geophysical fluid. Differences between the statistical properties of turbulence in a two-dimensional and a three-dimensional model of atmospheric flow are pointed out.

U. FRISCH shows the classical, statistical theory of turbulence to complement the newer, dynamical-system approach; the former describes spatial irregularity, mostly in a time-stationary manner, while the latter emphasizes irregularity in time within a relatively simple, orderly spatial structure. The intermittent character, in space and time, of fully developed turbulence is emphasized. This intermittency is linked to the appearance of preferred scales, which break the scaling self-similarity of the statistics for solutions of the governing equations. An appendix, written jointly with G. PARISI, discusses the intermittency in terms of nested sets of singularities for the solutions. Each set is characterized by a higher order of singularity and has a fractional dimension which could be verified experimentally.

C. W. VAN ATTA discusses what can be measured in fully developed turbulence, as opposed to what one would like to measure in order to verify turbulence theories. The effects of buoyancy and stratification in the laboratory, the atmosphere and the oceans are investigated. The oceanographically important transition in time and space between homogeneous, isotropic turbulence and a flow dominated by organized inertia-gravity waves is studied in the laboratory and in the field.

S. A. ORSZAG gives an introduction to spectral methods for the numerical study of the transition to turbulence and of fully developed turbulence. These methods are applied to the problem of transition in plane Poiseuille flow between two parallel walls. A nonlinear, secondary instability seems to account for the well-known fact that transition occurs at a much lower Reynolds number than linear stability of the parallel flow would suggest. Three flows at high Reynolds number are discussed, along with the question of computational subgrid-scale closure.

Part III addresses itself to the *turbulent aspects of geophysical flows*. R. SADOURNY gives an introduction to the statistical theory and numerical simulation of quasi-geostrophic turbulence, in which relatively rapid rotation, expressed by a small Rossby number, and high stratification, expressed by a small height-to-length aspect ratio, play a strongly constraining role. This is the situation for large-scale atmospheric and oceanic flows away from the equator, resulting in a nearly two-dimensional distribution of vorticity. The statistical theory is developed for barotropic models, in which vertical variations are neglected altogether or greatly simplified, and for baroclinic models, in which vertical variations are taken into account more carefully. The propagation, or « cascade », of kinetic energy and of enstrophy, the mean-squared vorticity, from one wave number to another is studied in detail assuming statistical equilibrium.

R. HIDE describes the classical experiments in a rotating annulus with differentially heated side walls. The transitions from an axisymmetric, Hadley-type flow regime, to a Rossby regime with periodic and doubly periodic wave motion, and on to a turbulent regime are discussed. The case of heating in the interior and cooling at the walls is shown to produce stable, closed, anti-cyclonic eddies possessing certain similarities with long-lived features of the atmospheric circulation on the large, outer planets, Jupiter and Saturn.

D. J. TRITTON presents first the experimental interaction between homogeneous, isotropic turbulence and rotation. Transitions in space between flow regions affected and unaffected by the rotation can be quite sharp, large organized vortices coexisting with small-scale grid-generated turbulence. Next, results are given on « thermals » or rising plumes in a very viscous fluid under the effect of convective instability at low Reynolds number. This phenomenon might be associated with certain aspects of irregularly distributed convection in the Earth's mantle.

S. CHILDRESS gives an introduction to flow in the Earth's core in the presence of electromagnetic forces. The kinematic and hydromagnetic aspects of the dynamo mechanism for the generation of the geomagnetic field are studied. Irregular reversals of this field and the role of turbulence are also discussed.

T. MAXWORTHY and J. L. MITCHELL outline new evidence on planetary flows obtained from recent space missions, emphasizing the atmospheric circulation of Jupiter. Characteristics of the latter's banded structure and local, persistent features, such as the Great Red Spot and the White Ovals, are presented. Energy conversions in and near these flow structures are analyzed, and comparisons are made with a theoretical solitary-wave model for the motion associated with the local features.

Part IV discusses the *predictability of geophysical flows* within the context of turbulence theories. E. N. LORENZ analyzes the growth of prediction errors, defined as the time evolution of the difference between two initially neighboring states of the same physical system, or the same mathematical model. Prediction errors are studied first in simple models with a small number of degrees of freedom, then in large numerical models of the atmospheric general circulation, and finally in the atmosphere itself, using twice-daily maps of the height of the isobaric surfaces at 850 mb, 500 mb and 200 mb, and the near-recurrences of these maps. The connections between the stability of a system, the periodicity or recurrence of its behavior, and its predictability are emphasized. These connections provide part of the content of the dynamical-system approach to turbulence and justify to a large extent its application to geophysical flows.

C. E. LEITH discusses two-dimensional, stable, coherent structures as a model for persistent, localized flow features in the atmosphere and in the oceans. Depending on the governing equation, solitons or modons are examples of one- and two-dimensional coherent structures, respectively. Such structures enhance predictability and are investigated from a unifying, variational point of view.

D. K. LILLY studies the predictability of small-scale and meso-scale atmospheric flows with the tools of statistical turbulence theory. The high intermittency of these flows has a marked effect on their predictability. This effect is analyzed in an idealized flow of Rankine vortices spaced wide apart.

G. D. ROBINSON provides an observational study of the variance present over a range of scales in atmospheric motions. Different methods of deriving a variance spectrum from raw data are presented. Implications of the method used for the derived energy conversion rates and spectra are discussed. These spectra in turn affect predictability estimates based on statistical turbulence theories.

In Part V we turn to the low-frequency, small–wave-number questions of *climate dynamics*. E. KALNAY and R. LIVEZEY consider theoretical and practical questions of weather prediction on the time scale of months to years. The actual performance of large general circulation models in extended-range weather prediction is reviewed. The operational practice of the U. S. National Meteorological Center in issuing monthly and seasonal mean forecasts is outlined. The influence of slow land and ocean surface changes on mean atmospheric flow is discussed. Persistent features and low-frequency, planetary-scale variability patterns are indicated as sources of potential predictability over extended periods.

M. GHIL provides an introduction to climate variability on the scale of hundreds to millions of years. Models of the atmosphere's radiative budget are introduced and the stability of stationary solutions to these energy balance models is reviewed. The phenomenology of glaciation cycles during the last two million years of the Earth's history, the Quaternary period, is sketched. Simple models of the coupled atmosphere-ocean-ice-crust-mantle system are shown to possess stable periodic solutions, and to respond resonantly to small variations of insolation due to the orbital changes occurring in the planetary system. The Fourier spectra of this response exhibit the continuous background and a number of the peaks apparent in power spectra of Quaternary proxy records. The complexity of these spectra precludes a detailed reconstruction of paleoclimatic history. The general implications of spectral properties of a deterministic time series for its predictability are discussed.

R. BENZI and A. SUTERA study the combined effects of periodic forcing and stochastic perturbations on an energy balance model with multiple stable equilibria. The main effect is to produce nearly periodic, abrupt transitions between the equilibria, with mean period equal to that of the forcing. The results are shown to hold for more general systems, given suitable ranges of values for the amplitude and period of the forcing, the variance of the noise, the deterministic relaxation time to each stable equilibrium and the distance between the equilibria.

T. L. BELL discusses the theoretical underpinning of estimates for climatic change due to natural and anthropogenic causes. The main tool used is the fluctuation-dissipation relation, which is based on the idea that the decay of internal fluctuations gives a measure of the system's sensitivity to external changes as well. Fluctuations are spontaneous and hence frequent, allowing one to accumulate reliable statistics, while external changes are rare and hard to isolate. The validity of the fluctuation-dissipation relation is studied for climate models of increasing complexity, and statistical sampling problems arising in its applications are reviewed.

In addition to the material included in this volume, four one-hour lectures were given, based on material published elsewhere. S. CHILDRESS talked about geometric models of turbulence (*Geophys. Astrophys. Fluid Dyn.*, **29**, 29 (1984)). M. GHIL delivered the inaugural address for the ceremony celebrating the 30th anniversary of the School, which had its first course in the Summer of 1953. The topic of this inaugural lecture was « Nonlinearity, complexity and Boolean delay equations » (D. DEE and M. GHIL: *SIAM J. Appl. Math.*, **44**, 111 (1984); M. GHIL and A. MULLHAUPT: *J. Stat. Phys.*, submitted). D. K. LILLY discussed the role of helicity in relatively long-lived, intensely rotating convective storms (*Intense Atmospheric Vortices*, edited by L. BENGTSSON and J. LIGHTHILL (Berlin, Heidelberg, New York, N.Y., 1982), p. 149; *J. Atmos. Sci.*, in press).

Finally, E. N. LORENZ and K. VENKATARAMANAIAH related the observed quasi-geostrophic character of large-scale geophysical flows to the existence of a « slow manifold », an invariant manifold for the full equations of motion on which the flow is quasi-geostrophic, and thus slower than other types of motion supported by the full equations (C. E. LEITH: *J. Atmos. Sci.*, **37**, 958 (1980); E. N. LORENZ: *J. Atmos. Sci.*, **37**, 1685 (1980)). The existence of the slow manifold for a simple model containing a Lorenz attractor was discussed, and its dependence on the model's parameters analyzed. The preliminary result is that this manifold is not always stable and that faster motions can appear for certain parameter values, as they do in the atmosphere (K. VENKATARAMANAIAH: Ph. D. Thesis, M.I.T., in preparation).

The one-to-three-hour lectures were supplemented by four workshop-type discussion sessions. These started with a number of brief presentations by two or three participants, and continued with an open discussion of these presentations, of the pertinent lectures and of the topic in general.

The first discussion had as its topic « Turbulence, theoretical questions ». P. CONSTANTIN made a presentation about the finite dimensionality of the attractor set for the phase-space flow of the Navier-Stokes equations (P. CON-

STANTIN and C. FOIAS: *Commun. Pure Appl. Math.*, in press). D. SCHERTZER presented empirical probability distributions of velocity and buoyancy fluctuations for the atmosphere extending over a large range of wave numbers and connected them with scaling self-similarity laws for the governing equations and boundary conditions (D. SCHERTZER and S. LOVEJOY: *Turbulence and Chaotic Phenomena in Fluids*, edited by T. TATSUMI (North-Holland, Amsterdam, 1984)).

The open discussion centered on the properties and definitions of turbulent flows, and on the complementarity of the dynamical-system approach and of the statistical approach to turbulence. The first seems to emphasize transition to weakly turbulent behavior and disorder in time, while the second emphasized disorder in space and appears to be more useful in studying fully developed turbulence. Other approaches, such as stability and regularity theory for the Euler and Navier-Stokes equations, and blow-up of solutions in finite time, were also discussed.

The second discussion centered on « Blocking and persistent anomalies of atmospheric flow ». R. BENZI gave a phenomenological introduction using video equipment to show synoptic patterns of blocking from 500 mb height maps. These patterns persist beyond the usual, mean atmospheric predictability time and have also been simulated in simple models (B. LEGRAS and M. GHIL: *J. Méc. Théor. Appl.*, no. spécial, 45 (1983); *J. Atmos. Sci.* (1984), in press).

B. LEGRAS presented results in which a flow with higher or lower predictability occurs in different regions of a simple model's extended phase-parameter time-space (B. LEGRAS and M. GHIL: *Predictability of Fluid Motions*, edited by G. HOLLOWAY and B. WEST (American Institute of Physics, New York, N.Y., 1984), p. 87). A. SPERANZA emphasized the role of baroclinic effects in the generation and maintenance of blocks (P. MALGUZZI and P. MALANOTTE-RIZZOLI: *J. Atmos. Sci.*, submitted; A. BUZZI, A. TREVISAN and A. SPERANZA: *J. Atmos. Sci.*, **41**, 637 (1984)).

The open discussion covered the temporal behavior of persistent anomalies, alternating with more irregular flow, their localized character in space, the relative effects of internal atmospheric variability and of external changes in boundary conditions. Various tools for the study and prediction of these anomalies, besides simple models, were discussed, including large general circulation models, nearly recurrent synoptic maps (« analogues ») and purely statistical methods (« teleconnection patterns »).

The third discussion revolved around « Climate and its predictability ». P. PESTIAUX presented methods of spectral analysis for time series of proxy data from deep-sea cores and results on some cores with high sedimentation rates (Ph. D. Thesis, Louvain, 1984). A. P. MULLHAUPT discussed formal conceptual models of long-term climatic change based on Boolean delay equations, and showed stationary and periodic solutions of such a model, depending on delay values (M. GHIL, A. MULLHAUPT and P. PESTIAUX: in preparation). N. DALFES talked about stochastic perturbations of spatially one-dimensional energy balance models and the dependence of variance spectra on the additive or multiplicative character of the random noise (H. N. DALFES, S. H. SCHNEIDER and S. L. THOMPSON: *J. Atmos. Sci.*, **40**, 1648 (1983)).

The open discussion dealt with the questions of climate modeling and prediction on the practically important time scales of months to decades, up to the historically and geologically interesting ones of hundreds to millions of years. The phenomena active on various time scales, their causing mechanisms and the spatial scales of most interest on each time scale, from regional to global, were touched upon. As a relatively new discipline, climate dynamics seems to be in a good position to benefit from intradisciplinary, as well as interdisciplinary exchanges of methods and results.

The fourth and final discussion concentrated on « Predictability of weather and extended-range prediction ». G. VALLIS discussed predictability from the point of view of the statistical mechanics of fluid flow, which provides analogies with the connection between thermodynamic entropy and information-theoretic entropy (G. F. CARNEVALE and G. HOLLOWAY: *J. Fluid Mech.*, **116**, 115 (1982)). Numerical results on entropy and predictability from a study of the barotropic vorticity equation with stochastic forcing and high-order dissipation were also presented (G. F. CARNEVALE and G. K. VALLIS: *Predictability of Fluid Motions*, edited by G. HOLLOWAY and B. WEST (American Institute of Physics, New York, N.Y., 1984), p. 577). Other concepts of predictability, based on the dynamical-system notions of sensitive dependence on initial data and characteristic exponents, on statistical notions of error cascades from small to large scales and on time series analysis notions of lagged correlations and their relation to the Fourier spectrum of the process were also discussed.

On the practical side, the enhancement of predictability due to the presence of persistent anomalies, local in physical space or in wave number space, was emphasized. The predictability of means due to slow changes in external parameters or boundary data was recognized. A broad approach to extended-range prediction, combining the theoretical advances provided by simple models with the empirical results of statistically based forecasting and with further development of large numerical models, seemed the most promising at the present state of the art.

All this activity would not have been possible without some preparation and a lot of help. The organizing committee of the Course included the Director and Scientific Secretaries, who also edited this volume, and the following members: Prof. S. CHILDRESS, Dr. C. E. LEITH (chairman) and Dr. A. SUTERA. We are all grateful to the Italian Physical Society, its President, Prof. R. A. RICCI, and its Council, for having put at our disposal its hospitality and the marvelous and stimulating surroundings of Villa Monastero.

Ms. E. MAZZI, Secretary of the Society, headed a local staff which made everything run smoothly and pleasantly.

The travel of some American participants and lecturers was supported by the National Science Foundation (NSF), Atmospheric Sciences and Ocean Sciences Divisions, and by the National Aeronautics and Space Administration (NASA), Atmospheric Dynamics and Radiation Branch. Ms. P. STEVENS from NSF and Dr. J. THEON from NASA were especially helpful in making the necessary arrangements. The Ministero della Pubblica Istruzione, Consiglio Nazionale delle Ricerche and IBM Italy also contributed funds.

Finally, this volume could not have been produced and distributed without the careful technical editing of P. PAPALI, Publication Secretary of the Society, and the interest and solicitude of Dr. G. WOLZAK, of North-Holland Publishing Company.

We hope that in a time of increasing specialization this volume will help increase contacts between applied mathematicians, meteorologists, oceanographers and physicists.

M. GHIL, R. BENZI and G. PARISI
New York and Rome, September 1983

PART I

THE ONSET OF TURBULENCE

The Onset of Turbulence: a Mathematical Introduction.

D. RUELLE

Institut des Hautes Etudes Scientifiques
35, route de Chartres, 91440 Bures-sur-Yvette, France

0. – Introduction.

The time evolutions which one observes in the study of natural systems often exhibit a complicated nonperiodic behaviour, with apparently stochastic character. It has become increasingly clear in recent years that such complicated behaviour is common in many deterministic systems and does not depend on external « stochastic forces » acting on the system. One speaks then of *chaotic behaviour*, of *deterministic noise*, or of the presence of *strange attractors*.

Remarkably, HADAMARD, DUHEM and POINCARÉ, almost a century ago, already knew the existence of deterministic time evolutions with stochastic character. They also understood some of the physical and philosophical implications of this fact. However, the implementation of these « chaotic » ideas in physical theories is recent and has largely depended on the availability of the modern fast computers and of sophisticated experimental techniques. It is indeed one thing to know that physical systems may exhibit a certain type of behaviour, and another to assert and prove that this behaviour is present and significant for specific phenomena. In fact, in the case of hydrodynamic turbulence, and later of « chemical turbulence », there was serious opposition to the new ideas, which gained respectability only after intensive experimental study.

From a mathematical point of view, the new ideas which we shall discuss belong to the theory of differentiable dynamical systems. While this theory contains difficult and profound results, it is only fair to say that its *predictive* value for physics is limited. The theory of differentiable dynamical systems is nevertheless quite helpful in *interpreting* the behaviour of physical systems at a low level of excitation. This in particular is true for hydrodynamical systems at the onset of turbulence. We shall also discuss some *ergodic* concepts which may be useful in the theory of developed turbulence, but this appears to be a considerably more difficult problem.

Sections **1** and **2** deal with the bifurcation theory of differentiable dynamical systems, in particular with strange attractors and routes to chaos. Section **3** describes the ergodic theory.

1. – Strange attractors and sensitive dependence on initial condition.

1˙1. *Differentiable dynamical systems.*

We shall consider time evolutions with either discrete or continuous time t. In the first case t is an integer and we assume that

$$x_{t+1} = f(x_t). \tag{1}$$

In the second case t is a real number and the evolution is given by a differential equation

$$\frac{\mathrm{d}x_t}{\mathrm{d}t} = F(x_t). \tag{2}$$

In applications, x_t will represent the state of a system at time t, for instance the instantaneous velocity field of a fluid. We take x_t to belong to a finite- or infinite-dimensional manifold M. We also assume that f or F are *differentiable* functions on M (several times differentiable may be required).

One might think that differentiability is just a technical assumption without direct physical significance. Surprisingly, the contrary turns out to be true. Differentiable dynamical systems (DDSs) exhibit a number of recognizable phenomena like Hopf or Feigenbaum bifurcations, or occurrence of strange attractors. These phenomena are observed in hydrodynamical or chemical systems, even though they would be hard to predict.

For simplicity, we shall usually take M finite dimensional in these lectures. More general situations can be treated with more effort. For instance, the Navier-Stokes time evolution corresponds—where defined—to a differentiable (in fact, real analytic) semi-flow in infinite dimension.

1˙2. *Types of DDSs.*

a) *Maps* (time integer $\geqslant 0$). A differentiable map f defines a DDS according to (1). Corresponding to the positive integer time n we have the map f^n which gives the state $f^n x$ of the system at time n, when the state x at time 0 is known.

b) *Diffeomorphisms* (time integer $\gtrless 0$). If f has a differentiable inverse f^{-1}, *i.e.* if f is a diffeomorphism, then for any integer k, positive or negative, there is a map f^k.

c) *Flows* (time real ≥ 0). A vector field F on the finite-dimensional manifold M defines a time evolution by solution of (2). We assume that an existence and uniqueness theorem holds for all times and write the solution with initial condition x as $x_t = f^t x$. The family (f^t) indexed by $t \in \mathbf{R}$ is called a flow. Note the group property $f^s \circ f^t = f^{s+t}$.

d) *Semi-flows* (time real ≥ 0). Time evolutions defined only for positive times are called semi-flows and occur in the solution of PDEs like the Navier-Stokes and the heat equation, for which backward continuation is ill posed.

1˙3. *Sensitive dependence on initial condition.* – Let δx be an infinitesimal change of the initial condition x. After time t, this has become

$$\delta x_t = \frac{\partial f^t x}{\partial x} \delta x . \tag{3}$$

If our manifold M is Euclidean space, or if pieces of M are parametrized by pieces of $\mathbf{R}^m$, δx and δx_t are m-dimensional vectors, and $\partial f^t x / \partial x$ is the $m \times m$ matrix of partial derivatives of f^t at the point x. Notice (for the discrete-time case) the chain rule formula

$$\frac{\partial f^n x}{\partial x} = f'(f^{n-1} x) \dots f'(fx) f'(x) , \tag{4}$$

where we have written $f'(x) = \partial fx / \partial x$. Formula (4) makes it plausible that an exponential rate of growth of δx_t can be defined:

$$\lambda(x, u) = \lim_{t \to \infty} \frac{1}{t} \log \left\| \frac{\partial f^t x}{\partial x} \cdot u \right\| .$$

We shall see in sect. **3** that such a limit indeed exists almost everywhere (for probability measures invariant under time evolution). For obvious reasons we have replaced the infinitesimal δx by a finite vector u. The quantity $\lambda(x, u)$ is called a *characteristic exponent*.

If $\lambda(x, u) > 0$, a small error δx on the initial condition will grow exponentially with time, at least while it remains small (when δx_t becomes large, formula (3) no longer applies). This phenomenon of error growth is called *sensitive dependence on initial condition*, and systems with this property are called *chaotic*.

1˙4. *Attractors.* Suppose that the closed set $A \subset M$ attracts nearby trajectories: if x is sufficiently close to A, x_t tends to A for $t \to \infty$ (with discrete or continuous time t). The set A is then called an *attractor* (it may be appropriate to impose also some irreducibility condition). Attractors are physically important because they describe the asymptotic behaviour of a system. On the other hand, a nonattracting invariant set B will usually be invisible because, if $x \in B$, small perturbations will send the orbit $(f^t x)$ of x away from B, never to come back. If the time evolution on an attractor A exhibits sensitive dependence on initial condition, we say that A is a *strange* attractor. This is not a precise mathematical definition, but will suffice for our purposes.

Notice that a *conservative* (*i.e.* Hamiltonian) time evolution conserves the volume in phase space (Liouville's theorem). For such time evolutions, the concept of attractors is, therefore, not useful. On the other hand, strange and nonstrange attractors are important in the description of the asymptotic behaviour of *dissipative* systems.

2. – Routes to chaos.

2˙1. *Some simple technical facts.*

a) Genericity. DDSs exhibit a variety of phenomena which is so enormous as to be unclassifiable. It is thus reasonable to eliminate special cases and to discuss only the « generic » situation. In particular, Hamiltonian systems have nongeneric behaviour in preserving volume, and we shall not discuss them in the present lectures. We concentrate rather on dissipative systems like viscous fluids. We shall not try to make the notion of genericity precise, but further examples will occur below.

b) Time-one map. It is often convenient to pass from the study of a system with continuous time to that of a system with discrete time and *vice versa*. Starting from a flow (f^t), we obtain a diffeomorphism simply by taking $t = 1$. Unfortunately, this time-one map f^1 has nongeneric behaviour: if f^1 has a period p point x, with $p > 1$ (*i.e.* $f^p x = x$, $fx \neq x$), then f^1 has a continuum of period p points through x (the (f^t) orbit of x). By comparison, the periodic points of period p of a diffeomorphism are generically isolated.

c) Poincaré map. There if often a better vay to associate a map with a flow (f^t). This is achieved by using a piece of hypersurface Σ (dimension 1 less than M) and for $x \in \Sigma$ defining Px to be the first point on the orbit of x (*i.e.* the point $f^t x$ with smallest $t > 0$) which is again in Σ. For instance, if (f^t) has a periodic orbit through a and Σ is transversal at a to this periodic orbit, P is defined for x close enough to x in Σ; P is called Poincaré map. It is invertible near x (*i.e.* it is a local diffeomorphism) (see fig. 1).

There is a construction which is inverse to that of the Poincaré map (the *suspension*) which produces a flow (f^t) in $m + 1$ dimensions from a diffeomorphism in m dimensions. As a consequence of these constructions the phenomena observed for flows in a certain dimension usually correspond to the phenomena observed in dimension 1 less for diffeomorphisms.

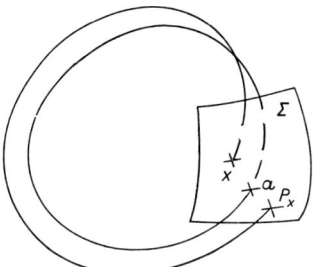

Fig. 1.

2'2. Generic bifurcations of an attracting fixed point. – A stable steady state of a physical system is represented by an attracting fixed point ξ of the corresponding dynamical system. The fixed-point condition is

$$\xi = f(\xi)$$

for discrete time (cf. (1)) or

$$F(\xi) = 0$$

for continuous time (cf. (2)). The attracting character is expressed in terms of the matrix of partial derivatives of f or F at ξ:

$(\partial_i f_j(\xi))$ has its eigenvalues inside the unit circle

or

$(\partial_i F_i(\xi))$ has its eigenvalues in the left half complex plane.

Suppose now that our dynamical system depends differentiably on a real parameter μ (bifurcation parameter). More precisely, we assume that $f = f_\mu$ or $F = F_\mu$ and that $f_\mu(x)$ or $F_\mu(x)$ is a differentiable function of μ, x. Suppose that the system has a fixed point ξ_0 for the value μ_0 of the bifurcation parameter, and that $\partial_i f_{\mu_0 j}(\xi_0) - \delta_{ij}$ or $(\partial_i F_{\mu_0 j}(\xi_0))$ is invertible (nonzero determinant, or no zero eigenvalue). Then the *implicit function theorem* implies that the system also has a unique fixed point $\xi = \xi(\mu)$ close to ξ_0 for μ close to μ_0. In particular, the invertibility condition is satisfied if ξ_0 is an attracting fixed point. As μ is varied, the invertibility condition may eventually break down in several ways.

a) Saddle node bifurcation. If $(\partial_i f_{\mu_0 j}(\xi_0))$ has an eigenvalue 1 or $(\partial_i F_{\mu_0 j}(\xi_0))$ an eigenvalue 0, the implicit function theorem no longer applies; ξ_0 is called a saddle node. Generically, the fixed point $\xi(\mu)$ which we assume to be present and attracting for $\mu < \mu_0$ collides for $\mu = \mu_0$ with a nonattracting fixed point $\xi'(\mu)$. For $\mu > \mu_0$ there is nothing attracting left near μ_0 (see fig. 2).

The inverted saddle node bifurcation occurs for $\mu > \mu_0$ (see fig. 3).

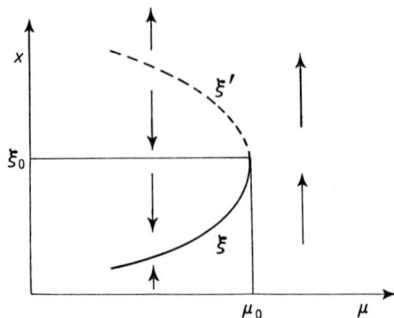

Fig. 2. – Saddle node bifurcation.

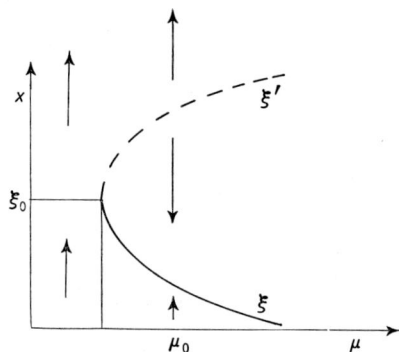

Fig. 3. – Inverted saddle node bifurcation.

Examples of *nongeneric* behaviour occur if the dynamical system has a symmetry group. For instance, the symmetry $x \to -x$ in $\mathbf{R}^m$ implies that 0 is always a fixed point and cannot disappear in a saddle node bifurcation.

b) Flip or pitchfork bifurcation (discrete times). If $(\partial_i f_{\mu j}(\xi))$ has eigenvalues inside the unit circle for $\mu < \mu_0$, and one of them crosses at -1 for $\mu = \mu_0$, then generically either an attracting periodic orbit (η_1, η_2) of period 2 is created for $\mu > \mu_0$ or a nonattracting periodic orbit of period 2 is present for $\mu < \mu_0$ (see fig. 4 and 5).

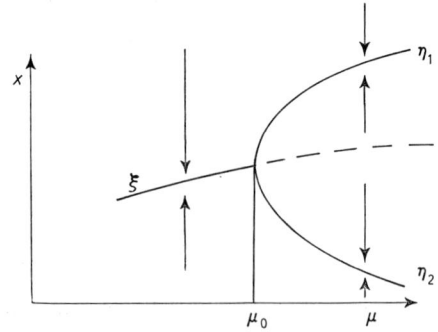

Fig. 4. – Flip.

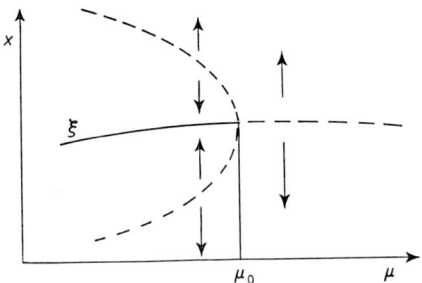

Fig. 5. – Inverted flip.

If the map f is the Poincaré map of a flow, the bifurcation replaces a periodic orbit of period T by a periodic orbit of period $\approx 2T$ for the flow (see fig. 6).

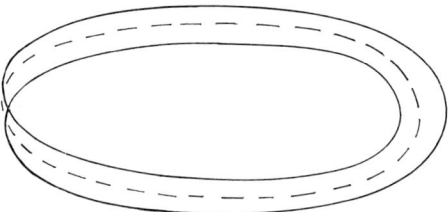

Fig. 6.

c) *Hopf bifurcation* (continuous or discrete time). Let $(\partial_i F_{\mu j}(\xi))$ have eigenvalues in the left half complex plane for $\mu < \mu_0$, one complex conjugate pair crossing to the right half-plane for $\mu > \mu_0$. Or let $(\partial_i f_{\mu j}(\xi))$ have eigenvalues inside the unit circle for $\mu > \mu_0$, one complex conjugate pair crossing to the outside for $\mu > \mu_0$. Then, generically, the dynamical system has a little invariant closed curve C either for $\mu > \mu_0$ (Hopf bifurcation, the closed curve is attracting) or for $\mu < \mu_0$ (inverted Hopf) (see fig. 7 and 8).

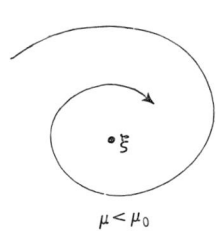
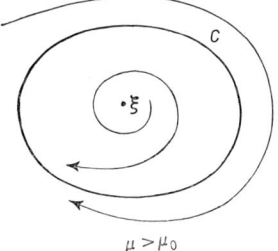

Fig. 7. Fig. 8.

Fig. 7. – Hopf bifurcation, $\mu < \mu_0$.

Fig. 8. – Hopf bifurcation, $\mu > \mu_0$.

In the continuous-time case, C is a periodic orbit. The Hopf bifurcation thus replaces a steady state by a periodic state. If f is the Poincaré map of a flow, the Hopf bifurcation produces an attracting invariant 2-torus (see fig. 9).

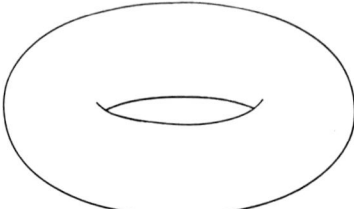

Fig. 9.

On this torus the time evolution often is quasi-periodic with 2 periods, *i.e.* of the form

(5) $$x(t) = \Phi(\omega_1 t, \omega_2 t),$$

where Φ has period 2π in each argument, and ω_1, ω_2 are not rationally related.

2`3. *Frequency locking.* – The quasi-periodic motion (5) is frequently replaced by a periodic motion where ω_1 and ω_2 are changed to become rationally related. Figure 10 shows the phenomenon on a torus drawn as a square with opposite sides identified.

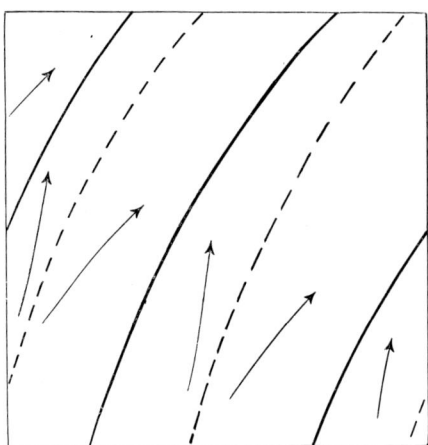

Fig. 10.

This phenomenon of *frequency locking* or *entrainment* is well known to engineers. It is a typically nonlinear phenomenon.

2`4. *When does sensitive dependence on initial condition occur?* – No criterion is known by which one could decide when there is a positive characteristic exponent. It is known, however, that this can occur only if the dimension of the manifold is sufficiently large, as the table I indicates.

TABLE I. – *Sensitive dependence on initial condition can occur on* M

for a noninvertible map	if dim $M \geqslant 1$
for a diffeomorphism	if dim $M \geqslant 2$
for a flow	if dim $M \geqslant 3$

We now give some examples.

The map $f: \theta \to 2\theta$ (mod 2π) of the circle is « chaotic » because $\delta\theta$ is replaced by $2^n \delta\theta$ after n applications of f, corresponding to a characteristic exponent $\log 2 > 0$.

A diffeomorphism of a compact manifold cannot expand everywhere in all directions (because it has to be invertible). There should be at least one contracting direction as well as an expanding one, and, therefore, dim $M \geqslant 2$ is needed for a strange attractor. Standard example: the Hénon attractor in $\mathbf{R}^2$ corresponding to the diffeomorphism

$$\begin{pmatrix} x \\ y \end{pmatrix} \mapsto \begin{pmatrix} y + 1 - ax^2 \\ bx \end{pmatrix}$$

with $b = 0.3$ and $a = 1.4$.

In a flow the direction of flow is neutral (neither expanding nor contracting exponentially, if the vector field does not vanish). Therefore, one more dimension is wasted, and dim $M \geqslant 3$ is needed here. Standard example: the Lorenz attractor in $\mathbf{R}^3$ corresponding to the equations

$$\dot{x} = -\sigma(x - y), \qquad \dot{y} = -y + (r - z)x, \qquad \dot{z} = -bz + xy$$

with $b = \frac{8}{3}$, $\sigma = 10$, $r = 28$.

It is true that we lack rigorous mathematical treatment of the Hénon and Lorenz attractors. There are, however, (more complicated!) examples in the same dimensions for which rigorous proofs are available.

2'5. *Scenarios leading to chaos.* – We consider again a DDS depending on a bifurcation parameter μ. In hydrodynamics one could take for μ the Reynolds number. Starting from the simple situation where the system has an attracting fixed point (physically, a steady state) one can reach turbulence in different manners. We shall describe three routes to turbulence which have been studied theoretically and observed experimentally. These routes—or scenarios as they have been called by ECKMANN—are related to the three bifurcations of subsect. **2'2**. There are many other possibilities, and even the scenarios described here are outlines of possible sequences of events rather than precise mathematical descriptions.

a) The intermittent scenario (MANNEVILLE-POMEAU). Let an attracting periodic orbit C, present for $\mu < \mu_0$, be replaced for $\mu > \mu_0$ by a strange attractor. Using a Poincaré map, we reduce to the situation of a fixed point ξ replaced by a strange attractor. The disappearance of ξ is assumed to take place by collision with a nonattracting fixed point ξ' (saddle node bifurcation). The situation is best visualized if f_μ is a 1-dimensional map (fig. 11 and 12).

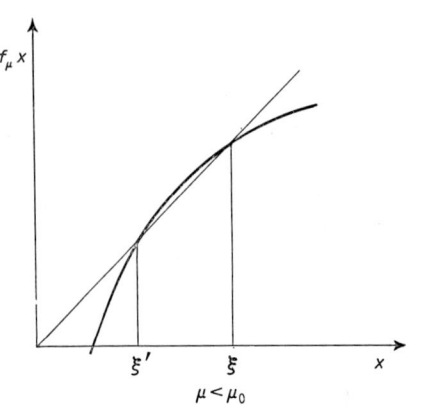
Fig. 11.
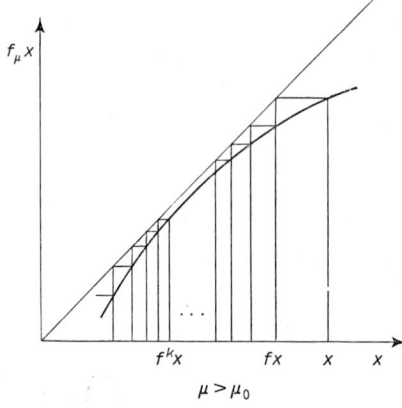
Fig. 12.

After the bifurcation, the attracting point ξ has disappeared, but the orbit $(x, fx, ..., f^k x, ...)$ still spends a lot of time near the saddle node ξ_0. For the original flow this means that x_t spends a lot of time circling near what was a periodic orbit C_0, with occasional excursions far away on the strange attractor. A signal u_t attached to the flow then has the aspect of fig. 13 with nearly pe-

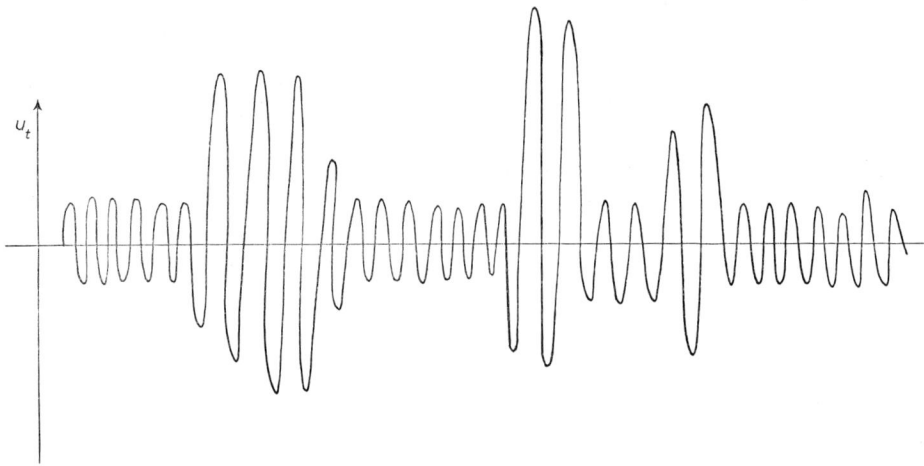
Fig. 13.

riodic sequences interrupted by *bursts of turbulence*. This *temporal intermittency* is not related in any clear manner to the spatial intermittency recognized in fully developed turbulence.

b) *The period-doubling cascade* (FEIGENBAUM). Here again we start with an attracting periodic orbit. Using a Poincaré map, we replace this orbit by an attracting fixed point ξ for a map f_μ. When μ increases, a pitchfork bifurcation may occur. It is a remarkable fact that successive pitchfork bifurcations may follow, leading to periods 4, 8, ..., $2k$, ... (fig. 14) and that the sequence

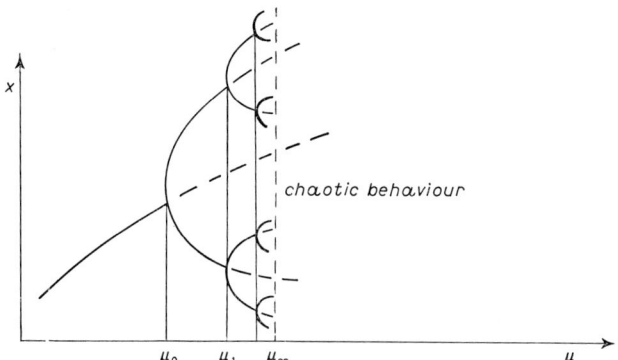

Fig. 14.

$\mu_0, \mu_1, \mu_2, \ldots$ of bifurcation values of the parameter tends geometrically to a limit μ_∞:

$$\lim_{k \to \infty} \frac{\mu_\infty - \mu_k}{\mu_\infty - \mu_{k+1}} = 4.669\,201\,609\,103\,\ldots\,.$$

The whole period-doubling cascade constitutes a new generic bifurcation. The cascade is beautifully visible in certain hydrodynamical experiments (LIBCHABER). For $\mu > \mu_0$, chaotic behaviour is known to occur, but it has not been analysed in detail.

c) *The quasi-periodic scenario* (RUELLE-TAKENS-NEWHOUSE). We have seen how a Hopf bifurcation can produce a periodic orbit in a continuous-time dynamical system. Using a Poincaré map and a second Hopf bifurcation, we have also obtained a quasi-periodic motion with two periods (see (5)). One can imagine that successive bifurcations could produce quasi-periodic motions of the form

$$x(t) = \Phi(\omega_1 t, \ldots, \omega_k t),$$

where Φ has period 2π in each argument, and $\omega_1, \ldots, \omega_k$ are not rationally related. With increasing k the motion would appear more and more « turbulent », and this is indeed the turbulence theory proposed by LANDAU and HOPF.

Quasi-periodic motions have, however, no sensitive dependence on initial condition and are also unstable to small perturbations of the equations of motion. This instability may lead to frequency locking, with creation of one or several attracting periodic orbits or quasi-periodic motions with less than k independent frequencies. If $k \geqslant 3$, the instability may also lead to the creation of strange attractors.

The scenario $1 \to 2 \to 3$ frequencies $\to$ strange attractor has been observed in hydrodynamical experiments (by GOLLUB and SWINNEY and others). The experiments involve frequency analysis of a time series: sharp frequencies correspond to periodic and quasi-periodic motions, then a *continuous spectrum* (broad-band spectrum) to a strange attractor. A variation of this scenario, where the 2-torus wrinkles and directly changes to a strange attractor without appearance of a third frequency, is much studied now (KADANOV, SCHENKER; RAND, OSTLUND, SETHNA, SIGGIA; MANTON, NAUENBERG).

3. – Ergodic theory of differentiable dynamical systems.

To define the characteristic exponents

$$\lambda(x, u) = \lim_{t \to \infty} \frac{1}{t} \log \left\| \frac{\partial f^t x}{x} \cdot u \right\|$$

it suffices to consider the discrete-time situation and to use the following result (first obtained by OSELEDEC (1968)).

3.1. Theorem (multiplicative ergodic theorem). – *Let ϱ be a probability measure on a space M, and $f: M \to M$ a measure-preserving map. Let also T be a (measurable) function M to the $m \times m$ matrices (real or complex), such that*

$$\int_M \varrho(\mathrm{d}x) \log^+ \|T(x)\| < +\infty,$$

where $\log^+$ is the positive part of the logarithm. Define $T_x^n = T(f^{n-1}x) \ldots T(fx) T(x)$ and let T_x^{n*} be the adjoint of T_x^n. Then

$$\lim_{n \to \infty} (T_x^{n*} T_x^n)^{1/2n} = \Lambda_x$$

exists for almost all x.

Let $\exp[\lambda_1(x)] \geqslant \ldots \geqslant \exp[\lambda_m(x)]$ be the eigenvalues of Λ_x, then $\lambda_1, \ldots, \lambda_m$ are called *characteristic exponents* (or Ljapounov exponents). If ϱ is *ergodic* (i.e. there is no invariant set S with $0 < \varrho(S) < 1$), then $\lambda_1, \ldots, \lambda_m$ are ϱ-almost everywhere constant. Notice that, if V is the subspace corresponding to the eigenvalues of Λ_x which are strictly less than λ_1, then

$$\text{(6)} \qquad \lim_{n\to\infty} \frac{1}{n} \log \|T_x^n u\| = \lambda_1 \qquad \text{if } u \in \mathbf{R}^m \setminus V.$$

There are similar statements involving the lower characteristic exponents.

If $M = \mathbf{R}^m$ and f is a differentiable map, we can apply the multiplicative ergodic theorem with $T(x) = \partial f/\partial x$. Then $T_x^n = \partial f^n/\partial x$ by the chain rule, and sensitive dependence on initial condition occurs if the largest characteristic exponent is strictly positive.

3˙2. *Choice of an invariant measure.* – On a strange attractor there are typically uncountably many ergodic invariant measures, and it is an important conceptual problem to decide which one is physically relevant. (Gaussianity assumptions, as used in the past, do not seem to be an appropriate answer.) In some cases which one has been able to handle the following was observed. For almost every initial point x *with respect to Lebesgue measure* $\mathrm{d}x$ in a neighbourhood of the attractor, the time average corresponds to a unique asymptotic measure ϱ on the attractor:

$$\lim_{n\to\infty} \frac{1}{n} \sum_{k=0}^{n-1} \varphi(f^k x) = \int \varrho(\mathrm{d}y)\varphi(y)$$

for every continuous function φ. The measure ϱ is stable under small stochastic perturbations, *i.e.*, if a bit of noise is added to the deterministic process f, the resulting random process has a stationary measure which tends to ϱ in the limit of zero noise. The measure ϱ is in general singular with respect to Lebesgue measure, but remains « continuous along expanding directions ». We shall not make this statement precise, but remark that the pictures of the Hénon attractor, for instance, show the continuous distribution of mass along curves (so-called unstable manifolds) oriented everywhere in the expanding direction. (The subspace V in (6) is, in the present case, the *contracting* direction. The *expanding* direction is the contracting direction for f^{-1}.)

3˙3. *Entropy* (Kolmogorov-Sinai invariant). – Sensitivity to initial condition implies that, if the initial state of a dynamical system is known only with finite precision, the state at a later time is known with less precision. In other words, the system acts as a source of information (or as some sort of random-number generator). The average rate of information measure ϱ is the *entropy* $h(\varrho)$, or Kolmogorov-Sinai invariant. (If the dynamical system is a fluid, the thermodynamic entropy of the fluid is also defined, but this is a totally different and unrelated physical concept.) The following general inequality holds for any ergodic probability measure:

$$h(\varrho) \leqslant \Sigma \text{ positive characteristic exponents (with respect to } \varrho\text{)}.$$

For the « asymptotic measures » of subsect. 3'2, the equality sign holds in this formula.

3'4. *Dimension of attractors.* – The attractors for the (incompressible) Navier-Stokes equation in a bounded region have finite Hausdorff dimension (for any finite Reynolds number). This means that the number of excited degrees of freedom of the fluid system is in some sense finite. The Hausdorff dimension of an attractor is larger than the number of positive characteristic exponents for any ergodic probability measure on the attractor. (This is because the expanding directions are contained in the attractor.)

4. – Some recommended reading.

The following set of references is modest, in the spirit of this «mathematical introduction». An elementary review of strange attractors and the onset of turbulence (with pictures) is given in [1]. The original papers of Lorenz [2] and Hénon [3] on the corresponding attractors are well worth reading. On pathways to turbulence see [4-9]. A review of the various scenarios is given by ECKMANN [10]. Smale's paper [11] on differentiable dynamical systems is recommended to the mathematically inclined reader. It is perhaps a bit outdated, but will remain a great classic of mathematical literature. At a more technical level also we mention the discussion of ergodic theory applied to hydrodynamics in [12].

REFERENCES

[1] D. RUELLE: *Math. Intell.*, **2**, 126 (1980).
[2] E. N. LORENZ: *J. Atmos. Sci.*, **20**, 130 (1963).
[3] M. HÉNON: *Commun. Math. Phys.*, **50**, 69 (1976).
[4] D. RUELLE and F. TAKENS: *Commun. Math. Phys.*, **20**, 167 (1971). Note concerning our paper *On the nature of turbulence*, *Commun. Math. Phys.*, **23**, 343 (1971).
[5] S. NEWHOUSE, D. RUELLE and F. TAKENS: *Commun. Math. Phys.*, **64**, 35 (1978).
[6] M. J. FEIGENBAUM: *J. Stat. Phys.*, **19**, 25 (1978).
[7] M. J. FEIGENBAUM: *J. Stat. Phys.*, **21**, 669 (1979).
[8] M. J. FEIGENBAUM: *Commun. Math. Phys.*, **77**, 65 (1980).
[9] Y. POMEAU and P. MANNEVILLE: *Commun. Math. Phys.*, **74**, 189 (1980).
[10] J. P. ECKMANN: *Rev. Mod. Phys.*, **53**, 643 (1981).
[11] S. SMALE: *Bull. Am. Math. Soc.*, **73**, 747 (1967).
[12] D. RUELLE: *Bull. Am. Math. Soc.*, **5**, 29 (1981).

The Onset of Weak Turbulence.
An Experimental Introduction.

A. LIBCHABER

The James Franck Institute, The Enrico Fermi Institute
The University of Chicago - Chicago, IL 60637

Since Benard early experiments [1] in 1900, thermal convection has been one of the main paradigms for the study of fluid instabilities. The theoretical approach has also moved a long way since the early work of Rayleigh [2] and Jeffreys [3]. The main architect for the general framework of thermal convection is BUSSE, whose review article [4] is the best general reference. A simple introduction can be found in a recent *Scientific American* article [5]. For the evolution of the experimental scene the articles by KRISHNAMURTI [6], GOLLUB [7], BERGE [8] and AHLERS [9] are good references.

Why are Rayleigh-Benard experiments such a paradigm for the observation of hydrodynamic instabilities?

The primary instability, transition from heat diffusion to heat convection, is well understood and documented. It is a second-order mean-field phase transition. The convective pattern may develop as an ordered two-dimensional flow, for low-aspect-ratio cell, or when an ordering field is applied [10, 11]. Cylindrical rolls, parallel to each other, will then define a new ordered state. Building upon it, and increasing the control parameter (Rayleigh number), a sequence of bifurcations will evolve up to a weak turbulent state. Let us emphasize that, in the more general case of a large-aspect-ratio cell and with no ordering field, the convective state above the transition will be disordered and the route to weak turbulence more complicated and not adequately approached by a dynamical-system theory [9].

It is also one of the rare cases, in nonlinear physics, where a detailed analysis of the secondary instabilities has been successfully performed [4], as a function of the Rayleigh number and of the Prandtl number. Finally, from a pure experimental point of view, the localization of the fluid flow in a small and closed vessel, with excellent thermal regulation of the boundaries, leads to very high signal-to-noise performance in the measurements [12]. In the oscillating state of the fluid, signal-to-noise ratio can be of the order of 60 dB.

1. – Basics of thermal convection.

The Oberbeck-Boussinesq approximation is used, this means that all fluid properties are assumed constant, with the exception of the temperature dependence of the density, taken into account in the gravity term. To get dimensionless numbers one uses d as a length scale, $d^2/\varkappa$ as a time scale, $(T_2 - T_1)/R$ as a temperature scale.

The governing equations become then

(1) $$P^{-1}\left(\frac{\partial}{\partial t}v + (v\cdot\nabla)v\right) = -\nabla p + \theta k + \nabla^2 v,$$

(2) $$\nabla\cdot v = 0,$$

(3) $$\frac{\partial}{\partial t}\theta + (v\cdot\nabla)\theta = Rk\cdot v + \nabla^2\theta,$$

where k is a vertical unit vector opposed to gravity, θ the deviation from the static temperature distribution, ∇p all terms that can be written in gradient form, and where R and P are the Rayleigh number and the Prandtl number.

The onset of convection defines a critical Rayleigh number, R_c, and a critical wave number, α_c, for a fluid layer of infinite horizontal extent:

$$R_c = 1707, \quad \alpha_c = 3.117.$$

Whereas for the onset of convection the Rayleigh number is the only relevant parameter, the Prandtl number enters insofar as the nonlinear properties of convection are concerned. Depending on the type of fluids used, *i.e.* the Prandtl number, various types of instabilities are observed as one increases the Rayleigh number. The main ones are called oscillatory instability (OS), skewed varicose (SV), zigzag (ZZ), cross rolls (CR). The so-called « Busse balloon » is shown in fig. 1: it indicates the region of stable rolls and the marginal stability curves. The experimental work, reported here, is related to two low-Prandtl-number fluids, helium ($P \simeq 0.5$) and mercury ($P = 0.025$). The two main instabilities of relevance are the oscillatory and the skewed varicose, which can be very easily deduced from fig. 1. Of all the instabilities present in the Busse diagram, the OS is the only time-dependent one. In an experiment it will lead to a time-dependent oscillation of the fluid velocity and temperature fields. This will represent in our experiment the first oscillator measured by our temperature probe [12].

From now on we shall deal only with the low-Prandtl-number fluid case. There, going back to the transport equations, the only relevant nonlinear term is the momentum advection one of the Navier-Stokes equation (1), $(v\cdot\nabla)v$, because it has a very large coefficient for small P. When this term

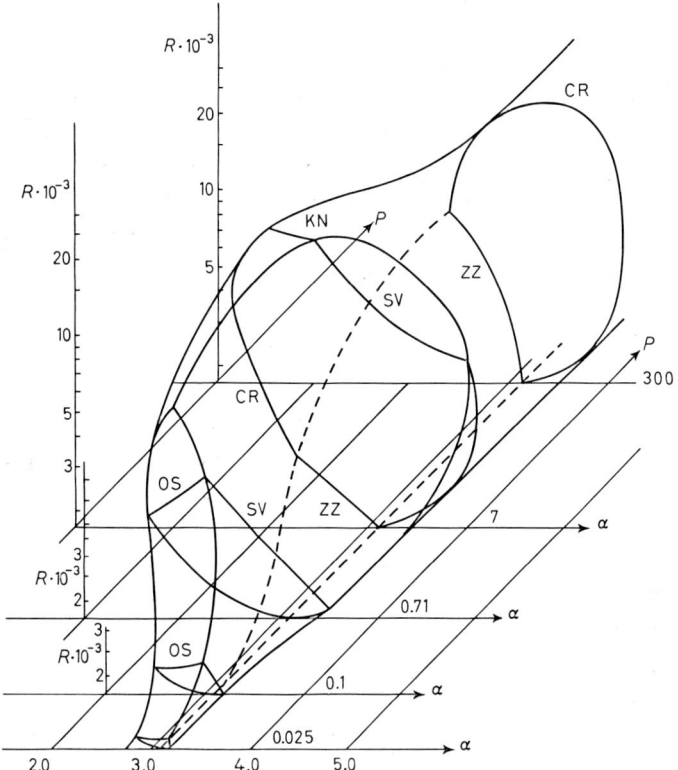

Fig. 1. – Busse stability analysis in the (α, P, R)-plane.

becomes larger than the diffusion one, $\nabla^2 \boldsymbol{v}$, the fluid will start to oscillate. From simple dimensional arguments, the onset is clearly related to a critical Reynolds number. So, whereas the onset of convection is related to a critical Rayleigh number, the onset of the oscillatory instability is related to a critical Reynolds number, given the roll wavelength as the length scale. The OS consists of a transverse, time-dependent oscillation of the convective rolls which propagates along the roll axis.

Let us now introduce another characteristic of one of our fluids, mercury. It is an electrical conductive fluid. Thus an applied magnetic field provides another experimental control parameter. Given the poor electrical conductivity of mercury, the main effect of an horizontal magnetic field will be to add a kind of anisotropic viscosity, inhibiting velocity variations along the roll axis. Shortly speaking, it introduces a magnetic rigidity. The dimensionless number associated with the magnetic-field control parameter is called the Chandrasekhar number [13]:

$$Q = \frac{\sigma B^2 d^2}{\varrho \nu}$$

(σ electrical conductivity, ϱ fluid density, d depth of the fluid layer, B magnetic field).

In the presence of an horizontal magnetic field, the onset of convection will not be affected, but the roll orientation will tend to align with the field direction. The onset of the oscillatory instability, on the other hand, will be strongly affected, the oscillatory motion being strongly inhibited by the field [11].

Recent experimental [14] and theoretical works [15] indicate that the oscillatory-instability onset is shifted towards larger Rayleigh numbers R in the presence of a field, following a power law:

$$R_Q - R_0 = Q^{1.2},$$

where R_0 is the R number without field and R_Q that with a field. Thus, by increasing the magnetic-field strength, one shifts the OS onset to higher Rayleigh numbers, thus testing stronger nonlinear regimes. In doing so we will show that various routes to a chaotic behavior will appear. Let us now show the relevance of such experiments to dynamical-system theory.

2. – Bifurcations from a limit cycle.

We are now entering into a new chapter, which we would intitulate « the mathematics of time ». It is related to the time behavior of the fluid. For low-P fluids, the secondary instability is a time-dependent one. This defines our first limit cycle. Referring now to Ruelle presentation, there are three generic ways a limit cycle can loose its stability, as a control parameter is increased:

  1) through a saddle node bifurcation, which leads to intermittency;

  2) through a pitchfork bifurcation, leading to a cascade of period-doubling bifurcations;

  3) through a Hopf bifurcation, which defines a quasi-periodic state.

Before developing the experimental observations of those three routes to a chaotic state, let us introduce an important nonlinear phenomenon, frequency locking of oscillators. Locking between oscillators will take us back from quasi-periodicity to periodicity.

3. – Frequency lock-in.

When two oscillators are present, of frequencies f_1 and f_2, the first nonlinear phenomena are related to frequency mixing. Due to quadratic nonlinear coupling, it leads to the appearance in the Fourier spectrum of the com-

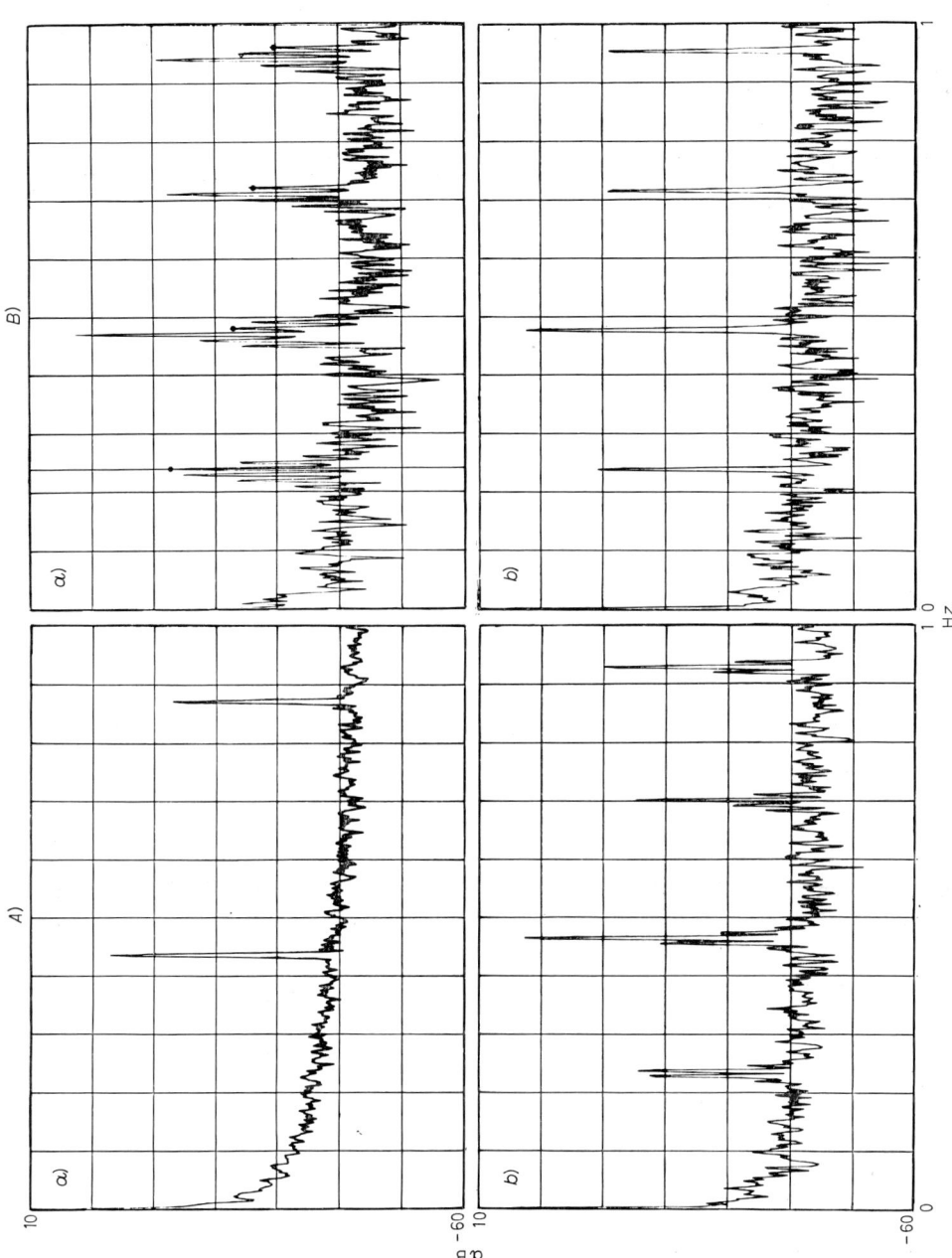

Fig. 2. – A) Onset of the oscillatory instability with harmonics, onset of the second oscillator and frequency mix-in; B) development of the mixing between modes, frequency lock-in. The fluid is liquid helium.

bination frequencies $mf_1 \pm nf_2$ [16]. For higher nonlinearities frequency lock-in phenomena may occur. There the low-amplitude oscillator becomes locked in frequency to the large-amplitude one, the frequency ratio f_1/f_2 taking a rational value. This value remains constant for some range value of the control parameter. A cascade of frequency lock-in states may eventually occur. In this case the various rational values follow the theory of Farey fractions [17]. For example, we show in fig. 2 the result of an experiment in liquid helium, where starting from a quasi-periodic state, a frequency lock-in state appears with a rational ratio of 2. A typical locking cascade is shown in fig. 3, again for the

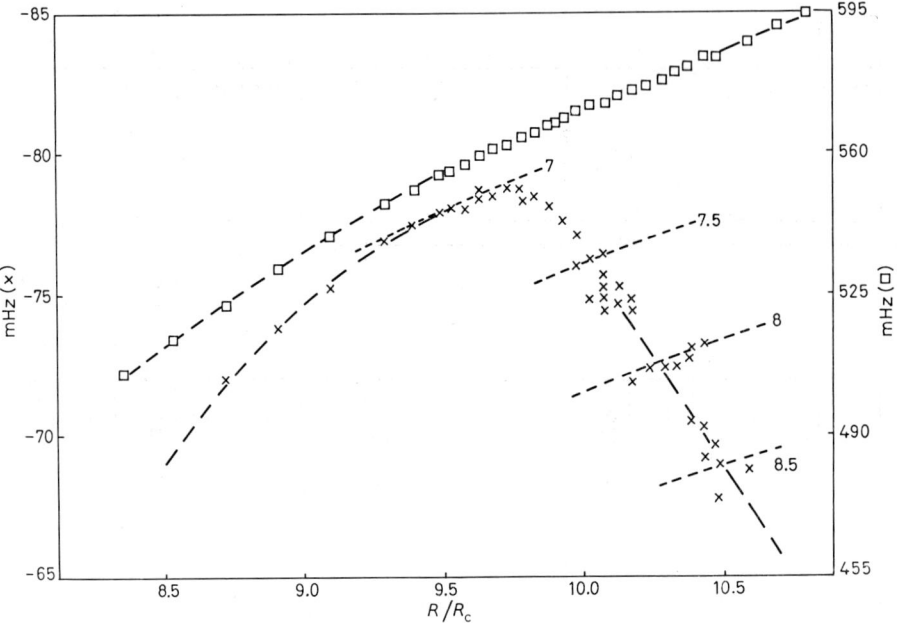

Fig. 3. – A cascade of locking states. One may observe the following ratios: 7, 22/3, 15/2, 23/3, 8, following the Farey fraction theory. The fluid is liquid helium.

case of liquid helium. The Farey theory goes as follows: if the main lock-in states, called parents, have rational values p_1/q_1 and p_2/q_2, the first term in the hierarchy will have a lock-in state with ratio $(p_1 + p_2)/(q_1 + q_2)$, and so forth for the succeeding ones. In fig. 3 the parents had values 7 and 8, so that the first daughter observed had a value 15/2, and the next one 22/3 and 23/3, which can be seen also [12].

4. – Saddle node bifurcation and intermittency.

We refer to subsect. 2`2a) of Ruelle lecture for the general presentation of the saddle node bifurcation. The experimental observation goes as follows:

starting from a limit cycle and increasing the control parameter, one reaches a critical value R_0, where occasionally a burst of noise appears in the time recording. For $R > R_0$ the bursts of noise become more frequent and finally the time recording becomes totally noisy. Figure 4 shows a typical recording in

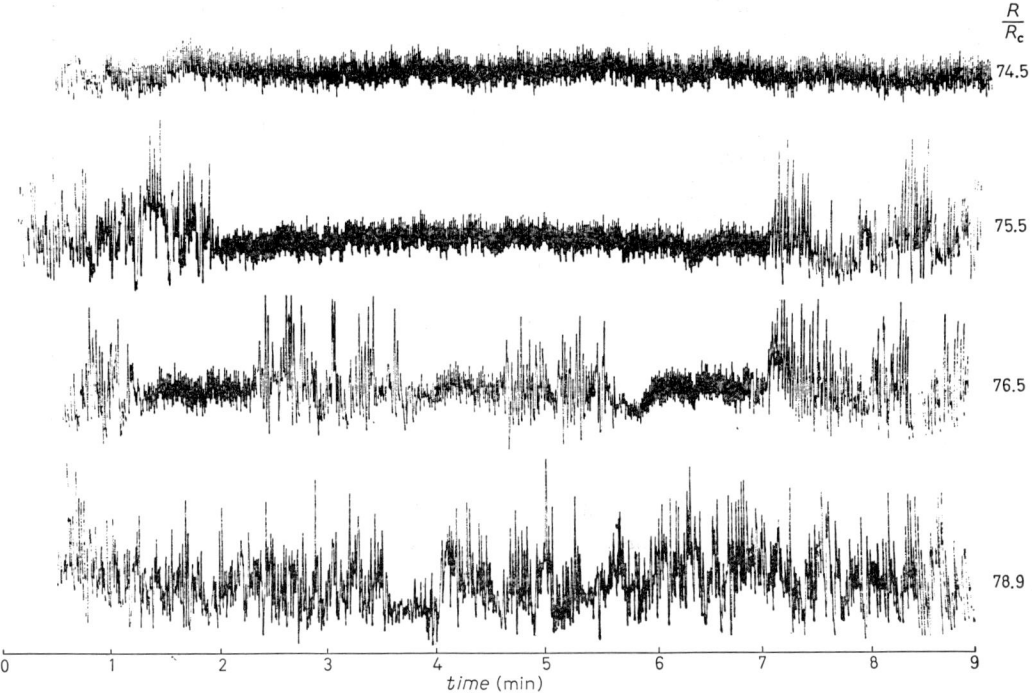

Fig. 4. – Intermittency from a torus, in liquid helium.

liquid helium [18]. In the model proposed by POMEAU and MANNEVILLE [19], the interesting features of this intermittent transition to chaos are their scaling properties. The laminar periods (period between two noise bursts) diverge as one approaches R_0, following a law

$$T \simeq (R - R_0)^{-\alpha}$$

for $R > R_0$, the value of α depending on the model. For the simplest case $\alpha = 1/2$.

Intermittency has also been observed in silicon oil [19] and water [7]. In the cascade of period-doubling bifurcations, discussed in the next section, intermittency is present any time one enters a stable band in the chaotic region [20].

An intermittent transition to chaos can also occur, starting from a quasi-periodic case, a two-torus, and we refer to Daido presentation [21].

5. – Pitchfork bifurcations and the period-doubling cascade.

The period-doubling scenario corresponds to a cascade of pitchfork bifurcations. The articles by MAY [22] and ECKMANN [23, 24] are two pedagogical presentations of the subject. Its relation with the Navier-Stokes equation has been studied by FRANCESCHINI. The three main features of the scenario are:

1) Starting from a limit cycle, one observes a cascade of period-doubling bifurcations, as the control parameter R is increased, with a geometrical progression for the successive R values (see Ruelle paper) up to the transition value R_∞. The amplitudes of the successive subharmonics are reduced by about 13 dB at each bifurcation [25].

2) Beyond R_∞ the flow is chaotic. Increasing R a reversed period-doubling cascade is observed, with a noisy background, and scaling properties also. Speaking in terms of phase transition, it is a second-order phase transition, the critical point being R_∞ and the order parameter the noise amplitude [26].

3) In the chaotic region, laminar windows are present everywhere, following the U sequence of Metropolis [27]. One enters a window through a saddle node bifurcation, thus intermittency. One leaves a window by another periodic doubling cascade, self-similar to the main one.

Experimental observations of part of the scenario were first performed in liquid helium [12] and water [28]. A more complete study was done in liquid mercury, with an horizontal magnetic field present [13]. The magnetic field appeared necessary, in order to introduce a large damping to the oscillatory instability. This, being analogous to large area contraction in phase space, allows to compare the results to one-dimensional maps. Otherwise, a two-dimensional mapping, such as the Henon one [29], seems necessary [30].

The cascade of period-doubling bifurcations is shown in fig. 5, for a mercury cell with four convective rolls. The value of δ measured was 4.4, and the subharmonic amplitude reduction 14 dB, thus close to the theoretical predictions.

The reversed cascade, in the chaotic regime, is shown on fig. 6, for the liquid-helium experiment. As one increases the Rayleigh number, the subharmonics disappear one by one, in reverse order from their appearance, each time replaced by a noisy band.

An example of a laminar window, in the chaotic regime, is shown in fig. 7 with a period-three order. More precisely, this figure is taken for a value of the control parameter where one leaves the period three, following a period-doubling cascade. Periods 6 and 12 and 24 can be seen, periods 12 and 24 being noisy, as the reversed cascade is already present.

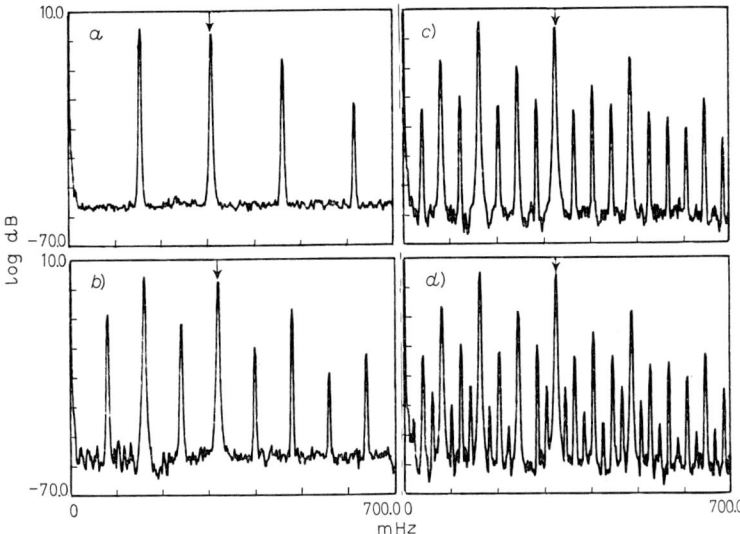

Fig. 5. – Cascade of period-doubling bifurcations, in mercury: a) $R/R_c = 3.47$, b) $R/R_c = 3.59$, c) $R/R_c = 3.64$, d) $R/R_c = 3.65$.

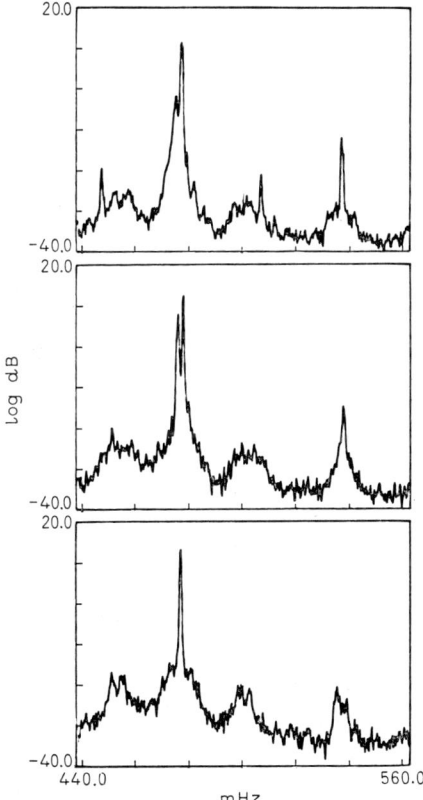

Fig. 6. – The reversed cascade in liquid helium.

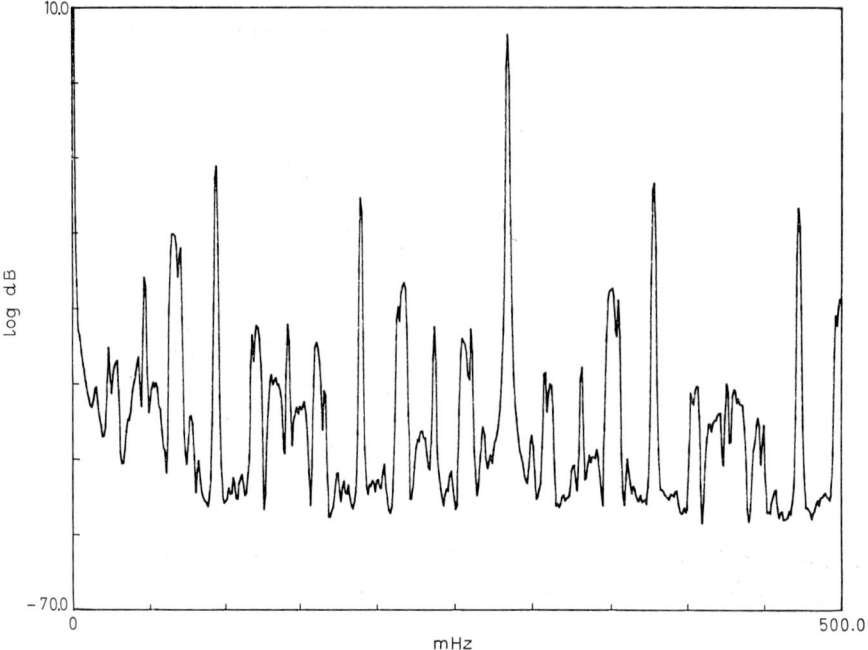

Fig. 7. – Period three in the chaotic region, in a mercury experiment.

6. – Hopf bifurcation and the Ruelle-Takens scenario.

The two preceding routes, intermittency and period doubling, were a well-defined scenario, which originate from the simple possible state, one limit cycle. Let us now try to describe a more general case, a succession of Hopf bifurcations. Starting from a limit cycle and increasing the control parameter, one reaches a quasi-periodic state through a Hopf bifurcation; will this process continue as LANDAU suggested [31], or will one reach a chaotic state after a small and finite number of Hopf bifurcations, as conjectured by RUELLE and TAKENS?

Up to now experiments in water, helium and mercury have shown that after three Hopf bifurcations at most the fluid reaches a chaotic state. The scenario for this transition is not uniquely defined, but the important result is that a finite and small number of bifurcations leads to a chaotic state.

Let us show, as an example, the results of an experiment in mercury. Beyond the onset of the oscillatory instability, a quasi-periodic state sets in, with two oscillators of frequencies f_1 and f_2 (fig. 8a)). Increasing R, one reaches a chaotic state (fig. 8b)), with three oscillators present, and a noise of very large amplitude, decreasing exponentially with frequency. Notice that the noise leaves unaffected the line shape of the frequency peaks. From the time recording a strange attractor can be reconstructed.

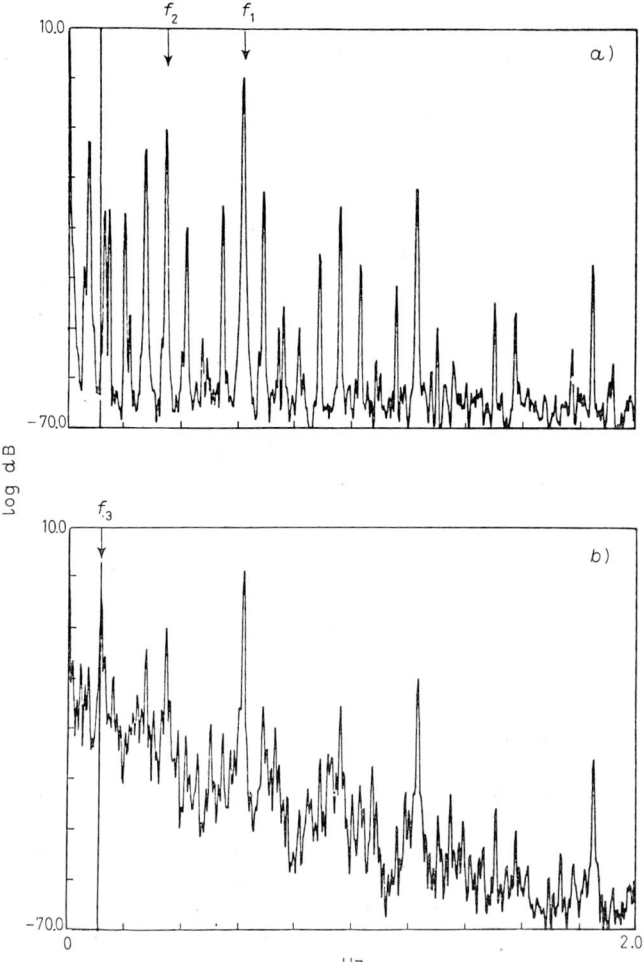

Fig. 8. – a) The quasi-periodic state for $R/R_c = 8.91$, b) the chaotic state for $R/R_c = 8.98$. The fluid is mercury.

There is much more to be said and done on this more general route to chaos, but it will need more experimental work.

REFERENCES

[1] H. BENARD: *Rev. Gen. Sci.*, **1**, 1261 (1900).
[2] Lord RAYLEIGH: *Philos. Mag.*, **32**, 529 (1916).
[3] H. JEFFREYS: *Philos. Mag.*, **2**, 823 (1926); *Proc. R. Soc. London, Ser. A*, **118**, 195 (1928).
[4] F. H. BUSSE: *Rep. Prog. Phys.*, **41**, 1929 (1978).

[5] M. Velarde and C. Normand: *Sci. Am.*, **243**, 78 (1980).
[6] R. Krishnamurti: *J. Fluid Mech.*, **42**, 295 (1970); **60**, 285 (1973).
[7] J. P. Gollub and S. V. Benson: *J. Fluid Mech.*, **100**, 449 (1980); J. P. Gollub, A. R. McCarriar and J. F. Steinman: *J. Fluid Mech.*, **125**, 259 (1982).
[8] P. Berge: in *Chaos and Order in Nature*, edited by H. Haken (Berlin, 1981).
[9] G. Ahlers and R. W. Walden: *Phys. Rev. Lett.*, **44**, 445 (1980).
[10] H. Benard and D. Avsec: *J. Phys. (Paris)*, **11**, 486 (1938).
[11] S. Fauve, C. Laroche and A. Libchaber: *J. Phys. (Paris) Lett.*, **42**, 455 (1981).
[12] A. Libchaber and J. Maurer: in *Non Linear Phenomena at Phase Transitions and Instabilities*, edited by T. Riste (New York, N. Y., 1981).
[13] A. Libchaber, C. Laroche and S. Fauve: *J. Phys. (Paris) Lett.*, **43**, 211 (1982); S. Chandrasekhar: *Hydrodynamic and Hydromagnetic Stability* (New York, N.Y., 1981).
[14] R. H. Busse and R. M. Clever: *The stability of convection rolls in the presence of an horizontal magnetic field*, preprint.
[15] J. Maurer and A. Libchaber: *J. Phys. (Paris) Lett.*, **40**, L419 (1979).
[16] M. Sargent, M. O. Scully and W. E. Lamb: *Laser Physics* (Reading, Mass., 1974).
[17] T. Allen: *Physica (Utrecht)*, **6**, 305 (1983).
[18] J. Maurer and A. Libchaber: *J. Phys. (Paris) Lett.*, **41**, 515 (1980).
[19] P. Berge, M. Dubois, P. Manneville and Y. Pomeau: *J. Phys. (Paris) Lett.*, **41**, 341 (1980).
[20] J. E. Hirsch, B. A. Huberman and D. J. Scalapino: *Phys. Rev. A*, **25**, 519 (1982).
[21] H. Daido: *Onset of intermittency from torus*, preprint.
[22] R. M. May: *Nature (London)*, **261**, 459 (1976).
[23] J. P. Eckmann: *Rev. Mod. Phys.*, **53**, 643 (1981).
[24] P. Collet and J. P. Eckmann: *Integrated Maps on the Interval as Dynamical Systems* (Basel, 1980); M. Feigenbaum: *J. Stat. Phys.*, **19**, 25 (1978); **21**, 669 (1979).
[25] M. Nauenberg and J. Rudnick: *Phys. Rev. B*, **24**, 493 (1981).
[26] J. P. Crutchfield, J. D. Farmer and B. A. Huberman: *Phys. Rep.*, **92**, 47 (1982).
[27] N. Metropolis, M. L. Stein and P. R. Stein: *J. Comb. Theory, Ser. A*, **15**, 25 (1973).
[28] M. Giglio, S. Muzatti and U. Perini: *Phys. Rev. Lett.*, **47**, 243 (1981).
[29] M. Henon: *Commun. Math. Phys.*, **50**, 69 (1976).
[30] A. Arneodo, P. Coullet, C. Tresser, A. Libchaber, J. Maurer and D. D. Humieres: *Physica (Utrecht) D*, **6**, 385 (1983).
[31] L. Landau and E. Lifshitz: *Fluid Mechanics* (London, 1959).

Small Stochastic Perturbations of Dynamical Systems.

G. Jona-Lasinio

Dipartimento di Fisica dell'Università « La Sapienza » - Roma, Italia

1. – Preliminary remarks.

The theory of small stochastic perturbations of dynamical systems is a branch in the theory of stochastic processes of considerable significance for both physics and mathematics. From the stand-point of physics it has almost universal applicability, as any system living in the real world is subject to disturbances of the most diverse nature which very often are adequately described by some stochastic process.

Of course, in many cases the effects of such perturbations are irrelevant and the idealized models of Galilean and Newtonian physics are sufficient. However, if one observes a system over a sufficiently long interval of time, the situation may be substantially different. Not only there may be deviations from the deterministic behaviour predicted by some differential equation: new phenomena may appear which are not possible in a deterministic scheme or, if in principle possible, require a higher level of sophistication in the mathematical modelling.

Small random perturbations of dynamical systems are also very interesting from a purely mathematical point of view. There is now a considerable body of results which may be considered as the core of a more general qualitative theory of stochastic differential equations. Such a theory, besides its intrinsic interest, is relevant for a deeper analysis of deterministic dynamical systems and has received important applications in the study of elliptic and parabolic linear partial differential equations.

From the probabilistic point of view the theory of small random perturbations of dynamical systems is a chapter in the so-called theory of large deviations. However, its strict connection with differential equations gives it a more intuitive and appealing flavour.

The purpose of these lectures is to give an introduction to some of the basic mathematical ideas in this field. The exposition will be largely descriptive and only in a few key points we will provide some details. The mathematically inclined reader may obtain a deeper picture going through the literature given at the end.

2. – Motion along the flow.

We begin with a simple comparison problem. Suppose we are given a stochastic differential system

$$(2.1) \qquad \dot{x}_t^\varepsilon = b(x_t^\varepsilon) + \varepsilon \dot{W}_t, \qquad x_0^\varepsilon = x \in R^d,$$

where ε is a small number and W_t is the Wiener process. A natural question is then: what is the relationship between the trajectories of the stochastic process x_t and the solutions of the deterministic problem

$$(2.2) \qquad \dot{x}_t = b(x_t), \qquad x_0 = x.$$

Intuitively we would expect that for ε small the trajectories of (1.1) and (1.2) should remain close to each other at least for some finite time T. Let us show that this is, in fact, the case. More precisely we want to estimate the probability $P\left(\sup_{0 \leq t \leq T} |x_t^\varepsilon - x_t| > \delta\right)$ and show that it can be made as small as we like by a convenient choice of ε. From (2.1) and (2.2) we obtain

$$(2.3) \qquad x_t^\varepsilon - x_t = \int_0^t [b(x_s^\varepsilon) - b(x_s)] \, ds + \varepsilon W_t.$$

If now $b(x)$ satisfies the usual Lipschitz condition which ensures existence and uniqueness of the solution, from (2.3) we obtain

$$(2.4) \qquad \sup_{0 \leq t \leq T} |x_t^\varepsilon - x_t| \leq \varepsilon \exp[KT] \sup_{0 \leq t \leq T} |W_t|,$$

where K is defined by $|b(x_1) - b(x_2)| \leq K|x_1 - x_2|$. Therefore,

$$(2.5) \qquad P\left(\sup_{0 \leq t \leq T} |x_t^\varepsilon - x_t| > \delta\right) \leq P\left(\sup_{0 \leq t \leq T} |W_t| > \frac{\delta}{\varepsilon} \exp[-KT]\right) \leq$$

$$\leq 2P\left(|W_T| > \frac{\delta}{2\varepsilon} \exp[-KT]\right) \approx$$

$$\approx \left(\frac{\delta^2}{\varepsilon^2 T} \exp[-2KT]\right)^{d/2-1} \exp\left[-\frac{\delta^2}{8\varepsilon^2 T} \exp[-2KT]\right].$$

d is the dimensionality of the space. In going from the first to the second line we have used a well-known property of the Brownian motion. We then see that, by keeping T and δ fixed, this probability is exponentially small when $\varepsilon \to 0$.

The meaning of this estimate is that, if any subset A of the space of continuous trajectories in R^d defined in the interval of time $[0, T]$ contains the function x_t together with its δ-neighbourhood (or δ-tube), the main contribution to the probability $P(x^\varepsilon \in A)$ comes from this δ-neighbourhood, the remaining trajectories giving only an exponentially small term.

In many problems, however, one is interested in a more systematic study of those events A which do not contain the trajectory x_t. Such events are important, for example, in connection with stability problems, where one is interested in calculating the exit time from a neighbourhood of an equilibrium position.

Fig. 1.

It is clear that any set of trajectories leading away from 0 does not contain any solution of (2.2). Similar problems are encountered in the study of the equilibrium measure of the process x_t generated by (2.1) when $\varepsilon \to 0$.

3. – Motion against the flow.

We refer again to eq. (2.1) and associate with it the so-called action functional

$$(3.1) \qquad I_T(\varphi) = \tfrac{1}{2} \int_0^T |\dot\varphi_t - b(\varphi_t)|^2 \, dt$$

defined on all φ_t continuous in $[0, T]$ and satisfying $\varphi_0 = x$. When the integral does not exist (a continuous function may be nondifferentiable), we set $I_T(\varphi) = \infty$. $I_T(\varphi)$ measures in some sense how much φ_t differs from a solution of (2.2). We define in addition

$$(3.2) \qquad I(A) = \inf_{\varphi \in A} I_T(\varphi),$$

where A is a set of trajectories.

All the calculations that follow are based on the fundamental relationship

$$(3.3) \qquad I(A) = -\lim_{\varepsilon \to 0} \varepsilon^2 \log P(x^\varepsilon \in A).$$

(3.3) says that for small ε we can write

$$(3.4) \qquad \exp\left[-\frac{I(A)+h}{\varepsilon^2}\right] < P(x^\varepsilon \in A) < \exp\left[-\frac{I(A)-h}{\varepsilon^2}\right],$$

where h is a small error. It establishes a relationship between the probability of the event A and a quantity $I(A)$ which is defined exclusively in terms of the deterministic system (2.2).

We want to estimate now the probability that the process x_t^ε, initially in a neighbourhood of an equilibrium position O of (2.2), leaves a domain D containing O, within a given time T. To simplify the calculation, we first suppose that the vector field b is a gradient field, that is

(3.5) $$b = -\nabla U.$$

This restriction will be removed later.

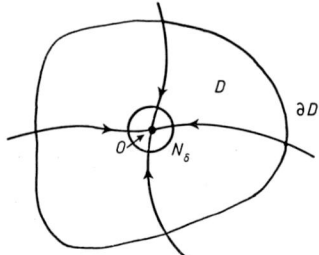

Fig. 2.

We define the set A

(3.6) $$A = \{\varphi_t : \varphi_0 \in N_\delta,\ \varphi_T \notin D\},$$

where N_δ is a small sphere of diameter δ centred at O. A contains all the trajectories starting within N_δ and leaving D before T. Because of (3.4), to obtain a lower bound on $P(x^\varepsilon \in A)$, we need an upper bound on $I(A)$. By definition we have

(3.7) $$I(A) \leqslant I_T(\varphi), \qquad\qquad \varphi \in A.$$

In order to get a good estimate, we have to choose φ carefully.

We split our trajectory in two pieces. The first $\varphi_t^{(1)}$ connects any initial point x in the interior of N_δ with any point x' on the boundary of N_δ and takes a time $\tau < T$. More precisely, we choose $\varphi_t^{(1)} = x(1 - t/\tau) + x' t/\tau$, the linear path joining x and x'. For this part of the trajectory we have the easy estimate

$$I_\tau(\varphi) \leqslant \int_0^\tau [|\dot\varphi_t|^2 + |\nabla U|^2]\, dt \leqslant \frac{|x - x'|^2}{\tau} + C\tau,$$

with C a positive constant. If we take $\tau = |x - x'|/C$, we obtain

(3.8) $$I_\tau(\varphi) \leqslant (C+1)|x-x'| \leqslant (C+1)\delta.$$

To construct the second part of the trajectory $\varphi_t^{(2)}$ which takes a time $\bar{T} = T - \tau$, we observe the identity valid for any φ

(3.9) $$I_T(\varphi) = \tfrac{1}{2}\int_0^T |\dot\varphi_t + \nabla U|^2 \, dt = \tfrac{1}{2}\int_0^T |\dot\varphi_t - \nabla U|^2 \, dt + 2[U(\varphi_T) - U(\varphi_0)].$$

(3.9) shows that, if φ is a solution of the system $\dot\varphi = \nabla U = -b$, the integral on the right-hand side vanishes. Now, if O was a sink for the original dynamical system, it will become a source when we change the sign of b or ∇U. Therefore, if we choose for $\varphi_t^{(2)}$ the solution of the above equation starting at x' for $t = 0$, the integral will give no contribution provided $\bar{T}$ and, therefore, T are large enough to reach the boundary ∂D. It is not difficult to see that typically we need $T \geqslant O(\ln(1/\delta))$. We arrive then at our optimal estimate

(3.10) $$I(A) \leqslant I_\tau(\varphi^{(1)}) + I_{\bar{T}}(\varphi^{(2)}) \leqslant 2\left[\inf_{x \in \partial D} U(x) - U(0)\right] + c(\delta) =$$
$$= 2[U(\bar{x}) - U(0)] + c(\delta),$$

where $c(\delta)$ is an error term and $\bar{x}$ is the point of ∂D where $U(x)$ reaches its minimum. We have used the continuity of U to replace $U(x')$ with $U(0)$.

Let us show that we can easily construct a lower bound for $I(A)$ starting again from (3.9). By choosing a $T' < T$ from (3.9), we have the obvious inequality

(3.11) $$I_T(\varphi) \geqslant \tfrac{1}{2}\int_0^{T'} |\dot\varphi + \nabla U|^2 \, dt =$$
$$= \tfrac{1}{2}\int_0^{T'} |\dot\varphi - \nabla U|^2 \, dt + 2[U(\varphi_{T'}) - U(\varphi_0)] \geqslant 2[U(\varphi_{T'}) - U(\varphi_0)].$$

We take T' as the time φ_t hits the boundary ∂D. Then, since $U(\varphi_{T'}) \geqslant U(\bar{x})$, we have

(3.12) $$I(A) = \inf_{\varphi \in A} I_T(\varphi) \geqslant 2[U(\bar{x}) - U(0)] - c'(\delta).$$

(3.4), (3.10) and (3.12) give us finally

(3.13) $$\exp\left[-\frac{2[U(\bar{x}) - U(0)] + h + c(\delta)}{\varepsilon^2}\right] <$$
$$< P(x^\varepsilon \in A) < \exp\left[-\frac{2[U(\bar{x}) - U(0)] - h - c'(\delta)}{\varepsilon^2}\right].$$

We would like to emphasize that our estimate would not have changed if we had started at a point $x \in D$ at a finite distance from O. In fact, the system would fall immediately in a small neighbourhood of O along a deterministic trajectory of (2.1) as discussed in sect. **2** and after a long time it will succeed in leaving D. Actually, with a calculation similar to the one we have done, one can show that the average time necessary to leave D is

$$(3.14) \qquad E(\tau) \approx \exp\left[\frac{2[U(\bar{x}) - U(0)]}{\varepsilon^2}\right],$$

where τ is the moment of first exit from D and $\approx$ means that error terms like in (3.13) must be included. One can also show that $\bar{x}$ is the point where the system will pass with high probability when leaving D.

The hypothesis that b is a gradient field can be easily removed. We introduce the so-called quasi-potential relative to the stable equilibrium point O

$$(3.15) \qquad V_0(x) = \inf_{\substack{\varphi_0 = 0 \\ \varphi_T = x \\ T}} I_T(\varphi).$$

The inf is taken over all possible trajectories connecting O and x in an arbitrary time T. $V_0(x)$ can be shown to be a continuous function of x. All the previous estimates then remain true provided one substitutes $2[U(\bar{x}) - U(0)]$ with $V_{0D} = \inf_{x \in \partial D} V_0(x)$.

This is an important generalization as it allows to include, for example, the case when the stable-equilibrium orbit is a limit cycle. In this case in (3.15) we can take any point O on the limit cycle as the initial point of the trajectory φ. Since any two points O and O' of the equilibrium orbit are connected by a solution of $\dot{\varphi} = b(\varphi)$, the value of $V_0(x)$ will not depend on which point is chosen.

4. – The invariant measure.

A prominent question in the study of stochastic perturbations of a dynamical system is the behaviour when t becomes very large. The interesting cases are those in which the process evolves towards a stationary-equilibrium situation. This will be the case, for example, if for large x the scalar product $\langle b \cdot x \rangle < 0$. We are then especially interested in knowing the structure of the equilibrium distribution. If $b = -\nabla U$, it is well known that the invariant measure or equilibrium distribution has a density which is simply given by

$$(4.1) \qquad \varrho(x) = C \exp\left[-\frac{2U(x)}{\varepsilon^2}\right],$$

where C is a normalization constant. From (4.1) we see that the measure for $\varepsilon \to 0$ will be sharply peaked near the minima of $U(x)$, i.e. near the stable-equilibrium points of system (2.2). Actually in the end it will be concentrated on the lowest minimum. This is an interesting fact, because it tells us that not all the stationary solutions of the unperturbed system (i.e. its equilibrium points) can be obtained as limits when the perturbation vanishes. The perturbation somehow selects the most stable configuration of the system. The natural question is then whether this type of conclusion continues to hold also for the nongradient case for which we have no explicit expression of the invariant measure as (4.1). The answer is affirmative and for a large class of systems can be obtained through a very interesting construction that we now describe. The class of systems that we consider here are the so-called Morse-Smale dynamical systems which exhibit a finite number of equilibrium positions or limit cycles. The conclusions, however, hold also for more general systems characterized by chaotic behaviour. The important thing is that the attractors be compact sets.

Consider first the simple case of a single stable-equilibrium position O. We surround O with two small spheres that we denote by Γ and γ.

Fig. 3.

Define the following sequence of increasing random times $\tau_0, \sigma_0, \tau_1, \sigma_1, \ldots$, $\tau_0 = 0$:

$$\sigma_n = \inf \{t > \tau_n : x_t^\varepsilon \in \Gamma\}, \qquad \tau_n = \inf \{t > \sigma_{n-1} : x_t^\varepsilon \in \gamma\}.$$

Therefore, σ_0 is the first time the process hits Γ, τ_1 is the first time the process hits γ after having touched Γ and so on. Define then the sequence $Z_n = X_{\tau_n}^\varepsilon$. One can easily show that, due to the strong Markov property of the process, the sequence Z_n is a Markov chain with state space γ. It has a unique invariant measure $\nu(\mathrm{d}x)$ defined on γ. A remarkable theorem due to HASMINSKI now asserts that the invariant measure μ of the process x_t^ε can be expressed in terms of ν through the following formula:

$$(4.2) \qquad \mu(D) = C \int_\gamma E_x \left(\int_0^{\tau_1} \chi_D(x_s^\varepsilon) \, \mathrm{d}s \right) \nu(\mathrm{d}x),$$

where χ_D is the indicator function of the set D in R^d, E_x is the expectation with respect to the process starting at x and C is a normalization constant. (4.2) expresses the equilibrium probability of D in terms of the time spent by the process in D before the first hit on γ. Starting from (4.2), we can obtain estimates for small ε. Defining $\tau_D = \inf\{t : x_t^\varepsilon \in D\}$, we have the straightforward estimate

$$(4.3) \qquad \mu(D) \leqslant C \max_{x \in \gamma} P(\tau_D < \tau_1) \max_{x \in \partial D} E_x \tau_1 .$$

From (4.3) one arrives, through calculations similar to those of sect. 3, at the estimate

$$(4.4) \qquad \mu(D) < \exp\left[-\frac{V_0 - h}{\varepsilon^2}\right],$$

where $V_0 = \inf_{x \in D} V_0(x)$; $V_0(x)$ is the quasi-potential defined by (3.15). It can be shown that also the opposite inequality holds, so that for small ε

$$(4.5) \qquad \mu(D) \approx \exp\left[-\frac{V_0}{\varepsilon^2}\right].$$

The previous construction and estimates can be extended to the case of several stable-equilibrium positions or stable limit cycles. The Markov chain now will have as state space the union $\bigcup_i \gamma_i$ of the spheres constructed for each attractor. The new feature appearing in this case is the possibility of jumps from one attractor to the other. It can be shown that the probability of transition from one point $x \in \gamma_i$ to some neighbourhood $B \subset \gamma_j$ depends very weakly on both x and B for ε small. In fact, it depends primarily only on the pair i, j, and the estimate holds

$$(4.6) \qquad P_{ij} \approx \exp\left[-\frac{V_{ij}}{\varepsilon^2}\right],$$

where $V_{ij} = \inf_{\varphi_0 = O_i, \varphi_T = O_j} I_T(\varphi)$, i.e. the minimum is taken over all trajectories connecting the attractors O_i and O_j in any interval of time T.

At this point we have obtained a drastic reduction of the phase space of the chain which consists of a finite number of states represented by the different attractors. The invariant measure associated with this chain can be easily calculated and, in general, it will be concentrated on a particular attractor. It is a theorem that *also* the invariant measure of the process is concentrated on the same attractor for ε small. In addition, if there is only one measure μ_0 on the attractor invariant for the dynamical system (2.2), the invariant measure of the process tends weakly to μ_0.

As we mentioned before, this type of conclusions hold also for systems with chaotic behaviour: a particular invariant measure is selected by the random perturbation.

To conclude our discussion on the invariant measure, we briefly analyse the so-called stabilization problem. Let us consider the probability that the process starting at x will belong to the set D at time t.

We denote this probability $P(x_t^\varepsilon \in D)$ and we want to study its behaviour when both $\varepsilon \to 0$ and $t \to \infty$. It is clear that, if D contains the limit set of the solution x_t of (2.2) starting at x, we expect $\lim_{t \to \infty} \lim_{\varepsilon \to 0} P = 1$. On the other hand, if D contains all the limit sets of (2.2), we obtain the same result if we invert the order of the limits, i.e. $\lim_{\varepsilon \to 0} \lim_{t \to \infty} P = 1$. The question we ask is what happens if $\varepsilon \to 0$ and $t \to \infty$ in some correlated way. One can prove the following. If D contains only one stable equilibrium and is entirely contained in its basin of attraction, when we let $t(\varepsilon) \to \infty$ as the characteristic exit time from D, we obtain

(4.7) $$\lim_{\varepsilon \to 0} P(x_{t(\varepsilon)}^\varepsilon \in D) = 1 .$$

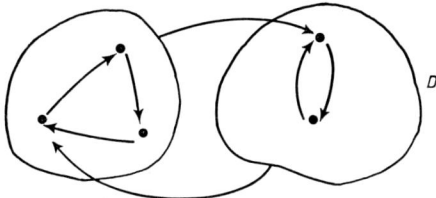

Fig. 4.

More precisely we take $0 < \lim_{\varepsilon \to 0} \varepsilon^2 \log t(\varepsilon) < V_{0D}$ (see (3.14) and (3.15)). The meaning of (4.7) is that, even if the invariant measure of the process is not concentrated on the attractor contained in D, on the scale $t(\varepsilon)$ everything goes as if it were the support of the stable invariant measure and the system is trapped in D. In general, i.e. with D containing several attractors, more complicated things can happen and the process can be trapped into subcycles (see fig. 4).

If one takes the limit $\varepsilon \to 0$, $t(\varepsilon) \to \infty$ in an appropriate way, (4.7) holds for D containing a subcycle.

5. – Infinite-dimensional dynamical systems.

In this section we discuss the generalization of the previous theory to an infinite-dimensional context. Since there is not yet a systematic treatment of infinite-dimensional dynamical systems, we describe immediately the model problem that we shall consider. Formally our equation can be written

(5.1) $$\frac{\partial u}{\partial t}(x, t) = -\frac{\delta S(u)}{\delta u(x, t)} + \varepsilon \alpha(x, t) ,$$

where $S(u)$ is the potential functional

(5.2) $$S(u) = \int_0^L \left[\frac{1}{2}\left(\frac{\partial u}{\partial x}\right)^2 + V(u)\right] dx,$$

$V(u)$ is a polynomial function of u and α is a two-dimensional Gaussian white noise characterized by the covariance

(5.3) $$E(\alpha(x, t)\alpha(x', t')) = \delta(x-x')\delta(t-t').$$

Therefore, (5.1) is an infinite-dimensional gradient system which can be written more explicitly

(5.4) $$\frac{\partial u}{\partial t} = \frac{\partial^2 u}{\partial x^2} - V'(u) + \varepsilon\alpha.$$

The model is specified further by assuming Dirichlet boundary conditions on the interval $[0, L]$, i.e. $u(0, t) = u(L, t) = 0$ $\forall t$ and taking

(5.5) $$V(u) = \frac{\lambda}{4}u^4 - \frac{\mu}{2}u^2, \qquad \lambda, \mu > 0.$$

Our model problem is inspired by the quantum-mechanical double-well anharmonic oscillator. In fact, the formal equilibrium density of (5.4) $\exp[-2S(u)/\varepsilon^2]$ is just the density in the functional integral description of quantum mechanics at imaginary time. However, other interpretations are possible. We may think of (5.4) as describing the motion of a nonlinear elastic string in a high-viscosity noisy environment (Smoluchowsky approximation). Heat diffusion in a nonlinear environment is another possible interpretation: u in this case would be the local temperature. Both are useful references for intuition.

We begin our analysis by discussing the equilibrium positions of the deterministic equation corresponding to $\varepsilon = 0$. They are solution of the equation

(5.6) $$\frac{d^2 u}{dx^2} - \lambda u^3 + \mu u = 0$$

with Dirichlet boundary conditions. It can be shown that (5.6) has a finite number of solutions u_i for sufficiently large values of $\mu^{\frac{1}{2}} L$. They are naturally classified by the increasing values of $S(u_i)$. Of course, due to the symmetry of the equation $u \to -u$, if u_i is a solution, also $-u_i$ is a solution. Only the lowest ones $\pm u_1$ are stable equilibria. All the others are saddle points. The shapes of u_1 and u_2 are roughly those in fig. 5 and 6. $\pm(\mu/\lambda)^{\frac{1}{2}}$ are the minima of the double-well potential (5.5). When i increases, the number of oscillations increases.

The intuition gained in the finite-dimensional case now suggests that the solutions of (5.5) will fluctuate for long times near the stable configurations $\pm u_1$. From time to time a large fluctuation will drive u from the vicinity

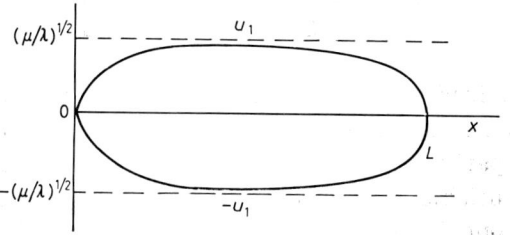

Fig. 5.

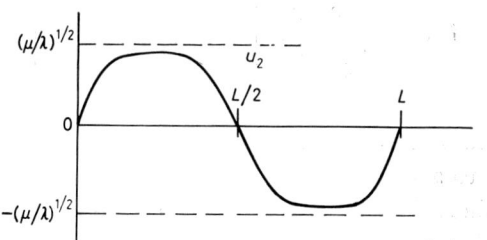

Fig. 6.

of u_1 to the vicinity of $-u_i$ or *vice versa*. Our problem then consists in evaluating the probability that such a jump takes place in a certain time T. Here one has to face a certain number of technical questions. To give a rigorous meaning to (5.4), we transform it into an integral equation

$$(5.7) \qquad u = -G * V'(u) + \varepsilon Z + \bar{u},$$

where $G(x, t)$ is the fundamental solution of the linear heat equation, $\bar{u}$ is the solution of the same equation satisfying the initial condition $\bar{u}(x, 0) = u_0(x)$, Z is the Gaussian process $G * \alpha$ and $*$ means convolution in both time and space.

The Gaussian process $Z(x, t)$ is completely defined by its covariance, that can be written explicitly

$$(5.8) \qquad E\big(Z(x, t) Z(x', t')\big) = G * G^+(x, t, x', t').$$

From (5.8), using well-known methods for the heat equation and theorems on Gaussian processes, one proves that the trajectories of Z are continuous with probability one. At this point (5.7) can be solved for each $Z(x, t)$ continuous.

The solution u will also be continuous. Here we meet a difficulty typical of the infinite-dimensional nature of the problem: the potential $S(u)$ is not continuous in the space of continuous functions! In fact, a continuous function u in general is nondifferentiable and $S(u) = \infty$. Nevertheless, one finds that the transition probability from one stable equilibrium to the other is of the order

$$(5.9) \qquad P \approx \exp\left[-\frac{2[S(u_2) - S(u_1)]}{\varepsilon^2}\right],$$

that is P is determined by the potential difference between u_1 and the lowest saddle point u_2 as in the finite-dimensional case. The reason why it works is the strongly dissipative character of the heat equation. Even if one starts from a nondifferentiable u, this relaxes very rapidly to a smooth function. To arrive at (5.9) one defines in analogy with (3.1) the action functional

$$(5.10) \qquad I(u) = \frac{1}{2} \int_0^T dt \int_0^L dx \left[\frac{\partial u}{\partial t} - \frac{\partial^2 u}{dx^2} + V'(u)\right]^2$$

and proves an estimate like (3.4).

It is interesting to ask which is the shape of the trajectory leading from one attractor to the other. This, like in the finite-dimensional case, will be a solution of the equation with inverted drift until it reaches the saddle point u_2. It then relaxes to the new equilibrium following the flow.

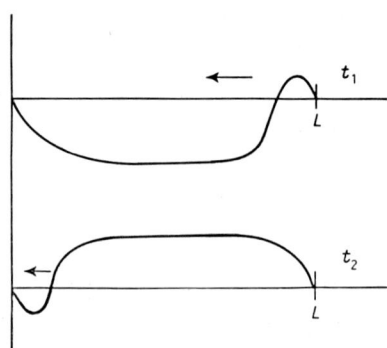

Fig. 7.

Preliminary computer calculations indicate an evolution of the form represented in fig. 7.

A small bump appears near one of the ends and then propagates until the whole solution is on the other side.

In conclusion we would like to comment on the possibility of extending the theory to higher space dimensions, that is to the equation

$$\frac{\partial u}{\partial t} = \Delta u - V'(u) + \varepsilon \alpha, \tag{5.11}$$

where Δ is the Laplacian. Difficulties now arise already at the level of defining what is a solution to (5.11). When we transform (5.11) into an integral equation like (5.7), the Gaussian process $Z(x, t)$ has typical trajectories which are distributions. Therefore, the equation itself has no meaning unless we interpret it in a suitable way. For $d = 2$, by adopting a suitable notion of probabilistic weak solution, one can reduce the problem to one in constructive field theory where the usual renormalization techniques can be applied. A large-deviation theory for this case along the lines of these lectures has yet to be constructed.

BIBLIOGRAPHY

The basic reference in this field is A. D. VENTZEL and M. I. FREIDLIN: *Fluctuations in Dynamical Systems under the Action of Small Random Perturbations* (Moskva, 1979). An English translation has appeared in the *Springer Series Grundlehren der Mathematischen Wissenschaften*, Vol. **260** (1984). In sect. **2**, **3** and **4** I have closely followed this book.

If the above book is unavailable, a good general reference is still the original article by A. D. VENTZEL and M. I. FREIDLIN: *On small random perturbations of dynamical systems*, Russ. Math. Surv., **25**, 1 (1970).

For the topics discussed in the last part of sect. **4** one may consult YU. KIFER: *On small random perturbations of some smooth dynamical systems*, Math. USSR Izv., **8**, 1083 (1974); M. I. FREIDLIN: *Subliminting distributions and stabilization of solutions of parabolic equations with a small parameter*, Sov. Math. Dokl., **18**, 1114 (1977).

Section **5** is based on W. G. FARIS and G. JONA-LASINIO: *Large fluctuations for a non-linear heat equation with noise*, J. Phys. A, **15**, 3025 (1982).

In connection with infinite-dimensional dynamical systems of the type discussed in sect. **5** the following is a good reference: D. HENRY: *Geometric Theory of Semilinear Parabolic Equations*, Lect. Notes Math., Vol. **840** (Berlin, 1981).

For the relevance of the theory described in these notes in connection with geophysical problems see R. BENZI and A. SUTERA: this volume, p. 403.

PART II

FULLY DEVELOPED TURBULENCE

A Comparative Pathology of Atmospheric Turbulence in Two and Three Dimensions.

H. TENNEKES

Royal Netherlands Meteorological Institute - de Bilt
Free University - Amsterdam

1. – Introduction.

My exposure to turbulence started in 1958, when I took graduate courses from Francis CLAUSER, Stanley CORRSIN, Lesley KOVASZNAY and Owen PHILLIPS at the John Hopkins University. Looking back at twenty-five years of turbulence research, I find that the concepts I associate with the word turbulence have changed radically over the years. Before 1970, I would not have dreamt of putting the words turbulence and predictability side by side, as in the title of this summer course. To me, turbulence was unpredictable by definition. Turbulence was the chaos that arises in fluids because of the innumerable instabilities associated with vortex stretching. I remember how upset I was when the terms two-dimensional turbulence and geostrophic turbulence were coined. Fluid motion without a vigorous energy cascade toward small scales should not claim the space I had reserved for my own ideas.

I sought refuge in my dictionary. It gives the following definitions of turbulence: 1) violent disorder or commotion, 2) haphazard secondary motion caused by eddies in a moving fluid, 3) irregular motion of the atmosphere, as that indicated by gusts and lulls in the wind. The references to utter lack of predictability abound: « disorder », « haphazard », « irregular », « gusts and lulls ». These terms certainly do not point in the direction of calculations of error growth in the chaotic evolution of deterministic systems. However, a dictionary reflects the wisdom of the past, and I have long ago given up my resistance to the evolution of the concepts associated with the label « turbulence ». These days, I tend to think of turbulent flow as flow in which deterministic calculations become useless in a finite time interval.

The term « useless » refers to a person who makes a pragmatic decision, based on the circumstances in which he operates. To a general-circulation researcher, all eddy motions in the atmosphere, including the synoptic eddies, are taken as turbulence. A numerical modeler, on the other hand, is interested in deterministic forecasts of synoptic-scale motions; he treats the small-scale stuff (clouds, boundary-layer eddies and the like) as turbulence. It all depends on the time and length scales of the phenomena one is interested in. Most people do not mind the lack of predictability of gusts and lulls in the wind, but a small-craft sailor watches patches of cats' paws eagerly, because a few seconds' advance warning is vital to him.

In this paper we study the statistical properties, the spectral characteristics and the predictability of fully developed turbulence in two and three dimensions. We shall discover that the statistical properties of two-dimensional chaotic flow differ considerably from its two-dimensional counterpart. Throughout the paper, we approach the subject from a classical, statistical point of view: we do not study bifurcations, solitons, modons, or strange attractors. However, we will use the opportunity to make a few comments on entropy and on the distribution of characteristic exponents. The main theme of this lecture will be the relation between the statistical properties of turbulence in two and three dimensions and the consequences of these properties for predictability. We will develop this theme by reviewing the vorticity dynamics of two- and three-dimensional flow and the spectra that result from the various cascade processes that are possible.

Before we start, a note on terminology is in order. When we speak of small-scale or microscale turbulence in the atmosphere, we do not necessarily think of micrometre or millimetre eddies. In fact, one millimetre is about the smallest scale encountered in the atmosphere; micrometre eddies do not occur. Small scale merely means that the dimensions of the eddy motion are substantially smaller than the scale height of the atmosphere. In the same way, mesoscale refers to scales of motion comparable to the scale height, and large scale refers to phenomena whose horizontal extent substantially exceeds the scale height. There are numerous refinements and exceptions to this rough division, but it suffices for our present purposes.

The term three-dimensional turbulence refers to chaotic motion in which vorticity amplification by fluid line extension (vortex stretching, for short) is of great dynamical significance, because a succession of instabilities over a wide range of scales causes an explosive growth of eddy enstrophy, in particular at the smallest scales of motion. Vortex stretching accounts for much of the vigour of hurricanes, tornadoes, dust devils and bath-tub vortices. It is not hard to imagine how unstable a flow containing a great number of closely packed tornadoes would be. By comparison, two-dimensional turbulence evolves gently. The many instabilities associated with vorticity amplification are absent in two-dimensional turbulence; they are severely constrained in geostrophic flow. Three-dimensional turbulence is not limited in vertical extent by the gravity-controlled thickness of the atmospheric shell around the planet, but geostrophic turbulence is. Speaking concisely, three-dimensional turbulence is unpredictable vortex-stretching chaos, while two-dimensional turbulence is barely predictable vorticity-advecting chaos. I use the adjectives advisedly; in the course of our analysis it will become clear why they are appropriate. The absence of vortex stretching somehow allows for a limited kind of predictability: after all, the weather is predictable up to about ten days ahead. If we want to define the difference in terms of enstrophy (enstrophy is one-half of the squared vorticity), we can state that three-dimensional turbulence needs no external supply of enstrophy, but two-dimensional turbulence does. In an inviscid fluid, two-dimensional turbulence conserves its total energy and its total enstrophy; geostrophic turbulence has similar invariants [1]. If the viscosity is finite, however, enstrophy is dissipated at a nonnegligible rate; therefore, the maintenance of a stationary state requires an external source. In three-dimensional turbulence, on the other hand, vortex stretching creates as much enstrophy as viscosity permits.

2. – Some properties of flow with friction.

We are concerned with incompressible fluids whose viscosity is very small. The Reynolds number characterizing the flow field then is very large. At first thought, the behaviour of a flow field in a fluid whose viscosity is nearly zero should differ only a little from that in a fluid whose viscosity is identically equal to zero. There is a massive amount of evidence, however, showing that this is not the case. In fluid dynamics, the limit of vanishing viscosity is in general a singular limit; nearly inviscid flow fields often differ dramatically from their counterparts in exactly inviscid fluids. The singular behaviour of flow fields in the limit as the Reynolds number approaches infinity was first realized clearly by PRANDTL, at the turn of this century. That event marks the beginning of modern fluid dynamics.

In the Navier-Stokes equation for an incompressible, viscous fluid the kinematic viscosity ν occurs as the coefficient of a term containing the Laplacian of the velocity field:

$$(1) \qquad \frac{\partial u_i}{\partial t} + u_j \frac{\partial u_i}{\partial x_j} = -\frac{1}{\varrho}\frac{\partial p}{\partial x_i} + \nu \frac{\partial^2 u_i}{\partial x_j \partial x_j}.$$

In the limit as $\nu \to 0$ (properly speaking, in the limit as $\mathrm{Re} = UL/\nu \to \infty$, where U is a characteristic velocity, L a characteristic length and Re the Reynolds number), one is tempted to discard the last term in (1). However, fluid flow at large Reynolds numbers is not characterized by weak viscous effects throughout the flow field, but by strong viscous effects in a very small fraction of the entire field. Viscosity continues to exert a significant influence on boundary layers and viscous vortex filaments, no matter how small the viscosity becomes. Inside the viscous boundary layer along a solid body, for example, the advective and viscous terms of (1) tend to balance. This is possible only if the boundary-layer thickness δ is of order

$$\delta \sim (\nu L/U)^{\frac{1}{2}}. \tag{2}$$

This estimate is obtained from (1) by dimensional analysis. It can be confirmed by equating the diffusion time δ^2/ν to the advection time L/U. If the viscosity goes to zero, the relative boundary-layer thickness δ/L also becomes vanishingly small ($\delta/L \sim (\nu/UL)^{\frac{1}{2}} \sim \mathrm{Re}^{-\frac{1}{2}}$). In other words, the viscous layer becomes very thin, without ever disappearing completely. Similar behaviour is observed in turbulence, where viscosity remains significant at the smallest scales of motion, in the thinnest vortex filaments, no matter how large the Reynolds number becomes.

The kinetic-energy equation associated with (1) is [2]

$$\frac{\partial}{\partial t}\frac{1}{2}u_i u_i + u_j \frac{\partial}{\partial x_j}\frac{1}{2}u_i u_i = -\frac{1}{\varrho}\frac{\partial p u_i}{\partial x_i} + 2\nu \frac{\partial}{\partial x_j}u_i e_{ij} - 2\nu e_{ij} e_{ij}. \tag{3}$$

Here $\partial u_i/\partial x_i = 0$ has been used.
The strain rate e_{ij} is defined by

$$e_{ij} = \frac{1}{2}\left(\frac{\partial u_i}{\partial x_j} + \frac{\partial u_j}{\partial x_i}\right). \tag{4}$$

The viscous dissipation term $2\nu e_{ij} e_{ij}$ can be written out as

$$2\nu e_{ij} e_{ij} = \nu \frac{\partial u_i}{\partial x_j}\frac{\partial u_i}{\partial x_j} + \nu \frac{\partial u_i}{\partial x_j}\frac{\partial u_j}{\partial x_i}. \tag{5}$$

For homogeneous incompressible turbulence the mean value of the last term in (5) is zero, so that the mean dissipation rate is equal to the mean value of the first term on the right-hand side.

In the energy equation, viscosity causes both diffusion and dissipation. Viscous diffusion of energy takes the form of the divergence of a viscous energy flux—the one but last term on the r.h.s. of (3). The dissipation term is negative definite, pointing to the irreversible character of dissipative processes. One cannot gain kinetic energy from friction! The viscous dissipation of energy can be ignored only if squared velocity gradients remain finite as $\mathrm{Re} \to \infty$. This is not always the case, however. In the laminar boundary layer we encountered above, for example, the viscous dissipation rate must be estimated as

$$2\nu e_{ij} e_{ij} \sim \nu \frac{\partial u_i}{\partial x_j}\frac{\partial u_i}{\partial x_j} \sim \nu \frac{U^2}{\delta^2} \sim \frac{U^3}{L}. \tag{6}$$

Since the gradients become steeper as viscosity decreases, the viscous loss of energy inside the boundary layer is locally *independent* of viscosity! Clearly, one has to be very careful before one neglects viscous dissipation. We shall meet several other examples of this pathology shortly. Time and again we shall discover that the limit of vanishing viscosity is singular.

The dissipation term in (3) corresponds to the conversion of mechanical energy into the thermal energy of molecular motion. Dissipation is an irreversible thermodynamic process, as is seen clearly from the entropy equation for incompressible, viscous flow [2]

$$(7) \qquad T\frac{\partial S}{\partial t} + u_j T \frac{\partial S}{\partial x_j} = 2\nu e_{ij} e_{ij} + \varkappa C_p \frac{\partial^2 T}{\partial x_j \partial x_j}.$$

Here, S is the entropy per unit mass, T the temperature, $\varkappa$ the thermal diffusivity and C_p the specific heat at constant pressure.

Fluid flow in which the dissipation rate is not negligible requires a source of mechanical energy and a sink for entropy if a stationary state is to be maintained. Dissipative flow systems need some appropriate kind of forcing, else they decay irreversibly.

In this lecture, we are talking about turbulence and predictability in forced, dissipative flows. There is good reason to wonder about possible relations between reversibility and predictability. I will present a brief discussion in the next section, though the subject is outside the scope of my competence. At this point, I venture the opinion that chaotic flows producing very little entropy are more likely to be predictable than those producing entropy at a significant rate. We shall see later that two-dimensional turbulence is almost nondissipative as $\nu \to 0$; is its strange predictability behaviour somehow associated with its thermodynamics?

3. – Entropy and information.

Most of us tend to associate entropy with chaos, loss of information, lack of predictability and irreversibility. Turbulent motion is chaotic and its range of predictability is limited. It is, therefore, natural to search for parameters that characterize the extent or amount of chaos in a turbulent flow field. Such parameters ought to have properties similar to that of entropy, in the sense that they should tend to increase in a monotonous manner as the amount of disorder increases.

In the language of dynamical-systems theory, turbulent flows are « sensitive to initial conditions ». The difference between two realizations that are initially very close together tends to increase in time. After a while the two realizations differ as much from one another as two randomly chosen states. If the initial difference is an error in one of the two realizations, the true state of the system, therefore, becomes unpredictable after some time. Error growth and loss of information are related, and it seems logical to define an eddy entropy that increases at a rate proportional to the error growth. This approach has been explored by EDWARDS, BETCHOV and others; recently it was taken up by CARNEVALE and HOLLOWAY ([3], see also [4]). Their entropy represents a « monotonic tendency to complete ignorance, in accord with the second law of thermodynamics ». Viscous dissipation tends to decrease the eddy entropy, because « viscosity drives the system toward the perfectly predictable state of zero motion ». The thermodynamic entropy, however, *increases* due to viscous dissipation, so that the rates of change of the two are opposite in sign. It is not clear how the two kinds of entropy are related; perhaps we need several kinds of entropy, just like we have several kinds of energy. Parenthetically, what is the information content of zero motion? Is zero motion not a perfectly uninformative flow?

A point of view opposite to that of Carnevale and Holloway is taken by RUELLE [5], who claims that a turbulent fluid acts as a *source* of information. He defines the production of information in terms of the distribution of characteristic exponents (which are related to the amplification rates of small disturbances at all possible scales of motion). RUELLE shows that the information production and the energy dissipation rate are related to each other; in the case of two-dimensional turbulence he proves that the information production is bounded from above by the dissipation rate (and thus is related to the production of thermodynamic entropy). Ruelle's position is in line with the concepts advocated by PRIGOGINE [6], who asserts that dissipative dynamical systems that are sensitive to initial conditions (*i.e.* those that have strange attractors) are a source of order, because they tend to create coherent structures, which contain new information. Prigogine's term for coherent structures is « dissipative structures »; he wants to emphasize that the creation of information must be accompanied by increased dissipation rates (« excess entropy production ») in order to maintain compliance with the second law of thermodynamics.

Instabilities in a flow field create eddies. Eddies are coherent structures, no doubt about that whatsoever. Coherent structures contain useful information: a mid-latitude storm, with its well-developed frontal structure and its well-defined life cycle, is a forecaster's delight.

However, in a chaotic flow field it is hard to predict (and impossible to predict in the long run) where and when these beautiful coherent structures can be found. New information, yes! But is it accessible? And how rapidly is it degraded? The information contained in any coherent structure will not last longer than its lifetime; after that, we have to wait for the birth of the next generation of eddies. Apparently, information is a perishable commodity.

In this context it seems useful to review briefly sowe of the other concepts introduced by PRIGOGINE, because they bear directly on the relations among reversibility, predictability, order, chaos and entropy that concern us here.

According to PRIGOGINE, the trajectories of a system with a strange attractor are no longer observables, because it is impossible to tell on which sheet or fold of the attractor one is located if the accuracy of measurement (*i.e.* the accuracy of determining the position of the system in phase space at a given time) is limited. In other words, « experiments performed with an arbitrary but finite accuracy lead us only to the identification of some *finite* region of phase space where the system may be located ». Through such a small, but finite volume uncountably many folds of the attractor pass. In that case one can do no more than compute the evolution of the probability distribution of a bundle of trajectories. These trajectories diverge in accordance with sensitivity to initial conditions, and before too long the evolution of the system has become unpredictable.

If the trajectories of systems with strange attractors are unobservables, there must be an uncertainty relation involved, just like the one between position and momentum in quantum mechanics. PRIGOGINE [6] hypothesizes a relation of the type $\Delta\lambda \cdot \Delta t = 1$, where λ is the amplification rate of small disturbances and t is the age of a coherent structure. If I interpret it correctly, this uncertainty relation states that one cannot simultaneously determine both the growth rate and the age of a newly created eddy. I do not think the relation applies to nonlinear stages in the evolution of a coherent structure, because the concept of characteristic exponents (rates of growth of disturbance amplitudes) then does not apply. Since amplification rates and characteristic exponents are believed to be associated with the evolution of entropy, Prigogine's uncertainty relation in fact deals with entropy and time as the two complementary variables, both of them in the form of operators (like the operators in quantum mechanics). However, I have not found a concrete example of the way this uncertainty relation is involved in the evolution of a bundle of trajectories. It would be of great interest to determine if these ideas can be applied to fluid dynamics.

This detour has been long enough now; let me try to summarize. Turbulence contains many coherent structures, ranging from blocking ridges, mid-latitude storms, hurricanes and tornadoes to dust devils, vortex streets and highly intermittent dissipative vortex filaments. Coherent structures arise in the flow field because of sensitivity to initial conditions (*i.e.* instabilities). Coherent structures contain newly created information, but the accessibility and usefulness of the information is limited because the eddies tend to be located chaotically in four-dimensional space-time and because their lifetimes are finite. Chaos can be a source of order, but it is hard to get hold of the order in the chaos. The very sign of eddy entropy may depend on one's point of view! We need a theory that connects irreversibility to lack of predictability, that acknowledges the presence of information in coherent structures and that generalizes the concept of entropy. A challenge for some of the students in this summer school, perhaps?

4. – Vorticity dynamics.

For incompressible flow in a Newtonian fluid, observed from an inertial frame of reference, the vorticity equation reads

$$\frac{\partial \zeta_i}{\partial t} + u_j \frac{\partial \zeta_i}{\partial x_j} = \zeta_j e_{ij} + \nu \frac{\partial^2 \zeta_i}{\partial x_j \partial x_j}. \tag{8}$$

Here, the vorticity ζ_i is defined by

$$\zeta_i = e_{ijk} \frac{\partial u_k}{\partial x_j}. \tag{9}$$

The term $\zeta_j e_{ij}$ represents vortex stretching and turning. A vertical vortex, such as a dust devil in which ζ_3 is the only nonzero vorticity component, suffers extension or contraction from $e_{33} = \partial u_3/\partial x_3$; its axis is turned by the off-diagonal components of e_{ij}. If the horizontal wind field is convergent, $\partial u_3/\partial x_3 > 0$; that leads to amplification of the angular velocity in the vortex (note that the vorticity is twice the local angular velocity).

The vortex-stretching term is zero in two-dimensional flow, because the vorticity vector is perpendicular to the plane of motion everywhere: the velocity component parallel to the vorticity is zero, and the horizontal velocity components do not depend on the vertical co-ordinate If the viscosity of the fluids equals zero, the vorticity equation of two-dimensional flow ($\zeta_i = \zeta_3 = \zeta$) reduces to

$$\frac{\partial \zeta}{\partial t} + u_j \frac{\partial \zeta}{\partial x_j} = \frac{d\zeta}{dt} = 0. \tag{10}$$

Two-dimensional, inviscid motion thus conserves the vorticity of all fluid particles. Two-dimensional flow merely advects vorticity (apart from the diffusive effects of viscosity). Vorticity drifts along in the flow field. This is the principal correspondence with quasi-geostrophic inviscid motion in planetary atmospheres: there not the vorticity itself, but some appropriate form of potential vorticity is conserved. In a homogeneous ocean, potential vorticity conservation is expressed by

$$\frac{d}{dt}\left(\frac{\zeta + f}{H}\right) = 0, \tag{11}$$

where f is the Coriolis parameter and H the depth of the water column. It is no exaggeration to claim that weather forecasting with numerical models is feasible because the dominant short-term process for synoptic-scale motion is adiabatic advection of potential vorticity. In three-dimensional turbulence, on the other hand, vortex stretching is also an effective short-term mechanism, with serious consequences for predictability.

Before we have a look at the equation that governs the evolution of enstrophy, two comments are in order. First, since the limit $\nu \to 0$ is likely to be singular, vorticity conservation in two-dimensional turbulence is a good short-term approximation, but not necessarily a good long-term approximation. A domain of size L is permeated by viscous diffusion in a time of order L^2/ν; conservation of vorticity is bound to be a valid approximation only for times $t \ll L^2/\nu$.

Second, the vorticity equation is strikingly similar to the equations that govern the evolution of passive scalars in a flow field. The temperature equation, for example, reads

$$(12) \qquad \frac{\partial T}{\partial t} + u_j \frac{\partial T}{\partial x_j} = \varkappa \frac{\partial^2 T}{\partial x_j \partial x_j}.$$

The similarity between (8) and (12) can be exploited to derive the spectral characteristics of the temperature field. One example is the k^{-1} temperature spectrum in the viscous-convective range of three-dimensional turbulence at large Prandtl numbers ($\nu/\varkappa \gg 1$), which is similar to the vorticity spectrum in the enstrophy-cascading inertial range of two-dimensional turbulence [7, 8].

Now we turn to the enstrophy, which, as we recall, is one-half of the vorticity squared. The enstrophy equation is obtained from the vorticity equation (8) by multiplication by ζ_i:

$$(13) \qquad \frac{\partial}{\partial t} \frac{1}{2} \zeta_i \zeta_i + u_j \frac{\partial}{\partial x_j} \frac{1}{2} \zeta_i \zeta_i = \zeta_i \zeta_j e_{ij} + \frac{1}{2} \nu \frac{\partial^2 \zeta_i \zeta_i}{\partial x_j \partial x_j} - \nu \frac{\partial \zeta_i}{\partial x_j} \frac{\partial \zeta_i}{\partial x_j}.$$

In two-dimensional flow, this reduces to

$$(14) \qquad \frac{\partial}{\partial t} \frac{1}{2} \zeta^2 + u_j \frac{\partial}{\partial x_j} \frac{1}{2} \zeta^2 = \nu \frac{\partial^2 \frac{1}{2}\zeta^2}{\partial x_j \partial x_j} - \nu \frac{\partial \zeta}{\partial x_j} \frac{\partial \zeta}{\partial x_j}.$$

The absence of e_{ij} from (14) does not imply that it equals zero or that two-dimensional flow does not suffer deformation. On the contrary: e_{ij} can increase vorticity gradients very effectively, for example.

In the literature on large-scale atmospheric flow, one often encounters the concept of enstrophy conservation. Numerical models, for example, can be made to conserve both their total energy (integrated over the entire globe) and their total enstrophy. It is evident from (14) that enstrophy conservation is possible only if the viscosity is exactly equal to zero. It is not good enough to require that the viscosity be vanishingly small, because (as we shall see shortly) two-dimensional turbulence tends to amplify vorticity gradients, much as colour gradients increase when paint is mixed by stirring. The mean square gradients become larger when the viscosity becomes smaller, and the mean value of the last term in (14) will remain finite as the viscosity tends to zero [8]. This is one of the many examples of the singular behaviour of a vanishingly small viscosity. The mean viscous diffusion of enstrophy, on the other hand, represented by the first term on the r.h.s. of (14), can be neglected if the viscosity is small enough (except near solid boundaries and inside viscous filaments).

5. – Energy and enstrophy in three-dimensional turbulence.

Having collected all of the tools required, we now can study vortex-stretching turbulence in more detail. We are interested in statistically steady states of forced, dissipative flows; shear in the mean flow is an appropriate source of energy for three-dimensional turbulence because it resembles the way smaller eddies are strained by their large brothers and sisters. Therefore, we divide the flow into a mean current and eddy motion and employ Reynolds' averaging. This process is formally identical with that of separating the resolved scales from the subgrid scale fluctuations in a numerical grid point model; we shall use this correspondence repeatedly.

The equation for the mean eddy enstrophy in three-dimensional turbulence reads

$$(15) \quad \frac{\partial}{\partial t}\frac{1}{2}\overline{\zeta'_i \zeta'_i} + \bar{u}_j \frac{\partial}{\partial x_j}\frac{1}{2}\overline{\zeta'_i \zeta'_i} = -\overline{u'_j \zeta'_i}\frac{\partial \bar{\zeta}_i}{\partial x_j} - \frac{\partial F_j}{\partial x_j} + \overline{\bar{\zeta}_i \zeta'_j e'_{ij}} + \overline{\bar{e}_{ij} \zeta'_i \zeta'_j} + \overline{\zeta'_i \zeta'_j e'_{ij}} - \nu \overline{\frac{\partial \zeta'_i}{\partial x_j}\frac{\partial \zeta'_i}{\partial x_j}}.$$

Here, F_j represents two spatial enstrophy fluxes that have little relevance in the present context. The first term on the r.h.s. of (15) represents the exchange of enstrophy between the mean flow and the turbulence (or between the resolved scales and the subgrid scale eddies). The third, fourth and fifth terms are associated with vortex stretching in one way or another. The fourth term, for example, is the amplification of subgrid scale enstrophy by the straining motion of the resolved scales. Since the amplification of vorticity causes very large strain rates at small scales of motion, the fifth term on the r.h.s. of (15) is by far the largest source term for enstrophy in three-dimensional turbulence at large Reynolds numbers [9]. In first approximation, the enstrophy budget of three-dimensional turbulence thus reads ([7], Chapt. 3)

$$(16) \quad \overline{\zeta'_i \zeta'_j e'_{ij}} = \nu \overline{\frac{\partial \zeta'_i}{\partial x_j}\frac{\partial \zeta'_i}{\partial x_j}}.$$

In three-dimensional turbulence, enstrophy is *not* a conservative quantity. In fact, in the absence of viscosity the enstrophy would increase without limit.

In fluids with a small viscosity, the mean enstrophy level is very high; an external source of enstrophy or transfer of enstrophy from the mean flow to the eddy motion is *not* needed to maintain a high enstrophy level in three-dimensional turbulence. The enstrophy level is as large as viscosity permits.

The autonomous and extremely vigorous creation of eddy enstrophy in (16) finds its counterpart in the spectral flux of energy. As vortex tubes are extended, the rotation rates increase because the angular momentum converges into narrow filaments. At the same time the kinetic energy converges into small-diameter motion. Enstrophy amplification and a cascade of energy towards small scales are two aspects of the same phenomenon, just like the dynamic processes occurring in dust devils, tornadoes and hurricanes.

In three-dimensional turbulence maintained by shear in the mean motion (or by the straining of the large-scale eddies), the kinetic-energy budget for the eddy motion reads ([7], Chapt. 3)

$$(17) \quad \frac{\partial}{\partial t}\frac{1}{2}\overline{u'_i u'_i} + \bar{u}_j \frac{\partial}{\partial x_j}\frac{1}{2}\overline{u'_i u'_i} = -\frac{\partial}{\partial x_j}\left(\frac{1}{\varrho}\overline{u'_j p'} + \frac{1}{2}\overline{u'_i u'_i u'_j}\right) + 2\nu \frac{\partial}{\partial x_j}\overline{u'_i e'_{ij}} - \overline{u'_i u'_j}\,\bar{e}_{ij} - 2\nu\overline{e'_{ij} e'_{ij}}.$$

In homogeneous turbulence, moreover, we have

$$(18) \quad 2\nu\overline{e'_{ij} e'_{ij}} = \nu\overline{\zeta'_i \zeta'_i}.$$

The only source term for energy in (17) is the mechanical (or shear) production term, $-\overline{u'_i u'_j}\,\bar{e}_{ij}$. All other terms on the right-hand side (except for the dissipation) are flux divergences, which integrate out to zero if no energy is added or removed at the boundaries of the domain. The overall energy level can be maintained only if production and dissipation of eddy kinetic energy are in balance:

$$(19) \quad -\overline{u'_i u'_j}\,\bar{e}_{ij} = \nu\overline{\zeta'_i \zeta'_i}.$$

Here we have used (18) in order to link energy and enstrophy. Even if the turbulence is not homogeneous, the difference between the l.h.s. and the r.h.s. of (18) is negligible.

Apart from the effects of viscosity, energy is a conservative quantity in three-dimensional turbulence. However, the enstrophy is maintained at an extremely high level by the source term in (16) if the viscosity is small enough; therefore, the product of viscosity and enstrophy in (19) need not become small in the limit as $\nu \to 0$. This is, in fact, the case: turbulence is strongly dissipative, even in fluids with vanishing viscosity. Energy is not conserved, no matter how small the viscosity. In current terminology, energy is a dissipated integral of three-dimensional flow. In order to avoid confusion, I should add that this behaviour is not caused by the way in which a small viscosity operates, but by the relentness nature of an energy cascade driven by a multitude of vortex-stretching instabilities. The spectral energy flux and, therefore, the dissipation rate are determined by the dynamics at the *large-scale* end of the spectrum.

Dissipation occurs almost exclusively at the very smallest scales of motion, because the enstrophy is concentrated there. The viscous sink in (19), therefore, occurs at the end of the spectral range. The source term in (19), on the other hand, represents the exchange of energy between the mean flow and the eddy motion, or from the resolved scales to the subgrid scale eddies. If the mean flow is the only source of energy and if the viscous losses are confined to the smallest scales, there must be a spectral flux of energy toward the small scales. This spectral flux is constant over a wide range of scales if the source term is confined to the largest scales of motion and viscous dissipation to the smallest. This is the celebrated energy cascade in three-dimensional turbulence. The cascade rate is a process whose rate is dictated by the instabilities in the larger eddies; viscosity merely mops up whatever winds up on the floor.

The energy cascade rate of three-dimensional turbulence is one of its principal parameters. If we take the l.h.s. of (19) to represent the transfer of energy from the resolved scales to the subgrid scale motion and assume that this spectral flux is constant, we can define

(20) $$\varepsilon = \overline{u_i' u_j'}\, \bar{e}_{ij},$$

so that

(21) $$\overline{\zeta_i' \zeta_i'} = \varepsilon/\nu.$$

Using these estimates in the enstrophy equation (16), we find that the smallest scale of motion, η_3, must be given by

$$\eta_3^2 = \nu (\nu/\varepsilon)^{\frac{1}{2}},$$

whence

(22) $$\eta_3 = (\nu^3/\varepsilon)^{\frac{1}{4}}.$$

This is the Kolmogorov scale of three-dimensional turbulence. In the atmosphere, it is typically one millimetre or larger.

Let me summarize the situation before giving a few examples. Vortex stretching creates the autonomous processes (no external forcing needed) of enstrophy amplification and the associated spectral flux of energy toward small scales. If the viscosity is zero, this leads to an unlimited enstrophy explosion at infinitely small scales. A nonzero viscosity, however, no matter how small, will dissipate the energy and constrain the enstrophy to a finite level. Since the spectral energy flux is driven by vortex stretching, the energy cascade rate ε and the ultimate dissipation rate of energy (which equals ε in stationary, homogeneous flow) are independent of viscosity, no matter how small the latter becomes.

The use of (19) and (20) can be illustrated profitably by deriving the logarithmic wind profile in a neutral surface layer. The vertical momentum flux, $-\overline{u'w'}$, is independent of z:

$$-\overline{u'w'} = u_*^2 , \tag{23}$$

where u_* is the friction velocity. The energy conversion rate from the mean wind to the eddy motion is $-\overline{u'w'}\,\partial U/\partial z$; this has to be balanced by the spectral flux ε. Therefore,

$$u_*^2 \frac{\partial U}{\partial z} = \varepsilon , \tag{24}$$

which requires an estimate for the cascade rate. Since the energy cascade is initiated by the instabilities of eddies with velocities proportional to u_* and diameters proportional to the distance from the surface, we estimate

$$\varepsilon = u_*^3 / kz , \tag{25}$$

where k, the von Kármán constant, is an empirical coefficient. If we now substitute (25) into (24), we find $\partial U/\partial z = u_*/kz$. This leads to the logarithmic law. Note, incidentally, the similarity between (25) and (6).

Another example is the conventional estimate of the intensity of convective turbulence. An upward vertical flux of sensible heat, H, leads to an energy production rate that is equal to $gH/\varrho C_p T$. If the dissipation rate is estimated as w_*^3/h, where h is the height of the convective mixed layer, then

$$w_*^3 = \frac{gH}{\varrho C_p T} h . \tag{26}$$

Turbulence theory abounds with estimates of this kind. Three-dimensional turbulence is as dissipative as the vigour of its energy cascade permits.

6. – The enstrophy cascade in two-dimensional turbulence.

We return to two-dimensional flow. The equation for the mean eddy enstrophy (or subgrid scale enstrophy) in two-dimensional turbulence is obtained from (15) by putting $\zeta_i = \zeta$ and discarding all vortex-stretching terms:

$$\frac{\partial}{\partial t} \frac{1}{2} \overline{\zeta' \zeta'} + \bar{u}_j \frac{\partial}{\partial x_j} \frac{1}{2} \overline{\zeta' \zeta'} = -\overline{u'_j \zeta'} \frac{\partial \bar{\zeta}}{\partial x_j} - \overline{\frac{\partial F_j}{\partial x_j}} - \nu \overline{\frac{\partial \zeta'}{\partial x_j} \frac{\partial \zeta'}{\partial x_j}}. \tag{27}$$

A stationary, homogeneous situation can be maintained only if

$$-\overline{u'_j \zeta'} \frac{\partial \bar{\zeta}}{\partial x_j} = \nu \overline{\frac{\partial \zeta'}{\partial x_j} \frac{\partial \zeta'}{\partial x_j}}. \tag{28}$$

This equation should be compared with its counterpart in three-dimensional flow (16) and with the energy equation (19). Enstrophy is a conservative quantity in two-dimensional, inviscid flow. However, when the viscosity is finite, the resulting enstrophy dissipation has to be balanced by a spectral flux of enstrophy if a stationary state is to be maintained. That, in turn, requires the transfer of enstrophy from the mean flow to the eddies that is represented by the left-hand side of (28).

In three-dimensional turbulence, (16) shows that there is an autonomous, internal source of eddy enstrophy. In two-dimensional turbulence, on the other hand, (28) shows that there must be a spectral flux of enstrophy from the resolved scales to the subgrid scale motion, similar to the spectral flux of energy described by (19).

How large is the spectral enstrophy flux in two-dimensional turbulence? This can be determined from the way the right-hand side of (28) behaves as the viscosity tends to zero [8]. In the energy equation for three-dimensional turbulence, (19), the product of viscosity and enstrophy becomes independent of ν as $\nu \to 0$, because the enstrophy increases beyond bound by virtue of (16). The question concerning the r.h.s. of (28) then can be solved by looking at the equation for the mean square vorticity gradients. This reads in first approximation [8]

$$\overline{\frac{\partial \zeta'}{\partial x_i} \frac{\partial \zeta'}{\partial x_j} e'_{ij}} = \nu \overline{\frac{\partial^2 \zeta'}{\partial x_i \partial x_j} \frac{\partial^2 \zeta'}{\partial x_i \partial x_j}}. \tag{29}$$

The structure of this equation is identical to that of (16). The vorticity gradients are amplified by straining motion, because on the average material lines are extended in two-dimensional turbulence. The amplification of vorticity gradients will lead to an autonomous spectral flux of enstrophy toward larger wave numbers, just like the amplification of enstrophy in (16) leads to a spectral flux of energy. In this way, it will become impossible to conserve enstrophy in the limit as $\nu \to 0$.

The amplification of vorticity gradients is an autonomous process; it does *not* require an external source mechanism. The vorticity gradients thus will become as large as viscosity permits, and the spectral flux of enstrophy does *not* depend on viscosity in the limit as $\nu \to 0$. Therefore, (28) and (29) together define an enstrophy cascade, just like (16) and (19) define the energy cascade of three-dimensional turbulence. In analogy with (20), we now define the spectral enstrophy flux in (28) as

$$\chi = -\overline{u_j' \zeta'} \frac{\partial \bar{\zeta}}{\partial x_j}. \tag{30}$$

Therefore,

$$\overline{\frac{\partial \zeta'}{\partial x_j} \frac{\partial \zeta'}{\partial x_j}} = \chi/\nu. \tag{31}$$

The smallest scale of motion, η_2, is obtained by straightforward dimensional analysis; it reads [8]

$$\eta_2 = (\nu^3/\chi)^{\frac{1}{6}}. \tag{32}$$

This is obtained most easily by requiring η_2 to be a function of ν and χ only. A careful estimate of the strain rate fluctuations in (29) will lead to the same result.

The enstrophy-cascading range of two-dimensional turbulence has a negligible spectral flux of energy if the viscosity is small enough. Since the enstrophy remains finite, the r.h.s. of (19) goes to zero as $\nu \to 0$. The spectral energy flux in the conventional sense of the word is zero in this case: all scales of motion lose a little energy directly to viscous friction, not to each other. If the viscosity is small enough, the eddy energy is very nearly conserved, even though the spectral enstrophy flux leads to nonnegligible losses of enstrophy, no matter how small the viscosity becomes. In current terminology, enstrophy is a dissipated integral, but energy a rugged integral of two-dimensional flow. Two-dimensional turbulence is almost nondissipative; therefore, it is almost reversible in the thermodynamic sense of the word. These characteristics are perhaps related to the favourable predictability behaviour of two-dimensional and geostrophic flow.

Two-dimensional turbulence also can have an energy-cascading range, in which there is a significant spectral flux of energy toward larger scales, but a negligible spectral flux of enstrophy [10]. The reverse energy cascade is of great significance to geostrophic flow, because it is related to the maintenance of the general circulation. However, I have not yet been able to develop an attractive, reasonably simple explanation of the reverse energy cascade in terms of the tools used in this paper.

7. – Spectral characteristics of three-dimensional turbulence.

The spectrum in the energy-cascading range of three-dimensional turbulence depends on the spectral energy flux ε, the wave number k and the viscosity ν. In the range $k\eta_3 \ll 1$ the local viscous effects are small, so that the spectral characteristics in the inertial subrange are determined by ε and k only. The typical velocity $u(k)$ of an eddy at wave number k then is estimated as

$$u(k) \sim \varepsilon^{\frac{1}{3}} k^{-\frac{1}{3}}. \tag{33}$$

Therefore,

$$u^2(k) \sim \varepsilon^{\frac{2}{3}} k^{-\frac{2}{3}}, \tag{34}$$

whence the inertial-range spectrum becomes

$$E_3(k) \sim k^{-1} u^2(k) \sim \varepsilon^{\frac{2}{3}} k^{-\frac{5}{3}}. \tag{35}$$

This is the well-known minus five-thirds spectrum. Another way to derive (33) or (35) is to assume that the energy spectrum is proportional to some unknown power of the wave number, to estimate the spectral flux at k in terms of $-\overline{u'_i u'_j} \bar{e}_{ij}$ and to require that the flux be independent of k. The same result is obtained. In a recent paper, BENZI et al. [11] derive the inertial-range spectrum from a maximum-chaos principle, by requiring that the predictability time be as small as possible consistent with the integral constraints.

The typical eddy vorticity and eddy strain rate at wave number k are

$$\zeta(k) \sim e_{ij}(k) \sim ku(k) \sim \varepsilon^{\frac{1}{3}} k^{\frac{2}{3}}. \tag{36}$$

These increase with wave number, making the vorticity fluctuations greatest at the smallest scales of motion. This is consistent with our statements on the amplification of enstrophy by the extension of material lines.

How can we see that no energy is lost while it moves down the cascade? On first inspection, something is amiss, because the eddy energy decreases with increasing wave number. However, it is not the energy *residing* at k that counts, but the energy *flux* toward the smaller scales. According to (20), this has the form of the product of the square of a typical eddy velocity at a wave number just above the value of k at which the spectral flux is measured and the strain rate at a wave number just below the toll gate. Therefore, the spectral energy flux is estimated as the product of (34) and (36):

$$-\overline{(u'_i u'_j} \bar{e}_{ij})_k \sim \varepsilon^{\frac{2}{3}} k^{-\frac{2}{3}} \cdot \varepsilon^{\frac{1}{3}} k^{\frac{2}{3}} = \varepsilon. \tag{37}$$

This proves consistency with (20) and confirms that the energy flux is constant. The energy level at large wave numbers is small because the «velocity» of the spectral flux increases as the wave number increases. Or, if the reciprocal of $e_{ij}(k)$ can be interpreted as a residence time at wave number k, we conclude that the energy flux is constant because the energy level and the residence time at wave number k decrease at the same rate as k increases.

The spectral vorticity estimate (36) can also be used to determine whether the overall enstrophy level is consistent with (21). Most of the enstrophy resides near the viscous cut-off; therefore, ζ^2 at $k\eta_3 = 1$ must be equal to ε/ν. Using the definition of η_3, we can easily confirm that this is the case.

How about the enstrophy cascade rate in vortex-stretching turbulence? The typical vorticity gradient at k is estimated as

$$\left(\frac{\partial \zeta}{\partial x}\right)_k \sim k^2 u(k) \sim \varepsilon^{\frac{1}{3}} k^{\frac{5}{3}}. \tag{38}$$

The spectral enstrophy flux is of the type given in (30), though it is by no means constant in this case. At wave number k, we estimate

$$-\overline{\left(u'_j \zeta'_i \frac{\partial \zeta_i}{\partial x_j}\right)} \sim \varepsilon^{\frac{1}{3}} k^{-\frac{1}{3}} \cdot \varepsilon^{\frac{1}{3}} k^{\frac{2}{3}} \cdot \varepsilon^{\frac{1}{3}} k^{\frac{5}{3}} \sim \varepsilon k^2. \tag{39}$$

The enstrophy flux increases very rapidly with k, because the enstrophy produced at each wave number adds to the flux coming from the larger scales. The spectral flux is greatest at the viscous cut-off. At $k\eta_3 = 1$, we obtain

$$-\overline{\left(u'_j \zeta'_i \frac{\partial \zeta_i}{\partial x_j}\right)}_{\eta_3} \sim \varepsilon \eta_3^{-2} \sim (\varepsilon/\nu)^{\frac{3}{2}}. \tag{40}$$

This is consistent with (16). The enstrophy flux at $k\eta_3 = 1$ must be proportional to the total enstrophy produced throughout the spectrum; this is represented by the left-hand side of (16), which is of order $(\varepsilon/\nu)^{\frac{3}{2}}$ because the vorticity fluctuations are of order $(\varepsilon/\nu)^{\frac{1}{2}}$ by virtue of (21). The enstrophy generation rate by vortex stretching becomes infinitely large as the viscosity tends to zero.

Now we turn to the consequences of this behaviour for the predictability of small-scale turbulence. The time scale at wave number k is the reciprocal of the vorticity:

$$t(k) \sim \zeta^{-1}(k) \sim \varepsilon^{-\frac{1}{3}} k^{-\frac{2}{3}}. \tag{41}$$

The time scale decreases as k increases. This makes the spectral cascade « local »: eddies of widely different wave numbers are tuned to widely different frequencies and cannot interact effectively. The time scale and, therefore, the dynamic interaction period and life span of the very smallest eddies are given by

$$t(\eta_3) \sim \varepsilon^{-\frac{1}{3}} \eta_3^{\frac{2}{3}} = (\nu/\varepsilon)^{\frac{1}{2}}. \tag{42}$$

A numerical example is illustrative. In the atmosphere, $\nu = 15 \cdot 10^{-6}$ m^2 s^{-1}; a typical value of ε above the surface layer is about 10^{-3} m^2 s^{-3}. The Kolmogorov time scale (42) thus is a little larger than 10^{-1} s. Since the useful forecast period of deterministic calculations is on the order of 10 eddy time scales [12], we cannot compute the evolution of the very smallest eddies more than *one second* ahead, even if the initialization and resolution problems were surmountable (they are not). After one second, errors have grown large enough to make the forecast purely stochastic. This is not the way to go.

The predictability of the small-scale structure of atmospheric turbulence remains extremely limited, even if we look at somewhat larger scales. If $\varepsilon = 10^{-3}$ m² s⁻³, the eddy time scale is 10 s at $k^{-1} = 1$ m, 46 s at $k^{-1} = 10$ m and 220 s at $k^{-1} = 100$ m. At the hundred-metre scale, about half an hour of predictability is the best one can ever hope for. Moreover, since $t(k)$ decreases with decreasing scale, deterministic microstructure calculations are impossible *in principle* when the required forecast period is of the order of the life span of the larger eddies. It helps to *reduce* the resolution (to increase the mesh size); the small-scale motion can better be handled statistically. If forecasts make no sense, the small-scale stuff is best treated as subgrid scale motion. This is why most turbulence research is done on a wholesale statistical basis. Turbulence research does not resemble numerical forecasting, but studies of the general circulation, where all weather is treated as noise.

The dependence of typical accelerations on wave number sheds more light on the problem of choosing an appropriate resolution. The acceleration at k is estimated as

$$a(k) \sim k u^2(k) \sim \varepsilon^{\frac{2}{3}} k^{\frac{1}{3}}. \tag{43}$$

This increases with wave number: the smallest eddies have the largest accelerations. The acceleration at the Kolmogorov microscale is

$$a(\eta_3) \sim \varepsilon^{\frac{2}{3}} \eta_3^{-2} \sim (\varepsilon^3/\nu)^{\frac{1}{4}}, \tag{44}$$

which goes to infinity as $\nu \to 0$.

For $\varepsilon = 10^{-3}$ m² s⁻³ and $\nu = 15 \cdot 10^{-6}$ m² s⁻¹, the accelerations in the smallest eddies are of order 10^{-1} ms⁻², which is one-thousand times as large as the net accelerations associated with large-scale atmospheric motion.

Accelerations are at the heart of the equations of motion. According to (43), the acceleration field is dominated by high-frequency noise. If we are mainly interested in the small accelerations associated with the evolution of the large eddies, we have to be very careful, else our calculations will get lost in the subgrid scale noise. Drastic averaging will improve the noise problem; this is in line with our conclusions concerning the wave number dependence of the characteristic time. The problem of parametrizing subgrid scale fluxes may become more difficult, however, because at scales larger than those in the inertial subrange one cannot rely on local isotropy and other special features of the small-scale motion.

Suppose we take the wave number representing the numerical grid size Δ inside the inertial subrange. The acceleration associated with the divergence of the unresolved eddy flux of momentum (which is the contribution of the unresolved motion to the acceleration field) then is estimated as $\varepsilon^{\frac{2}{3}}\Delta^{-\frac{1}{3}}$. The acceleration at the largest scale of motion is of order $\varepsilon^{\frac{2}{3}}L^{-\frac{1}{3}}$ (L is the mixing height, for example). This means that the subgrid scale fluxes have to be parametrized extremely carefully if one wants a deterministic calculation of the evolution of the large-scale features. I do not know of any investigation that deals with this problem. In Deardorff's famous numerical simulations [13], for example, for the subgrid scale parametrization is chosen such that the spectral energy flux and other statistical properties are represented correctly. I doubt very much that the parametrization is capable of producing reliable individual realizations of the turbulence field. Deardorff's simulations were general-circulation experiments, not forecasting runs.

The picture I paint is too bleak. The effective forecast range at $k\Delta = 1$ is of order $\varepsilon^{-\frac{1}{3}}\Delta^{\frac{2}{3}}$; it is small compared to $\varepsilon^{-\frac{1}{3}}L^{\frac{2}{3}}$. Deterministic forecasts of the barely resolved eddies become useless after a relatively short time. One might just as well forget about them altogether, and start out with a relatively crude subgrid scale parametrization. The only thing that matters is that the statistical features are parametrized accurately.

We have arrived at the point where we can summarize. Three-dimensional turbulence has a spectrum in which characteristic frequencies increase with increasing wave number. The microstructure of turbulence is unpredictable in principle when calculations have to be extended over the entire life span of the larger eddies. The problem of the parametrization of subgrid scale fluxes is severe, because the acceleration field is contaminated by high-frequency noise. Three-dimensional turbulence is *very* dissipative and *very* unpredictable. It belongs in Lorenz' third category: « for any particular future time, there is a limit below which the error involved in a forecasting calculation cannot be reduced, no matter how small the initial error is made » [14]. The rapid growth of errors at small scales of motion is responsible for the rapid loss of predictability.

8. – Characteristic exponents.

In the preceding section we have encountered the characteristic time or time scale of turbulence in the inertial subrange. We have, perhaps rather naively, equated the dynamic interaction time and the life span of an eddy with the reciprocal of the local vorticity (local both in the spectral and in the spatial sense). We have also assumed that error propagation rates are proportional to the characteristic frequency. Can some of these assumptions be justified? It is evident that turn-over time and vorticity are geometrically related, because the vorticity is twice the local angular velocity. However, what about turn-over times and lifetimes? Also, the growth rate of small errors is determined by the dominant characteristic exponent; how is the latter related to the vorticity at the same scale of motion?

Over periods of time that are small compared to the lifetime of an eddy, the viscous term in the vorticity equation may be neglected, providing we look at eddies that are large compared to the Kolmogorov microscale. In a frame of reference aligned with the local vorticity vector, the vorticity equation (8) then can be approximated temporarily by

$$\frac{d\zeta}{dt} = e_{pp}\zeta, \tag{45}$$

where e_{pp} is the appropriate principal component of the strain rate along the vortex axis ([7], Chapt. 8). It follows that

$$\frac{d}{dt}(\ln \zeta) = e_{pp},$$

which integrates to

$$\ln(\zeta/\zeta_0) = \langle e_{pp}\rangle t. \tag{46}$$

In (46) the angular brackets denote an average over the integration interval. The growth rate of small errors, on the other hand, is defined by

$$\ln(D/D_0) = \lambda_M t, \tag{47}$$

where D is the difference between two realizations and λ_M is the maximum characteristic exponent [5, 11]. Clearly, (46) is also valid for any error in vorticity, so that the strain rate at each wave number acts as the characteristic exponent in the early evolution of a vortex tube.

According to (18), the mean square strain rate in three-dimensional turbulence equals one-half the mean square vorticity. There is no reason to suspect that this relation does not hold, at least qualitatively, at each wave number. Therefore, the characteristic vorticity is comparable to the characteristic strain rate at each scale of motion. From that point on it is easy: (46) and (47) relate characteristic exponents to strain rates, whence we can conclude that the maximum characteristic exponent is comparable to the local vorticity, and thus inversely proportional to the local turn-over time (see also [11]). This implies that (36) or (41) can be used as an estimate of the spectral distribution of characteristic exponents in three-dimensional turbulence:

$$\lambda(k) \sim \zeta(k) \sim \varepsilon^{\frac{1}{3}} k^{\frac{2}{3}}. \tag{48}$$

A detailed analysis of the energy transfer associated with (45) and (46) shows that the strain rate is also a measure for the rate at which a vortex tube absorbs energy from its environment ([7], Chapt. 8). The energy-cascading rate, therefore, is also proportional to the turn-over rate. The relation between the lifetime and the turn-over time cannot be determined with arguments of this kind, however. I speculate that the strain rate does not remain aligned with a vortex tube for more than a few eddy revolutions (perhaps even less than one turn-over time). As soon as vorticity and strain rate are out of alignment, the eddy begins to lose its energy to its surroundings. Only if the decay process were under the exclusive effect of friction could one expect a long decay time; in fact, the environment is highly chaotic and the decay consists primarily of loss of coherence. There is, therefore, no reason to claim that lifetime and turn-over time should be widely different.

Vorticity amplification of the kind expressed by (45) and (46) is absent in two-dimensional turbulence. Therefore, the approach used here cannot help us to determine the distribution of characteristic exponents in two-dimensional flow. I have not yet found a more general line of reasoning, however.

9. – Spectral characteristics of two-dimensional turbulence.

Our study of the distinguishing features of turbulence in two and three dimensions now brings us to a discussion of the spectrum and the predictability properties of the enstrophy-cascading range in two-dimensional flow.

The spectrum in the enstrophy-cascading range depends on the spectral enstrophy flux χ, the viscosity ν and the wave number k. In the range $k\eta_2 \ll 1$ the local effects of viscosity should be small. The spectral characteristics in the inertial range then are determined by χ and k only [8]. The typical velocity $u(k)$ of eddies at wave number k then may be estimated as [15]

$$(49) \qquad u(k) \sim \chi^{\frac{1}{3}} k^{-1} .$$

Therefore,

$$(50) \qquad u^2(k) \sim \chi^{\frac{2}{3}} k^{-2} ,$$

whence the inertial-range spectrum becomes

$$(51) \qquad E_2(k) \sim k^{-1} u^2(k) \sim \chi^{\frac{2}{3}} k^{-3} .$$

This is the familiar minus-third power law of the enstrophy-cascading range. Equations (49)-(51) can be obtained also by assuming that the spectrum obeys an arbitrary power law, estimating the spectral enstrophy flux $-\overline{u_i' \zeta'\, \partial \zeta / \partial x_i}$

at wave number k on the basis of the assumed spectral behaviour and requiring that the flux be independent of k. That procedure gives the same results.

The vorticity and strain rate fluctuations at wave number k must have the following r.m.s. amplitudes:

$$\zeta(k) \sim e_{ij}(k) \sim ku(k) \sim \chi^{\frac{1}{3}}. \tag{52}$$

This seems to confirm that vorticity and enstrophy are conserved as long as viscosity does not enter the picture: the eddy vorticity is neither amplified nor attenuated as k increases. The real issue is a bit more delicate though, as we shall see in a moment.

The vorticity gradients associated with (52) are

$$\left(\frac{\partial \zeta}{\partial x}\right)_k \sim k\zeta(k) \sim \chi^{\frac{1}{3}}k. \tag{53}$$

These increase with increasing k, showing that the mechanism implied in (28) and (29) indeed leads to amplification of vorticity gradients and a nonvanishing dissipation of enstrophy as $\nu \to 0$. In fact, it is easy to verify that (53) is consistent with (31). Squaring (53), we have

$$\left(\frac{\partial \zeta}{\partial x}\right)_k^2 \sim \chi^{\frac{2}{3}}k^2. \tag{54}$$

Because the largest vorticity gradients occur at the smallest scales of motion, the mean square vorticity gradient in (31) can be estimated by taking $k\eta_2 = 1$ in this expression and employing definition (32) of η_2. This yields

$$\overline{\frac{\partial \zeta'}{\partial x_j}\frac{\partial \zeta'}{\partial x_j}} \sim \left(\frac{\partial \zeta}{\partial x}\right)^2_{k\eta_2=1} \sim \chi^{\frac{2}{3}} \cdot (\nu/\chi^{\frac{1}{3}})^{-1} = \chi/\nu, \tag{55}$$

which agrees with (31).

With the aid of (53), we can also perform a consistency check on the spectral enstrophy flux. This is done by interpreting the l.h.s. of (30) as the enstrophy flux from the resolved scales to the unresolved eddy motion, with the resolution put at k. Using (49), (52) and (53), we find that this flux is indeed independent of k and equal to χ (apart from numerical coefficients). The point here is that ζ is independent of k because the enstrophy transfer time is also independent of k.

Two other aspects of the k^{-3} range require attention: the characteristic time at wave number k and the characteristic acceleration. These are estimated as follows:

$$t(k) \sim \zeta^{-1}(k) \sim \chi^{-\frac{1}{3}}, \tag{56}$$

$$a(k) \sim ku^2(k) \sim \chi^{\frac{2}{3}} k^{-1}. \tag{57}$$

The characteristic evolution time of eddies in the enstrophy-cascading range is independent of wave number. This means that temporal changes in the flow field are not caused primarily by rapid evolution of small-scale systems. This view is confirmed by the wave number dependence of the characteristic acceleration. Small eddies in two-dimensional turbulence have small accelerations. The application of Newton's second law of motion thus is not contaminated excessively by high-frequency noise. Specifically, (57) represents the effects of the subgrid scale eddy flux of momentum on the evolution of the resolved motion. Clearly, these effects are small if the wave number of the numerical grid is large compared to that of the most energetic eddies. The spectral behaviour of two-dimensional turbulence permits a deterministic approach to the evolution of the large eddies, and the parametrization of subgrid scale fluxes is not a nearly insurmountable issue.

The fact that the time scale is independent of wave number requires further thought. The period over which the evolution of an eddy can be predicted deterministically is comparable with its life span. On the average, we cannot predict the time and place of birth and the subsequent evolution of the next generation of eddies. In general, the range of predictability is on the order of 10 time scales of the eddies concerned [12, 16]. This suggests that small eddies in two-dimensional turbulence have the same range of predictability as their brothers and sisters. That conclusion is not correct if the initial errors reside primarily at large wave numbers. It takes time before errors propagate up the spectrum; smaller eddies lose predictability before the larger ones do. In three-dimensional turbulence, the errors propagate so fast at high wave numbers that the net range of predictability of the large-scale motion is not appreciably extended by the period it takes the errors to reach the large eddies. In two-dimensional turbulence, however, we have to be a bit more careful.

Errors in a numerical forecast propagate from the unresolved scales toward the larger scales of motion. The contamination of the solution in the next octave takes about one time scale [12]. In the atmosphere, $\chi^{-\frac{1}{3}}$ equals about one day, because χ is on the order of 10^{-15} s^{-3}. The effective forecast period thus increases by a day if the resolution of a numerical model is improved by a factor of two. Also, if motions on a scale of $2^{12} \simeq 4000$ km are predictable ten days ahead, say, then those at the $2^{11} \simeq 2000$ km scale are contaminated within nine days, those at $2^{10} \simeq 1000$ km within eight, those at $2^9 \simeq 500$ km within seven and so on, even though their time scales are independent of wave number.

In these circumstances the total forecast period is a function of the number of octaves in the minus-three spectrum. If the viscosity were infinitely small and if the flow remained exactly two-dimensional at all scales, no matter how small, the range of predictability could be made as large as desired by selecting a sufficiently fine-mesh numerical grid. Two-dimensional turbulence belongs to Lorenz' second category [14]: « at any particular future time the error may be made arbitrarily small by making the initial error sufficiently small, but (...) the error becomes large in the sufficiently distant future ». In this asymptotic sense (the argument requires an arbitrarily long enstrophy-cascading range), two-dimensional turbulence is not nearly as unpredictable as its two-dimensional counterpart. The total gain obtainable in three-dimensional turbulence is a few seconds at best once the resolution is adequate, but in two-dimensional turbulence one can gain days as long as one can add octaves at the small-scale end.

In the atmosphere, however, the enstrophy cascade contains a finite number of octaves. There are 10 octaves between $2^3 = 8$ km and $2^{13} = 8192$ km; all motion at scales smaller than the scale height is likely to have a different spectral behaviour, with very little additional predictability.

The k^3 spectrum has other curious properties, too. Since the characteristic time is independent of wave number, the strain rate of large eddies is comparable to that of the smallest. Therefore, the cascade is not « local »: the spectral flux is not primarily between spectral neighbours. Also, it takes an infinitely long time to *initiate* a fully developed spectrum in a nearly inviscid flow driven by random forcing at a fixed wave number.

Let me summarize this discussion with a note on reversibility and predictability. In the limit as $\nu \to 0$, two-dimensional turbulence is almost non-dissipative and thermodynamically nearly reversible because its enstrophy is not amplified. The total amount of enstrophy in an infinitely long inertial range remains finite only if the enstrophy spectrum decreases at least as fast as k^{-1}. In that case, the characteristic time does not decrease with decreasing scale, and the range of predictability can be extended at will by appropriate resolution improvements.

10. – Epilogue.

Having arrived at the end of our trip through the classical theory of turbulence, I wonder how we should proceed. The classical theory deals with statistical properties of chaotic flows, such as mean energy and enstrophy, dissipation rates, spectra and inertial ranges. It can, and does, provide a frame of reference for further explorations, but it is not capable of answering some of the questions that occupy researchers today. Sensitivity to initial conditions is a case in point. The classical theory of turbulence tacitly assumes that individual realizations of a turbulent flow are sensitive to initial conditions and that averages do not exhibit such sensitivity because the fluctuations become uncorrelated before too long. The classical theory ignores the issue of predictability and proceeds to treat turbulence with stochastic methods. However, we now know that there is a profound difference between stochastic (« random », statistical) behaviour and the chaotic behaviour exhibited by a deterministic system with a strange attractor [17, 18]. Also, systems that are sensitive to initial conditions are capable of creating coherent structures, which are a kind of order within chaos. We cannot continue to ignore the presence of coherent structures, whether they are called solitons, modons, cyclones or merely « eddies », and we will have to come to grips with the relation between predictability and reversibility in forced, dissipative flow systems. I suspect that the predictability of chaotic solutions of the Navier-Stokes equations will someday be explained in terms of attractors; perhaps Prigogine's uncertainty relation between time and entropy will play a prominent role.

The classical theory has another serious drawback as far as I am concerned: it expands the velocity field in Fourier coefficients, which are the eigenfunctions of linear systems. Fourier coefficients, however, are not particularly suited for the study of nonlinear systems. Turbulence theory ought to be framed in terms of eddies that are reasonably local, both in physical space and in wave number space. This is a second reason to be interested in coherent structures; I hope that the current research on modons will lead to a workable definition of individual eddies. From that point, we can venture in the direction of multiple interactions between eddies, and on toward chaotic behaviour. Finally, we may arrive at a new theory of turbulence, in which coherent structures, strange attractors and a generalized concept of entropy are the cornerstones of understanding.

* * *

During the preparation of this paper, I have enjoyed several discussion with J. D. OPSTEEGH and W. VERKLEY. K. VAN GASTEL and F. NIEUWSTADT provided useful comments on an early draft. Sections 3 and 8 are based on discussions with several participants at the summer school; they were added afterwards.

REFERENCES

[1] J. CHARNEY: *J. Atmos. Sci.*, **28**, 1087 (1972).
[2] G. K. BATCHELOR: *An Introduction to Fluid Dynamics* (Cambridge, 1967).
[3] G. F. CARNEVALE and G. HOLLOWAY: *J. Fluid Mech.*, **116**, 115 (1982).
[4] G. F. CARNEVALE: *J. Fluid Mech.*, **122**, 143 (1982).
[5] D. RUELLE: *Commun. Math. Phys.*, **84**, 287 (1982).
[6] I. PRIGOGINE: *From Being to Becoming* (San Francisco, Cal., 1980).
[7] H. TENNEKES and J. L. LUMLEY: *A First Course in Turbulence* (Cambridge, Mass., 1972).
[8] G. K. BATCHELOR: *Phys. Fluids*, **12**, Suppl. 2, 233 (1969).
[9] G. I. TAYLOR: *Proc. R. Soc. London, Ser. A*, **164**, 15 (1938).

[10] R. H. KRAICHNAN: *Phys. Fluids*, **10**, 1417 (1967).
[11] R. BENZI, M. VITALETTI and A. VULPIANI: *J. Phys. A*, **15**, 883 (1982).
[12] C. E. LEITH and R. H. KRAICHNAN: *J. Atmos. Sci.*, **29**, 1041 (1972).
[13] J. W. DEARDOFF: *J. Atmos. Sci.*, **29**, 91 (1972).
[14] E. N. LORENZ: *Tellus*, **21**, 289 (1969).
[15] II. TENNEKES: *Bull. Am. Meteorol. Soc.*, **59**, 22 (1978).
[16] C. E. LEITH: *J. Atmos. Sci.*, **28**, 145 (1971).
[17] E. N. LORENZ: *J. Atmos. Sci.*, **20**, 130 (1963).
[18] O. E. LANFORD: *Annu. Rev. Fluid Mech.*, **14**, 347 (1982).

Fully Developed Turbulence and Intermittency.

U. FRISCH

CNRS, Observatoire de Nice - BP 139, 06003 *Nice Cedex, France*

1. – Simple ideas and misconceptions about turbulence.

Viscous incompressible 3-D flow can become turbulent when the Reynolds number R is sufficiently large. The latter is expressible, in terms of a typical scale L of the flow, a typical velocity V and the kinematic viscosity ν, as the ratio of the viscous diffusion time L^2/ν to the circulation (turn-over) time L/V. Information about turbulent flows comes from experiments, observations of nature and, increasingly, from computer simulations (cf. other lectures in this volume).

We now list and discuss some outstanding features of turbulent flows. In each case we begin with naive widely accepted statements and show that they can lead to misconceptions. We shall assume that the reader is at least moderately familiar with dynamical-system concepts (cf. the lectures by LIBCHABER, LORENZ and RUELLE in this volume and ref. [1-3]).

1'1. *Sharp transitions can occur when the Reynolds number is varied.* – Transitions from laminar to turbulent flows are discussed elsewhere in this volume. Very carefully controlled experiments on, *e.g.*, Rayleigh-Bénard convection have revealed a great variety of scenarios for the transition. In such experiments, when the flow becomes turbulent, it is often chaotic only in time and highly organized in space. In shear flows the transition may lead to much stronger chaos in both time and space via the 3-D destabilization of 2-D coherent structures (cf. the computer experiments by ORSZAG, PATERA and BRACHET reported in ref. [4] and [5] and the lectures by ORSZAG).

1'2. *The flow is unstable and unpredictable.* – A very weak perturbation introduced at some time t_0 may rapidly result in a complete distortion of the detailed flow pattern. Thus the flow may not be predictable in deterministic terms for more than a short time. It is conceivable, however, that statistical properties (involving, *e.g.*, time averages) are stable and can be predicted (cf. the difference between predicting weather and climate).

What do we understand by a «very weak perturbation»? The simplest way is to take an infinitesimal perturbation; such a perturbation is governed by the linearized dynamical equations. The growth of infinitesimal perturbations is controlled by the maximal Lyapounov characteristic exponent (LCE) λ_m (ref. [6, 7] and the lectures of Ruelle). Loosely stated, the fastest growing perturbation goes for very large times like $\exp[\lambda_m t]$, where $\lambda_m \geqslant 0$. Chaotic (or intrinsically stochastic) dynamical systems have $\lambda_m > 0$. In practice this does not necessarily mean complete lack of predictability. Indeed, i) small but finite perturbations may at first grow exponentially but eventually saturate at rather low levels, ii) the positive LCEs correspond to those directions in which the separation of neighbouring trajectories grows exponentially. There are also negative LCEs corresponding to shrinking separations and hence to no lack of predictability. In fact, it is known that chaotic dynamical systems can have strongly predictable features; one example is the «noisy periodicity» discussed by LORENZ [8]. Dissipative dynamical systems (e.g. viscous Navier-Stokes flow) may have a set that attracts trajectories for $t \to \infty$. The phase space can be infinite-dimensional (it usually is if the flow is governed by a partial differential equation), still the attractor can be finite-dimensional. If the dimension is small, the flow has in a sense only a finite number of unpredictable features. There is experimental evidence that many turbulent flows have strongly coherent quite predictable structures (cf. other lectures in this volume and ref. [9]). It is of interest to try to measure the dimension of their attractors.

Dynamical-system theory is also telling us that a turbulent flow with prescribed geometry and Reynolds number need not have unique statistical properties. Just like a body resting on a table may have several positions of stable equilibrium, a dynamical system can have several attractors with distinct basins. For example, in transition experiments of the Rayleigh-Bénard type the statistical regime that is obtained may depend on the way the system is prepared (cf. the lectures by LIBCHABER).

1`3. *Trajectories of marked particles are unstable and unpredictable.* – We now consider not the flow in some abstract phase space but the trajectories of individual marked particles following the fluid motion in (2-D or 3-D) physical space. Let us denote by $\boldsymbol{v}(t, \boldsymbol{r})$ the velocity field in Eulerian co-ordinates. Particle trajectories are determined by the following dynamical system

(1.1) $$\mathrm{d}\boldsymbol{r}(t, \boldsymbol{a})/\mathrm{d}t = \boldsymbol{v}(t, \boldsymbol{r}(t, \boldsymbol{a})), \quad \boldsymbol{r}(0, \boldsymbol{a}) = \boldsymbol{a}.$$

If the flow is incompressible, $\nabla \cdot v = 0$ and the dynamical system is conservative, *i.e.* the mapping $\boldsymbol{a} \mapsto \boldsymbol{r}(t, \boldsymbol{a})$ conserves the Lebesgue measure.

The interesting observation is that the dynamical system (1.1) may be chaotic without the velocity field being so. One simple known example is the spatially periodic steady Arnold-Beltrami-Childress flow (A, B, C real)

(1.2) $$V_1 = A \sin x_3 + C \cos x_2,$$

(1.3) $$V_2 = B \sin x_1 + A \cos x_3,$$

(1.4) $$V_3 = C \sin x_2 + B \cos x_1,$$

introduced by ARNOLD [10] and CHILDRESS [11] and studied by HÉNON [12]. It is easily checked that $\boldsymbol{v} \times \operatorname{curl} \boldsymbol{v} = 0$. Hence this flow is an exact solution of the 3-D Euler equations. This solution being steady hardly qualifies as «turbulent». Still the numerical evidence is that streamlines (here identical with the trajectories of marked particles) are chaotic when $ABC \neq 0$. We thus have a sort of «Lagrangian turbulence»: drops of dye introduced into the flow will develop extremely intrincate structures. Lagrangian turbulence of this sort can also take place in two dimensions provided that the velocity field becomes time-dependent (periodic is enough). A very simple example is given by AREF [13]: in a circular shallow tank the fluid is agitated T-periodically by a stirrer (assumed to act like a point vortex); during a time interval $T/2$ the stirrer is placed at some point M and then for another half-period at the diametrically opposite point M'. This simple device can give very efficient mixing.

The above observations about Lagrangian turbulence may be relevant to atmospheric predictability: the motion of an advected long-lived small-scale structure (say a radioactive cloud) in a large-scale 2-D flow can be unpredictable as soon as the large-scale flow is not steady.

1˙4. *Turbulent flow enhances transport.* – In a nonturbulent fluid the transport of heat, momentum, etc. is caused by molecular motion and collisions: at scales much larger than the mean free path λ, a molecular-diffusion process takes place with a diffusion coefficient $k_{\text{mol}} \sim \lambda v_{\text{th}}$, where v_{th} is the r.m.s. thermal velocity. In a 3-D turbulent fluid with integral scale l_0 and r.m.s. turbulent velocity v_0 one can similarly construct a turbulent diffusion coefficient $k_{\text{turb}} \sim l_0 v_0$ usually $\gg k_{\text{mol}}$. The enhancement of transport by turbulence is sometimes the only evidence that a flow is turbulent (*e.g.* in the interiors of stars).

Steady 3-D flow with chaotic streamlines (such as the flow discussed above) will not usually produce turbulent diffusion. The reason is that the dynamical system defined by the streamlines will have Kolmogorov-Arnold-Moser (KAM) invariant surfaces within which marked particles remain trapped. Introducing some molecular diffusion amounts to letting the marked particles perform an additional Brownian motion by which they can escape from the KAM surfaces. Diffusive behaviour is then recovered at large scales.

We also stress that turbulent-transport coefficients are appropriate only for phenomena on scales much larger than l_0. This restriction is frequently ignored in empirical modelling of turbulent flows. One example is the use of eddy viscosities in subgrid scale modelling in connection with computer simulations when it is not practical to resolve all the relevant scales. Such procedures may yield reasonable results when used in a dimensionally consistent way. They may also badly fail. The most obvious pitfall is that by increasing the viscosity the flow can be made laminar. The opposite can also happen: in Taylor-Couette flow and mixing layers transitions can occur such that, when R is increased, the large-scale flow becomes *more* coherent, while the small scales remain chaotic.

This may be somewhat similar to the existence of laminar windows for the control parameter within the chaotic domain observed, *e.g.*, in iterated maps [14]. In such a situation subgrid scale calculations may be unable to predict the coherence of the flow.

There are circumstances in which turbulence on a scale l_0 acts to *enhance* rather than to diffusively damp the amplitude of the quantity being transported on a scale $\gg l_0$. In 2-D incompressible flow negative eddy viscosities can be used to explain that the energy cascade proceeds from small to large scales (cf. ref. [15, 16] and the lectures by SADOURNY and TENNEKES). In a 3-D flow with nonvanishing helicity (= integral over space of $v \cdot \mathrm{curl}\, v$) large-scale magnetic fields are amplified due to the «α-effect» (cf. ref. [17] and the lectures by CHILDRESS).

1˙5. *High-Reynolds-number turbulence has a wide range of scales.* – Fourier analysis of velocity signals from a probe in high-Reynolds-number flow (*e.g.* a turbulent jet) reveals the kind of spectrum shown in fig. 1. The energy spectrum follows a power law $\propto k^{-m}$ ($m \approx 5/3$) over a range of scales (*i.e.* inverse wave numbers) extending from the integral scale l_0 to the dissipation

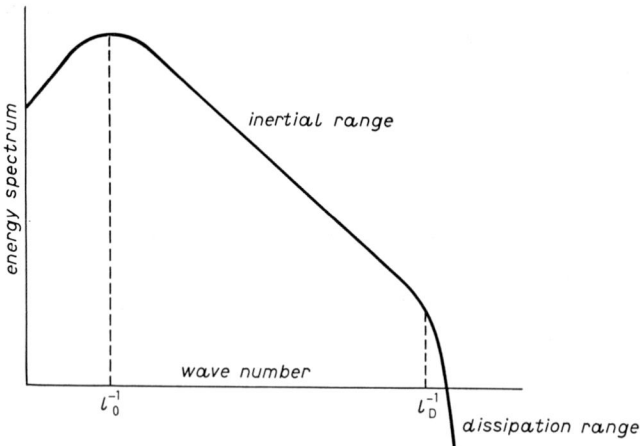

Fig. 1. – Energy spectrum of high-Reynolds-number turbulence. Double logarithmic co-ordinates. l_0 is the integral scale and l_D is the dissipation scale.

scale l_D. The ratio l_0/l_D increases with R like R^n ($n \approx 3/4$) and the small-scale motion is approximately isotropic. All this was predicted by KOLMOGOROV in 1941 [18]. Simple phenomenological interpretations can be given [19, 20]. The range of scales $l \sim l_0$ is called the energy-carrying or production range, because that is where most of the turbulent energy is produced (usually by some instability mechanism). The range $l_0 \gg l \gg l_D$ is the inertial range, because the dynamics are here dominated by the inertial terms in the Navier-Stokes equations (direct production and dissipation are negligible). The range $l \lesssim l_D$ is called the dissipation range because both inertial and dissipation terms are relevant. A frequent misconception is that only dissipation is relevant (cf. ref. [21]). The increasing range of scales as R is increased does not itself imply that the number of basic degrees of freedom (those governing the attractor of the flow) is increasing. As stressed, for example, in Tennekes' lectures,

Fig. 2. – Intermittent velocity signal. The plot shows high-pass filtered output of a hot-wire probe measuring the velocity in grid-generated turbulent flow (Y. GAGNE, Institut de Mécanique de Grenoble).

turbulent flows are full of coherent structures. Presumably a chaotic process (that may have an attractor of relatively low dimension) controls the formation of such structures. It is conceivable that the formation of small scales by subsequent flattening of the structures into, e.g., ribbons or sheets is governed by a predictable deterministic process. This picture is consistent with a low-dimensional attactor at high Reynolds numbers. Alternatively, and more likely, there may be additional small-scale instabilities (inviscid or viscous) leading to increasing chaos.

High-pass filtering of turbulent signals reveals that the small-scale activity is intermittent; it comes in bursts, as shown in fig. 2. This is already conspicuous in the inertial range and even more so in the dissipation range. Intermittency was discovered by BATCHELOR and TOWNSEND [22]. Inertial-range intermittency has not yet received any systematic explanation; it is not consistent with the original 1941 Kolmogorov theory (often referred to as K 41) and has led to various modified theories [20, 23-25]; we shall come back to this in sect. 2 and the appendix. Dissipation range intermittency is much better understood and can be related to singularities of the solutions of the Navier-Stokes equations at complex times (cf. ref. [21, 26]); an interesting early interpretation of this intermittency may be found in ref. [27]. The essence of the explanation is that high-pass filtering of an analytic function with complex-time singularities produces bursts centred at the real part of the singularities and with an overall amplitude proportional to $\exp[-\Omega|\tau|]$, where Ω is the filter frequency and τ the imaginary part of the singularity. The (rare) singularities closest to the real axis are strongly favoured for large Ω.

2. – Fully developed turbulence: intermittency as a broken symmetry.

By fully developed turbulence (FDT) we understand the asymptotic regime that is obtained by letting the Reynolds number tend to infinity. We shall here consider only the 3-D case; for the quite different 2-D case, see the lectures by SADOURNY and TENNEKES and ref. [15, 16]. In geophysical flows inertial ranges of three decades and more are not uncommon [28]. Laboratory experiments can display substantial inertial ranges [29]. Computer simulations of the full Navier-Stokes on a grid of 256^3 points have now reached the point where features of fully developed turbulence begin to be conspicuous [30, 31].

Attempts to construct a theory of FDT go back more than 40 years: a long history of frustrated attempts which we shall not review here [32, 33].

One central difficulty of FDT appears to be a broken symmetry. This has already been discussed in ref. [26]. The argument will here be just summarized. Consider the Navier-Stokes equations

$$(2.1) \quad \begin{cases} \partial_t \boldsymbol{v} + \boldsymbol{v} \cdot \nabla \boldsymbol{v} = -\nabla p + \nu \nabla^2 \boldsymbol{v}, \\ \nabla \cdot \boldsymbol{v} = 0 + \text{boundary and initial conditions}. \end{cases}$$

In the infinite-Reynolds-number limit ($\nu \to 0$) the Navier-Stokes equations are formally invariant under all the groups of scaling transformations $\mathscr{D}_h$

$$(2.2) \quad \boldsymbol{r} \to \lambda \boldsymbol{r}, \quad \boldsymbol{v} \to \lambda^h \boldsymbol{v}, \quad t \to \lambda^{1-h} t, \qquad \lambda > 0,$$

with arbitrary similarity exponent h. Since the production of turbulence singles out the scale l_0, the flow cannot be globally invariant under $\mathscr{D}_h$ (neither in a deterministic nor in a statistical sense). Still, the invariance may hold asymptotically at scales $l \ll l_0$. This is precisely one of the postulates of the K 41 theory [18], the other one being the assumption that there is a finite rate of energy dissipation per unit mass; it then follows that the flow is statistically self-similar with exponent $h = 1/3$. This has important consequences for the (longitudinal) structure functions. The latter are defined by

$$\langle (\delta v(l))^p \rangle \equiv \langle (v(\mathbf{r} + \mathbf{l}) - v(\mathbf{r}))^p \rangle, \tag{2.3}$$

i.e. the p-th-order moment of the velocity increments over a distance l (the velocity component being measured parallel to $\mathbf{l}$). For homogeneous isotropic turbulence the r.h.s. of (2.3) is a function only of $l = |\mathbf{l}|$ and of p. (Asymptotic) self-similarity with $h = 1/3$ implies that

$$\langle (\delta v(l))^p \rangle \propto l^{\zeta_p}, \qquad \zeta_p = p/3, \quad l_0 \gg l \gg l_D. \tag{2.4}$$

For $p = 2$ this is just another way of writing the celebrated $k^{-5/3}$ law. The trouble is that eq. (2.4) is only marginally supported by experiment for small p's and not at all for large p's. Power law behaviour appears to hold, but not with $\zeta_p = p/3$. Fairly accurate measurements of the ζ_p for p up to about 10 have been reported recently in ref. [29]. The experimental values of ζ_p for three flows are given in table I (taken from ref. [29]). R_λ is the Reynolds number based on the Taylor microscale. Clearly the results indicate that self-similarity is broken. In ref. [26] we show that, since scale invariance corresponds to a noncompact group, it is more likely to be broken than, say, isotropy, which corresponds to the compact group of rotations.

TABLE I. – *Exponents of structure functions (taken from ref. [29]).*

p	2	3	4	5	6	7	8
$R_\lambda = 515$ (duct)	0.71	1	1.33	—	1.8	—	2.27
$R_\lambda = 536$ (jet)	0 71	1	1.33	1.54	1.8	2.06	2.28
$R_\lambda = 852$ (jet)	0.71	1	1.33	1.65	1.8	2.12	2.22

p	9	10	12	14	16	18
$R_\lambda = 515$ (duct)	—	2.64	2.94	3.32	—	—
$R_\lambda = 536$ (jet)	2.41	2.60	2.74	—	—	—
$R_\lambda = 852$ (jet)	2.52	2.59	2.84	3.28	3.49	3.71

A weaker assumption that may be consistent with the data (but not necessarily with the Navier-Stokes equations) is *conditional* self-similarity: one assumes, for example, a hierarchical embedding of bursts within bursts as one proceeds to ever smaller scales (cf. ref. [20, 25]). A number of such hierarchical intermittent models has been constructed with probabilistic elements introduced in an *ad hoc* way; according to such models, in FDT the energy dissipation is concentrated in a fractal set [34]. The fractal dimension D of this set can be inferred from the sixth-order structure function [20]. According to the recent data reported in ref. [29] $D \approx 2.8$; this would indicate that dissipative structures are so extremely convoluted that they are nearly space filling ($D = 3$). In the appendix, written with PARISI, we show that the data are consistent with a picture of intermittency involving more than one fractal set.

Explaining the broken self-similarity by the intermittency is just displacing the problem. There are recent indications that the broken-symmetry–intermittency problem of FDT has to do with chaotic dynamical systems. This is based on results obtained with the « two-component shell model » of ref. [35, 36]. Earlier shell models have been introduced in ref. [37] and studied in ref. [38-40]. The general idea is as follows. Let us start with 3-D Navier-Stokes turbulence. Fourier space for the space variable is divided into octave shells comprised between successive wave numbers k_n defined dy

$$(2.5) \qquad k_{-1} = 0, \qquad k_0 = l_0^{-1}, \qquad k_n = 2^n k_0 \qquad (n \geqslant 1).$$

The total velocity field $v(t, r)$ can be decomposed into contributions $v_n(t, r)$ each of which involves $O(k_n^3)$ Fourier components. In shell models it is assumed that each shell has only a small number of degrees of freedom which does not increase with n. The models are chosen to have many structural properties in common with the Navier-Stokes equations: form of linear and nonlinear terms, conservation of energy for $v = 0$, conservation of volume in phase space for $v = 0$, etc. Interactions are only between neighbouring shells (motivated by the observation that in 3-D FDT energy transfer involves mostly interactions between comparable scales). The two-component shell model reads (ref. [36])

$$(2.6) \qquad \partial_t u_n = \alpha(k_n u_{n-1}^2 - k_{n+1} u_n u_{n+1} - k_n b_{n-1}^2 - k_{n+1} b_n b_{n+1}) + \\ + \beta(k_n u_{n-1} u_u - k_{n+1} u_{n+1}^2 - k_n b_{n-1} b_n + k_{n+1} b_{n+1}^2) - \nu k_n^2 u_n + \delta_{1n},$$

$$(2.7) \qquad \partial_t b_n = \alpha k_n (u_{n+1} b_n - u_n b_{n+1}) + \beta k_n (u_n b_{n-1} - u_{n-1} b_n) - \nu k_n^2 b_n,$$

$$(2.8) \qquad n = 1, 2, 3, \ldots, \quad u_0 = b_0 = 0.$$

α and β are real parameters. The Kronecker δ_{1n} provides a driving force for the first shell; higher-order shells are excited by nonlinear interactions. The following properties are easily checked: 1) the nonlinear interactions conserve the total energy

(2.9) $$E = \tfrac{1}{2} \sum_{n=1}^{\infty} (u_n^2 + b_n^2);$$

ii) when $\nu = 0$ and only a finite number N of shells are kept (with $u_{N+1} = b_{N+1} = 0$), the flow in the $2N$-dimensional space defined by eqs. (2.6), (2.7) is conservative, i.e.

(2.10) $$\sum_{1}^{N} \left(\frac{\partial}{\partial u_n} (\partial_t u_n) + \frac{\partial}{\partial b_n} (\partial_t b_n) \right) = 0 \qquad \text{(Liouville theorem)}.$$

Similar properties hold for the Navier-Stokes equations (ref. [32, 33]). The two-component shell model was introduced by GRAPPIN, LÉORAT and POUQUET [35] as a generalization to MHD (conducting flow with magnetic field) of the model of Desnyansky and Novikov [37], to which it reduces when $b_n \equiv 0$. It is also possible to consider the two-component shell model as a model of the Navier-Stokes equations that retains two rather than one mode per shell. It is noteworthy that the Liouville theorem (eq. (2.10)) does not hold for the Desnyansky-Novikov model.

The main results obtained for the two-component shell model (ref. [36]) in the FDT limit ($\nu \to 0$) are as follows:

a) If all the b_n's are set equal to zero and if $2^{1/3}\alpha - \beta > 0$, the solutions of eqs. (2.6), (2.7) tend for $t \to \infty$ to a « K 41 fixed point ».

b) For nonvanishing b_n's intermittent chaotic solutions have been obtained (so far in a limited number of numerical experiments).

The precise meaning of statement a) is

(2.11) $$\lim_{n\to\infty} \lim_{\nu\to 0} \lim_{t\to\infty} u_{n+1}/u_n = 2^{-1/3},$$

and similarly for the b_n's. It is easily checked that all nonlinear terms in eqs. (2.6), (2.7) vanish when $u_{n+1}/u_n = b_{n+1}/b_n = 2^{-1/3}$. Stability of this K 41 solution is demonstrated by a linear stability analysis and by computer experiments. For nonvanishing b_n's at high Reynolds numbers chaotic solutions (with positive maximum LCE) are observed, the high-shell u_n's and b_n's being conspicuously intermittent in time. Time averages of the quantities $\langle (u_n(t))^p \rangle$, analogous to structure functions, show power law dependence (in k_n) with significant deviations from K 41 scaling (ref. [36]). Note that intermittent behaviour has been previously observed in shell models in which the energy dissipation is not through a linear term but through a nonlinear eddy viscosity [40, 41].

To understand how the intermittent chaotic behaviour comes about in shell models, there are at least two strategies. One can truncate the model and study bifurcations controlled by the Reynolds number. This has already been done for the cases of two and three shells [36]. Alternatively one can work directly with infinitely many shells and infinite Reynolds number. Indeed, it is possible to construct a one-parameter family of models which in some ranges have K 41 solutions and in other ranges chaotic intermittent solutions. As the parameter is varied, bifurcations are expected which lead to symmetry breaking and chaos.

3. – Turbulence with a spectral gap and predictability (based in part on ref. [42]).

In the atmosphere of the Earth there is a number of instability mechanisms acting on very different scales: for example, the large-scale baroclinic instability and the small-scale convective instability. Under what conditions can the resulting turbulent flows coexist and be separated by a spectral gap (*i.e.* be spectrally segregated as depicted in fig. 3)? Many atmospheric scientists believe that there is such a gap in the atmosphere and it has been argued by LORENZ ([43] and this volume) that this can result in increased predictability of the weather.

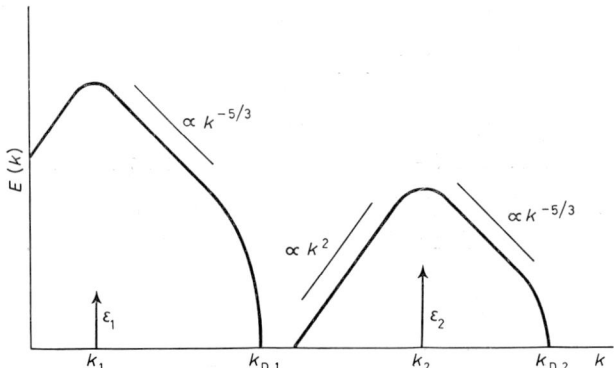

Fig. 3. – 3-D turbulence with a spectral gap.

The simplest model on which the possibility of a spectral gap can be examined is homogeneous isotropic turbulence with two distinct production mechanisms acting at scales $l_1 = k_1^{-1}$ and $l_2 = k_2^{-1} \ll l_1$. To make the problem (somewhat) tractable it is essential to assume a wide separation of scales.

Before putting the two turbulences together, let us recall some additional facts about the spectrum of high-Reynolds-number homogeneous isotropic turbulence driven at a single scale, as shown in fig. 4. Subtleties such as intermittency corrections to K 41 will be mostly ignored, since they are probably not very relevant for what will follow. We denote by ε_1 the rate at which energy is injected per unit mass into the turbulent flow. Stationarity is assumed; hence ε_1 is equal to the rate at which energy cascades to smaller scales and

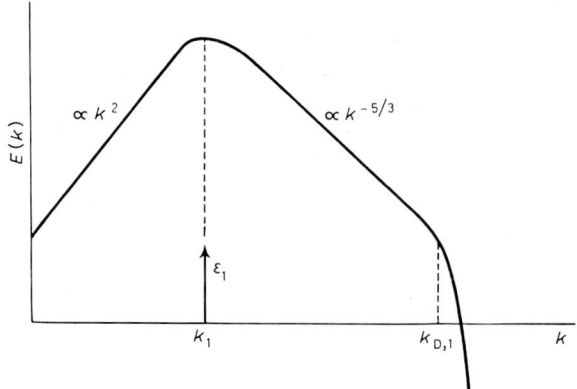

Fig. 4. – Inertial and equilibrium ranges of 3-D turbulence.

also equal to the rate of dissipation. The latter is $2\nu\Omega_1$, where ν is the viscosity and Ω_1 the enstrophy (mean square vorticity). The former can be estimated in terms of l_1 and the r.m.s. turbulent velocity v_1: the amount of energy in scales $\sim l_1$ is $\sim v_1^2$; the characteristic time for transferring this to smaller scales is the eddy turn-over time $t_1 \sim l_1/v_1$; hence $\varepsilon_1 \sim v_1^2/t_1 \sim v_1^3/l_1$. The energy spectrum follows the Kolmogorov law

$$(3.1) \qquad E(k) \sim \varepsilon^{2/3} k^{-5/3}$$

in the inertial range

$$(3.2) \qquad k_1 = l_1^{-1} \ll k \ll k_{D,1} \sim k_1 R^{3/4},$$

where $R \sim l_1 v_1/\nu$. At small wave numbers, $k \ll k_1$, there is an « equilibrium range » in which

$$(3.3) \qquad E(k) \propto k^2.$$

In this range there is an equipartition of kinetic energy between all Fourier modes (there are $\sim k^2 \, dk$ modes with wave number between k and $k + dk$). The equilibrium range is fed by beating-type interactions between two eddies in the energy-containing range, this being balanced by an eddy viscosity also coming mostly from the energy range [44]. In field-theoretic jargon the equilibrium range has « asymptotic infra-red freedom » (this is a school of physics!) as can be demonstrated using renormalization group methods [45, 46].

We are now in a position to understand the interactions of two turbulent motions separated by a gap (fig. 3). Let us denote by $\varepsilon_i \sim v_i^3/l_i$ the energy production rates and by $k_i = l_i^{-1}$ the energy-carrying wave number for the large-scale flow ($i = 1$) and the small-scale flow ($i = 2$). The main effect fo the small scales on the large ones is to replace the molecular viscosity ν by an eddy viscosity $\nu_t \sim l_2 v_2$ (cf. subsect. 1`4). Thus the large-scale Reynolds number is reduced to an effective $R_1^{\text{eff}} \sim l_1 v_1/l_2 v_2$. We denote by $E_2(k)$ the energy spectrum that would be established with only small-scale production and by $E_1(k)$ the spectrum that would be established with only large-scale production but with ν_t used instead of ν. The viscous cut-off for $E_1(k)$ is at $k_{\text{D},1} \sim k_1 (R_1^{\text{eff}})^{3/4}$. A necessary condition for spectral segregation is that $k_{\text{D},1} \ll k_2$. This is easily seen to be equivalent to

(3.4) $$\varepsilon_1 \ll \varepsilon_2 \quad \text{or} \quad v_1^3/l_1 \ll v_2^3/l_2 .$$

We have assumed up to this point that the small-scale turbulence is mostly unaffected by the presence of the large-scale turbulence. This requires that the turn-over time t_2 be small compared to the characteristic time for distortion by large-scale shears, namely $t_s \sim \Omega_1^{-\frac{1}{2}}$. It is easily checked that the second condition gives the same relation eq. (3.4) as before. Finally we shall require that the large-scale turbulence should have a large effective Reynolds number $R_1^{\text{eff}} \gg 1$. Putting everything together, we get the conditions for spectral segregation of two fully developed 3-D turbulent flows

(3.5) $$(l_1/l_2)^{-1} \ll v_1/v_2 \ll (l_1/l_2)^{1/3} .$$

If the left inequality becomes violated, the large-scale flow can become laminar (by the action of the small-scale turbulence!). A somewhat more quantitative study of this problem can be done using closure theory [42].

When intermittency is included in the above analysis (in the sense of ref. [20]), it is found that the exponent 1/3 in the r.h.s. of eq. (3.5) is lowered to $(D-2)/3$, where D is the fractal dimension of dissipation.

McLaughlin, Papanicolaou and Pironneau [47] have studied by asymptotic methods a variant of the spectral-gap problem in which the large- and small-scale flows are not driven and decay. So does the eddy viscosity and, eventually, molecular viscosity dominates. This may, however, take a very long time if the initial small-scale Reynolds number is large.

The above 3-D analysis is also easily extended to the coexistence of large-scale 2-D turbulence with small-scale 3-D turbulence. For this we must, of course, assume that there is a mechanism that keeps the large-scale flow 2-D. The situation is as shown in fig. 5. β_1 is the rate of enstrophy production $\beta_1 \sim (v_1^2/l_1^2)/(l_1/v_1) \sim v_1^3/l_1^3$. The enstrophy inertial range extends from $\sim k_1 = l_1^{-1}$ to $\sim k_{\text{D},1} \sim k_1 (R_1^{\text{eff}})^{1/2}$, where R_1^{eff} is defined as before. By going through the same kind of analysis as before, we find the following conditions for spectral segregation of the two fully developed 2-D and 3-D flows

(3.6) $$(l_1/l_2)^{-1} \ll v_1/v_2 \ll l_1/l_2 .$$

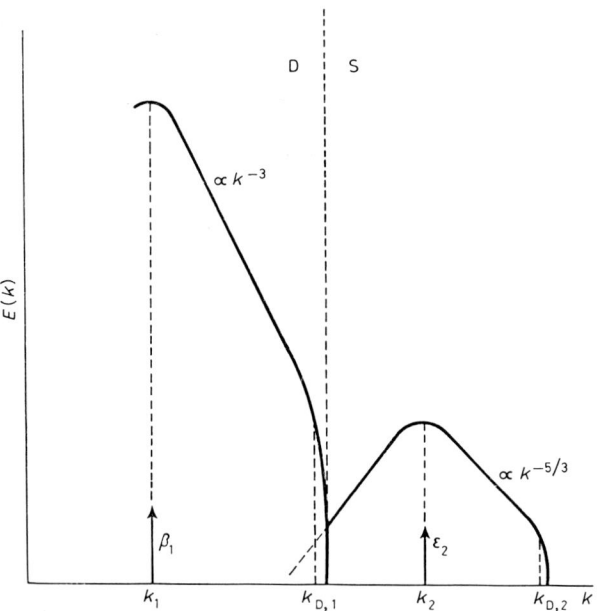

Fig. 5. – Spectral gap between large-scale 2-D turbulence and small-scale 3-D turbulence.

We now show that a spectral gap between small-scale 3-D turbulence and large-scale 2-D turbulence may lead to increased predictability of the large-scale motion (as suggested by LORENZ ([43] and this volume)). We assume that the flow can be resolved deterministically (D) up to a wave number in the spectral gap (indicated by a vertical dashed line in fig. 5) and that beyond the flow is only known statistically (S). According to ref. [48, 49] errors migrate from larger to smaller wave numbers. In a range where nonlinear interactions are mostly local (e.g. an inertial range), the characteristic time for error migration over, say, one octave of wave numbers is the local eddy turn-over time (what else could it be?). Across the spectral gap, however, interactions are highly nonlocal. The mechanism for error generation at wave numbers $k \ll k_2$, due to the small-scale 3-D turbulence, is basically the same as for the generation of the equilibrium range: errors are generated by beating interactions between wave numbers $\sim k_2$ and are damped by eddy viscosity. We can, therefore, estimate the error spectrum (in the absence of large-scale motion) to be

(3.7) $$E^{\mathrm{err}}(k) \sim (k/k_2)^2 E(k_2) \sim (k/k_2)^2 l_2 v_2^2.$$

Errors generated by the small-scale turbulence will be enhanced by the instability of the large-scale flow, and most efficiently at wave numbers of the order of the dissipation wave number $k_{D,1}$. The relative error is

$$(3.8) \qquad r \sim E^{\text{err}}(k_{D,1})/E(k_{D,1}) \sim k_{D,1}/k_2 \,.$$

We now assume exponential amplification of the error with a characteristic time equal to the turn-over time l_1/v_1 in the enstrophy inertial range (roughly wave number independent); we find that the relative error has become of order unity after a time $\sim t_1 \log(1/r) \sim t_1 \log(k_2/k_{D,1})$. To this we must add the time for errors to migrate along the enstrophy inertial range from $k_{D,1}$ to k_1 which is $\sim t_1 \log(k_{D,1}/k_1)$. Hence the total predictability time is $\sim t_1 \log(k_2/k_1)$. In other words, thanks to the spectral gap, the predictability is the same as if we had a 2-D flow with full resolution of all scales down to the integral scale of the 3-D turbulence. The increased predictability can be rather important (days?) if a spectral gap exists at scales ~ 100 km. This, of course, is debatable. It could be that the gap in the energy spectrum is filled by rare but violent meteorological events and does not exist in the mean; some increase in the predictability is then nevertheless expected.

Finally, we observe that coherent structures may play an important part in the dynamics of atmospheric turbulence. If this is so, predictability estimates based on turbulence phenomenology (*à la* Kolmogorov) may be very misleading.

* * *

We have greatly benefitted from discussions with C. GLOAGUEN (two-component shell model) and with J. C. ANDRÉ, M. LESIEUR and O. THUAL (spectral gap and atmospheric predictability).

APPENDIX

On the singularity structure of fully developed turbulence.

with

G. PARISI

Dipartimento di Fisica, Università di Roma II « Tor Vergata » - Roma, Italia

A simple way of explaining power law structure function is to invoke singularities of the Euler equations considered as limit of the Navier-Stokes equations as the viscosity tends to zero. For Burgers' equation we know that such singularities exist (shocks) and that they provide the required explanation of scaling. For the 2-D Euler equations we know that singularities do not exist (see, *e.g.*, ref. [26] and references therein). For the 3-D Euler equations the numerical evidence is inconclusive [26, 31]. MANDELBROT [24, 25] and others [20] have considered models with singularities concentrated on a set $\subset \mathbf{R}^3$ having noninteger (fractal) Hausdorff dimension. We shall here show that the data suggest the existence of a hierarchy of such sets (a « multifractal »).

Since the Navier-Stokes equations (in the zero-viscosity limit) are invariant under the group of scaling transformations (defined in eq. (2.2)) for *any* value of h, singularities of arbitrary exponents (and mixtures thereof) are consistent with the equations. Specifically, we start with a definition, the velocity field at a given time $v(x)$ is said to have a singularity of order $h > 0$ at the point x if

(A.1) $$\overline{\lim_{x \to y}} |v(x) - v(y)|/|x-y|^h \neq 0 .$$

For negative h eq. (A.1) is modified by not subtracting $v(y)$.

We call $S(h)$ the set of points for which the velocity field has a singularity of order h. It is obvious that

(A.2) $$S(h') \supset S(h) \qquad \text{if } h' > h .$$

Roughly speaking, $S(h)$ is the region where the velocity field is not an Hölder function of order h. We denote by $d(h)$ the Hausdorff dimension of $S(h)$ (see ref. [34] and [50] for definitions). It follows from eq. (A.2) that $d'(h) > 0$; we also make the concavity assumption $d''(h) < 0$.

If such singularities exist, then, in the fully developed turbulence regime, $d(h)$ has a nontrivial dependence on h: different kind of singularities are associated with sets having different Hausdorff dimensions. Note that the opposite phenomenon happens for the solutions of stochastic differential equations with white noise (like those studied in Jona-Lasinio's contribution to this volume): there the one-dimensional trajectories are (with probability one) Hölder functions of order $\frac{1}{2}$, so that

(A.3) $$d(h) = \theta(h - \tfrac{1}{2}) \qquad (\theta = \text{step function}) .$$

It is useful to connect the function $d(h)$ with the exponents ζ_p introduced in eq. (2.4) which control the asymptotic behaviour of the longitudinal structure functions. We can try to rephrase the previous statements on the Hausdorff dimensions of $S(h)$ by saying that the probability of having $|v(x) - v(y)|$ of order $|x-y|^h$ goes to zero like $|x-y|^{3-d(h)}$ when $|x-y| \to 0$. We thus arrive to the following integral representation for the moments:

(A.4) $$\langle (\delta v(l))^p \rangle \sim \int d\mu(h) \, l^{(ph+3-d(h))} ,$$

where $d\mu(h)$ is a measure concentrated on the region where $d(h) > 0$.

In the K 41 [18] picture and in the β-model [20], we have, respectively,

(A.5) $$\begin{cases} \zeta_p = p/3 , \\ \zeta_p = \lambda(p-3) + 1 \end{cases} \qquad (\lambda < \tfrac{1}{3}) .$$

Consequently we have, respectively,

(A.6) $$\begin{cases} d(h) = 3\theta(h - \tfrac{1}{3}) , & \text{K 41}, \\ d(h) = (2 + 3\lambda)\theta(h - \lambda) , & \beta\text{-model}. \end{cases}$$

In more sophisticated models, and also in actual turbulence, according to ref. [29], ζ_p is a nonlinear function of p. Evaluating the integral (A.4) using the saddle point method, we easily find

(A.7) $$\zeta_p = \min_h [ph + 3 - d(h)].$$

We have thus found that ζ_p is the Legendre transform (see ref. [51], sect. 14) of the codimension ($c(h) \equiv 3 - d(h)$) of the set $S(h)$. This is assuring that the convexity properties of ζ_p are automatically preserved by eq. (A.7).

If eqs. (A.4) and (A.7) are correct, the dimensions $d(h)$ are experimentally well-defined quantities: they can be extracted from the ζ_p's by using the inverse Legendre transform

(A.8) $$d(h) = 3 - \min_p (\zeta_p - ph).$$

We shall not try to do this using the data displayed in table I, although this is clearly possible, at least in the range of h for which the value of p minimizing eq. (8) falls in the experimentally observed interval: it is, however, likely that $d(h)$ will not be a step function because ζ_p appears to significantly deviate from a linear function of p. The function $d(h)$ is thus nontrivial and singularities of different kinds, if they exist, are concentrated on sets having different Hausdorff dimensions.

The function $d(h)$ (or, equivalently, ζ_p) has a clear dynamical meaning because it contains most of the relevant information on the scaling laws for fully developed turbulence. It would be rather important to measure accurately $d(h)$ and to find good evidence for its universality, i.e. its independence on the initial conditions and on all the other parameters which should become irrelevant in the fully developed turbulence regime.

If the multifractal model is basically correct, accurate measurements of the ζ_p's may be quite difficult. Indeed, the structure functions are a mixture of power laws (eq. (A.4)), so that very small scales (i.e. very high Reynolds numbers) may be needed before the contribution with the smallest exponent clearly dominates; where exactly this happens depends on the distribution $d\mu(h)$.

Note that consistency of the multifractal model with the data is by no means evidence for real singularities of the Euler equations. There is certainly more than one way to obtain scaling, otherwise scaling would not be observed in two dimensions, where singularities are ruled out [26].

We note two interesting consequences of the inversion formula (A.8). First, if ζ_p vanishes for $p \to 0$, then the weakest singularities, which has the exponent ζ'_0, are space filling ($d = 3$). It is clearly of interest to measure ζ_p for small noninteger p's. Second, the multifractal model is not completely consistent with Kolmogorov's [23] lognormal model for which $\zeta_p = p/3 + \mu p(3-p)/18$ (μ, if it exists, is somewhere between 0.2 and 0.5; see ref. [29]). Indeed, with this choice of ζ_p we find from eq. (A.8) that beyond $p_{max} = 9(2/3\mu)^{\frac{1}{2}}$ a negative dimension is obtained. Accurate measurements of very-high-order structure functions are required to test for a possible inconsistency of the multifractal model.

Finally, one may wonder how the above «multifractal» model relates to the models of ref. [20, 25, 34, 52]. In Mandelbrot's [25, 34, 52] probabilistic models for the dissipation a random weighting factor W appears at each stage of the cascade. The case when W has a binomial distribution («absolute curdling») corresponds to a single fractal in our approach (it is also equivalent to the β-model). For more general W-distributions (« weighted curdling ») one obtains exponents ζ_p that depend nonlinearly on p like in the multifractal model. There is a single fractal for the energy dissipation, but it is conceivable that other fractals will be uncovered by investigating all possible singularities of the dissipation. Still the multifractal model appears to be somewhat more restrictive than Mandelbrot's weighted-curdling model which does include the lognormal case.

REFERENCES

[1] *Proceedings of « Nonlinear Dynamics »*, edited by H. G. HELLEMAN, Ann. N. Y. Acad. Sci., **357** (1980).
[2] *Proceedings of « Dynamical Systems and Chaos »*, edited by L. GARRIDO, Lecture Notes in Physics, Vol. **179** (Berlin, 1983).
[3] *Proceedings of Les Houches, Session XXXVI, 1981*, edited by G. IOOSS, H. G. HELLEMAN and R. STORA (Amsterdam, 1983).
[4] S. ORSZAG and A. PATERA: *Phys. Rev. Lett.*, **45**, 989 (1980).
[5] M. BRACHET and S. ORSZAG: *Secondary instability of free shear flows*, to be published (1983).
[6] D. RUELLE: *Publ. Math. IHES*, **51** (1980).
[7] G. BENETTIN, L. GALGANI, A. GIORGILLI and J. M. STRELCYN: *Meccanica* (Bologna, 1980), p. 9 and 21.
[8] E. LORENZ: *Ann. N. Y. Acad. Sci.*, **357**, 282 (1980).
[9] *Proceedings of « The Role of Coherent Structures in Modelling Turbulence and Mixing*, edited by J. JIMENEZ, Lecture Notes in Physics, Vol. **136** (Berlin, 1981).
[10] V. ARNOLD: *C. R. Acad. Sci.*, **261**, 17 (1965).
[11] S. CHILDRESS: *J. Math. Phys. (N. Y.)*, **11**, 3063 (1970).
[12] M. HÉNON: *C. R. Acad. Sci.*, **262**, 312 (1966).
[13] H. AREF: *Stirring by chaotic advection*, J. Fluid Mech., in press (1984).
[14] P. COLLET and J. P. ECKMANN: *Iterated Maps on the Interval as Dynamical Systems* (Boston, 1980).
[15] R. H. KRAICHNAN: *Phys. Fluids*, **10**, 1417 (1967).
[16] R. H. KRAICHNAN: *J. Fluid Mech.*, **67**, 155 (1975).
[17] H. K. MOFFATT: *Magnetic Field Generation in Electrically Conducting Fluids* (Cambridge, 1978).
[18] A. N. KOLMOGOROV: *Dokl. Akad. Nauk SSSR*, **30**, 301 (1941).
[19] G. K. BATCHELOR: *The Theory of Homogeneous Turbulence* (Cambridge, 1953).
[20] U. FRISCH, P.-L. SULEM and M. NELKIN: *J. Fluid Mech.*, **87**, 719 (1978).
[21] U. FRISCH and R. MORF: *Phys. Rev. A*, **23**, 2673 (1981).
[22] G. K. BATCHELOR and A. A. TOWNSEND: *Proc. R. Soc. London, Ser. A*, **199**, 238 (1949).
[23] A. N. KOLMOGOROV: *J. Fluid Mech.*, **13**, 82 (1962).
[24] R. H. KRAICHNAN: *J. Fluid Mech.*, **62**, 305 (1974).
[25] B. MANDELBROT: in *Turbulence and Navier-Stokes Equation*, edited by R. TEMAM, Lecture Notes in Mathematics, Vol. **565** (Berlin, 1976), p. 121.
[26] U. FRISCH: in *Les Houches, Session XXXVI, Chaotic Behaviour in Deterministic Systems, 1981*, edited by G. IOOSS, H. G. HELLEMAN and R. STORA (Amsterdam, 1983).

[27] R. H. KRAICHNAN: *Phys. Fluids*, **10**, 2080 (1967).
[28] H. L. GRANT, R. W. STEWART and A. MOILLIET: *J. Fluid Mech.*, **12**, 241 (1962).
[29] F. ANSELMET, Y. GAGNE, E. J. HOPFINGER and R. A. ANTONIA: *High order velocity structure functions in turbulent shear flows*, preprint Institut de Mécanique de Grenoble (1983).
[30] M.-E. BRACHET: *C. R. Acad. Sci.*, **294**, 537 (1982).
[31] M.-E. BRACHET, D. I. MEIRON, S. A. ORSZAG, B. G. NICKEL, R. H. MORF and U. FRISCH: *J. Fluid Mech.*, **130**, 411 (1983).
[32] S. A. ORSZAG: in *Fluid Dynamics, Les Houches Summer School 1973*, edited by R. BALIAN and J. L. PEUBE (New York, N.Y., 1977), p. 235.
[33] H. A. ROSE and P.-L. SULEM: *J. Phys. (Paris)*, **39**, 441 (1978).
[34] B. MANDELBROT: *Fractals: Form, Chance and Dimension* (San Francisco, Cal., 1977).
[35] R. GRAPPIN, J. LÉORAT and A. POUQUET: unpublished (1981).
[36] C. GLOAGUEN: Thèse de 3ème Cycle, Observatoire de Meudon (1983).
[37] V. N. DESNYANSKY and E. A. NOVIKOV: *Prikl. Mat. Mekh.*, **38**, 507 (1974).
[38] T. L. BELL and M. NELKIN: *Phys. Fluids*, **20**, 345 (1977).
[39] T. L. BELL and M. NELKIN: *J. Fluid Mech.*, **88**, 369 (1978).
[40] E. SIGGIA: *Phys. Rev. A*, **17**, 1166 (1978).
[41] R. M. KERR and E. SIGGIA: *J. Stat. Phys.*, **19**, 543 (1978). See also J. LEE: *J. Fluid Mech.*, **101**, 349 (1980).
[42] A. POUQUET, U. FRISCH and J. P. CHOLLET: *Phys. Fluids*, **26**, 877 (1983).
[43] E. LORENZ: in *Predictability of Fluid Motions*, edited by G. HOLLOWAY and B. WEST (American Institute of Physics, New York, N.Y., 1984), p. 133.
[44] M. LESIEUR and D. SCHERTZER: *J. Méc.*, **17**, 610 (1978).
[45] D. FORSTER, D. R. NELSON and M. STEPHEN: *Phys. Rev. A*, **16**, 732 (1977).
[46] J.-D. FOURNIER, P.-L. SULEM and A. POUQUET: *J. Phys. A*, **15**, 1393 (1982).
[47] D. MCLAUGHLIN, G. PAPANICOLAOU and O. PIRONNEAU: *Convection of microstructure*, preprint, Courant Institute, New York (1983).
[48] E. LORENZ: *Tellus*, **21**, 289 (1969).
[49] C. E. LEITH and R. H. KRAICHNAN: *J. Atmos. Sci.*, **29**, 1041 (1972).
[50] J. P. KAHANE: in *Turbulence and Navier-Stokes Equation*, edited by R. TEMAM, Lecture Notes in Mathematics, Vol. **565** (Berlin, 1976), p. 94.
[51] V. ARNOLD: *Mathematical Methods of Classical Mechanics* (Russian edition, Moscow, 1974; French edition, Moscow, 1976).
[52] B. B. MANDELBROT: *J. Fluid Mech.*, **62**, 331 (1974).

Stratified-Turbulence Experiments.

C. W. VAN ATTA

Scripps Institution of Oceanography and
Department of Applied Mechanics and Engineering Sciences
University of California - San Diego, La Jolla, CA 92093

1. – Introduction.

The present lectures are intended to provide some insight into the interaction between experiments and theory for some basic questions involving buoyancy-influenced turbulence in geophysical fluid dynamics. Effects of the Earth's rotation will not be included. The discussion will focus on laboratory experiments to study the physical processes dominating turbulence encountered both in the open ocean and in tidally driven flows in inlets and straits, where stabilizing buoyancy forces are acting.

The study of the influence of vertical density stratification on geophysical turbulent flows has a long history in meteorology and oceanography. Good descriptions of the physical processes at work in « boundary » turbulence like that in the atmospheric boundary layer or in « interior » turbulence far from boundaries may be found in the texts of Turner [1] and Phillips [2]. The success of Monin-Obukhov scaling and the usefulness of parameters such as the flux and gradient Richardson numbers is well established for the atmospheric boundary layer and to a lesser extent for ocean bottom boundary layer flows. Two decades have elapsed since the splendid measurements of Grant, Stewart and Moilliet [3] in a tidal channel, which exhibited nearly three decades of inertial subrange in the spectrum of streamwise velocity, providing crucial evidence for the ideas of an energy cascade and local isotropy for homogeneous flows. Stable stratification of nonhomogeneous flows in the atmosphere and oceans, as parameterized by the Brunt-Väisälä buoyancy frequency $N = (-g(\partial \bar{\varrho}/\partial z)/\bar{\varrho})^{1/2}$, allows the existence of internal waves and alters the energy transfer between different scales, modifying the energy cascade to smaller scales of homogeneous turbulence. However, atmospheric spectral measurement [4] show that local isotropy still holds for the largest wave numbers. As N decreases, the $-5/3$ range in energy spectra extends to lower wave numbers (larger scales), as does the range over which the spectra of orthogonal velocity components conform to the local isotropy relations. Very similar findings for buoyancy-influenced tidally driven inlet flows have been recently reported [5]. Ocean microstructure measurements of the smaller scales of variation of temperature and velocity made in the past decade have raised a number of new questions about how to describe what is happening in the ocean which have no parallel in atmospheric studies, including the connection between turbulent motions, double diffusive processes, intrusions and the ubiquitous oceanic internal wave field. Recent measurements [6-8] indicate that turbulent motions, having many characteristics of those found in atmospheric and laboratory measurements with stabilizing buoyancy forces, do exist in and below the thermocline. Ocean microstructure measurements are usually highly localized, most often consisting of vertical sections of temperature and conductivity, and sometimes small-scale velocity fluctuations. The overall dynamical state of the flow in the area of the drop is not normally measured or known. It is difficult enough to make good microstructure measurements, and a determination of the complete flow field including mean velocity profiles, shear stresses and other quantities important for the turbulent dynamics, which might normally be measured in laboratory experiments, is not feasible. This leaves an undesirable amount of freedom in the interpretation of microstructure measurements, and oceanographers could very profitably use some sound experimentally founded principles to apply to their data. Because of the difficulty of laboratory measurements in stratified flows, there have been few fundamental experiments in stratified turbulence of general application for ocean turbulence, and some of the basic concepts have not been properly tested.

The present renaissance in interest in fluid turbulence among physicists seems in part prompted by the hope that recently ripening ideas for the theoretical treatment of simpler mechanical systems can be eventually adapted to provide new theoretical insight and experimental diagnostic techniques for the understanding of turbulent flows. While the idea of attractors has been found to be useful in describing some aspects of the transition to turbulence in convection and in the flow between rotating cylinders, they have not been shown to be a universal ingredient in laminar-turbulent transition and their relevance for fully developed turbulent flows has not been established. The present state of understanding in this area is reviewed in these lectures by LIBCHABER [9]. In what follows I shall employ the conventional moment equation approach to describe turbulence dynamics [1-3], a description particularly well suited for nearly homogeneous turbulent flows in which attractors or any organized coherent large-scale structure may be difficult to define or detect.

As a prelude to discussion of the experimental data, for the present lectures it seems appropriate to first briefly discuss what can actually be measured and how this relates to what can be theoretically predicted, *i.e.* what can an experimentalist measure and how are these measurable quantities related to the fundamental quantities appearing in the basic differential field equations or in the equations for statistically defined quantities derivable from the basic equations? After this introduction, a detailed discussion will be given of the results of some laboratory and field studies on buoyancy effects in ocean turbulence.

2. – What should be measured?

Detailed comparison of experiments with theory is severely limited by available experiment techniques. In many cases, the measurable quantities are not simply related to the variables of primary interest and an experimentalist must be content to examine only measurable « symptoms » and to deduce reasonable physical suggestions about the behavior of quantities he cannot measure. Theoretical development is limited and often impeded by incomplete, misleading, or incorrectly interpreted experimental results.

We assume that the flow satisfies the Navier-Stokes equations with the Boussinesq approximation, *i.e.*

$$\frac{\partial \tilde{u}_i}{\partial t} + \tilde{u}_j \frac{\partial \tilde{u}_i}{\partial x_j} = -\frac{1}{\varrho_0}\frac{\partial \boldsymbol{p}}{\partial x_i} + g\frac{\boldsymbol{\varrho}}{\varrho_0} + \nu \frac{\partial^2 \tilde{u}_i}{\partial x_j^2}, \tag{1}$$

$$\frac{\partial \tilde{u}_i}{\partial x_i} = 0, \tag{2}$$

$$\frac{\partial \varrho}{\partial t} + \tilde{u}_i \frac{\partial \varrho}{\partial x_i} + w \frac{\partial \bar{\varrho}}{\partial z} = \varkappa \nabla^2 \varrho. \tag{3}$$

Here $\boldsymbol{x} = (x, y, z)$ is a right-handed Cartesian co-ordinate system with z directed in the vertical, $\tilde{u} = (u, v, w)$ is the corresponding velocity, ϱ_0 a constant reference density, $\boldsymbol{p}$ the pressure deviation from the ambient pressure, $\boldsymbol{g} = (0, 0, -g)$ the gravity vector, ϱ the density deviation from the ambient $\bar{\varrho}(z)$, ν the kinematic viscosity and $\varkappa$ the mass diffusivity. One obtains statistical moment equations from these by multiplying them by any of the dependent variables, and time, ensemble, or phase averaging.

If we decompose the velocity, density and pressure fields into mean and fluctuating components, *i.e.*

$$\tilde{u}_i = U_i + u_i, \quad \boldsymbol{\varrho} = \bar{\varrho} + \varrho, \quad \boldsymbol{P} = P + p,$$

multiply through each equation by appropriate dependent variables and average, the statistical moment equations describing the evolution of the turbulent kinetic energy per unit mass $\frac{1}{2}\overline{u_i^2}$, mean square density fluctuation $\overline{\varrho^2}$ (proportional to the fluctuating potential energy) and vertical buoyancy flux $\overline{\varrho w}$ are

$$(4) \quad \left(\frac{\partial}{\partial t} + U_j \frac{\partial}{\partial x_j}\right)\frac{1}{2}\overline{u_i^2} + \frac{\partial}{\partial x_j}\overline{\left\{u_j\left(p/\varrho_0 + \frac{1}{2}u_i^2\right)\right\}} = -\overline{u_i u_j}\frac{\partial U_i}{\partial x_j} - \frac{g}{\overline{\varrho}}\overline{\varrho w} - \nu \overline{\left(\frac{\partial u_i}{\partial x_j}\right)^2},$$

$$(5) \quad \left(\frac{\partial}{\partial t} + U_j \frac{\partial}{\partial x_j}\right)\overline{\varrho^2} = -2\overline{\varrho w}\frac{\partial \overline{\varrho}}{\partial z} - 2\varkappa \overline{\left(\frac{\partial \varrho}{\partial x_j}\right)^2} + \frac{\partial}{\partial x_i}\overline{[u_i \varrho^2]},$$

$$(6) \quad \left(\frac{\partial}{\partial t} + U_j \frac{\partial}{\partial x_j}\right)\overline{\varrho w} = -\overline{w^2}\frac{\partial \overline{\varrho}}{\partial z} - \frac{g}{\varrho_0}\overline{\varrho^2} - \frac{1}{\varrho_0}\overline{\varrho \frac{\partial p}{\partial z}} - (\varkappa + \nu)\overline{\frac{\partial w}{\partial x_j}\frac{\partial \varrho}{\partial x_j}},$$

where some molecular diffusion and other divergence terms have been dropped in anticipation of the simplified situation to be discussed in sect. 4. Similar equations may be constructed for the Reynolds stress $\overline{u_i u_j}$ and other moments. Useful locally isotropic forms of the turbulent dissipation of kinetic energy ε and destruction of density variance χ are [10]

$$(7) \quad \varepsilon = \nu \overline{\left(\frac{\partial u_i}{\partial x_j}\right)^2} = 15\nu \overline{\left(\frac{\partial u}{\partial x}\right)^2},$$

$$(8) \quad \chi = 2\varkappa \overline{\left(\frac{\partial \varrho}{\partial x_j}\right)^2} = 6\varkappa \overline{\left(\frac{\partial \varrho}{\partial x}\right)^2} = 6\varkappa \overline{\left(\frac{\partial \varrho}{\partial z}\right)^2}.$$

3. – What can be measured.

Consider each of the terms in the above equations. The velocity u_i can be measured with adequate time resolution and with spatial resolution of about a millimeter using hot films and (in the laboratory) laser-Doppler anemometers. Practical problems make it very difficult to measure all three components simultaneously at the same point with good resolution. In the ocean airfoil probes [11] as well as heated films and thermistors are used to measure velocity fluctuations. In the laboratory, measurement problems are compounded near solid or fluid boundaries by unwanted heat transfer to the boundaries from heated sensors or refraction effects on laser beams. There appears to be no reliable way to measure the fluctuating static pressure at an interior fluid point, so the pressure-velocity correlations are usually beyond reach. Static pressure fluctuations can be measured on solid boundaries using, for example, sensitive condenser microphones [12].

Velocity time derivatives $\partial u_i/\partial t$ can be accurately measured utilizing the usually adequate frequency response. Measurements of velocity streamwise spatial derivatives $\partial u_i/\partial x$ are not practical using heated sensors because of probe intereference effects, unavoidable when one sensor is downstream of the other. Multiple laser-Doppler systems may some day be useful for laboratory measurements of spatial gradients, but have not yet been sufficiently developed.

Cross-stream gradients $\partial u/\partial y$, $\partial u/\partial z$ are much easier to measure, but gradients in v and w require too many sensors, which degrades the spatial resolution. Such problems have precluded the measurement of most of the terms in the turbulent dissipation rate. Usually only the term $\overline{(\partial u/\partial x)^2}$ is measured, and it is approximated using the time derivative and Taylor's hypothesis, since the time derivative is well resolved. In vertically sampled ocean microstructure measurements the component $\overline{(\partial u/\partial z)^2}$ is obtained [13].

Fine-scale temperature fluctuations can be measured both in the laboratory and in the field using resistance thermometers or thermistors with spatial resolution equal to or better than that achievable for the velocity field [14]. Density fluctuations in salt-stratified laboratory flows are measured with conductivity probes, which involves a trade-off between spatial resolution and drift of the output level with time [15]. In the ocean density depends upon temperature and salinity. The spatial resolution of oceanic-conductivity instruments is approaching, but has not yet equalled, that of the thermistors used for temperature microstructure measurements.

Since point measurement of scalars like density or temperature requires only one sensor, lateral gradients of scalars can often be measured with sufficient resolution, and all three terms in the temperature dissipation rate have been simultaneously measured for some laboratory flows [16].

Time spectra of all quantities can be readily determined, but spatial wave number spectral measurements are routinely obtained in only one preferred wave number direction in laboratory flows. With a single fixed probe one can get only the wave number in the direction of the mean motion, and this again requires the assumption of Taylor's hypothesis. It is often not practical to move probes in lateral directions at high enough speed to obtain wave number spectra in other directions. The eventual use of crossed laser-Doppler beams swept through the flow by rotating mirrors might be feasible, but there are many other problems with laser systems which presently severely limit their use. Spatial wave number spectra are commonly obtained in ocean measurements by towing or dropping instruments, taking advantage of the relatively slow time evolution of the spatial variations when possible. For example, it is possible to perform a microstructure cast over a considerable depth in a time small compared with a Brunt-Väisälä period, so that microstructure can be measured without contaminating internal-wave effects.

The second-derivative terms in the basic equations are for all practical purposes not measurable in laboratory experiments, because of the problems already discussed, except by approximation as second-order time derivatives. However, the corresponding viscous and diffusive terms in the equations for the statistically defined properties of mean square turbulent kinetic energy, potential energy, heat flux or buoyancy flux involve only first-order derivatives and can often be measured, when the diffusivity is not so small that the gradients contributing the major part of the dissipation rates are smaller than the spatial resolution of the probe. The latter situation is frequently encountered in the ocean and in laboratory salt-stratified turbulence, because of the relatively small molecular diffusivity of salt compared with that of heat or momentum.

A principal difficulty in interpreting ocean turbulence measurements is that usually data are not available for enough variables to intelligently discuss the dynamics of the flow. To do this it is important to know whether or not the turbulent kinetic energy is decaying or being maintained by the production terms.

It is very difficult to measure the vertical velocity component w in the ocean, as it is often smaller than uncertainties caused by mooring or vehicle-related motions. Direct measurements of the buoyancy flux are, therefore, extremely difficult and considerable speculation exists regarding its behavior in oceanic turbulence [17].

In the ocean one finds examples both of transition of initially fully developed turbulence into an internal wave field and the development of turbulence from instabilities caused by the breakdown of sufficiently large-amplitude isolated internal waves or other nonlinear interactions. The second type of transition can in turn be followed by the occurrence of the first type, and *vice versa*. For example, tidal flow over a sill may generate lee waves which produce overturning turbulence. The turbulence evolves under the influence of buoyancy forces and redistributes its kinetic energy by production of mean and fluctuating potential energy, production of new internal-wave motions and loss of energy by direct viscous dissipation.

The events giving rise to observed ocean turbulence are usually not identifiable and experimental observations clearly indicating the mechanisms of generation are rare. This is not just a matter of bad timing or luck in observations. The violent sea states associated with conditions likely to produce both large turbulence and wave fields often preclude the gathering of suitable data. Turbulence-generating events have, however, been beautifully documented in flows closer to shore. Measurements of tidally generated flow in inlets [18], wind-driven layers in lochs [19] and in outflows from interior seas [20] have provided good evidence for the generation of turbulence by the breaking of internal waves and other instabilities.

4. – Buoyancy effects in vertically homogeneous unsheared turbulent flow.

Let us first consider the spatial (x) development of the simplest vertically homogeneous (in a statistical sense) turbulent flow, initially uncoupled from internal-wave motions due to an initial disparity in turbulence and buoyancy length scales. That is, with initial turbulent scales much smaller than the buoyancy scales. Buoyancy forces and internal waves will then become of importance at a crucial later stage in the evolution of the flow.

We shall simplify the discussion by restricting it initially to the case of spatially decaying homogeneous turbulence in which there is no shear and hence no production, a case which exhibits many principal features of buoyancy effects on turbulent flow with minimal complication. In this case eqs. (4)-(6) become

$$U \frac{\partial}{\partial x} \frac{\overline{u_i^2}}{2} = -\frac{g}{\bar\varrho} \overline{\varrho w} - \varepsilon \,, \tag{9}$$

$$U \frac{\partial}{\partial x} \frac{\overline{\varrho^2}}{2} = -\overline{\varrho w} \frac{\partial \bar\varrho}{\partial z} - \chi \,, \tag{10}$$

$$U \frac{\partial}{\partial x} \overline{\varrho w} = -\overline{w^2} \frac{\partial \bar\varrho}{\partial z} - \frac{g}{\bar\varrho} \overline{\varrho^2} - \frac{1}{\bar\varrho} \overline{\varrho \frac{\partial p}{\partial z}} - (\varkappa + \nu) \overline{\frac{\partial w}{\partial x_j} \frac{\partial \varrho}{\partial x_j}}. \tag{11}$$

All terms in these equations are in principle measurable except for the pressure-velocity gradient correlation term in the buoyancy flux evolution equation.

Nondimensionalizing eq. (9) using l and l_z as horizontal and vertical length scales, and q as a turbulent velocity scale, and observing that the kinetic-energy dissipation rate is of order q^3/l for a turbulent flow, we have

$$\frac{U}{q} \frac{\partial}{\partial \tilde x} \left(\frac{\tilde q^2}{2} \right) = N^2 \frac{w' l l_z}{q^3} \frac{\overline{\varrho w}}{\varrho' w'} \frac{\varrho'}{(\partial \bar\varrho/\partial z) l_z} - 1 \,, \tag{12}$$

where $\tilde x = x/l$, $\tilde q = (\overline{u_i^2})^{1/2}/q$, and ϱ' and w' are the root mean square values of ϱ and w. Assuming q/U to be of order one, we find that both the advection and dissipation terms are of order one. Assume that the density fluctuation ϱ' is proportional to the mean gradient times the length scale l_z and that the normalized buoyancy flux correlation is of order one, which is appropriate for a fully turbulent field but not for internal waves. Then, if $w' \sim q$ and $l_z \sim l$, the buoyancy flux term is of order $(l_z/L_b)^2$, where $L_b = w/N$ is the buoyancy scale and L_b can be shown to be proportional to the vertical distance a fluid particle will move while converting all its vertical kinetic energy into potential energy. Another useful buoyancy length scale, which under certain assumptions [21] is equivalent to w/N, is the Ozmidov scale $L_R = (\varepsilon/N^3)^{1/2}$. The buoyancy flux term in (12) is of order $(l_z/L_R)^{4/3}$.

For a fully turbulent flow in which l is initially considerably less than the buoyancy length scale, the buoyancy flux term will initially be negligible. If, as the flow evolves, the ratio of the two scales approaches unity, buoyancy forces will become important and kinetic energy will be converted into fluctuating potential energy by the buoyancy flux. This will be the case for decaying turbulence in which the length scale l increases with time or distance. The equivalent derivation for a time-dependent flow (rather than a spatially dependent one) shows that buoyancy effects first become important when the turbulent overturning time scale l/q becomes of the same order as N^{-1}. The behavior of the fluctuating potential energy described by eq. (10) depends on the behavior of the buoyancy flux in a somewhat more subtle fashion than does the kinetic energy, as no term directly involving the acceleration of gravity g appears in eq. (10). A passive homogeneous scalar field is described by the same equations with the term involving g omitted. Note that the buoyancy flux term will be absent from the kinetic-energy equation, but still appears in the equation describing the evolution of $\overline{\varrho^2}$. The kinetic energy simply decays by energy transfer to higher wave numbers and viscous dissipation. Experiments [22, 23] show that in the competition between production of mean square density fluctuations by the buoyancy flux term and dissipation by the scalar dissipation term in eq. (10), the production term dominates and $\overline{\varrho^2}$ grows steadily without bound with increasing time or distance. The normalized buoyancy flux $\overline{\varrho w}/\varrho' w'$ increases to an asymptotically constant value of 0.7. A physical explanation of this behavior is that, as the turbulent kinetic energy decays, the smaller scales are more rapidly damped than larger ones and the turbulent integral length scale l grows during decay, producing vertical mixing of fluid particles with larger vertical excursions from their mean positions. This produces larger density fluctuations at a fixed point as either time or x increases. Under passive conditions l is equal to l_z, and all three terms in eq. (10) are of the same order. If we scale the buoyancy term in eq. (10) as we did for eq. (9), the order of the buoyancy term relative to the other two terms will be l/l_z, or of order one before buoyancy forces act. When buoyancy forces become important, the buoyancy flux no longer scales with ϱ' and w' as it does in a fully turbulent flow, but the magnitude of $\overline{\varrho w}$ is decreased relative to those of ϱ' and w'. For an internal wave field $\overline{\varrho w}$ is nearly zero, while ϱ' and w' are not, and the normalized density flux is small.

When the buoyancy flux term in the fluctuating-potential-energy equation becomes small enough, destruction of the density fluctuations by the diffusive term dominates and the fluctuations will cease to grow and begin to decrease. Since the buoyancy flux must change in order for the density fluctuation to be affected, one might expect to see changes first in the buoyancy flux and later in the fluctuating potential energy as a flow evolves in either time or space.

The influence of buoyancy forces on the buoyancy flux itself can be seen from eq. (11), similar to the discussion for the kinetic-energy equation. The terms involving g become as large as the other terms when the length scale l and the buoyancy scale become of the same order. When this happens, the buoyancy flux is reduced from its robust value during the non-buoyancy-influenced stage, and this in turn then reduces the rate of increase of $\overline{\varrho w}$. The balance then swings in favor of the $\overline{\varrho w}$ destruction terms and $\overline{\varrho w}$ decreases.

For spectrum measurements, it is useful to define a buoyancy wave number k_b as the reciprocal of the buoyancy length. One might anticipate that wave numbers of the order of k_b or smaller would be affected by buoyancy forces.

5. – A laboratory experiment.

Laboratory experiments to investigate the transition of homogeneous turbulence into internal waves in a stratified fluid have been carried out by STILLINGER, HELLAND and VAN ATTA [21]. These experiments differ in several basic aspects from previous studies of decaying stratified turbulence by other investigators. A unique closed-loop gravity-driven water channel designed for investigation of density stratified shear flows [24] was employed, whereas all previous experiments had been carried out in conventional towing tanks [25-28]. Direct density flux $\overline{\varrho w}$ measurements were obtained using a very-low-drift microscale conductivity instrument developed by HEAD [15].

Simultaneous point measurements of the horizontal and vertical velocity and density fluctuations were obtained at a number of downstream locations as unsheared grid-generated turbulence evolved under the influence of stable stratification.

For these experiments the initial vertical overturning length scale $L_T =$

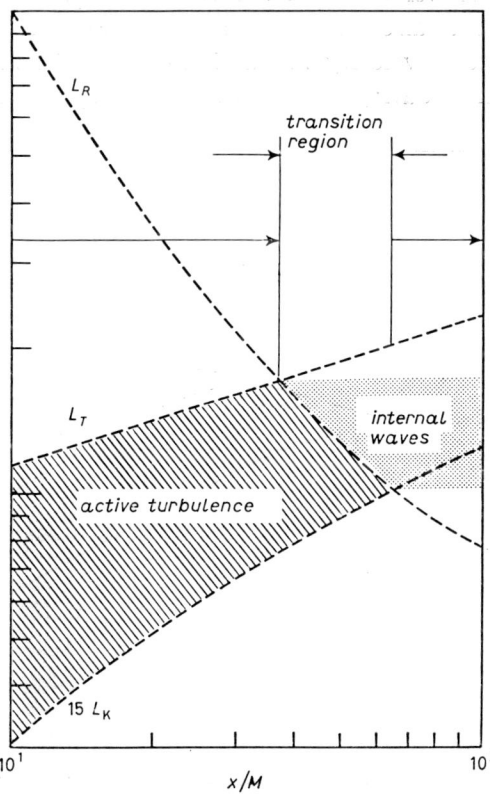

Fig. 1. – Length scale evolution diagram for the transition of decaying homogeneous turbulence to internal waves, after [21].

$= 2(\overline{\varrho^2})^{1/2}/(\partial\bar{\varrho}/\partial z)$ of the turbulence was from two to six times smaller than the initial buoyancy scales, the ratio being smaller for larger values of N, and the measured turbulent dissipation rate was fairly insensitive to N. In the following figures x is the distance downstream of the grid and M is the biplane grid mesh length.

Figure 1 is a length scale evolution map which summarizes some of the main aspects of the observed behavior. The upper dashed curve shows the decrease in the buoyancy length scale L_R as the turbulence decays. Buoyancy forces are dynamically important for scales larger than the buoyancy length scale. The middle dashed line gives the increase in the overturning length scale L_T observed during the evolution of a passive scalar field and in an active scalar field in the stratified cases before buoyancy effects become dynamically important. The lower dashed curve shows the evolution of a multiple of the Kolmogorov viscous length scale $L_K = (\nu^3/\varepsilon)^{1/4}$. Active turbulence is confined to the hatched region bounded by these scales. As the turbulent velocity field decays, the overturning scale and buoyancy scale approach one another, heralding the onset of dynamically important buoyancy effects. This onset is manifested in a slower decay of the velocity fluctuations, compared with the unstratified case, as also observed in some other experiments [28]. Because a fixed probe measures velocity fluctuation contributions from both internal waves and turbulence during the time interval required for these measure-

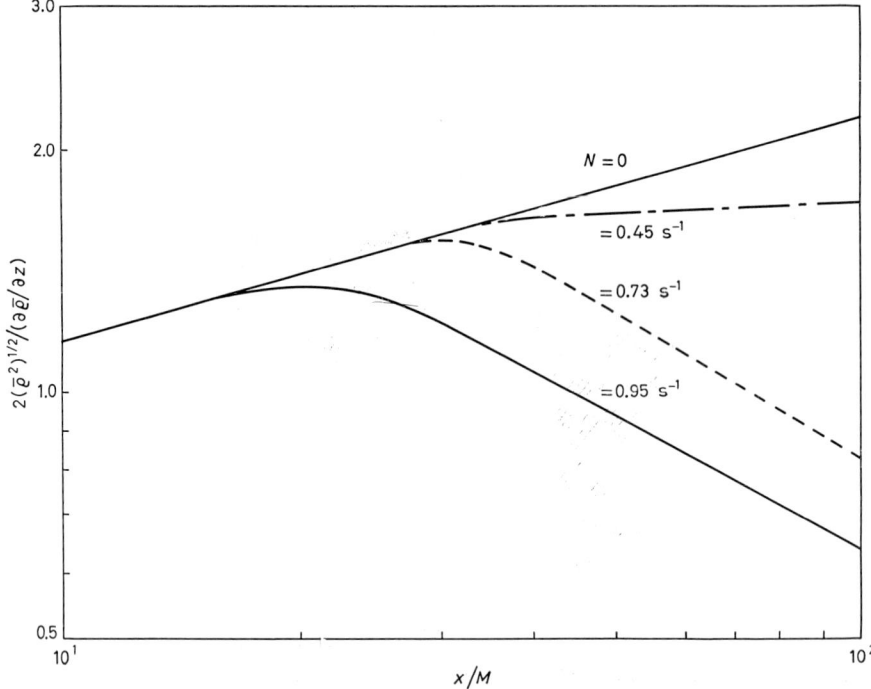

Fig. 2. – Mean squared density fluctuation during transition, after [21].

ments, the actual decay rate of the « turbulent » component cannot be directly determined, although it may be recoverable from the viscous-dissipation rate measured directly from the high-wave-number portion of the energy spectra. Another dramatic effect of stratification is the cessation of growth and eventual decay of the mean square density fluctuation, which occurs further downstream as N decreases, as shown in fig. 2. The onset of buoyancy effects in $\overline{\varrho^2}$ occurs when the buoyancy and overturning lengths are roughly equal, with average values of $L_T/L_b = 1.6$ and $L_T/L_R = 1.4$.

As $\overline{\varrho^2}$ and consequently the overturning scale L_T decay due to the buoyancy effects, the overturning scale « locks in » on the buoyancy scale and remains nearly proportional to it, as shown in fig. 3. This behavior is similar to that observed in ocean microstructure measurements by DILLON [29]. DILLON measured an overturning scale first employed by THORPE [19], which is obtained by resorting the discrete samples of a vertical temperature profile to produce a monotonically decreasing temperature with depth. The root mean square of the vertical displacements required is called the Thorpe scale, which DILLON suggests is equal to one-half of L_T. Dillon's best-fit line to his data is also shown in fig. 3. DILLON found no points far from this fit, unlike the laboratory data before the onset of buoyancy effects. If the oceanic flow dynamics at the time of the observations were very similar to those of the decaying laboratory grid

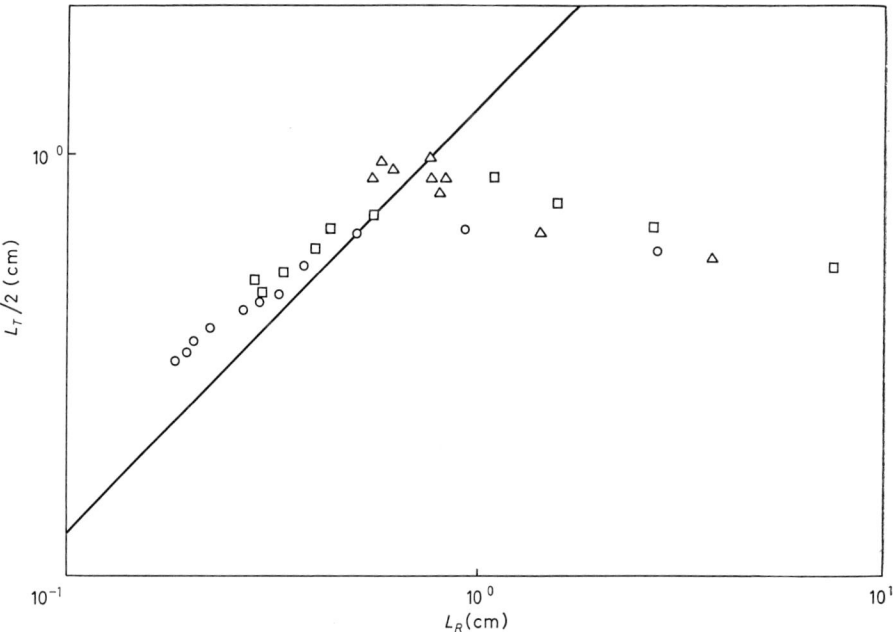

Fig. 3. – Turbulent overturning scale vs. buoyancy length scale during transition [21] and for ocean microstructure [29]. Laboratory L_T data: ○ $N = 0.95$ rad/s, △ $N = 0.73$, □ $N = 0.45$, ——— ocean Thorpe scale data.

data (negligible shear and vertical homogeneity), then one might hypothesize that all of his observations were of decaying buoyancy-influenced turbulence. This does not seem probable, but it poses a difficult question to assess with available data. It is reasonable to alternatively suggest [29] that the proportionality of buoyancy and Thorpe scales is a general feature of stratified turbulent flows, including shear flows which are being maintained by Reynolds stress production as well as decaying turbulence.

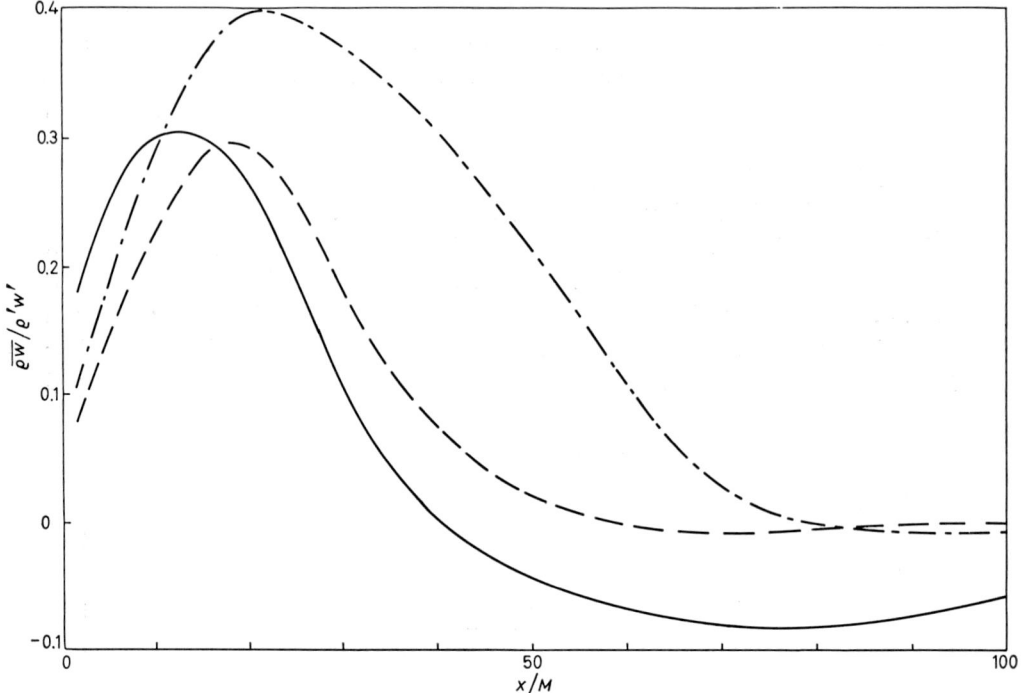

Fig. 4. – Vertical mass (buoyancy) flux during transition, after [21]. Notation same as in fig. 2.

As illustrated in fig. 4, the normalized buoyancy flux increases from near-zero values immediately behind the grid to a maximum value of about 0.3 at a location somewhat upstream of that at which the density fluctuation reaches its maximum value, in agreement with the earlier conjecture that $\overline{\varrho w}$ would show signs of buoyancy influence before the fluctuating potential energy did. The buoyancy flux then decreases to zero as the turbulence decays, with $\overline{\varrho w}$ becoming negative for the largest values of N studied and eventually returning toward zero. This countergradient flux or restratification process involving a downward density flux is also seen in the simulations of Riley et al. [30] for $N = 1.57$ and 3.14 rad/s and the «ringing» proceeds through several cycles of positive and negative $\overline{\varrho w}$ for $N = 3.14$.

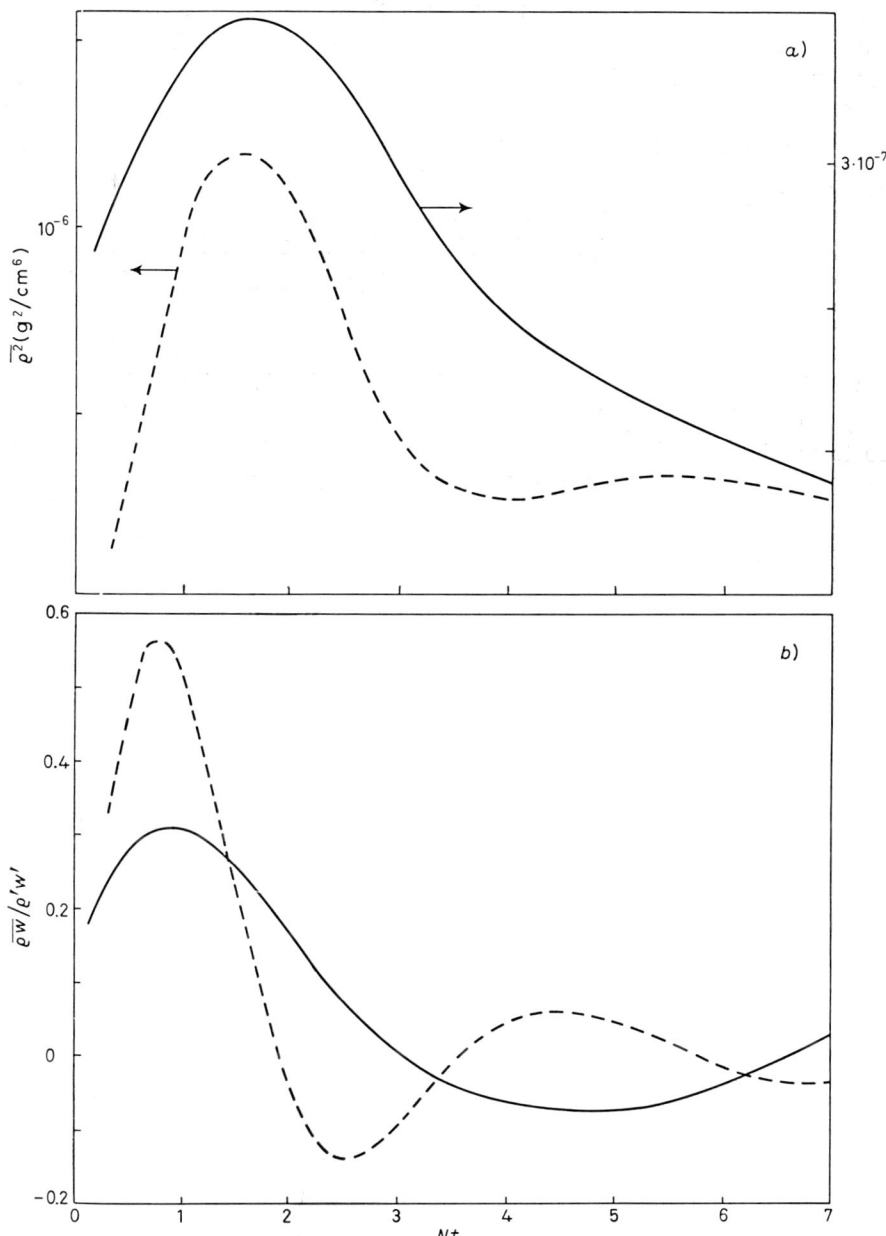

Fig. 5. – Comparison of experimental [21] and numerical simulation [30] results: a) mean square density fluctuations; b) buoyancy flux; ——— $\overline{\varrho w}/\varrho' w'$, experiment, $N = 0.95$ rad/s; ——— $\overline{\varrho w}$ g cm^{-2} s^{-1}, simulation, $N = 1.57$.

When the normalized $\overline{\varrho^2}$ data are plotted vs. a « time » equal to x/U made dimensionless with the Väisälä frequency, the data collapse toward a single curve for large Nt [21]. If we use the condition when $\overline{\varrho w}$ first reaches zero as an operational definition for the final transition to an internal wave field, then the transition time required to establish such a field from the first onset of buoyancy effects is found to be only about 0.14 Brunt-Väisälä periods, where the period is defined as $2\pi/N$. This transition time is about two-thirds the large eddy turn-over time, defined as L_T/q. The turbulent field thus « collapses » rather abruptly. This behavior is also found in the simulations of Riley et al. [30], with the collapse occurring in about 0.18 B-V periods for $N = 1.57$ rad/s.

Some comparisons of the laboratory data with the direct numerical simulations [30] are shown in fig. 5. The physical behavior is quite similar. In both cases, the density fluctuations and buoyancy flux increase smoothly from near zero to maximum values, with the peak value of the buoyancy flux occurring somewhat earlier than that for the mean square density fluctuation.

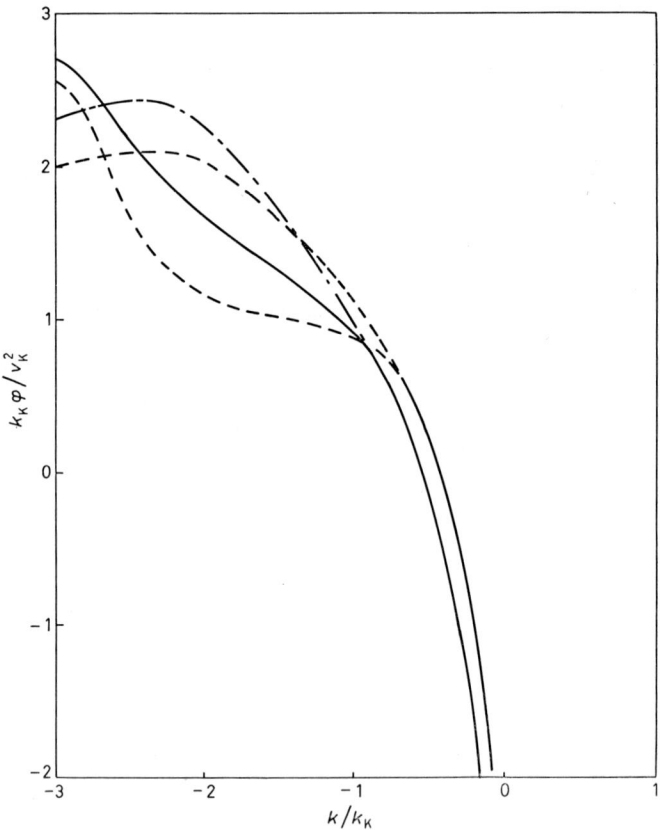

Fig. 6. – Comparison of velocity energy spectra for stratified and unstratified laboratory experiments. $N = 0.95$: ——— φ_u, – – – φ_w. $N = 0$: – – – φ_u, – – – φ_w.

Measured velocity spectra [21] collapse well with the universal viscous Kolmogorov similarity curve for the largest wave numbers, but stratification depletes the energy in lower wave numbers relative to the unstratified case, as shown in fig. 6. The wave number below which buoyancy effects are apparent in the velocity energy spectra is approximately equal to $k_b = (N^3/\varepsilon)^{1/2}$, and this wave number increases as the turbulence decays.

6. – Some geophysical measurements.

The influence of buoyancy forces on turbulent velocity energy spectra has also recently been observed in a convincing fashion in a very-high-Reynolds-number flow in Knight Inlet by GARGETT, OSBORN and NASMYTH [5]. These measurements were made from a research submersible instrumented for measuring fine-scale turbulent velocity and temperature fluctuations, as well as some parameters of the larger-scale motions. The longitudinal turbulent velocity component was measured using heated film sensors, and the cross-stream components with two single-axis airfoil probes [11]. It was found that the energy spectra of the velocity could be classified according to observed departures from the homogeneous case [3] and could be parameterized by the ratio of the Kolmogorov wave number k_K to the buoyancy wave number k_b, which is also the reciprocal of the ratio of the Ozmidov length to the Kolmogorov length. The observed behavior is shown schematically in fig. 7a). For large k_K/k_b the spectra exhibit classical homogeneous inertial subrange $-5/3$ power law behavior. For decreasing k_K/k_b the spectral levels at the lower wave numbers are lowered. The spectra of the vertical velocity component rejoin the unstratified universal spectrum at a wave number roughly equal to $2\pi k_b$. The spectra of the horizontal component exhibit the influence of buoyancy forces only at somewhat lower wave numbers and the systematic effects are not as clear. This behavior is generally similar to that previously found for the laboratory grid turbulence data [21] and atmospheric measurements [4]. One can also define a buoyancy velocity scale as $u_b = (\varepsilon/N)^{1/2}$. The same data replotted using buoyancy scales are shown schematically in fig. 7b), and it is clear that buoyancy scaling successfully collapses the data for low wave numbers. GARGETT et al. [5] note that the levels of the vertical velocity spectra are successfully collapsed to a «universal» level over a k^{-1} range and the junction of this range with the isotropic curve is collapsed to $k = 10k_b$. As the turbulence decayed away from the generating sill, the relation $\overline{u_i^2} = 1.2 u_b^2$ was roughly satisfied. For the largest values of k_K/k_b approximate isotropy of energy was found for all three velocity components. As k_K/k_b decreased, the energies of the cross-stream components decreased relative to that in the horizontal component, with $\overline{u_2^2} \simeq \overline{u_3^2} \simeq 0.7 u_b^2$ for $k_K/k_b \sim 200 \div 300$. As the turbulence further decayed, the difference increased, suggesting perhaps an eventual approach to a state of «two-dimensional» turbulence. Further comparisons

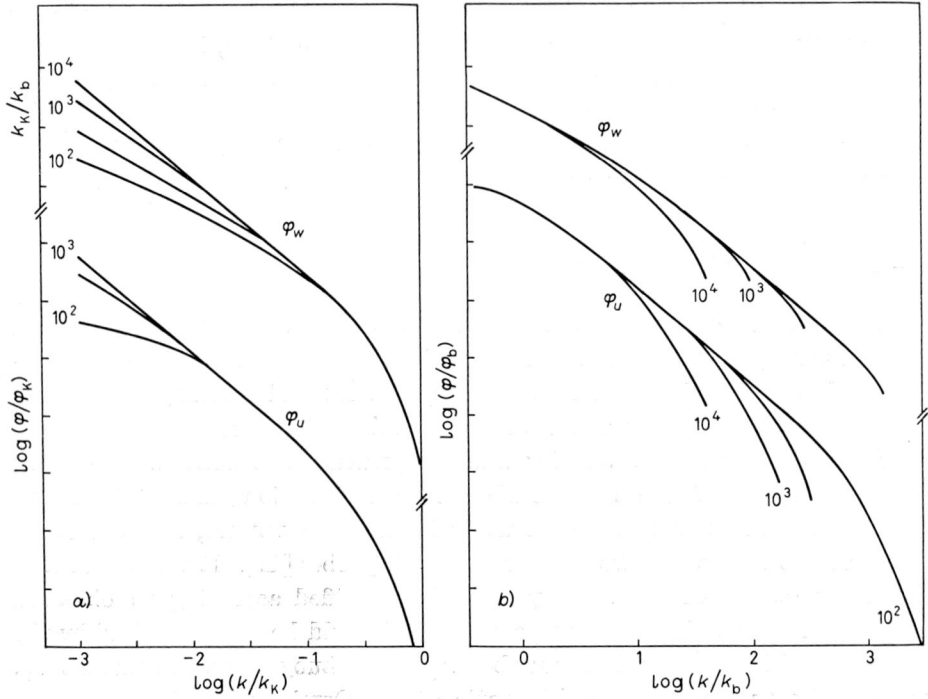

Fig. 7. – Schematic representation of behavior of velocity energy spectra for tidally driven Knight inlet turbulence [5]: a) Kolmogorov scaling, b) buoyancy scaling.

of these data with laboratory measurements, particularly of the scalar-field behavior, will be very interesting.

Scalar fluctuations, mainly temperature, have been measured in the atmosphere and oceans for some time. One important respect in which the ocean differs from the atmosphere is the relatively high Prandtl number of water compared with air. Numerous comparisons of the theoretical universal scalar spectral forms predicted by BATCHELOR [31] for large Prandtl and Reynolds numbers have been made with spectra obtained in natural waters [32-36]. Since the applicability of a Batchelor spectrum allows one to make an estimate of the kinetic-energy dissipation rate from a temperature measurement alone, it is important for oceanographers to know under what conditions such a spectrum will be obtained in a stratified fluid. This would be of special value for microstructure measurements, for which the dynamical state of the turbulence may only be inferred in many cases. It has been found [7] that the observed viscous-convective subrange spectrum of vertical temperature gradients approaches the Batchelor spectrum as the Cox number increases, where the Cox number, $\overline{(\partial T/\partial z)^2}/(\partial \overline{T}/\partial z)^2$, is a measure of the strength of the turbulent mixing. For low Cox numbers the spectrum may be broader and flatter than the Batchelor spectrum [7].

7. – Résumé.

There are many areas in which fruitful research on stratified turbulence could be pursued. Several are evident from the present brief and somewhat specialized discussion. On the experimental side laboratory measurements in flows with controlled shear and stratification are needed to investigate whether generalizations [37] of the simple length scale arguments found to be appropriate for decaying, unsheared grid turbulence may confidently be applied to more complex oceanographic situations. Direct numerical simulations of the same flows under as nearly the same conditions as possible should be performed and compared with the experimental data. As we have seen, there is still a need for more direct numerical simulations of the simplest case of decaying unsheared homogeneous turbulence with parameters chosen to correspond to those conditions for which the most complete set of measurements is available.

A good candidate flow for adding a controlled shear to the stratification would be a uniform shear flow with constant velocity gradient. In the homogeneous case, while turbulence intensities, Reynolds stresses, etc. become independent of downstream co-ordinate, integral length scales continue to increase until they reach the size of the apparatus. In the stratified case the vertical length scale would be limited by the buoyancy length and could remain constant. The horizontal scale, which continues to grow in wakes and jets in stratified environments even after vertical scales have « collapsed », might continue to grow as in the unstratified case. The increasing disparity in vertical and horizontal length scales would produce an interesting type of quasi–two-dimensional turbulent motion.

Laboratory measurements made using both vertical and horizontal sampling under identical stratification and other flow conditions would be most helpful in providing comparisons between statistical data obtained in these two ways which could be used in interpreting oceanic data. In the ocean measurements usually only one type of sampling is employed in a given experiment, and questions on the effect of the degree of anisotropy or of different sampling periods compared with the mean flow velocity or Väisälä period normally go unasked or unanswered.

Direct simulation of the high-Reynolds-number geophysical situation is clearly not yet feasible with existing computing capacities. Efforts to develop closure models which can accurately reproduce both the experiments and the direct numerical simulations should, therefore, be encouraged at the present time.

* * *

My research on stratified turbulence is supported by the National Science Foundation under Physical Oceanography Grant OCE82-05946 and Mechanical Engineering and Applied Mechanics Grant MEA81-00431.

REFERENCES

[1] J. S. Turner: *Buoyancy Effects in Fluids* (Cambridge, 1973).
[2] O. M. Phillips: *The Dynamics of the Upper Ocean* (Cambridge, 1977).
[3] H. L. Grant, R. W. Stewart and A. Moilliet: *J. Fluid Mech.*, **12**, 241 (1962).
[4] J. C. Kaimal, J. C. Wyngaard, Y. Izumi and O. R. Cote: *Q. J. R. Meteorol. Soc.*, **98**, 563 (1972).
[5] A. E. Gargett, T. R. Osborn and P. W. Nasmyth: preprint (April, 1983).
[6] M. C. Gregg: *J. Phys. Oceanogr.*, **10**, 915 (1980).
[7] T. M. Dillon and D. R. Caldwell: *J. Geophys. Res.*, **85**, 1910 (1980).
[8] T. R. Osborn: *J. Phys. Oceanogr.*, **10**, 83 (1980).
[9] A. Libchaber: this volume, p. 17.
[10] H. Tennekes and J. L. Lumley: *A First Course in Turbulence* (Cambridge, Mass., 1972).
[11] T. R. Osborn and T. Siddon: *Proceedings of the III Symposium on Turbulence in Liquids* (Rolla, Miss., 1973).
[12] T. S. Mautner and C. W. Van Atta: *J. Fluid Mech.*, **118**, 59 (1982).
[13] N. S. Okey: *J. Phys. Oceanogr.*, **12**, 256 (1982).
[14] D. R. Caldwell and T. M. Dillon: Ref. 81-10, School of Oceanography, Oregon State University (1981).
[15] M. J. Head: Ph. D. Thesis, University of California, San Diego (1983).
[16] K. R. Sreenivasan, K. R. Danh and R. A. Antonia: *Proceedings of the IUTAM Symposium on Structure of Turbulence and Drag Reduction, Phys. Fluids Suppl.* (1977).
[17] G. Holloway: *Atmospheres and Oceans* (Canadian), **21**, 107 (1983).
[18] D. M. Farmer and J. D. Smith: in *Hydrodynamics of Estuaries and Fjords*, edited by J. Nihoul (New York, N. Y., 1978).
[19] S. A. Thorpe: *Philos. Trans. R. Soc. London, Ser. A*, **286**, 125 (1977).
[20] J. D. Woods: *J. Fluid Mech.*, **32**, 791 (1968).
[21] D. C. Stillinger, K. N. Helland and C. W. Van Atta: *J. Fluid Mech.*, **131**, 73 (1983).
[22] R. D. Montgomery: Ph. D. Thesis, University of Michigan (1974).
[23] A. Sirivat and Z. Warhaft: *J. Fluid Mech.*, **128**, 323 (1983).
[24] D. C. Stillinger, M. J. Head, K. N. Helland and C. W. Van Atta: *J. Fluid Mech.*, **131**, 91 (1983).
[25] M. Tao: Ph. D. Thesis, University of Iowa (1971).
[26] R. E. Lange: Ph. D. Thesis, University of California, San Diego (1974).
[27] T. J. Lin and S. D. Veenhuizen: Flow Research Note No. 85 (1975).
[28] T. D. Dickey and G. L. Mellor: *J. Fluid Mech.*, **99**, 13 (1980).
[29] T. M. Dillon: *J. Geophys. Res.*, **87**, 9601 (1982).
[30] J. J. Riley, R. W. Metcalf and M. S. Weissman: in *Nonlinear Properties of Internal Waves*, edited by B. J. West, AIP Conf. Proc. No. 76 (1981).
[31] G. K. Batchelor: *J. Fluid Mech.*, **5**, 113 (1959).
[32] H. L. Grant, B. A. Hughes, N. M. Vogel and A. Moilliet: *J. Fluid Mech.*, **34**, 423 (1968).
[33] M. C. Gregg: *J. Phys. Oceanogr.*, **6**, 528 (1976).
[34] P. W. Nasmyth: Ph. D. Thesis, Institute of Oceanography, University of British Columbia, Vancouver (1970).
[35] J. A. Elliot and N. S. Oakey: *J. Fish. Res. Board Can.*, **33**, 2296 (1976).
[36] C. O. Marmorino and D. R. Caldwell: *Deep-Sea Res.*, **25**, 1073 (1978).
[37] C. H. Gibson: in *Marine Turbulence*, edited by J. Nihoul (New York, N. Y., 1980).

Lectures on Spectral Methods for Turbulence Computations.

S. A. ORSZAG (*)

Massachusetts Institute of Technology - Cambridge, MA 02139

1. – Introduction to spectral methods.

Spectral methods are based on representing the solution to a problem as a truncated series of smooth functions of the independent variables. Whereas finite-element methods are based on expansions in local basis functions, spectral methods are based on expansions in global functions. Spectral methods are the extension of the standard technique of separation of variables to the solution of arbitrarily complicated problems.

Let us begin by illustrating spectral methods for the simple one-dimensional heat equation. Consider the mixed initial-boundary-value problem

(1.1) $$\frac{\partial u(x,t)}{\partial t} = K \frac{\partial^2 u(x,t)}{\partial x^2} \qquad (0 < x < \pi, \, t > 0),$$

(1.2) $$u(0, t) = u(\pi, t) = 0 \qquad (t \geq 0),$$

(1.3) $$u(x, 0) = f(x) \qquad (0 \leq x \leq \pi).$$

The solution to this problem is

(1.4) $$u(x, t) = \sum_{n=1}^{\infty} a_n(t) \sin nx,$$

(1.5) $$a_n(t) = f_n \exp[-Kn^2 t],$$

where

(1.6) $$f_n = \frac{2}{\pi} \int_0^{\pi} f(x) \sin nx \, dx$$

are the coefficients of the Fourier sign series expansion of $f(x)$.

(*) Present address: Princeton University, Princeton, NJ 08540.

A spectral approximation to (1.1)-(1.3) is gotten by simply truncating (1.4) to

$$u_N(x, t) = \sum_{n=1}^{N} a_n(t) \sin nx \tag{1.7}$$

and replacing (1.5) by the evolution equation

$$\frac{da_n}{dt} = -Kn^2 a_n \qquad (n = 1, ..., N) \tag{1.8}$$

with the initial conditions $a_n(0) = f_n$ $(n = 1, ..., N)$.

The spectral approximation (1.7), (1.8) to (1.1)-(1.3) is an exceedingly good approximation for any time t greater than zero as $N \to \infty$. In fact, the error $u(x, t) - u_N(x, t)$ satisfies

$$u(x, t) - u_N(x, t) = \sum_{n=N+1}^{\infty} f_n \exp[-Kn^2 t] \sin nx = O(\exp[-KN^2 t]) \tag{1.9}$$
$$(N \to \infty)$$

for any $t > 0$. In contrast to (1.9), finite-difference approximations to the heat equation using N grid points in x lead to errors that decay only algebraically with N as $N \to \infty$. Furthermore, this spectral method for the solution of the heat equation is efficiently implementable by the fast Fourier transform (FFT) in $O(N \log N)$ operations.

There are several significant difficulties in extending the simple spectral method employed for (1.1)-(1.3) to more general problems. Among these difficulties are those caused by imposition of nontrivial boundary conditions, nonlinear and nonconstant coefficient terms, and complex geometries. These difficulties and their solutions will be discussed below (see also [1, 2]).

The Fourier series (1.4) converges fast if $u(x, t)$ is infinitely differentiable and $u(x, t)$ satisfies the boundary conditions

$$\frac{\partial^{2n} u(x, t)}{\partial x^{2n}} = 0 \qquad (x = 0, \pi) \tag{1.10}$$

for all nonnegative integers n. Under these conditions, the error after N terms

$$\varepsilon_N(x, t) = u(x, t) - \sum_{n=1}^{N} a_n(t) \sin nx$$

goes to zero uniformly in x faster than any power of $1/N$ as $N \to \infty$. On the other hand, if $u(x, t)$ is not infinitely differentiable *or* if any of the conditions (1.10) is violated, then $\varepsilon_N(x, t) = O(1/N^p)$ as $N \to \infty$ for some finite p. For example,

$$(1.11) \qquad 1 = \sum_{n=0}^{\infty} (-1)^n \frac{\sin(2n+1)x}{2n+1} \qquad (0 < x < \pi),$$

but the error incurred by truncating after N terms is of order $1/N$ for any fixed x, $0 < x < \pi$. Furthermore, the convergence of (1.11) is not uniform in x; (1.11) exhibits Gibbs' phenomenon, namely

$$\varepsilon_N(\xi/N) = O(1) \qquad (N \to \infty, \, \xi \text{ fixed}).$$

For any fixed N, there are points x at which the error after N terms of (1.11) is not small. The poor convergence of (1.11) is due to the violation of (1.10) for $n = 0$.

More generally, most eigenfunction expansions of a function $f(x)$ converge faster than algebraically (*i.e.* the error incurred by truncating after N terms goes to zero faster than any finite power of $1/N$ as $N \to \infty$) only if $f(x)$ is infinitely differentiable *and* $f(x)$ satisfies an infinite number of special boundary conditions. For example, the Fourier-Bessel expansion

$$f(x) = \sum_{n=0}^{\infty} a_n J_0(\lambda_n x) \qquad (0 \leqslant x \leqslant 1),$$

where λ_n is the n-th smallest root of $J_0(\lambda) = 0$, converges faster than algebraically only if f is infinitely differentiable and

$$(1.12) \qquad \left[\frac{1}{x}\frac{d}{dx} x \frac{d}{dx}\right]^k f(x) = 0 \qquad \text{at } x = 1$$

for $k = 0, 1, 2, \ldots$.

When a spectral expansion converges only algebraically fast, spectral methods based on these eigenfunction expansions cannot offer significant advantages over more conventional (finite-difference, finite-element) methods. Eigenfunction expansions of this kind should not normally be used *unless* the boundary conditions of the problem imply all the extra boundary constraints like (1.10) or (1.12). For example, if periodic boundary conditions are compatible with the differential equation to be solved, complex Fourier series are suitable to develop efficient spectral approximations.

In the development of spectral methods for general problems, it is important that the rate of convergence of the eigenfunction expansion being used does not depend on special properties of the eigenfunctions, like boundary conditions, but rather depend only on the smoothness of the function being expanded. Of course, if the solution to the problem being solved is not smooth, one should not expect errors that decrease faster than algebraically with $1/N$ when global eigenfunction expansions are used. Faster than algebraic rates of convergence may be achieved for these problems by either patching the solution at discontinuities or pre- and post-processing of the solution (see [2]).

There is an easy way to ensure that the rate of convergence of a spectral expansion of a function $f(x)$ depends only on the smoothness of $f(x)$, not on its boundary properties. The idea is to expand in terms of suitable classes of orthogonal polynomials, including Chebyshev and Legendre polynomials for all those problems in which constraints like (1.10) and (1.12) are unrealistic. These polynomial expansions avoid all difficulties associated with the Gibbs phenomenon provided the solution $f(x)$ is smooth.

From the mathematical point of view, the classical orthogonal polynomials are eigenfunctions of singular Sturm-Liouville problems. It is not hard to show [1] that expansions using eigenfunctions of such singular Sturm-Liouville problems converge at a rate that depends only on the smoothness of $f(x)$, in contrast to eigenfunction expansions based on nonsingular Sturm-Liouville problems that lead to additional boundary constraints like (1.10) on $f(x)$.

These results for orthogonal polynomial expansions are easily demonstrated in the case of Chebyshev polynomial expansions. The n-th-degree Chebyshev polynomial $T_n(x)$ is defined by

$$(1.13) \qquad T_n(\cos\theta) = \cos n\theta .$$

Therefore, if

$$(1.14) \qquad f(x) = \sum_{n=0}^{\infty} a_n T_n(x) ,$$

then

$$(1.15) \qquad g(\theta) = f(\cos\theta) = \sum_{n=0}^{\infty} a_n \cos n\theta .$$

Thus the Chebyshev polynomial expansion coefficients a_n of $f(x)$ are just the Fourier cosine expansion coefficients of the even, periodic function $g(\theta)$. A simple integration-by-parts argument then shows that

$$n^p a_n \to 0 \qquad (n \to \infty) ,$$

provided $g(\theta)$ (or, equivalently, $f(x)$) has p continuous derivatives. Since

$$\left| f(x) - \sum_{n=0}^{N} a_n T_n(x) \right| \leqslant \sum_{n=N+1}^{\infty} |a_n| \qquad (|x| \leqslant 1) ,$$

it follows that the rate of convergence of (1.14) is faster than algebraic if f is smooth.

In summary, spectral expansions should be made using series of orthogonal polynomials unless the boundary conditions of the problem are fully compatible with some other class of eigenfunctions. In practice, Chebyshev and Legendre polynomial expansions are recommended for most applications, supplemented by Fourier series and surface harmonic series when boundary conditions permit.

Another difficulty with general kinds of spectral methods is their application to problems with nonlinear and nonconstant coefficient terms. Before explaining the solution to this problem, let us illustrate the difficulty.

Suppose we wish to solve the partial differential equation

$$\frac{\partial u}{\partial t} = \mathcal{N}(u, u) + \mathcal{L}u, \tag{1.16}$$

where $u = u(\mathbf{x}, t)$ and $\mathcal{N}$ is a bilinear (nonlinear) operator that involves only spatial derivatives and $\mathcal{L}$ is a linear operator that involves only spatial derivatives. The operators $\mathcal{N}$ and $\mathcal{L}$ may depend on both $\mathbf{x}$ and t. A spectral method for the solution of (1.16) is obtained by seeking the solution as a finite spectral expansion:

$$u(\mathbf{x}, t) = \sum_{n=1}^{N} a_n(t) \psi_n(\mathbf{x}), \tag{1.17}$$

where we assume for now that $\psi_n(\mathbf{x})$ ($1 \leq n \leq \infty$) are a complete set of orthogonal functions. If we introduce the re-expansion coefficients c_{nmp} and d_{nm} so that

$$\mathcal{N}(\psi_m, \psi_p) = \sum_{n=1}^{\infty} c_{nmp}(t) \psi_n,$$

$$\mathcal{L}(\psi_m) = \sum_{n=1}^{\infty} d_{nm}(t) \psi_n$$

and equate coefficients of $\psi_n(\mathbf{x})$ ($n = 1, \ldots, N$) in (1.16), we obtain

$$\frac{da_n}{dt} = \sum_{m=1}^{N} \sum_{p=1}^{N} c_{nmp}(t) a_m(t) a_p(t) + \sum_{m=1}^{N} d_{nm}(t) a_m(t) \qquad (n = 1, \ldots, N). \tag{1.18}$$

Equations (1.18) are the spectral evolution equations for the solution of (1.16). They have one very serious drawback. In general c_{nmp} and d_{nm} are nonzero for typical n, m, p, so that evaluation of da/dt from (1.18) for all $n = 1, \ldots, N$ requires $O(N^3)$ arithmetic operations for the bilinear term and $O(N^2)$ operations for the linear term. Thus solution of (1.18) requires order N^3 operations per time step. Since operational spectral calculations now involve $N \geq 10^6$, the computational cost of the direct solution of (1.18) is prohibitive (even if only linear terms are present).

The problem here is one of computational complexity. Finite-difference methods for the solution of (1.16) on N grid points may require only order N operations per time step. If the spectral method really requires order N^3 operations per time step, it cannot compete when N is large.

Another example illustrating the computational complexity of spectral methods is given by the nonlinear diffusion equation

$$(1.19) \qquad \frac{\partial u(x,t)}{\partial t} = \exp[u] \frac{\partial^2 u}{\partial x^2}(x,t).$$

If we seek the solution as

$$(1.20) \qquad u(x,t) = \sum_{n=1}^{N} a_n(t) \psi_n(x)$$

in terms of the orthonormal functions $\psi_n(x)$, then

$$(1.21) \qquad \frac{da_n}{dt} = \int \psi_n(x) \exp\left[\sum_{m=1}^{N} a_m(t)\psi_m(x)\right] \sum_{p=1}^{N} a_p \psi_p''(x) \, dx$$

for $n = 1, \ldots, N$. These evolution equations for $\{a_n(t)\}$ have an exponential degree of computational complexity as they are expressed as an integral functional of $\{a_n(t)\}$.

The solution to the problem of computational complexity is to use the author's transform methods. Let us illustrate the technique for a pseudo-spectral (or collocation) approximation to (1.19). First, we introduce N suitable collocation points $x_1, x_2, \ldots, x_N$ lying within the computational domain. Then the approximate solution (1.20) is forced to satisfy the partial differential equation (1.19) (or its boundary conditions) exactly at these discrete points at every time t. More specifically, the following three steps are done at each time step t:

i) Determine N coefficients $a_n(t)$ $(n-1, \ldots, N)$ so that

$$(1.22) \qquad u(x_j, t) = \sum_{n=1}^{N} a_n(t) \psi_n(x_j) \qquad (j=1, \ldots, N).$$

ii) Evaluate $u_{xx}(x_j, t)$ by

$$(1.23) \qquad u_{xx}(x_j, t) = \sum_{n=1}^{N} a_n(t) \psi_n''(x_j) \qquad (j=1, \ldots, N).$$

iii) Finally, evaluate $\partial u(x_j, t)/\partial t$ by

$$(1.24) \qquad \frac{\partial u(x_j, t)}{\partial t} = \exp[u(x_j,t)] u_{xx}(x_j,t) \qquad (j=1, \ldots, N)$$

and march forward to the next time step.

The idea of the pseudospectral transform method can be restated as follows: Transform freely between physical (x_j) and spectral (a_n) representations, evaluating each term in whatever representation that term is most accurately, and simply, evaluated. Thus, in (1.24), we evaluate $\exp[u]$ in the physical representation while we compute u_{xx} in the spectral representation by (1.22) because it is most accurately done there.

It should be apparent to the reader that pseudospectral transform methods can be applied to any problem that can be treated by finite-difference methods regardless of the technical complexity of nonlinear and nonconstant coefficient terms.

For the expressions of interest, computation of derivatives of a N-term spectral expansion requires order N arithmetical operations. For the Fourier series (1.7), this fact is obvious:

$$\frac{d}{dx} \sum_{n=1}^{N} a_n \sin nx = \sum_{n=1}^{N} n a_n \cos nx ,$$

$$\frac{d^2}{dx^2} \sum_{n=1}^{N} a_n \sin nx = -\sum_{n=1}^{N} n^2 a_n \sin nx .$$

For the Chebyshev polynomial expansion (1.14), the computational complexity of differentiation is a little less apparent. Since $T_n(\cos\theta) = \cos n\theta$,

$$\frac{T'_{n+1}(x)}{n+1} - \frac{T'_{n-1}(x)}{n-1} = \frac{2}{c_n} T_n(x) \qquad (n \geqslant 0) ,$$

where $c_0 = 2$, $c_n = 1$ ($n \geqslant 1$) and $T'_0 = T'_{-1} = 0$. Therefore, if

$$\frac{d}{dx} \sum_{n=0}^{N} a_n T_n(x) = \sum_{n=0}^{N} b_n T_n(x) ,$$

then

$$2 \sum_{n=1}^{N} a_n T'_n(x) = \sum_{n=0}^{N} c_n b_n \left[\frac{T'_{n+1}}{n+1} - \frac{T'_{n-1}}{n-1} \right] = \sum_{n=1}^{N+1} [c_{n-1} b_{n-1} - b_{n+1}] T'_n(x)/n .$$

Equating coefficients of $T'_n(x)$ for $n = 1, ..., N+1$ gives the recurrence relation

(1.25) $$\begin{cases} c_{n-1} b_{n-1} - b_{n+1} = 2n a_n & (1 \leqslant n \leqslant N) , \\ b_n = 0 & (n \geqslant N) . \end{cases}$$

The solution of (1.25) for b_n given a_n requires only order N arithmetic operations. Similar recurrence relations can be obtained for differentiation of spectral series based on other sets of orthogonal polynomials and functions.

In the case of Fourier series, the transform (1.7) and its inverse can be computed in $O(N \log_2 N)$ operations if $N = 2^p$ using the fast Fourier transform. However, most of the computational efficiency of transform methods comes not from the FFT but from the separability of multidimensional transforms. Thus a three-dimensional discrete Fourier transform can be expressed as three one-dimensional Fourier transforms

$$(1.26) \quad \sum_{j=0}^{J-1} \sum_{k=0}^{K-1} \sum_{l=0}^{L-1} a(j, k, l) \exp\left[2\pi i \left(\frac{jm}{J} + \frac{kn}{K} + \frac{lp}{P}\right)\right] =$$
$$= \sum_{j=0}^{J-1} \exp[2\pi ijm/J] \sum_{k=0}^{K-1} \exp[2\pi ikn/K] \sum_{l=0}^{L-1} a(j, k, l) \exp[2\pi ilp/L].$$

The left-hand side of (1.26) requires roughly $(JKL)^2$ operations to evaluate at all the points $0 \leqslant m < J$, $0 \leqslant n < K$, $0 \leqslant p < L$. On the other hand, even without the FFT, the right-hand side of (1.26) requires only about $(JKL)\cdot(J+K+L)$ operations to evaluate at all the points. When the FFT is applied to the one-dimensional transforms on the right-hand side of (1.26), the number of operations necessary to evaluate (1.26) is reduced further to $(JKL)\cdot(\log_2 J + \log_2 K + \log_2 L)$ if J, K, L are powers of 2.

Spectral approximations to general boundary-value problems lead to full $N \times N$ matrix equations for the N expansion coefficients a_n. It would seem that solution of these equations requires $O(N^3)$ arithmetic operations, while storage of the matrix requires $O(N^2)$ memory locations. Since typical problems now involve $N \sim 10^6$, the direct solution (or even the direct formulation) of such problems would seem unworkable now.

Consider the solution of a general linear differential equation $Lu = f$. Let a N-term spectral approximation to this problem be given by

$$(1.27) \quad L_{\rm sp} u_N = f_N,$$

where f_N is a suitable N-term approximation to f. As mentioned several times earlier, the matrix representation of (1.27) is generally a full $N \times N$ matrix, so that direct solution of (1.27) by Gauss elimination methods would require order N^2 storage (for the matrix representation of $L_{\rm sp}$) and order N^3 arithmetic operations.

Here we shall describe a method that permits the solution of (1.27) using order N storage locations with the number of arithmetic operations of order the larger of $N \log N$ and the number of operations required to solve $Lu = f$ by a *first-order* finite-difference method. The important conclusion is that *spectral methods for general problems in general geometries can be implemented efficiently with operation costs and storage not much larger than that of the simplest finite-difference approximation to the problem with the same number of degrees of freedom.* Since spectral methods require many fewer degrees of freedom to achieve given accuracy (or, nearly equivalently, spectral methods achieve much higher accuracy for a given number of degrees of freedom) than required by finite-order finite-difference approximations, important computational efficiencies result from the new method.

The idea of the iteration method is as follows: Suppose we are able to construct an approximation L_{ap} to the spectral operator L_{sp} that has the following properties:

i) L_{ap} has a sparse matrix representation so that it can be represented using only $O(N)$ storage locations.

ii) L_{ap} is efficiently invertible in the sense that the equation

(1.28)
$$L_{\mathrm{ap}} u_N = f_N$$

is solvable as efficiently as a first-order finite-difference approximation to the problem.

iii) L_{ap} approximates L_{sp} in the sense that

(1.29)
$$0 < m \leqslant \|L_{\mathrm{ap}}^{-1} L_{\mathrm{sp}}\| \leqslant M < \infty$$

for suitable constants m, M as $N \to \infty$. Roughly speaking, (1.29) requires that the eigenvalues of $L_{\mathrm{ap}}^{-1} L_{\mathrm{sp}}$ be bounded from above and below as $N \to \infty$.

We propose to construct L_{ap} from L_{sp} by changing the discretization operator either in addition to or in place of approximating the differential operator. Thus we construct L_{ap} by a suitable *low-order finite-difference approximation to L*.

A simple example is given by the second-order differential equation

(1.30)
$$Lu = f(x) u''(x) + g(x) u'(x) + h(x) u(x) = v(x) \qquad (0 \leqslant x < 2\pi)$$

with periodic boundary conditions $u(x + 2\pi) = u(x)$ and $f(x) > 0$. A spectral approximation is approximately sought as the finite Fourier series

(1.31)
$$u(x) \sim \sum_{|k| < K} a_k \exp[ikx].$$

If the Fourier coefficients of $f(x)$, $g(x)$, $h(x)$, $v(x)$ are denoted f_k, g_k, h_k, v_k, respectively, then the spectral (Galerkin) equations for a_k are

(1.32)
$$L_{\mathrm{sp}} u = \sum_{\substack{|p| < K \\ |k-p| < K}} [-p^2 f_{k-p} + ipg_{k-p} + h_{k-p}] a_p = v_k.$$

Clearly, these equations have, in general, a full matrix representation that requires $O(K^2)$ storage locations and $O(K^3)$ operations to invert.

A suitable approximate operator L_{ap} is constructed using the collocation points $x_j = 2\pi j/N$ ($j = 0, 1, \ldots, N-1$), where $N = 2K$. In the physical space representation, we use the finite-difference approximation

$$(1.33) \qquad L_{ap} u|_{x_j} = f(x_j) \frac{u_{j+1} - 2u_j + u_{j-1}}{(\Delta x)^2} + g(x_j) \frac{u_{j+1} - u_{j-1}}{2\Delta x} + h(x_j) u_j,$$

where $u_j = u(x_j)$ and $\Delta x = 2\pi/N$. Obviously, L_{ap} is sparse and efficiently invertible. To verify (1.29) we use the following elementary argument (that may be made more rigorous but no more correct by more involved WKB-like arguments). If λ is an eigenvalue of $L_{ap}^{-1} L_{sp}$, then there exists a function $u(x)$ such that

$$(1.34) \qquad L_{sp} u = \lambda L_{ap} u.$$

If $u(x)$ is a smooth function of x (in the limit $N \to \infty$), then both $L_{ap} u$ and $L_{sp} u$ should be good approximations to $Lu(x)$, so (1.34) implies $\lambda \sim 1$. On the other hand, if $u(x)$ is a highly oscillatory function of x (in the limit $N \to \infty$), then

$$(1.35) \qquad u'' \gg u' \gg u \qquad (N \to \infty).$$

Therefore,

$$(1.36) \qquad L_{ap} u \sim f(x_j) \frac{u_{j+1} - 2u_j + u_{j-1}}{(\Delta x)^2}$$

and, if transform (pseudospectral) methods are used to evaluate $L_{sp} u$,

$$(1.37) \qquad L_{sp} u \sim f(x_j) \sum_{|k|<K} (-k^2) a_k \exp[ikx_j],$$

so (6.18) gives

$$(1.38) \qquad f(x_j) \sum_{|k|<K} (-k^2) a_k \exp[ikx_j] \sim \lambda f(x_j) \frac{u_{j+1} - 2u_j + u_{j-1}}{(\Delta x)^2}.$$

The eigenfunctions of (1.38) are

$$u_j = \exp[iqx_j] \qquad (|q|<K)$$

and the associated eigenvalue is

$$\lambda = \frac{(q\Delta x)^2}{4 \sin^2 \tfrac{1}{2} q \Delta x}.$$

Since $|q| < K$ with $K = \frac{1}{2}N = \pi/\Delta x$, we obtain

$$1 \leqslant \lambda \leqslant \frac{\pi^2}{4}.$$

Thus (1.29) holds with $m = 1$ and $M = \pi^2/4 \approx 2.5$.

There are several extensions of the above method for constructing L_{ap} that are important in practice. First, in the case of Chebyshev spectral methods, it is appropriate to construct L_{ap} using finite-difference approximations based on the collocation points $x_j = \cos \pi j/N$. In this case, the operator bounds (1.29) continue to hold with $M = 2.5$, $m = 1$ for a wide variety of operators L. Second, higher-order equations are best treated by writing them as a system of lower-order equations. Thus direct construction of L_{ap} for $L = \nabla^4$ gives

$$1 \leqslant \|L_{\mathrm{ap}}^{-1} L_{\mathrm{sp}}\| \leqslant 6 \approx \left(\frac{\pi^2}{4}\right)^2.$$

However, if we introduce $v = \nabla^2 u$ and define the second-order operator K by

$$K\begin{pmatrix} u \\ v \end{pmatrix} = \begin{cases} \nabla^2 u - v, \\ \nabla^2 v, \end{cases}$$

then direct construction of K_{ap} as a finite-difference operator gives

$$1 \leqslant \|K_{\mathrm{ap}}^{-1} K_{\mathrm{sp}}\| \leqslant 2.5.$$

Third, odd-order operators, initial-value problems and problems of mixed type are best treated by constructing L_{ap} on a grid that is roughly 50% finer than that used in construction of L_{sp} by collocation. In this case the spectral bounds (1.29) with $M \leqslant 2.5$ continue to hold for most problems. For example, the operator $\partial/\partial x$ with periodic boundary conditions has spectrum ik, while its centered finite-difference approximation has spectrum $i \sin(k\Delta x)/\Delta x$, so

$$\|L_{\mathrm{ap}}^{-1} L_{\mathrm{sp}}\| = O(k\Delta x/\sin k\Delta x),$$

which is unbounded for $|k\Delta x| < \pi$, but bounded by $4\pi/3\sqrt{3} \approx 2.4$ if $|k\Delta x| < 2\pi/3$.

2. – Applications.

2`1. Introduction. – Over the last few years, there has been progress in understanding fundamental nonlinear processes in shear flows. In this section, I shall survey some results that have emerged from numerical studies of transition and turbulence. I shall review for you three different aspects of these problems. First, I shall summarize results on the basic instabilities that seem to be responsible for the onset of chaos in these flows. These instabilities appear to be universal in character and may explain many of the unifying features of transition. Second, I shall give some examples of progress in the numerical simulation of high-Reynolds-number flows. Finally, I will give a synopsis of new ideas for subgrid scale closures of huge-Reynolds-number turbulence.

Full details of the ideas discussed here are given in the references.

2˙2. *A transitional instability.* – The processes by which laminar flows undergo transition to turbulence remain basically unsolved. However, recent numerical studies have provided some insights into transition, including:

2˙2.1. Nonclassical character of transitional instabilities. The primary linear (exponential) instability of classical plane parallel shear flows with noninflectional velocity profiles, as described by the Orr-Sommerfeld (or related) equations, is much too weak to describe transition. For example, linear instability of plane Poiseuille flow $\left(U(z) = 1 - z^2, |z| < 1\right)$ occurs for Reynolds numbers $R_c > 5778$, while Squire's theorem implies that the critical disturbance is two-dimensional. The fact that this instability is induced by a subtle interplay of viscosity and shear implies that its growth rates are quite small on convective time scales. For example, the most rapidly growing exponential mode of the Orr-Sommerfeld equation is obtained at $R_{opt} = 48\,000$; its growth rate is only 0.0076; it is so feeble that perturbations grow by a factor 10 in a time of about 300, in which time a point on the centerline moves about 150 channel widths. In contrast, transition is observed to occur explosively over a few channel widths at Reynolds numbers as low as roughly 1000. A transitional instability that affects noninflectional plane parallel shear flows must have a characteristic convective time scale.

2˙2.2. Three dimensionality of transition. Two-dimensional fluids do not appear to exhibit the kind of strong chaos that is characteristic of turbulent shear flows. In thermal convection, CURRY et al. [3] show that two-dimensional flows do not appear to act in a strongly chaotic way, but three-dimensional flows may be strongly chaotic at large enough Reynolds number. Even for inflectional free shear flows, in which there are strong inviscid two-dimensional instabilities, BRACHET and ORSZAG [4] show that the flows that develop from two-dimensional finite-amplitude disturbances are not strongly chaotic, in contrast to the flows that develop three-dimensionally.

2˙2.3. Instability of two-dimensional nonlinear travelling waves. Perhaps the simplest instability that has the character of a transitional instability is the linear three-dimensional instability of two-dimensional finite-amplitude flows. ORSZAG and KELLS [5] and ORSZAG and PATERA [6]

Fig. 1. – Streamlines of the steady (stable) finite-amplitude two-dimensional travelling wave for plane Poiseuille flow at $R = 4000$, plotted in the rest frame of the wave (from [6]).

show how such an instability fits the basic features of transition in classical shear flows, including their convective growth rates, inherent three-dimensionality, onset at Reynolds numbers in accord with experimental observations and flow features in accord with early transitional flows. These instabilities have been analyzed both by direct numerical simulation of the evolving three-dimensional flow and by a linear perturbation analysis of the nonparallel two-dimensional (nonlinear travelling wave) flow. In fig. 1, we show the streamlines of a typical two-dimensional base state (here for plane Poiseuille flow at $R = 4000$). The nonparallel character of the base flow leads to considerable complication in its linear stability analysis (see [6] for the formulation of these large-matrix eigenvalue problems). A topic of much current research interest is the development of efficient numerical methods for finding eigenvalues of the very large matrices encountered in problems of this sort. In fig. 2, we give a stability diagram for this transitional instability; here we plot contours of constant growth rate as a function of the amplitude of the two-dimensional base state and the Reynolds number. The growth rates of this instability are $1 \div 2$ orders of magnitude larger than those of Orr-Sommerfeld modes. The

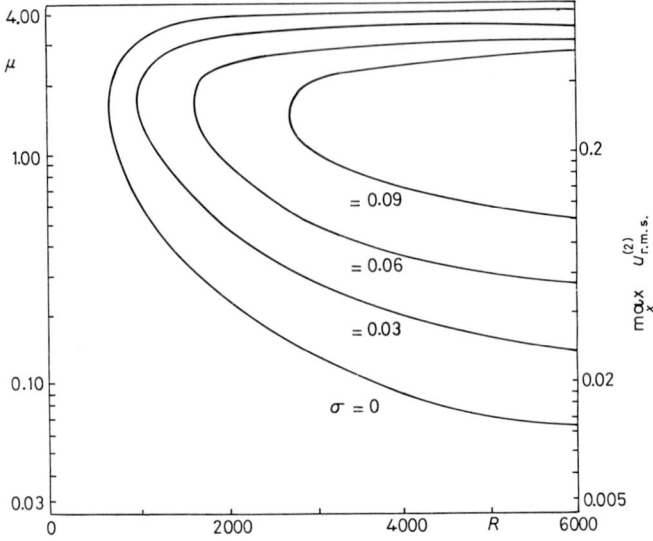

Fig. 2. – Contours of constant growth rate (labelled by growth rate) as a function of R and the amplitude of the background two-dimensional nonlinear wave (see right-hand scale).

development of this three-dimensional secondary instability seems to be consistent with available experimental data on early transitional flows. In fig. 3, we compare contours of the x velocity at the so-called one-spike stage of transition in plane Poiseuille flow obtained a) experimentally by NISHIOKA, IIDA and KANBAYASHI [7] and b) numerically by KLEISER and SCHUMANN [8]. The

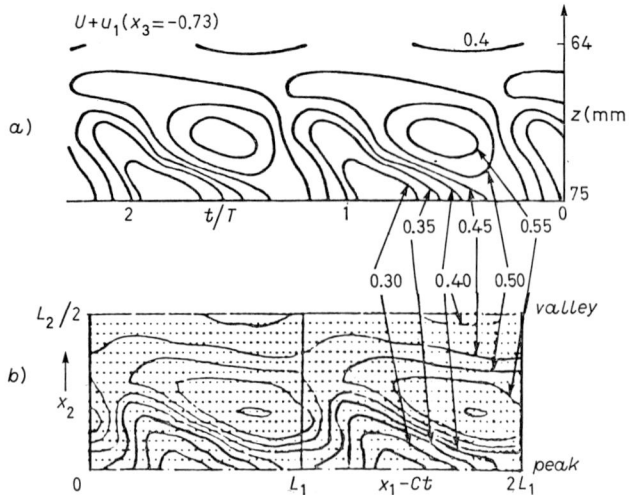

Fig. 3. – Contours of x velocity in the (x, y)-plane at the one-spike stage in the laboratory experiments of Nishioka et al. (a)) and in the numerical simulation of Kleiser and Schumann (b)) (from [8]).

flows that develop from the initial linear instability appear to lead directly to chaos and turbulence and not to saturate into ordered, laminar flow states. Similar instabilities have been found in boundary layers, plane Couette flow, pipe Poiseuille flow (see [6]) and in free shear flows (see [4]).

2˙2.4. *Competition between two-dimensional pairing and three-dimensional instabilities.* Inflectional free shear flows, like mixing layers and jets, are inviscidly unstable to two-dimensional disturbances. Squire's theorem implies that these instabilities are strongest when two-dimensional; when these two-dimensional instabilities evolve in time, they saturate into ordered laminar-flow states characterized by large-scale vortical flow structures. These vortical flows may themselves be unstable to subharmonic (pairing) instabilities, in which two (or more) vortices are paired and generate a new larger-scale vortex motion [9]. In these flows, the three-dimensional instability discussed above is also present [10], but it is not necessarily stronger than the pairing instability. However, the three-dimensional secondary instability is effective at much smaller spanwise spatial scales than is the inviscid primary instability and seems to lead directly to chaotic flows [4].

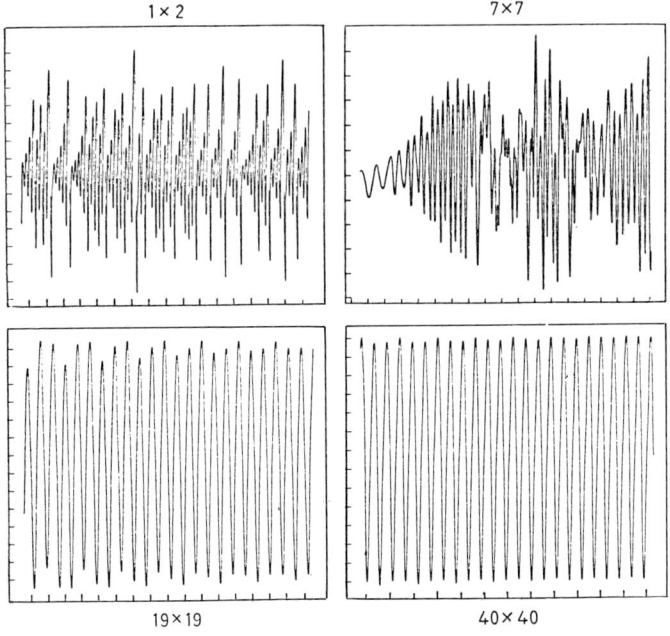

Fig. 4. – Time evolution of the Fourier component b_{02} of the temperature field in two-dimensional Bénard convection at $R_a = 120\, R_{ac}$, $\mathrm{Pr} = 20$ [3]. The numbers labelling each plot give the wave number cut-off used to derive the Galerkin approximation to the Boussinesq equations. Thus 1×2 gives the Lorentz equations, while the higher-order models are higher-order Galerkin approximations. Observe that as the resolution increases the chaos disappears.

2˙2.5. Spurious (numerical) turbulence. CURRY et al. [3] show that, while low-order dynamical systems derived by Galerkin approximation to the two-dimensional Boussinesq equations may exhibit chaotic solutions, this chaos typically disappears as the dimension of the projection space increases (see fig. 4). Similarly, it was shown by ORSZAG and KELLS [5] that under-resolved numerical calculations of transitional planar shear flows may be spuriously chaotic. Under-resolved computations do not have degrees of freedom associated with small spatial scales available to act as an eddy viscosity on well-resolved large scales.

2˙3. Computer simulations of turbulence. – In this subsection, I shall give three examples of numerical simulations of turbulent flows. The first two examples, turbulent channel flow and the simulation of a turbulent spot, are of the nature of numerical experiments in which the numericist uses the computer in much the same way as the experimentalist uses the laboratory, namely as a source of data about flows in a controlled environment. The final example, the Taylor-Green vortex, is an example in which the computer is being used to try to uncover fundamental physical laws of turbulence.

2˙3.1. Turbulent channel flow. Turbulent channel flows have been simulated numerically three ways: *a*) large-eddy simulation with a subgrid scale turbulence closure for eddies outside the wall layer and a heuristic boundary condition applied at the edge of the viscous sublayer by DEARDORFF [11] and SCHUMANN [12], *b*) large-eddy simulation with a subgrid scale turbulence closure applied to eddies of all scales including those in the wall layer by MOIN and KIM [13] and *c*) full numerical solution of the Navier-Stokes equations by ORSZAG and PATERA [14]. The really crucial differences are, as we again note in subsect. **2˙4** below, between *a*) and *b*)-*c*). Simulations of type *a*) have much smaller computational requirements at a given Reynolds number R than either of types *b*) or *c*), the latter requiring asymptotically similar computational work at large R. The deficiency of simulations of type *a*) is that they require modelling of wall layer effects in terms of an over-simplified boundary condition; the deficiencies of types *b*) and *c*) are that, with currently available computer resolution (say $64 \times 64 \times 65$ on a Cray-1 computer), Reynolds numbers are limited to about 10 000 (type *b*)) or 5000 (type *c*)). For simulations of types *b*) or *c*), the computational work scales as R^3, so future increases in computer power do little to increase the effective Reynolds number of the computations.

Nevertheless, it is possible to achieve interesting results with full numerical solutions of the Navier-Stokes equations. In fig. 5, we plot the mean velocity profile found in the channel flow computations of Orszag and Patera [14]. The fit to a logarithmic wall layer velocity profile is only marginal, but the resulting von Kármán constant 0.45 is within experimental bounds, so this calculation does give the first computation of a wall layer from the basic prin-

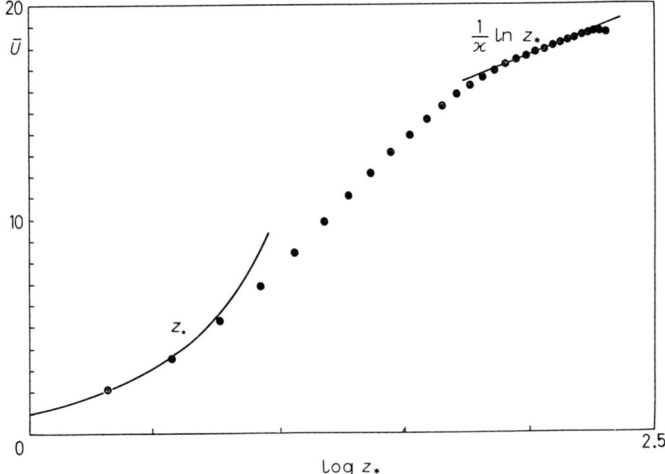

Fig. 5. – Mean turbulent profile obtained by full numerical simulation of plane Poiseuille flow at $R = 5000$ using a $64 \times 64 \times 65$ spectral simulation. Note the viscous sublayer, buffer region and logarithmic layer of $8 \div 9$ data points (from [14]).

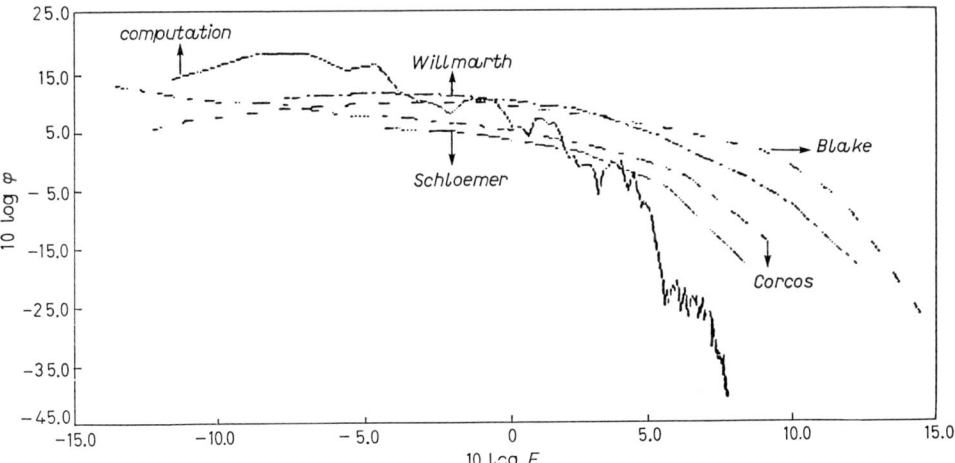

Fig. 6. – A plot of the turbulent wall pressure spectrum as a function of frequency (from [15]).

ciples of fluid dynamics. Another more recent result from computations of this type is given in fig. 6, in which we plot the wall pressure spectrum in a moderate-resolution $(32 \times 32 \times 33)$ run compared with available experimental data (see [15]). Despite the moderately low Reynolds number $(R = 5000)$ of the simulation, agreement is achieved because flow features that do not depend explicitly on the boundary wall layer structure tend to be Reynolds number independent.

2'3.2. Turbulent spot. There has been much recent interest in the evolution of localized «spots» in turbulent flows (see [16]). The first numerical simulation of a turbulent spot was reported by LEONARD [17], who used three-dimensional vortex filament techniques to compute the (inviscid) flow. More recently, we have begun a study of spots using full numerical solutions of the Navier-Stokes equations at moderate Reynolds numbers [18]. The latter simulations are performed by forcing the initial flow using a localized force to drive a jet of fluid vertically, then allowing the disturbance to evolve naturally. In fig. 7 and 8, we plot contours of maximum vertical-z velocity in the (x, y) and (x, z) planes at various times of evolution of plane Poiseuille flow. The character of this spot evolution is similar to that observed experimentally: the spot seems to spread in the spanwise direction by «transverse contamination», in agreement with the dye injection experiments of Gad-el-Hak et al. [19]; the greatest turbulent activity is near the edges of the spot, the «spreading» angle of the spot relative to its source is about 10°, in agreement with the channel flow experiments of Carlson et al. [20]; the vertical structure of the spot is in qualitative agreement with that observed experimentally. Further numerical experiments are under way that should elucidate details of the flow in spots and the surrounding fluid.

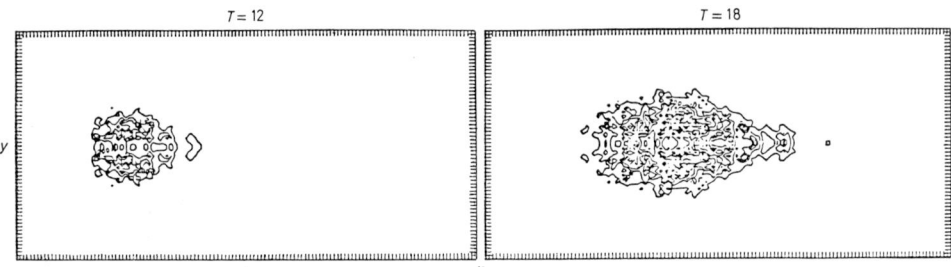

Fig. 7. – Contours of the maximum z-velocity in the (x, y)-plane at $t = 12, 18$ after initializing a turbulent-spot computation by imposed vertical forcing. These computations are performed using a spectral code with $128 \, (x) \times 32(y) \times 32(z)$ resolution. Fourier series are used in x and y; Chebyshev polynomial expansions are used in z. Here $R = 6000$.

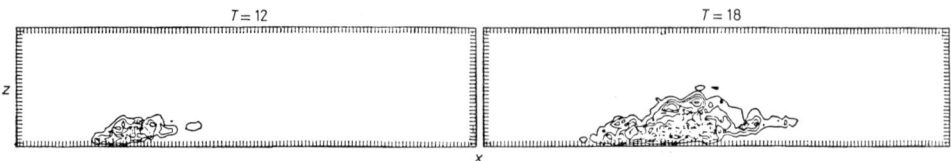

Fig. 8. – Same as fig. 7, except x-z contours of $\max_y |v_z|$.

2'3.3. Taylor-Green vortex. In order to gain understanding of the basic physics of the generation of small-scale turbulent flow features, a nice model problem is the Taylor-Green (TG) vortex flow [21, 22]. Here the flow is that which develops in time from initial conditions that consist of excitation in basically a single Fourier mode. Because of nonlinear interaction, the flow becomes strongly three-dimensional and develops excitation at all spatial scales. The TG vortex has been used to study such fundamental questions as the enhancement of vorticity by vortex line stretching, the approach to isotropy of the small scales, possible singular behavior of the Euler equations, formation of an inertial range and analysis of the geometry and intermittency of high-vorticity regions. The TG flow is advantageous for these studies because its special symmetry has allowed the development of numerical algorithms that are a factor 64 more efficient in both memory and storage than conventional periodic-geometry spectral methods. For a three-dimensional flow, this factor 64 translates into a factor 4 increased range of spatial scales—it is now possible to compute the TG vortex flow with $512 \times 512 \times 512$ Fourier modes for each velocity component on the Cray-1 computer (or more than $4 \cdot 10^8$ effective degrees of freedom!).

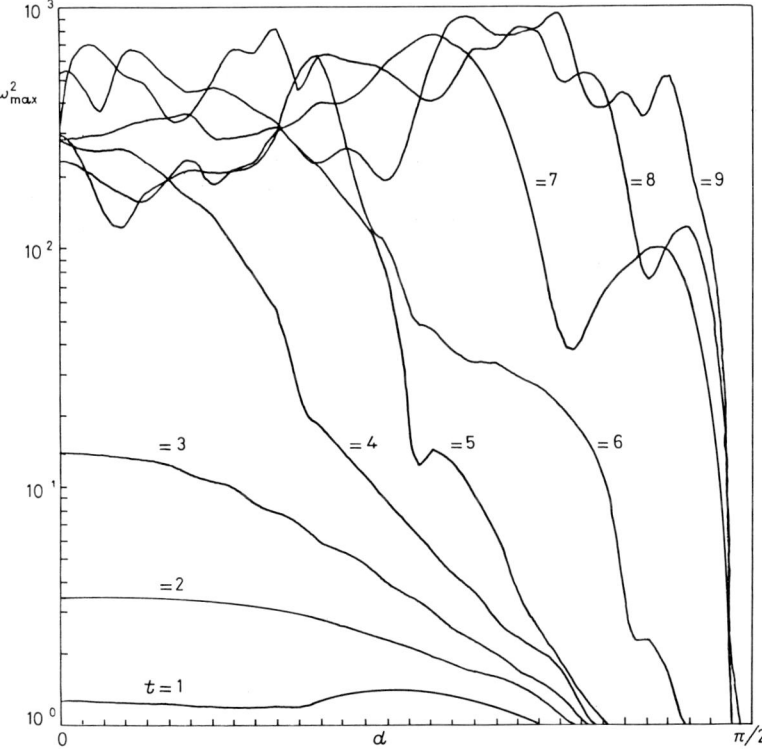

Fig. 9. – A plot of the distribution of large-vorticity regions in the TG vortex flow as a function of time t and distance d away from the side-walls of the impermeable cube in which the flow takes place. Observe how vorticity explodes in towards the center of the cube between $t = 4$ and $t = 8$ (from [22]).

One of the more exciting results to emerge from our studies of the TG flow is the suggestion that viscosity may play an essential role in the development of small-scale turbulence, not just acting as a sink of turbulent kinetic energy. Indeed, we find that the development of the turbulent flow seems to require viscosity to induce instabilities of vortical structures in which the initial large-scale nonturbulent vorticity undergoes an explosive redistribution in space (see fig. 9). These viscosity-induced instabilities are probably effective because viscosity allows vortex line reconnections prohibited in inviscid flow. Similar diffusional instabilities have now been shown to be responsible for the generation of small-scale structures in two-dimensional magnetohydrodynamic [23] and kinetic [24] turbulence. Further study of viscosity-induced instabilities should clarify the development of intermittent flow structures in turbulence.

2'4. *Subgrid scale turbulence closures.* – Perhaps the most distinguishing characteristic of high-Reynolds-number turbulent flows is their large range of excited space and time scales. In homogeneous turbulence, dissipation scale eddies are of order $R^{\frac{3}{4}}$ times smaller than energy-containing eddies. Including the effect of this range of spatial scales on the allowable time step in a numerical solution of the Navier-Stokes equations gives the estimate that order R^3 operations are required to simulate a turbulent flow. This is the reason for interest in the large-eddy simulation method in which excitations on scales smaller than those resolvable numerically are modelled, usually by an eddy viscosity coefficient (see [11, 12]). The basic action of an eddy viscosity on large eddies is reasonable, although it cannot reproduce the random character of the action of small-scale eddies. However, in order to model properly wall turbulence, it is necessary to extend the subgrid modelling ideas of Deardorff and Schumann and treat the turbulence all the way up to a rigid wall, as in recent work by MOIN and KIM [13]. Unfortunately, in order for MOIN and KIM to resolve motions down to the scale of turbulent bursts, which is necessary in order to capture the mechanism producing the turbulence, the work restriction $O(R^3)$ remains. Thus the Reynolds-number restrictions are similar for large-eddy and full numerical solutions of the Navier-Stokes equations that attempt to integrate all the way through the wall layer region.

In recent work, YAKHOT and ORSZAG [25] have used dynamic renormalization group (RNG) methods to treat wall-bounded turbulence. The idea of the infra-red RNG method is to use perturbation methods based on the direct-interaction approximation [26] to eliminate all small spatial scales up to the resolvable grid scale from the Navier-Stokes equations. This is done perturbatively by eliminating narrow bands of wave vectors from the dynamics (see fig. 10), renormalizing the resulting reduced dynamical equation to have the form of the Navier-Stokes equation with modified viscosity and random forcing terms, and then repeating the process iteratively until all the required small scales are removed. The resulting dynamical equations involve a modified eddy viscosity and a random force, both induced by renormalization. The eddy viscosity is modified from the Smagorinsky viscosity used by DEARDORFF, SCHUMANN and MOIN and KIM in the wall regions in which there is interference between the eddy and molecular viscosities. This interference effect is the key to obtaining a faithful representation of the wall region. Also, the induced

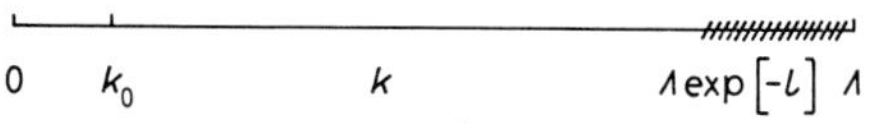

Fig. 10. – A schematic representation of the modal structure of the dynamic renormalization group. Here k_0 represents wave numbers within the energy-containing range, while Λ gives the high-wave-number (viscous) cut-off. Modes in the hatched band are removed at each step of the RNG procedure.

random force is large in the buffer layer between the viscous sublayer and the logarithmic layer, giving a turbulence source in this region. Further work is now under way applying these RNG-based closures to both large-eddy simulations of turbulent shear flows and to the derivation of new classes of turbulence transport (Reynolds averaged) equations that should be useful in engineering applications.

3. – Conclusion.

I have reviewed several areas of activity in the numerical simulation of transition and turbulence in which I have been intimately involved recently. In this short space, it has not been possible to do justice to all of the large number of researchers involved in these fields; the references do a more complete job of surveying the literature. The principal conclusions from our studies are:

i) Numerical methods now provide essential information complementary to that available from experiment and mathematical analysis.

ii) Computational fluid mechanics has now matured, so that there are techniques that can be reliably applied to the most difficult of fluid-mechanical problems. In contrast to 10 years ago, it is no longer mainly a question of how to compute a complicated flow, rather, now, it is a question of which flow to compute in order to extract the most useful information.

iii) It is crucial, especially in our studies of transitional flows, that we have used spectral numerical methods (see sect. **1** above). Spectral methods are so accurate for these problems that we can confidently conclude that properly tested numerical results are true fluid-mechanical results. In contrast to finite-difference or finite-element methods in which an increase in spatial resolution by a factor 2 leads to an error decrease by a factor 4 or 8 or so, with spectral methods a factor 2 increase in resolution typically decreases the error by several orders of magnitude. This permits accurate verification of results. For example, in recent studies of transition in circular Couette flow, MARCUS et al. [27] and MARCUS [28] have been able to achieve at least three-decimal-place agreement with experiment on wave speeds. The confidence in these results has permitted new analytical insights into the character of the onset of wavy instabilities of Taylor vortices in Couette flow [29].

iv) New generations of bigger and faster computers can most profitably be used to extend the range of application of computational fluid dynamics. Transition and turbulence problems in complex geometries with complex physics, like multiphase flows, will surely be the subject of studies in the near future.

* * *

This work was supported by the Office of Naval Research under Contracts N00014-82-C-0451 and N00014-83-K-0227, by the Air Force Office of Scientific Research under Contract F49620-83-C-0064 and by the National Science Foundation under Grants ATM-8310210 and MEA-8215695.

REFERENCES

[1] D. GOTTLIEB and S. A. ORSZAG: *Numerical Analysis of Spectral Methods: Theory and Applications* (Philadelphia, Penn., 1977).
[2] D. GOTTLIEB, M. Y. HUSSAINI and S. A. ORSZAG: *Theory and applications of spectral methods*, in *Proceedings of the Symposium on Spectral Methods* (Philadelphia, Penn., 1984), p. 1.
[3] J. H. CURRY, J. R. HERRING, J. LONCARIC and S. A. ORSZAG: *Order and disorder in two- and three-dimensional Benard convection*, in *J. Fluid Mech.*, to appear.
[4] M. E. BRACHET and S. A. ORSZAG: *Secondary instability of free-shear flows*, submitted to *J. Fluid Mech.*
[5] S. A. ORSZAG and L. C. KELLS: *J. Fluid Mech.*, **96**, 159 (1980).
[6] S. A. ORSZAG and A. T. PATERA: *J. Fluid Mech.*, **128**, 347 (1983).
[7] M. NISHIOKA, S. IIDA and S. KANBAYASHI: *An experimental investigation of the subcritical instability in plane Poiseuille flow*, in *Proceedings of the X Turbulence Symposium* (Tokyo, 1978), p. 55.
[8] L. KLEISER and U. SCHUMANN: *Laminar-turbulent transition process in plane Poiseuille flow*, in *Proceedings of the Symposium in Spectral Methods* (Philadelphia, Penn., 1984), p. 141.
[9] P. C. PATNAIK, F. S. SHERMAN and G. M. CORCOS: *J. Fluid Mech.*, **73**, 215 (1976).
[10] R. T. PIERREHUMBERT and S. E. WIDNALL: *J. Fluid Mech.*, **114**, 59 (1982).
[11] J. W. DEARDORFF: *J. Fluid Mech.*, **41**, 453 (1970).
[12] U. SCHUMANN: *J. Comput. Phys.*, **18**, 376 (1975).
[13] P. MOIN and J. KIM: *J. Fluid Mech.*, **118**, 341 (1982).
[14] S. A. ORSZAG and A. T. PATERA: *Phys. Rev. Lett.*, **47**, 832 (1981).

[15] R. A. HANDLER, R. J. HANSEN, L. SAKELL, E. T. BULLISTER and S. A. ORSZAG: *Phys. Fluids*, **27**, 579 (1984).
[16] I. J. WYGNANSKI, M. SOKOLOV and D. FRIEDMAN: *J. Fluid Mech.*, **78**, 785 (1976).
[17] A. LEONARD: *Vortex simulation of three-dimensional, spotlike disturbances in a laminar boundary layer*, in *Turbulent Shear Flows*, II, edited by L. J. S. BRADBURY *et al.* (Berlin, 1980), p. 67.
[18] E. T. BULLISTER and S. A. ORSZAG: *Numerical simulation of turbulent spots*, to be published.
[19] M. GAD-EL-HAK, R. F. BLACKWELDER and J. J. RILEY: *J. Fluid Mech.*, **110**, 73 (1981).
[20] D. R. CARLSON, S. E. WIDNALL and M. F. PEETERS: *J. Fluid Mech.*, **121**, 407 (1982).
[21] G. I. TAYLOR and A. E. GREEN: *Proc. R. Soc. London, Ser. A*, **158**, 499 (1937).
[22] M. E. BRACHET, D. I. MEIRON, S. A. ORSZAG, B. G. NICKEL, R. H. MORF and U. FRISCH: *J. Fluid Mech.*, **130**, 411 (1983).
[23] U. FRISCH, A. POUQUET, P. L. SULEM and U. MENEGUZZI: to be published.
[24] M. E. BRACHET: to be published.
[25] V. YAKHOT and S. A. ORSZAG: *Renormalization group formulation of large eddy simulation*, submitted to *J. Fluid Mech.*
[26] R. H. KRAICHNAN: *J. Fluid Mech.*, **5**, 497 (1959).
[27] P. S. MARCUS, S. A. ORSZAG and A. T. PATERA: *Simulation of circular Couette flow*, in *Proceedings of the VIII International Conference on Numerical Methods in Fluid Dynamics*, edited by E. KRAUSE (Berlin, 1982), p. 371.
[28] P. S. MARCUS: to be published.
[29] B. J. BAYLY and P. S. MARCUS: *Analytic computation of Taylor-Couette wave speeds*, to be published.

Part III

TURBULENCE IN GEOPHYSICAL FLOWS

Quasi-Geostrophic Turbulence: An Introduction.

R. SADOURNY

Laboratoire de Météorologie Dynamique
Ecole Normale Supérieure - 24, rue Lhomond, 75231 Paris Cedex 05

1. – Introduction.

Quasi-geostrophic turbulence, a term originally introduced by CHARNEY [1], deals with the nonlinear behaviour of large-scale geophysical flows. Quasi-geostrophic flows are quasi–two-dimensional, quasi-nondivergent flows which are a good approximation of large-scale oceanic of atmospheric flows away from the equator; the quasi-geostrophic approximation, again originally due to CHARNEY [2], is derived in general terms in the appendix. A detailed review of quasi-geostrophic dynamics is to be found in [3]; here only basic aspects will be considered. We shall leave aside the important problem of the interaction of nonlinearity with Rossby waves, which has been partially investigated in [4-6] and reviewed with some detail in [3]. For the problem of quasi-geostrophic turbulence above topography, the reader is referred to [7-9] and again to [3] for review.

Section 2 deals with pure two-dimensional, or barotropic, flows. The basic properties of nonlinear transfers are first described, followed by a somewhat detailed discussion of statistical equilibria. After a brief description of stability theorems, the classical phenomenology of forced stationary two-dimensional turbulence is exposed and followed by a discussion of how well the theory is matched by direct numerical simulations.

Section 3 deals with three-dimensional, or baroclinic, quasi-geostrophic flows. This is basically an unsolved problem, presented here with less detail than the barotropic case and under severely restrictive simplifying assumptions.

For completeness, the governing equations (the quasi-geostrophic vorticity equation and the corresponding upper- and lower-boundary conditions) are derived in an appendix.

2. – Barotropic turbulence.

2`1. *Nonlinear transfers*. – The motion of a barotropic fluid is nondivergent on isentropic surfaces (see appendix). Therefore, on isentropic surfaces, it is governed by a single two-dimensional equation, the barotropic vorticity equation

$$\left(\frac{\partial}{\partial t} + \frac{\partial(\psi,\ )}{\partial(x, y)}\right)\nabla^2\psi = \nu\nabla^4\psi\ . \tag{1}$$

Here ψ is the streamfunction, $\nabla^2\psi$ the relative vorticity, x and y longitude and latitude, respectively, times the radius of the Earth. For the moment we shall concentrate on *pure* turbulence, disregarding, among other things, its interaction with Rossby waves: the planetary vorticity f, or its variation with latitude, has, therefore, been omitted. Dissipation has been included under the usual form of a viscosity term with viscosity coefficient ν—an artificial but convenient formulation. At vanishing viscosity, eq. (1) is characterized by an infinite number of invariants (any function of vorticity is obviously an invariant of the motion), amongst which two are quadratic: the *energy*, which reduces to kinetic energy

$$\mathscr{E} = \tfrac{1}{2}\iint_\mathscr{D}(\nabla\psi)^2\,\mathrm{d}x\,\mathrm{d}y\ , \tag{2}$$

and the *enstrophy*, or vorticity squared

$$\mathscr{Z} = \tfrac{1}{2}\iint_\mathscr{D}(\nabla^2\psi)^2\,\mathrm{d}x\,\mathrm{d}y\ . \tag{3}$$

The domain $\mathscr{D}$ of the flow can be either the whole sphere or a basin surrounded by rigid walls—in the latter case, ψ will have to be assumed constant over the boundary—or even, for simplicity, a simple periodic square.

The natural modes of the problem are the eigenmodes $\varLambda_k$ of the Laplacian operator on $\mathscr{D}$ (constrained to be constant on the boundary if $\mathscr{D}$ is an ocean basin), corresponding to real negative eigenvalues $-\lambda_k^2$ or *scales* λ_k^{-1}; we shall suppose that the $\varLambda_k$, λ_k have been so ordered that the scale decreases with increasing k. Expanding the streamfunction on the eigenmodes yields the spectral form of the equation of motion (1)

$$\frac{\mathrm{d}}{\mathrm{d}t}\hat{\psi}_k + \sum_{p,q} A_{kpq}\hat{\psi}_p\hat{\psi}_q = -\nu\lambda_k^2\hat{\psi}_k\ , \tag{4}$$

where the caret refers to expansion coefficients. We shall suppose that the convolution term has been symmetrized: $A_{kpq} = A_{kqp}$. The spectral forms of energy and enstrophy read

$$\mathscr{E} = \tfrac{1}{2}\sum_k \mathscr{E}_k \qquad \text{with } \mathscr{E}_k = \lambda_k^2\hat{\psi}_k^2\ , \tag{5}$$

$$\mathscr{Z} = \tfrac{1}{2}\sum_k \mathscr{Z}_k \qquad \text{with } \mathscr{Z}_k = \lambda_k^4\hat{\psi}_k^2 = \lambda_k^2\mathscr{E}_k\ . \tag{6}$$

In (5), (6), we have supposed that we use real eigenmodes and, therefore, real expansion coefficients.

The conservation of energy and enstrophy by the convolution term plays a major role in the organization of energy transfers among modes, yielding very special properties for two-dimensional, barotropic fluids, as first observed by FJØRTOFT in ref. [10]. It is readily seen from (4)-(6) that the root-mean-square scale relative to the spectral energy distribution,

$$\bar{\lambda}^{-1} = (\mathscr{E}/\mathscr{Z})^{1/2}, \tag{7}$$

is conserved in the absence of viscosity and increases with positive ν. This means that systematic transfer of energy from large scales to small scales is forbidden in barotropic fluids. In fact, enstrophy is so heavily weighted towards the smaller scales—remember (6)—that any such transfer cannot occur without violating the enstrophy conservation constraint. This type of argument is most clear if we consider an elementary *triad interaction*, say (k, p, q), within which *detailed* energy and enstrophy conservation are equivalent to

$$\lambda_k^2 A_{kpq} + \lambda_p^2 A_{pqk} + \lambda_q^2 A_{qkp} = 0, \tag{8}$$

$$\lambda_k^4 A_{kpq} + \lambda_p^4 A_{pqk} + \lambda_q^4 A_{qkp} = 0, \tag{9}$$

which, for instance, yields

$$\frac{\lambda_p^2 A_{pqk}}{\lambda_q^2 A_{qkp}} = \frac{\lambda_k^2 - \lambda_q^2}{\lambda_p^2 - \lambda_k^2}. \tag{10}$$

The meaning of (10) is that the larger-scale and the smaller-scale modes of a triad either gain energy from the middle one simultaneously, or simultaneously feed energy into it. If we suppose, say, $\lambda_q < \lambda_k < \lambda_p$, another consequence of (8)-(10) is that the (q, k)-transfer will be more efficient at transferring energy than enstrophy, and *vice versa* for the (k, p)-transfer. A schematic diagram showing these properties of the triad interaction is displayed in fig. 1.

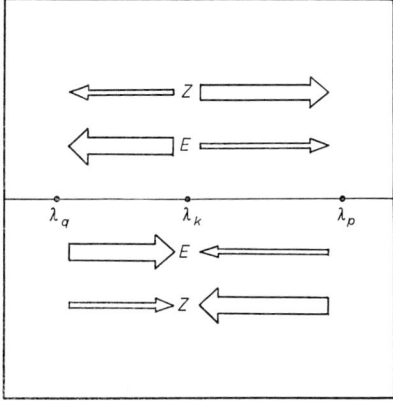

Fig. 1. – Schematic energy (E) and enstropy (Z) fluxes in a triad interaction.

2'2. *Statistical equilibria.* – A more elaborate viewpoint on the nonlinear trends of the flow is provided by the statistical mechanics of truncated systems involving only a finite number of modes—of which the triad interaction is just the simplest example. For instance, if we discard all modes whose scale is smaller than a given *cut-off scale l*, we are left with an analogue of (4) restricted to, say, N degrees of freedom, the evolution of which can be described by the motion of its image $\mathcal{M} = \{\lambda_k \hat{\psi}_k\}$ in a N-dimensional phase space. At zero viscosity, energy and enstrophy are conserved, and $\mathcal{M}$ follows the intersection $\mathscr{C}$ of the energy hypersphere (5) with the enstrophy hyperellipsoid (6). In addition, its motion is nondivergent, because A_{kpq}, obtained from the spectral transform of a Jacobian, is such that $A_{kok} = A_{kko} = 0$. This means that, if we consider $\mathscr{C}$ together with the measure induced by Lebesgue's measure in phase space, then the stochastic process defined by (4) is stationary.

We shall suppose that (4) has only a finite memory of its initial conditions, or, in other words, that two successive states of the system become uncorrelated for large enough time lags. If this is true, the time-averaged modal energies

(11) $$\bar{\mathscr{E}}_k^T = \frac{1}{T} \int_0^T \mathscr{E}_k(t) \, dt$$

converge in quadratic mean towards the corresponding expectations $\bar{\mathscr{E}}_k$; and in fact, from Birkoff's first theorem [11], the convergence is even almost sure since the process is stationary.

Now the modal-energy expectations $\bar{\mathscr{E}}_k$ remain to be evaluated. This can be done directly by numerical calculations on the *microcanonical* ensemble—the hypercurve $\mathscr{C}$ with its associated measure $\mathcal{M}$—as in ref. [12]. The simplest way (although less accurate for very small N), however, consists in substituting to $\mathcal{M}$ the approximate Boltzmann distribution

(12) $$\mathfrak{I}(\lambda_k \hat{\psi}_k) = \left(\frac{a + b\lambda_k^2}{2\pi}\right)^{1/2} \exp\left[-\frac{1}{2}(a + b\lambda_k^2) \mathscr{E}_k\right],$$

which yields

(13) $$\bar{\mathscr{E}}_k = (a + b\lambda_k^2)^{-1},$$

a form originally derived by KRAICHNAN in ref. [13] (se also [14, 15]). The two parameters a, b in (13) are uniquely determined by the (initial) values of $\mathscr{E}$ and $\mathscr{Z}$. To prove this statement, we proceed as follows. From (5)-(7), (13) we get, after eliminating a and setting $X = b\mathscr{E}$ as the unknown,

(14) $$\sum_{k=0}^{N-1} (N - X(\bar{\lambda}^2 - \lambda_k^2))^{-1} = 1.$$

The rational function on the left-hand side has $M \leqslant N$ distinct poles defining $M-1$ bounded consecutive intervals, one of which contains zero since $\lambda_0 < \bar{\lambda} < \lambda_{N-1}$. It increases within each bounded strictly positive interval and decreases within each bounded strictly negative interval, which yields $M-2$ zeros, all to be discarded since $\mathscr{E}_k$ must be positive for all k. On the other hand, it has M zeros, and $X = 0$ is another solution to be discarded. We are finally left with one and only one acceptable solution for X in the interval containing zero, which means one and only one equilibrium state.

Looking now at the derivative of (14) at $X = 0$, we see that

i) $\quad b < 0, \ a > 0 \quad$ corresponds to $\quad \bar{\lambda}^2 > \dfrac{1}{N} \sum\limits_{k=0}^{N-1} \lambda_k^2$.

In this case (high ratio of enstrophy to energy), the equilibrium spectrum exhibits a cusp at the smallest scale, where enstrophy is essentially concentrated, and a near-equipartition of energy among medium- to large-scale modes.

ii) $\quad b = 0, \ a > 0 \quad$ corresponds to $\quad \bar{\lambda}^2 = \dfrac{1}{N} \sum\limits_{k=0}^{N-1} \lambda_k^2$,

which corresponds to an equipartition of energy amongst all modes.

iii) $\quad b > 0, \ a > 0 \quad$ corresponds to $\quad \dfrac{1}{N} \sum\limits_{k=0}^{N-1} \lambda_k^{-2} < \bar{\lambda}^2 < \dfrac{1}{N} \sum\limits_{k=0}^{N-1} \lambda_k^2$.

In this case we get a tendency to equipartition of energy in the larger scales and equipartition of enstrophy in the smaller scales.

iv) $\quad b > 0, \ a = 0 \quad$ corresponds to $\quad \bar{\lambda}^{-2} = \dfrac{1}{N} \sum\limits_{k=0}^{N-1} \lambda_k^{-2}$.

Here we have equipartition of enstrophy amongst all modes.

v) $\quad b > 0, \ a < 0 \quad$ corresponds to $\quad \bar{\lambda}^{-2} > \dfrac{1}{N} \sum\limits_{k=0}^{N-1} \lambda_k^{-2}$.

In this final case, the equilibrium spectrum exhibits a cusp at the largest scale where energy is essentially concentrated, with a near-equipartition of enstrophy in medium to small scales. Note that half of the preceding discussion proceeds from the symmetry of the problem with respect to the exchange $(\mathscr{E}, \mathscr{L}; \lambda_k, \lambda_k^{-1})$.

To summarize, the trend of truncated inviscid systems towards statistical equilibrium generalizes the behaviour already observed on single-triad interactions, according to which enstrophy tends to accumulate among smaller scales, while energy accumulates at larger scales. Equipartitions, or, in the most extreme cases, cusps, are produced by physical ($\lambda \geqslant \lambda_0$) or artificial ($\lambda \leqslant \lambda_{N-1}$) scale limitations, which frustrate the trend of energy to reach even larger scales, or the trend of enstrophy to reach even smaller scales. The five cases i)-v) are illustrated by fig. 2, extracted from ref. [12].

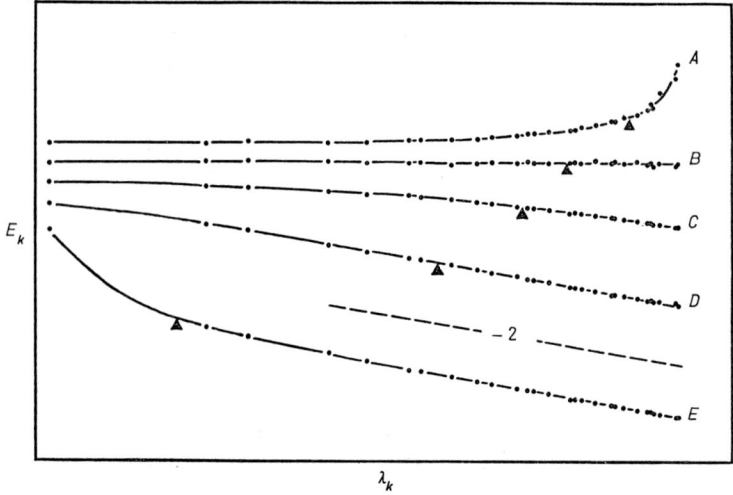

Fig. 2. – Equilibrium spectra for various values of λ (triangles) (from ref. [12]).

An interesting quantity to consider in connection with statistical equilibria is the *entropy* of the truncated flow

$$\mathscr{S} = \tfrac{1}{2} \sum_k \log \mathscr{E}_k,$$

a measure of the information necessary to completely specify the motion when the spectrum $\mathscr{E}_k$ is given. It has been shown that the equilibrium spectra (13) obtain by maximizing $\mathscr{S}$ for given $\mathscr{E}$ and $\mathscr{Z}$. Therefore, equilibrium spectra are those which, for given $\mathscr{E}$ and $\mathscr{Z}$, provide the least information on the actual state of the system (see [16, 17]).

2'3. *Stationary solutions: stability.* – At zero viscosity, the stationary solutions of (1) correspond to a vanishing Jacobian. This means that the streamfunction ψ^0 and vorticity $\zeta^0 = \nabla^2 \psi^0$ of a stationary solution are dependent on each other; or, in other words, that there exists a function F such that, everywhere on $\mathscr{D}$, $F(\psi^0(x, y), \zeta^0(x, y)) = 0$. A sufficient condition for stability of this type of motion is as follows:

The stationary solution is stable in each of the following two cases:

(15) $\quad\quad\quad\quad\quad\quad\quad d\psi^0/d\zeta^0 < 0 \quad\quad\quad\quad\quad\quad$ *everywhere on $\mathscr{D}$,*

(16) $\quad\quad\quad\quad\quad\quad\quad d\psi^0/d\zeta^0 \leqslant -\lambda_0^{-2} \quad\quad\quad\quad\quad$ *everywhere on $\mathscr{D}$,*

where λ_0^{-1} is the largest eigenscale.

Criterion (15) has been stated by ARNOLD in ref. [18], criterion (16) by BENZI, PIERINI, VULPIANI and SALUSTI in ref. [19]. A simple demonstration can be sketched as follows. ANDREWS in ref. [20] remarks that the space integral of

(17) $$I = \frac{d\psi^0}{d\zeta^0}\zeta'^2 + (\nabla\psi')^2$$

is an invariant of the linearized perturbation equation

(18) $$\frac{\partial \zeta'}{\partial t} + \frac{\partial(\psi^0, \zeta')}{\partial(x, y)} + \frac{\partial(\psi', \zeta^0)}{\partial(x, y)} = 0.$$

Here primes refer to perturbation quantities. More precisely, it is straightforward to show that

(19) $$\frac{d\mathscr{I}}{dt} = -\iint_{\mathscr{D}}\left(\frac{\partial(\zeta^0, \psi'^2/2)}{\partial(x, y)} - \frac{\partial(\psi^0, \psi'\zeta')}{\partial(x, y)} + \frac{\partial(\zeta^0, \zeta'^2/2)}{\partial(x, y)}\right) dx\, dy,$$

where $\mathscr{I}$ is the space integral of I

(20) $$\mathscr{I} = \iint_{\mathscr{D}} I\, dx\, dy,$$

and ξ^0 is defined from the local specification $d\xi^0/d\psi^0 = d\psi^0/d\zeta^0$. The basic flow is stable if I keeps a given sign (either positive: criterion (15), or negative: criterion (16)) throughout $\mathscr{D}$.

Note that equality is allowed in (16). The fact that the largest-scale eigenmode Λ_0 is stable follows directly from the nonlinear interaction properties described in subsect. **2'1, 2'2**. Instability of Λ_0 would mean that its energy can be transferred to smaller scales in violation of the enstrophy conservation property.

2'4. *Forced stationary turbulence: phenomenology.* – We now consider the simplest case of an inviscid fluid on the infinite plane, forced by a stationary, homogeneous, isotropic source acting at unit wave number: $k_F = 1$; therefore, it injects energy and enstrophy into the fluid at the same constant rate ε.

Denoting by $E(k, t)$ the expectation of *one-dimensional* energy density at wave number k and time t $\left(\text{with } \mathscr{E}(t) = \int_0^\infty E(k, t)\, dk\right)$, we define the spectral energy and enstrophy fluxes

(21)
$$\Pi_E(k, t) = \begin{cases} \int_0^k \frac{\partial E}{\partial t}\, dp & \text{for } k < 1, \\ \int_k^\infty \frac{\partial E}{\partial t}\, dp & \text{for } k > 1, \end{cases}$$

$$\Pi_Z(k, t) = \begin{cases} \int_0^k p^2 \frac{\partial E}{\partial t}\, dp & \text{for } k < 1, \\ \int_k^\infty p^2 \frac{\partial E}{\partial t}\, dp & \text{for } k > 1. \end{cases}$$

This yields

(22)
$$\begin{cases} \Pi_Z(k, t) < k^2 \Pi_E(k, t) & \text{for } k < 1, \\ \Pi_E(k, t) < k^{-2} \Pi_Z(k, t) & \text{for } k > 1. \end{cases}$$

If we now admit that the turbulence evolves towards a pseudostationary regime—stationary on an increasing-wave-number band $(k_E(t), k_Z(t))$, $0 \leftarrow k_E(t) < 1 < k_Z(t) \to \infty$,—we get from (22)

(23)
$$\begin{cases} \Pi_E(k, t) = \Pi_E(k_E(t), t) > \varepsilon(1 - k_Z^{-2}(t)), \\ \Pi_Z(k, t) = \Pi_Z(k_E(t), t) < \varepsilon k_E^2(t), \end{cases} \quad \text{for } k_E(t) < k < 1;$$

$$\begin{cases} \Pi_E(k, t) = \Pi_E(k_Z(t), t) < \varepsilon k_Z^{-2}(t), \\ \Pi_Z(k, t) = \Pi_Z(k_Z(t), t) > \varepsilon(1 - k_E^2(t)), \end{cases} \quad \text{for } 1 < k < k_Z(t).$$

Besides, simple arguments based on the assumption of a self-similar shape in the evolution of each spectral tail ($k < k_E$, $k < k_Z$) yield

(24)
$$\begin{cases} \dfrac{dk_E}{dt} \sim - \varepsilon / E(k_E(t), t), \\ \dfrac{dk_Z}{dt} \sim \varepsilon / k_Z^2(t)\, E(k_Z(t), t). \end{cases}$$

As $t \to \infty$, (23) yields

(25)
$$\begin{cases} \Pi_E(k, t) \to \varepsilon, \quad \Pi_Z(k, t) \to 0 & \text{for } k < 1, \\ \Pi_E(k, t) \to 0, \quad \Pi_Z(k, t) \to \varepsilon & \text{for } k > 1. \end{cases}$$

The asymptotic regime is, therefore, characterized by two contiguous inertial ranges: an *energy inertial range* (EIR) where energy cascades from the injection scale to ever increasing scales at a constant rate equal to the injection rate ε and an *enstrophy inertial range* (ZIR) where enstrophy cascades from the injection scale to ever decreasing scales again at a constant rate equal to the injection rate ε. In the EIR, the enstrophy flux vanishes, and, conversely, the energy flux vanishes in the ZIR.

The phenomenology of the two barotropic inertial ranges, originally derived by KRAICHNAN [13, 21] along the line of Kolmogorov's [22] arguments, has been most clearly illustrated by closure model simulations [23]. The derivation can be rapidly sketched as follows. The distortion of an eddy at wave number k is mainly due to straining by scales p^{-1} larger than k^{-1}, a measure of which is the enstrophy contained in these scales. This leads to the *eddy turn-over time* estimate

$$\tau(k) \sim \left(\int_0^k p^2 E(p) \, dp \right)^{-1/2}. \tag{26}$$

On the other hand, the energy and enstrophy cascade rates can be estimated respectively as

$$\begin{cases} \varepsilon \sim k E(k)/\tau(k) & \text{for } k < 1, \\ \varepsilon \sim k^3 E(k)/\tau(k) & \text{for } k > 1. \end{cases} \tag{27}$$

Combining (26) and (27) and solving for $E(k)$ yields

$$\begin{cases} E(k) = C \varepsilon^{2/3} k^{-5/3} & \text{for } k < 1, \\ E(k) = C \varepsilon^{2/3} k^{-3} (\log k)^{-1/3} & \text{for } k > 1, \end{cases} \tag{28}$$

where a *universal* constant C has been introduced. Looking back at the integrand of (26) with respect to the self-similar measure $d \log k$, we verify that this integrand is dominated by the wave numbers which are closest to k in the EIR, while, in the ZIR, wave numbers close to one dominate: we say that nonlinear interactions are essentially local in scale within the EIR and non-local within the ZIR.

From (26)-(28), we also readily obtain estimates of $k_E(t)$, $k_Z(t)$ or, equivalently, of the inverse functions $t_E(k)$, $t_Z(k)$, respectively, the times needed for the EIR or the ZIR to reach a given wave number k:

$$\begin{cases} t_E(k) \sim \varepsilon^{-1/3} k^{-2/3} & \text{for } k < 1, \\ t_Z(k) \sim \varepsilon^{-1/3} (\log k)^{2/3} & \text{for } k > 1. \end{cases} \tag{29}$$

It, therefore, takes an infinite time for each cascade to reach infinitely large or small scales: in fact, the regularity of initially smooth solutions of (1) at finite times has been mathematically demonstrated [24-26]. Another phenomenological interpretation of $t_E(k)$, $t_Z(k)$ is the time taken by a perturbation initially introduced at wave number one to induce perturbations at wave number k, or *vice versa*. In that sense, these times are also estimates of predictability times—for a more quantitative approach of predictability times using closure modelling, see ref. [27]. To say that barotropic flows have *infinite regularity times* is phenomenologically equivalent to say they have *infinite predictability times*, in the sense that it takes an infinite time for a perturbation of infinitely small initial scale to affect finite scales.

2`5. *Direct or semi-direct simulations*. – The above phenomenology is difficult to verify in practice. Pure two-dimensional turbulence is, to a large extent, a theoretician's abstraction. Some evidence that large-scale atmospheric energy spectra, for instance, do follow approximately k^{-3} laws has indeed been produced by WIIN-NIELSEN [28], DESBOIS [29], MOREL and LARCHEVÊQUE [30]; contradictory evidence, however, has also been produced recently by GAGE [31], then LILLY and PETERSEN [32]. In any case the complexity of atmospheric flow makes it a poor candidate to be used as a benchmark for testing the theory. It seems that there is slightly better hope on the side of laboratory experiments: hydrodynamic experiments on rotating tanks do generate turbulent flows which are approximately two-dimensional (*e.g.*, [33]), and it even seems feasible to generate flows which are *exactly* two-dimensional in magneto-hydrodynamics experiments [34].

The other useful benchmark, of course, is numerical experimentation. With the advent of bigger and faster computers, direct numerical simulations of (1) can be performed over wave number bands wide enough to allow, for instance, precise evaluation of spectral slopes in the inertial ranges. The problem with numerical experimentation, however, is that the simulation of fully developed turbulence with the discretized Navier-Stokes equation requires a phenomenological parametrization of the nonlinear interactions between explicit scales and subgrid scales. Such numerical simulations can no longer be referred to as *direct*, but rather as *semi-direct*. The contamination of the resulting solution by phenomenological *a priori*s in the small scales makes the comparison of simulations with theory a not so straightforward matter: as an extreme standpoint, we would think of numerical experimentation as absolutely useless if we had reasons to believe that the large-scale statistics of the flow depend crucially on the details of subgrid scale dynamics.

Our current experience with semi-direct simulations is that they have not—at least up to now—completely sustained the (probably too simple) phenomenology described above. The two interesting features to mention about them is the establishment of energy spectra generally steeper than expected and the emergence of quasi-stationary isolated, almost circular vortices.

2˙6. Spectral slopes and ZIR intermittency. – The logarithmic slope of the energy spectra obtained in semi-direct simulations of freely decaying turbulence is close to -3. On the other hand, forced turbulence seems to be associated with much steeper spectra [35, 36]. There is apparently no universal law, logarithmic slopes around -4, -5 or even -6 have been obtained. Steepness seems to be depend on the scale and the nature of the source: for example, sources defined as instability yield steeper slopes than white-noise forcing.

A plausible explanation of this behaviour is the *intermittency* argument developed in ref. [35] as an extension to two-dimensional flows of an argument initially derived by KOLMOGOROV [37] and further developed in [38, 39]. Intermittency in the ZIR would be related to space-time fluctuations of the enstrophy transfer; these fluctuations may occur in relation to confinement of the enstropy transfer at wave number k within small portions of the space-time domain $\hat{\mathscr{D}}(k, t)$, the measure of which—say, $\hat{V}(k)$—is supposed to be stationary in time and decreasing with scale. In order to derive a simple phenomenological theory of intermittency, we may assume that the *active* domain $\hat{\mathscr{D}}(k, t)$ is self-similar in space, its persistence in time being naturally measured by the local eddy turn-over time. Therefore, we take

$$(30) \qquad \hat{V}(k) \sim (k/k_0)^{-\alpha} \hat{\tau}(k)/\theta \,,$$

where k_0 and θ are normalizing constants. Instead of speaking of transfers, we may say that all the active turbulent energy is concentrated in $\mathscr{D}(k, t)$. This leads us to estimate the *specific* energy of active structures—what we call the *active* energy spectrum—as

$$(31) \qquad \hat{E}(k) \sim E(k)/\hat{V}(k) \,,$$

which yields an analogue of (26) for the present case

$$(32) \qquad \hat{\tau}(k) \sim \left(\int_0^k p^2 \hat{E}(p) \, \mathrm{d}p \right)^{-1/2} .$$

Combining (30)-(32) with

$$(33) \qquad \varepsilon \sim k^3 E(k)/\hat{\tau}(k) \,,$$

the analogue of (27) for the present case, and solving for $\hat{E}(k)$ or $E(k)$ yields an active spectrum of the form

$$(34) \qquad \hat{E}(k) = C\varepsilon\theta k^{-3+\alpha}$$

and an *observed* spectrum of the form

(35) $$E(k) = \begin{cases} C(\varepsilon/\theta)^{1/2}(k/k_0)^{-3-\alpha/2} & (\alpha > 0), \\ C(\varepsilon/\theta)^{1/2} \log(k/k_0)^{-1/2} k^{-3} & (\alpha = 0). \end{cases}$$

The active spectrum (34) is rather flat, which yields the unexpected result that, contrary to the nonintermittent case, the transfers within the intermittent ZIR are indeed *local* in wave number space. The observed spectrum itself (35) is, of course, steeper than in the nonintermittent case. Although assumption (30) is here just as an example, it has one interesting implication: the fact that $\hat{\tau}(k)$, a natural measure of the duration of active phases at wave number k, decreases with scale induces an irreducible, minimum level of intermittency, which makes (35) at $\alpha = 0$ slightly steeper than (28). At the other extreme α is bounded by two for obvious geometrical reasons; then the maximum steepeness authorized by (30) is a logarithmic slope of -4, a value insufficient to explain the whole set of numerical results.

2˙7. *Emergence of quasi-stationary vortices.* – The other striking feature of semi-direct simulations is the emergence of quasi-stationary vortices, equally observed in randomly forced flows [35] and in decaying turbulence [40]. Although their level of excitation is markedly higher than the level of excitation of their more turbulent surroundings, their structure is difficult to analyse because of this surrounding noise. The most striking example of isolated vortex we have found is illustrated in fig. 3: its streamfunction is very close to a Gaussian, with zero total vorticity. These vortices are quasi-stationary in the sense that they remain almost unperturbed until they get in the vicinity of one another and eventually interact. Sporadic interaction of isolated vortices can be seen as one manifestation of intermittency. Note that in [35] a hierarchy of isolated vortices with decreasing levels of excitation and scales can be observed throughout the ZIR. It is tempting to associate these vortices and their excitation level with the active energy spectrum defined in 2˙6: the separation between such active regions and the surrounding quiet domain is, however, difficult to realize in a quantitative way.

The stability theory summarized in subsect. 2˙3 is insufficient to explain the relative stability of isolated vortices observed in our semi-direct simulations. Of course, a Gaussian vortex has a circular core ($\varrho < \varrho_0$) where $\mathrm{d}\Psi^0/\mathrm{d}\zeta^0$ is negative, surrounded by an infinite region ($\varrho > \varrho_0$) where $\mathrm{d}\Psi^0/\mathrm{d}\zeta^0$ is positive: it is tempting to consider the surrounding region as relatively stable and acting as a shield to protect the core from being destabilized by small perturbations coming from outside. This is, however, not entirely convincing, since the stability of the external annulus has been demonstrated only for perturbations which vanish at $\varrho = \varrho_0$ (*).

(*) At this point the reader may refer to C. E. LEITH: this volume, p. 266.

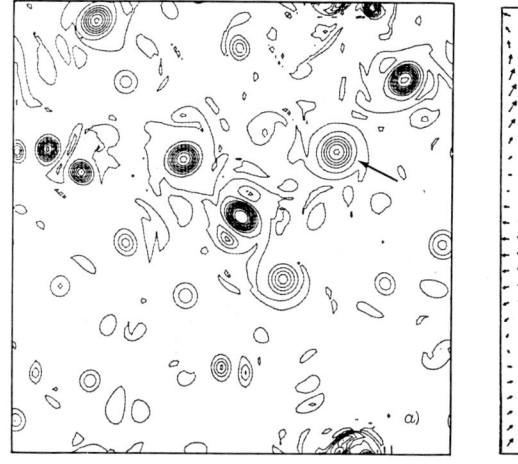
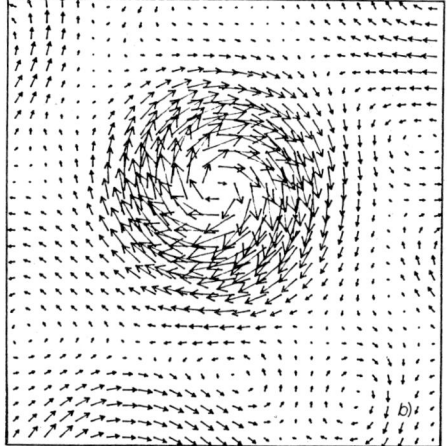
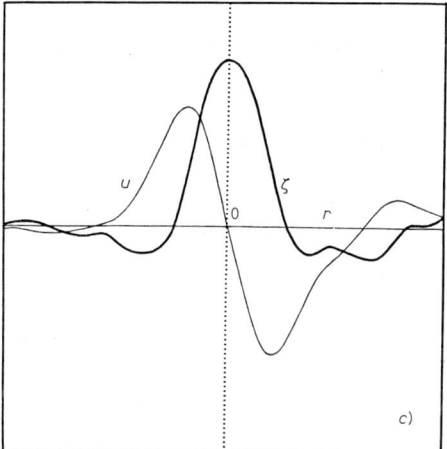

Fig. 3. – a) Vorticity field in a 128×128 point forced semi-direct simulation; b) enlargement of the vortex indicated by an arrow in a); c) radial structure of the vortex shown in b); u is tangential velocity, ζ is vorticity.

The other problem is how these organized vortices emerge from initial chaos or chaotic stirring. MC WILLIAMS [40], in his simulations of decaying turbulent flow, traces them back to initial vorticity maxima which persist for a longer time than others. Somewhat heuristically, we can envision the emergence of a quasi-stationary vortex as arising from the tendency of two-dimensional turbulent structures to aggregate into larger scales; random accretions would tend to generate circular shapes as the most probable shape; and the growth of the isolated vortex would end when the immediate surroundings which feed their energy into it eventually become exhausted.

3. – Baroclinic turbulence.

3'1. *Simplified formulations.* – The most general formulation of baroclinic quasi-geostrophic turbulence leads to a rather complicated problem, which, for the sake of simplicity, can be split into two partial but simpler ones: the *internal problem* ($\mathcal{J}$) of a quasi-geostrophic turbulence within isentropic boundaries at top and bottom, and the *external problem* ($\mathcal{E}$) of a quasi-geostrophic turbulence with vanishing potential vorticity everywhere and thus entirely driven by boundary dynamics (see the appendix for details). Problem $\mathcal{E}$ is a two-boundary, hence a two-vertical-mode problem. Problem $\mathcal{J}$, on the contrary, has a denumerable set of vertical eigenmodes, which makes it somewhat more complex.

A simplified formulation of (A.7), (A.14), (A.15) involving constant coefficients in the vertical is the following [1]:

$$(36) \quad \begin{cases} \left(\dfrac{\partial}{\partial t} + \dfrac{\partial(\Psi,\ )}{\partial(x,y)}\right)(\nabla^2 \Psi + \mu_1^2 \Psi_{zz}) = 0 & \text{for } -\dfrac{1}{2} < z < \dfrac{1}{2}, \\ \left(\dfrac{\partial}{\partial t} + \dfrac{\partial(\Psi,\ )}{\partial(x,y)}\right)\Psi_z = 0 & \text{for } z = \pm\dfrac{1}{2}. \end{cases}$$

Here z is a nondimensional vertical co-ordinate, μ_1^{-1} is equal to π times the first internal radius of deformation. In (36) the external-radius-of-deformation effect has been neglected, which means that we are considering the fluid as bounded by isobaric surfaces at both top and bottom: this assumption of vertical symmetry, although somewhat unphysical, will help simplify the problem.

We thus get a simplified formulation of $\mathcal{J}$ [41]

$$(37) \quad \begin{cases} \left(\dfrac{\partial}{\partial t} + \dfrac{\partial(\Psi,\ )}{\partial(x,y)}\right)(\nabla^2 \Psi + \mu_1^2 \Psi_{zz}) = 0 & \text{for } -\dfrac{1}{2} < z < \dfrac{1}{2}, \\ \Psi_z = 0 & \text{for } z = \pm\dfrac{1}{2}, \end{cases}$$

and of $\mathcal{E}$

$$(38) \quad \begin{cases} \nabla^2 \Psi + \mu_1^2 \Psi_{zz} = 0 & \text{for } -\dfrac{1}{2} < z < \dfrac{1}{2}. \\ \left(\dfrac{\partial}{\partial t} + \dfrac{\partial(\Psi,\ )}{\partial(x,y)}\right)\Psi_z = 0 & \text{for } z = \pm\dfrac{1}{2}. \end{cases}$$

The internal problem has been studied originally by CHARNEY [1] from a phenomenological standpoint, then by HERRING [42], who expanded a closure model around three-dimensional isotropy. Here we shall rather follow the approach of Salmon [43, 44] and Hoyer and Sadourny [45], who took the opposite approach of expanding around barotropy: if we limit ourselves to the first two vertical modes with horizontal eigenmodes as in subsect. 2˙1, the streamfunction can be expanded into the alternative forms

$$\Psi = \begin{cases} \sum_k (\Psi^0 + \Psi^1 \sqrt{2} \sin \pi z) \Lambda_k(x,y) & (\mathfrak{I}), \\ \sum_k \left(\Psi^0 \cosh \frac{\lambda_k z}{\mu_1} + \Psi^1 \sinh \frac{\lambda_k z}{\mu_1}\right) \Lambda_k(x,y) & (\mathfrak{E}). \end{cases} \qquad (39)$$

From now on we shall refer to modes 0 and 1 as to (pseudo)barotropic and baroclinic modes, respectively.

The two problems can be reduced to a single formalism by making the change of variables

$$(40) \qquad \tilde{\lambda}_k = \begin{cases} \lambda_k/\pi\mu_1, \\ \lambda_k/2\mu_1, \end{cases} \qquad \psi = \begin{cases} \mu_1^2 \Psi/\pi & (\mathfrak{I}), \\ \mu_1^2 \Psi/4 & (\mathfrak{E}). \end{cases}$$

If we set

$$(41) \qquad \lambda_k^0 = \begin{cases} \tilde{\lambda}_k, \\ (\tilde{\lambda}_k \operatorname{tgh} \tilde{\lambda}_k)^{1/2}, \end{cases} \qquad \lambda_k^1 = \begin{cases} (\tilde{\lambda}_k^2 + 1)^{1/2} & (\mathfrak{I}), \\ (\tilde{\lambda}_k/\operatorname{tgh} \tilde{\lambda})^{1/2} & (\mathfrak{E}), \end{cases}$$

(37), (38) both read

$$(42) \qquad \begin{cases} d\eta_k^0/dt = \sum_{p,q} A_{kpq}\left(\eta_p^0 \eta_q^0 (\lambda_q^0)^{-2} + \eta_p^1 \eta_q^1 (\lambda_q^1)^{-2}\right), \\ d\eta_k^1/dt = \sum_{p,q} A_{kpq}\left(\eta_p^1 \eta_q^0 (\lambda_q^0)^{-2} + \eta_p^0 \eta_q^1 (\lambda_q^1)^{-2}\right), \end{cases}$$

with the definition

$$(43) \qquad \eta_k^i = -(\lambda_k^i)^2 \psi_k^i,$$

which represent spectral coefficients of potential vorticity ($\mathfrak{I}$) or pressure fluctuations on boundaries ($\mathfrak{E}$). Both problems have three quadratic invariants (remember we have truncated to two vertical modes):

$$(44) \qquad \mathscr{E} = \sum_k \mathscr{E}_k, \qquad \mathscr{P} = \sum_k \mathscr{P}_k, \qquad \mathscr{Q} = \sum_k \mathscr{Q}_k,$$

with $\mathscr{E}_k = \mathscr{E}_k^0 + \mathscr{E}_k^1$, $\mathscr{P}_k = \mathscr{P}_k^0 + \mathscr{P}_k^1$, and

$$(45) \qquad \begin{cases} \mathscr{P}_k^i = \tfrac{1}{2}(\eta_k^i)^2 = (\lambda_k^i)^2 \mathscr{E}_k^i, \\ \mathscr{Q}_k = \eta_k^0 \eta_k^1; \end{cases}$$

$\mathscr{E}$ is total energy, $\mathscr{P}$ is total potential enstrophy ($\mathfrak{I}$) or available potential energy on both boundaries ($\mathfrak{E}$), $\mathscr{Q}$ is the correlation between the barotropic and baroclinic parts of the flow.

3`2. Statistical equilibria.** – Equations (42) generalize the two-layer baroclinic model whose statistical mechanics has been studied in [46]. Like in subsect. **2`2, we follow the motion of the image $\mathcal{M} = (\eta_k^i)$ in the $2N$-dimensional real phase space of the problem. $\mathcal{M}$ follows the intersection of three hypersurfaces $\mathcal{E} = \mathcal{E}_0$, $\mathcal{P} = \mathcal{P}_0$, $\mathcal{Q} = \mathcal{Q}_0$; a *mixing* hypothesis, *i.e.* the assumption of a finite correlation time for second-order moments, leads to almost sure convergence of time averages to microcanonical expectations. These in turn are approximated for large enough Ns by using Boltzman's probability distribution

$$(46) \qquad \mathfrak{F}(\eta_k^i) = \frac{\sqrt{d_k}}{2\pi} \exp\left[-\frac{1}{2}\left(a_k^0(\eta_k^0)^2 + a_k^1(\eta_k^1)^2 + 2c\eta_k^0\eta_k^1\right)\right],$$

with the definitions

$$(47) \qquad a_k^i = a(\lambda_k^i)^{-2} + b, \qquad d_k = a_k^1 a_k^0 - c^2,$$

which depend on three parameters a, b, c, corresponding to specified values of $\mathcal{E}_0$, $\mathcal{P}_0$, $\mathcal{Q}_0$. The corresponding expectations of modal variances read

$$(48) \qquad \begin{cases} \overline{\mathcal{P}_k^0} = a_k^1/d_k, & \overline{\mathcal{P}_k^1} = a_k^0/d_k, \\ \overline{\mathcal{Q}_k} = 2c/d_k. \end{cases}$$

The equilibrium spectra (48) can be studied along the same lines as already done in subsect. **2`2**. Without going to comparable detail, we shall consider only the simple case in which $\mathcal{Q}_0 = 0$ (the statistics are then invariant in a change of orientation of the vertical axis). Then $c = 0$ and the energy distribution is formally equivalent to (13). It is then easy to show that energy mostly accumulates in the larger barotropic scales (small λs), with potential enstrophy mostly equipartitioned amongst all scales smaller than the radius of deformation. A complete discussion of (47), (48) would be of interest in giving some insight on the role of the invariant $\mathcal{Q}$. This role—or, more generally, the property that potential enstrophy is actually conserved at every level and not only as a three-dimensional integral—has been systematically overlooked in all the theory of quasi-geostrophic turbulence, because it is not easily described in terms of vertical spectral expansions.

3`3. Turbulence with large-scale thermal forcing. – We consider now an inviscid fluid on the infinite plane, forced by stationary, homogeneous, isotropic large-scale thermal sources ($\lambda_F = 1$), which, therefore, inject $\mathcal{E}$ and $\mathcal{P}$ into the fluid at the same rate ε. For simplicity we further assume that no correlation is injected into the fluid: in other words, the thermal sources are invariant in a reversal of the vertical co-ordinate axis. Then they will remain identically zero, and the evolution of the fluid will be governed by the interplay of the two invariants $\mathscr{E}$, $\mathscr{P}$ only, as in the barotropic case. The functions $\lambda^0(k)$, $\lambda^1(k)$, which correspond, in the present case of an infinite domain, to our former discrete set $\{\lambda_k^0, \lambda_k^1\}$, are displayed in fig. 4. There we see that baroclinic and

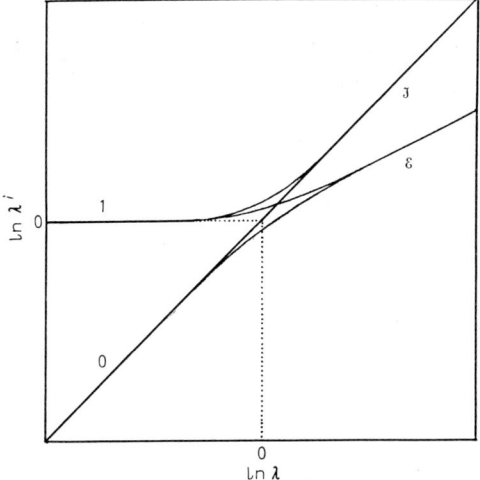

Fig. 4. – Dependence of eigenvalues λ^i on the normalized horizontal wave number λ for problems $\mathfrak{J}$ and $\mathscr{E}$.

barotropic modes differ only at large scales, while problems $\mathfrak{J}$ and $\mathscr{E}$ differ only at smaller scales. Like in subsect. 2`4, we denote by $E(k,t)$, $E^0(k,t)$, $E^1(k,t)$ the expectations of *one-dimensional* energy density at the normalized wave number k and time t, and we define the energy and enstrophy fluxes

(49)
$$\left| \begin{array}{l} \Pi_E^0(\lambda, t) = \displaystyle\int_{\lambda^0 < \lambda} \frac{\partial E^0}{\partial t}(p, t)\,\mathrm{d}p\,, \\[2pt] \Pi_P^0(\lambda, t) = \displaystyle\int_{\lambda^0 < \lambda} (\lambda^0(p))^2 \frac{\partial E^0}{\partial t}(p, t)\,\mathrm{d}p\,, \\[2pt] \Pi_E^1(\lambda, t) = \displaystyle\int_{\lambda^1 < \lambda} \frac{\partial E^1}{\partial t}(p, t)\,\mathrm{d}p\,, \\[2pt] \Pi_P^1(\lambda, t) = \displaystyle\int_{\lambda^1 < \lambda} (\lambda^1(p))^2 \frac{\partial E^0}{\partial t}(p, t)\,\mathrm{d}p\,, \\[2pt] \Pi_E(\lambda, t) = \displaystyle\int_{\lambda^0 > \lambda} \frac{\partial E^0}{\partial t}(p, t)\,\mathrm{d}p + \int_{\lambda^1 > \lambda} \frac{\partial E^1}{\partial t}(p, t)\,\mathrm{d}p\,, \\[2pt] \Pi_P(\lambda, t) = \displaystyle\int_{\lambda^0 > \lambda} (\lambda^0(p))^2 \frac{\partial E^0}{\partial t}(p, t)\,\mathrm{d}p + \int_{\lambda^1 > \lambda} (\lambda^1(p))^2 \frac{\partial E^1}{\partial t}(p, t)\,\mathrm{d}p\,, \end{array} \right\} \begin{array}{l} \lambda < 1; \\[80pt] \lambda > 1. \end{array}$$

This yields

(50) $$\begin{cases} \Pi_P^0(\lambda, t) < \lambda^2 \Pi_E^0(\lambda, t), \quad \Pi_P^1(\lambda, t) = \Pi_E^1(\lambda, t) = 0 & \text{for } \lambda < 1, \\ \Pi_E(\lambda, t) < \lambda^{-2} \Pi_P(\lambda, t) & \text{for } \lambda < 1. \end{cases}$$

If we now admit (like in subsect. 2´4) that the turbulence evolves towards a pseudostationary regime—stationary within a steadily increasing wave number band $(\lambda_E(t), \lambda_E(t), 0 \leftarrow \lambda_E(t) < 1 < \lambda_P(t) \to \infty)$—, we get from (50)

(51) $$\begin{cases} \Pi_E^0(\lambda, t) = \Pi_E^0(\lambda_E(t), t) > \varepsilon(1 - \lambda_P^{-2}(t)), \\ \Pi_E^1(\lambda, t) = \Pi_E^1(\lambda_E(t), t) = 0, \\ \Pi_P^0(\lambda, t) = \Pi_P^0(\lambda_E(t), t) < \varepsilon \lambda_E^2(t), \\ \Pi_P^1(\lambda, t) = \Pi_P^1(\lambda_E(t), t) = 0, \end{cases} \quad \text{for } \lambda_E(t) < \lambda < 1; \\ \begin{cases} \Pi_E(\lambda, t) = \Pi_E(\lambda_P(t), t) < \varepsilon \lambda_P^{-2}(t), \\ \Pi_P(\lambda, t) = \Pi_P(\lambda_P(t), t) > \varepsilon(1 - \lambda_E^2(t)), \end{cases} \quad \text{for } 1 < \lambda < \lambda_P(t).$$

As $t \to \infty$, (51) yields

(52) $$\begin{cases} \Pi_E^0(\lambda, t) \to \varepsilon, \quad \Pi_P^0(\lambda, t) \to 0 & \text{for } \lambda < 1, \\ \Pi_E(\lambda, t) \to 0, \quad \Pi_P(\lambda, t) \to \varepsilon & \text{for } \lambda > 1. \end{cases}$$

The asymptotic regime is, therefore, characterized by two inertial ranges: a *barotropic energy inertial range* (EIR) at scales larger than the internal radius of deformation, where energy cascades in barotropic (*i.e.* kinetic) form from the internal radius of deformation to ever increasing scales at a constant rate equal to the thermal injection rate ε, and a *potential enstrophy inertial range* (case J) or an *available potential energy inertial range* (case δ) (PIR) at scales smaller than the internal radius of deformation, where P cascades to ever decreasing scales at the same constant rate ε. In the EIR, the P flux vanishes, and the E flux vanishes in the PIR. Energy is thus trapped in the large scales; thermal energy must be converted into kinetic energy in the vicinity of the radius of deformation to feed the barotropic EIR. This argument demonstrates the *baroclinic instability* [47, 48] of thermally forced geophysical flows and interprets it as a process which keeps energy in the large scales while extracting potential enstrophy or available potential energy.

The phenomenological theory of the inertial ranges proceeds as before, although it is, of course, slightly more involved than for barotropic flows. The eddy turn-over time at wave number k is related to kinetic straining; therefore, it must be evaluated from, not potential enstrophy, but rather enstrophy:

(53) $$\tau(k) \sim \left[\int_0^k p^4 \left(\frac{E^0(p)}{\{\lambda^0(p)\}^2} + \frac{E^1(p)}{\{\lambda^1(p)\}^2} \right) dp \right]^{-1/2}.$$

In the EIR we have reasons to believe that $E^0(p) \gg E^1(p)$, since the reverse cascade is feeding the barotropic modes; and, furthermore, we have $\lambda^0(p) \ll \lambda^1(p)$, so that we can neglect baroclinic straining. In the ZIR, baroclinic and barotropic modes are *a priori* equivalent. We can thus restate (53) as

(54)
$$\left| \begin{array}{ll} \tau(k) \sim \left(\int_0^k p^2 E^0(p) \, dp \right)^{-1/2} & \text{for } k < 1, \\ \tau(k) \sim \left(\int_0^k p^4 \frac{E(p)}{(\lambda(p))^2} \, dp \right)^{-1/2} & \text{for } k > 1. \end{array} \right.$$

Cascade rates are

(55)
$$\left\{ \begin{array}{ll} \varepsilon \sim k E^0(k) / \tau(k) & \text{for } k < 1, \\ \varepsilon \sim k (\lambda(k))^2 E(k) / \tau(k) & \text{for } k > 1. \end{array} \right.$$

Combining (54) and (55), using the asymptotic behaviour of λ and solving for energy finally yields

(56)
$$\left| \begin{array}{lll} E^0(k) = C \varepsilon^{2/3} k^{-5/3} & & \text{for } k < 1, \\ E^i(k) = \frac{C}{2} \varepsilon^{2/3} k^{-3} (\log k)^{-1/3} & (\mathfrak{J}) & \\ E^i(k) = \frac{C}{2} \varepsilon^{2/3} k^{-8/3} & (\mathcal{E}) & \end{array} \right\} \text{for } k > 1;$$

note that the PIR of problem $\mathcal{E}$ can also be interpreted as an EIR for available potential energy.

If the scale of the thermal source is large enough, there is one more inertial range: the *baroclinic energy inertial range*, where thermal energy has to cascade from the injection scale to the radius of deformation, where it is converted to kinetic energy. Since the kinetic straining there is purely barotropic, this EIR is, in fact, a passive scalar inertial range:

(57)
$$E^1(k) = C \varepsilon^{2/3} k^{-5/3} \qquad \text{for } k < 1.$$

It has been shown in [49] that C^1 is probably a universal constant too for the 2-vertical-mode problem, with $C^1 \simeq C/30$. The establishment of (56)-(57) has been simulated in [45] using a statistical closure model formulation.

Like in subsect. 2`4, assuming self-similarity of the spectral tails yields the estimates

(58)
$$\left\{ \begin{array}{l} \dfrac{dk_E}{dt} \sim - \varepsilon / E^0(k_E(t), t), \\ \dfrac{dk_P}{dt} \sim \varepsilon / P^i(k_P(t), t). \end{array} \right.$$

Using (56), we evaluate again the times $t_E(k)$, $t_P(k)$ needed for the EIR and the PIR, respectively, to reach wave number k: for the EIR and the $\mathcal{E}$ PIR we get

(59) $$\begin{cases} t_E(k) \sim \varepsilon^{-1/3} k^{-2/3} & (k \to 0), \\ t_P(k) \sim \varepsilon^{-1/3} (\log k)^{2/3} & (k \to \infty) \; (\mathcal{E}), \end{cases}$$

exactly like in the barotropic case. The latter estimate means that internal quasi-geostrophic turbulence has infinite regularity and predictability times. On the other hand, the available potential-energy inertial range of external quasi-geostrophic turbulence yields a function $t_P(k)$ which remains bounded as $t \to \infty$. Therefore, it is likely that external quasi-geostrophic turbulence has finite regularity and predictability times (and, indeed, the atmosphere, for example, is observed to be less predictable near the ground and the tropopause than in-between). The apparition of singularities after a finite time in $\mathcal{E}$ is the quasi-geostrophic interpretation of *frontogenesis*. Although the quasi-geostrophic approximation is unable to physically describe fronts [50, 51], it still seems to provide a reasonably accurate description of frontogenesis statistics. ANDREWS and HOSKINS [52], using the *semi-geostrophic* approximation well adapted to describing fronts, have numerically simulated the establishment of an $\mathcal{E}$ PIR exactly similar to the quasi-geostrophic inertial range we have described.

APPENDIX

The quasi-geostrophic approximation.

The easiest way for deriving the quasi-geostrophic approximation is to start from the primitive equations using a function of entropy (say σ) as vertical co-ordinate; then the vertical velocity vanishes: $D\sigma/Dt = 0$, for adiabatic motion. The specific volume α is a function of σ and pressure p only (*), whose indefinite integral with respect to p will be denoted by $H(\sigma, p)$. For the two partial derivatives of H, we shall use the notations H_σ, H_p. Then the equation of motion

$$\frac{DV}{Dt} + fN \times V + \operatorname{grad} G + \alpha \operatorname{grad} p = 0,$$

where V refers to horizontal velocity, f to Coriolis parameter, N to the vertical unit vector and G to geopotential, can be rewritten as

(A.1) $$\frac{DV}{Dt} + fN \times V + \operatorname{grad} S = 0,$$

(*) The effect of moisture in the atmospheric case—or salinity in the oceanic case—is thereby excluded.

where $S = G + H$. For the atmosphere H is enthalpy, S is dry static energy. Elementary manipulations show that the hydrostatic equation

$$\frac{\partial G}{\partial \sigma} + a \frac{\partial p}{\partial \sigma} = 0$$

becomes

(A.2) $$\frac{\partial S}{\partial \sigma} = H_\sigma .$$

The pseudodensity in σ co-ordinate being denoted by $\imath = -\partial p/\partial \sigma$, the continuity equation reads

$$\frac{\partial \imath}{\partial t} + \operatorname{div}(\imath V) = 0 ,$$

which, combined to the curl of (A.1), yields the hydrostatic form of Ertel's potential vorticity theorem:

(A.3) $$\frac{DZ}{Dt} = 0 ,$$

where $Z = (f + \operatorname{rot} V)/\imath$ is the hydrostatic potential absolute vorticity.

We regard the state of the fluid as a perturbation around a basic stratified state of rest, defined, for instance, by a function $\overline{p}(\sigma)$. From this specification we can infer $\overline{a}(\sigma)$, $\overline{H}(\sigma)$, $\overline{S}(\sigma)$, etc. We shall also denote by $\overline{f}$ the average value of f over the domain of fluid we consider. All perturbations will be referred to with primes. We shall consider large-scale flows such that:

i) *The Lagrangian time scale associated to the material derivative* D/Dt *is large compared to the mean inertial time scale* $\overline{f}^{-1}$. This means, firstly, that the solutions of (A.1)-(A.3) we consider belong to the subspace of slow eigensolutions of the linearized equations (mainly, we exclude gravitational modes), and, secondly, that the nonlinear advective terms are small compared to $\overline{f}$, or, in other words, that the Rossby number is small. Then, to first order, (A.1) reduces to geostrophic equilibrium:

(A.4) $$f N \times V + \operatorname{grad} S = 0 .$$

ii) *The relative variation of f remains small over the flow domain.* By this condition we exclude planetary-scale flows as well as equatorial dynamics—for a detailed discussion of this point, see [53]. Then, to first order, we may replace f by $\overline{f}$ in (A.4): the velocity divergence vanishes and the motion is defined in terms of a streamfunction $\psi = S'/\overline{f}$:

(A.5) $$V = N \times \operatorname{grad} \psi .$$

iii) *The fluctuations of the pseudodensity are small compared to its mean value,* say

$$|\imath'|/\overline{\imath} = O(\mathrm{Ro}) ,$$

where Ro is the Rossby number (obviously, we need consider stable stratifications only: $\bar{\imath} > 0$). This requirement is equivalent to the classical condition $(\text{Ro Ri})^{-1} = O(\text{Ro})$, where Ri refers to Richardson number. It allows an expansion of Ertel's potential vorticity to first order in Ro:

$$Z = (f + \text{rot } V - \bar{f}\imath'/\bar{\imath})/\bar{\imath},$$

which may be called the *quasi-geostrophic potential absolute vorticity* (QGPAV). We may express Z in terms of the streamfunction only. Firstly, from (A.5), vorticity reduces to the horizontal Laplacian $\nabla^2 \psi$; further, we get from the hydrostatic equation (A.2)

(A.6) $$\frac{\partial S'}{\partial \sigma} = H'_\sigma = \bar{H}_{\sigma\mu}\mu',$$

since perturbations are taken at constant σ. Consequently,

$$\imath' = -\frac{\partial \mu'}{\partial \sigma}$$

can be rewritten

$$\imath' = -\frac{\partial}{\partial \sigma}\left(\frac{\partial S'}{\partial \sigma}\bigg/\bar{H}_{\sigma\mu}\right).$$

Finally, the Lagrangian derivative in (A.3) can be expressed in Jacobian form with respect to co-ordinates x (longitude times the radius of the Earth) and y (sine of latitude times the radius of the Earth). The result is the quasi-geostrophic potential absolute vorticity equation

(A.7) $$\left(\frac{\partial}{\partial t} + \frac{\partial(\Psi, \cdot)}{\partial(x, y)}\right)\left(f + \nabla^2\psi + \frac{\bar{f}^2}{\bar{\imath}}\frac{\partial}{\partial \sigma}\left(\frac{\partial \psi}{\partial \sigma}\bigg/\bar{H}_{\sigma\mu}\right)\right) = 0.$$

Note that the QGPAV is an exact invariant of the motion, contrary to the usual *pseudopotential vorticity* derived in height or pressure co-ordinate (although the two concepts are indeed equivalent in practice to the order of our approximations). This is a slight improvement due to the σ-co-ordinate formulation.

Let us now turn our attention towards the upper and lower boundaries of the flow domain; in the σ-co-ordinate frame, these boundaries are moving surfaces. We specify the upper boundary as a *free surface* subject to a given pressure $P(x, y)$ and the lower boundary as a *rigid surface* subject to a given geopotential $G(x, y)$. This choice yields

(A.8) $$\frac{D}{Dt}(G - \mathbf{G}) = 0 \qquad \text{at } G = \mathbf{G}(x, y),$$

(A.9) $$\frac{D}{Dt}(\mu - \mathbf{P}) = 0 \qquad \text{at } \mu = \mathbf{P}(x, y).$$

From (A.6), the pressure perturbation can be written

(A.10) $$\mu' = \bar{\imath}\frac{\partial \psi}{\partial \sigma}\bigg/\bar{H}_{\sigma\mu}.$$

Similarly, we write the perturbation of geopotential

(A.11)
$$G' = S' - \bar{H}_{\not{p}}\not{p}' = \bar{f}\left(1 - \bar{H}_{\not{p}}/\bar{H}_{\not{sp}}\frac{\partial}{\partial \not{s}}\right)\psi.$$

Developing **G** and **P** in terms of mean and fluctuation, $\mathbf{G} = \bar{\mathbf{G}} + \mathbf{G}'(x, y)$, $\mathbf{P} = \bar{\mathbf{P}} + \mathbf{P}'(x, y)$, and remembering that $D\bar{G}/Dt = D\bar{\not{p}}/Dt = 0$ since $\bar{G}$ and $\bar{\not{p}}$ are functions of $\not{s}$ only, we can rewrite (A.8), (A.9) in terms of perturbations:

(A.12)
$$\frac{D}{Dt}(G' - \mathbf{G}') = 0 \qquad \text{at } G = \mathbf{G}(x, y),$$

(A.13)
$$\frac{D}{Dt}(\not{p}' - \mathbf{P}') = 0 \qquad \text{at } \not{p} = \mathbf{P}(x, y).$$

If the order of magnitude of **G'** and **P'** is small compared to the vertical scale of the motion, (A.12), (A.13) can be approximated along isentropic boundaries $\bar{G} = \bar{\mathbf{G}}$, $\bar{\not{p}} = \bar{\mathbf{P}}$. Taking advantage of (A.10), (A.11), we then get the final form for the boundary conditions:

(A.14)
$$\left(\frac{\partial}{\partial t} + \frac{\partial(\psi, \ )}{\partial(x, y)}\right)\left(\bar{f}\left(1 - \bar{H}_{\not{p}}/\bar{H}_{\not{sp}}\frac{\partial}{\partial \not{s}}\right)\psi - \mathbf{G}'\right) = 0 \qquad \text{at } \bar{G} = \bar{\mathbf{G}},$$

(A.15)
$$\left(\frac{\partial}{\partial t} + \frac{\partial(\psi, \ )}{\partial(x, y)}\right)\left(\bar{f}\frac{\partial \psi}{\partial \not{s}}\Big/\bar{H}_{\not{sp}} - \mathbf{P}'\right) = 0 \qquad \text{at } \bar{\not{p}} = \bar{\mathbf{P}}.$$

In these last approximations, we have assumed

iv) $\dfrac{\not{s}' \partial \not{p}}{\not{p} \partial \not{s}} \ll 1$ at $\bar{G} = \bar{\mathbf{G}}$, \qquad v) $\dfrac{\not{s}' \partial G}{\bar{G} \partial \not{s}} \ll 1$ at $\bar{\not{p}} = \bar{\mathbf{P}}$,

$\not{s}'$ referring to the perturbation of boundary locations.

Multiplying (A.7) by $-\psi$ and integrating by parts, then using boundary conditions (A.14), (A.15) yield the *energy* invariant

(A.16)
$$\begin{cases} \dfrac{D\mathscr{E}}{Dt} = 0, \\[2mm] \mathscr{E} = \dfrac{1}{2}\iiint\left(\imath(\operatorname{grad}\psi)^2 + \dfrac{\bar{f}^2}{\bar{H}_{\not{sp}}}\left(\dfrac{\partial \psi}{\partial \not{s}}\right)^2\right)dx\,dy\,d\not{s} + \dfrac{1}{2}\iint_{\bar{G}=\bar{\mathbf{G}}}\dfrac{\bar{f}^2}{\bar{H}_{\not{p}}}\psi^2\,dx\,dy. \end{cases}$$

The total energy $\mathscr{E}$ is the sum of kinetic energy (the first term in the volume integral) and potential energy (the second term in the volume intergal plus a rigid boundary contribution). Another invariant is the *potential enstrophy*:

(A.17)
$$\begin{cases} \dfrac{D\mathscr{L}(\not{s})}{Dt} = 0, \\[2mm] \mathscr{L}(\not{s}) = \dfrac{1}{2}\iint\left(f + \nabla^2\psi + \dfrac{\bar{f}^2}{\imath}\dfrac{\partial}{\partial \not{s}}\left(\dfrac{\partial \psi}{\partial \not{s}}\Big/\bar{H}_{\not{sp}}\right)\right)^2 dx\,dy. \end{cases}$$

Finally, boundary conditions (A.14), (A.15) yield two further invariants, the *available potential energies* on top and bottom boundaries

(A.18) $$\mathscr{P} = \frac{1}{2} \iint \left(\bar{\imath} \frac{\partial \psi}{\partial s} \Big/ \bar{H}_{s/\mu} - \mathbf{P}' \right)^2 dx\,dy \qquad \text{at } \bar{\mu} = \bar{\mathbf{P}},$$

(A.19) $$\mathscr{G} = \frac{1}{2} \iint \left(\bar{\imath} \left(1 - \frac{\bar{H}_\mu}{\bar{H}_{s/\mu}} \frac{\partial}{\partial s} \right) \psi - \mathbf{G}' \right)^2 dx\,dy \qquad \text{at } \bar{G} = \bar{\mathbf{G}}.$$

The situation is much simpler if we consider the motion of a barotropic fluid. A barotropic fluid has only one independent variable of state, say s. Then $\imath$ is a function of s and the continuity equation reduces to the non-divergence of the velocity field over isentropic surfaces. Velocity is then defined from a streamfunction ψ, and the curl of (A.1) yields the *barotropic vorticity equation*

(A.20) $$\left(\frac{\partial}{\partial t} + \frac{\partial(\psi,\ )}{\partial(x,y)} \right) (f + \nabla^2 \psi) = 0.$$

Note that, in a barotropic fluid, all isentropic layers are decoupled. Now, instead of considering pure barotropic motion, we may look for a two-dimensional approximation of (A.7), (A.14), (A.15), based on vertical averaging with respect to $\bar{\imath}$. This yields

(A.21) $$\left(\frac{\partial}{\partial t} + \frac{\partial(\bar{\Psi},\ )}{\partial(x,y)} \right) (f + h(x,y) + (\nabla^2 - \mu_0^2) \bar{\Psi}) = 0,$$

where $\mu_0^2 = \bar{f}^2 / \bar{H}_\mu \bar{P}$ and $h(x,y) = \bar{\imath} (\mathbf{P} + \mathbf{G}/\bar{H}_\mu) / \bar{P}$. Here $\bar{\Psi}$ is the vertical average of ψ with respect to $\bar{\imath}$, $\bar{P}$ is the vertical pressure scale or the integral of $\bar{\imath}$ over the fluid depth, μ_0^{-1} is the external radius of deformation associated with the presence of a rigid bottom boundary, and $h(x,y)$ is the inhomogeneous term induced by the imposed bottom orography and top pressure. We may refer to (A.21) as to the *pseudobarotropic approximation* of the quasi-geostropic equations. The corresponding invariants are the pseudobarotropic energy

(A.22) $$\mathscr{E} = \frac{1}{2} \iint ((\operatorname{grad} \bar{\Psi})^2 + \mu_0^2 \bar{\Psi}^2)\,dx\,dy$$

and the pseudobarotropic potential enstrophy

(A.23) $$\mathscr{L} = \frac{1}{2} \iint (f + h(x,y) + (\nabla^2 - \mu_0^2) \bar{\Psi})^2 dx\,dy.$$

Other simplified cases of (A.7), (A.14), (A.15) are useful to consider. For example, assuming that the top and bottom boundaries are isentropic surfaces makes boundary dynamics trivial; in that case (A.14), (A.15) are replaced by

(A.24) $$\bar{\imath} \left(1 - \bar{H}_\mu / \bar{H}_{s/\mu} \frac{\partial}{\partial s} \right) \psi - \mathbf{G}' = 0 \qquad \text{at } \bar{G} = \bar{\mathbf{G}},$$

(A.25) $$\bar{\imath} \frac{\partial \psi}{\partial s} \Big/ \bar{H}_{s/\mu} - \mathbf{P}' = 0 \qquad \text{at } \bar{\mu} = \bar{\mathbf{P}}.$$

We shall refer to (A.7), (A.24), (A.25) as to the *internal* problem (ℑ). On the contrary, we may assume vanishing potential vorticity

$$(A.26) \qquad f + \nabla^2 \psi + \frac{\bar{f}^2}{\bar{i}} \frac{\partial}{\partial \mathfrak{z}} \left(\frac{\partial \psi}{\partial \mathfrak{z}} \Big/ \bar{H}_{\mathfrak{z}/\mathfrak{z}} \right) = 0$$

inside the domain; then internal dynamics become trivial and we are left with a problem driven by boundary dynamics only—although, of course, the top and bottom boundaries remain coupled through (A.26). We shall refer to (A.14), (A.15), (A.26) as to the *external* problem ($\mathcal{E}$). The invariants of ℑ are $\mathscr{E}$ and $\mathscr{Z}(\mathfrak{z})$ already stated in (A.16), (A.17). The invariants of $\mathcal{E}$ are $\mathscr{E}$, $\mathscr{P}$ and $\mathscr{G}$ (A.16), (A.18), (A.19).

REFERENCES

[1] J. G. Charney: *J. Atmos. Sci.*, **28**, 1087 (1971).
[2] J. G. Charney: *Geofys. Publ.*, **17**, 1 (1948).
[3] P. B. Rhines: *Annu. Rev. Fluid Mech.*, **11**, 401 (1979).
[4] P. B. Rhines: *J. Fluid Mech.*, **69**, 417 (1975).
[5] G. Holloway and M. C. Hendershott: *J. Fluid Mech.*, **82**, 747 (1977).
[6] G. Holloway: *Geophys. Astrophys. Fluid Dyn.*, **11**, 271 (1979).
[7] F. P. Bretherton and D. Haidvogel: *J. Fluid Mech.*, **78**, 129 (1976).
[8] G. Holloway: *J. Phys. Oceanogr.*, **8**, 414 (1978).
[9] J. R. Herring: *J. Atmos. Sci.*, **32**, 2254 (1977).
[10] R. Fjørtoft: *Tellus* **5**, 225 (1953).
[11] A. I. Khinchin: *Mathematical Foundations of Statistical Mechanics* (New York, N.Y., 1949).
[12] C. Basdevant and R. Sadourny: *J. Fluid Mech.*, **69**, 673 (1975).
[13] R. H. Kraichnan: *Phys. Fluids*, **10**, 1417 (1967).
[14] J. S. Frederiksen and B. L. Sawford: *J. Atmos. Sci.*, **37**, 717 (1980).
[15] D. G. Fox and S. A. Orszag: *Phys. Fluids*, **16**, 169 (1973).
[16] I. Cook: *J. Plasma Phys.*, **12**, 501 (1974).
[17] G. F. Carnevale: *J. Fluid Mech.*, **122**, 143 (1982).
[18] V. I. Arnold: *Dokl. Akad. Nauk SSSR*, **162**, 975 (1965).
[19] R. Benzi, S. Pierini, A. Vulpiani and E. Salusti: *Geophys. Astrophys. Fluid Dyn.*, **20**, 293 (1982).
[20] D. G. Andrews: *J. Atmos. Sci.*, **40**, 85 (1983).
[21] R. H. Kraichnan: *J. Fluid Mech.*, **47**, 525 (1971).
[22] A. N. Kolmogorov: *C. R. Acad. Sci. URSS*, **30**, 301 (1941).
[23] A. Pouquet, M. Lesieur, J. C. Andre and C. Basdevant: *J. Fluid Mech.*, **72**, 305 (1975).
[24] J. Leray: *J. Math. Pures Appl.*, **12**, 1 (1933).
[25] W. Wolibner: *Math. Z.*, **37**, 727 (1933).
[26] T. Kato: *Arch. Ration. Mech. Anal.*, **25**, 302 (1967).
[27] C. E. Leith and R. H. Kraichnan: *J. Atmos. Sci.*, **29**, 1041 (1972).
[28] A. Wiin-Nielsen: *Tellus*, **19**, 540 (1967).
[29] M. Desbois: *J. Atmos. Sci.*, **32**, 1838 (1975).
[30] P. Morel and M. Larchevêque: *J. Atmos. Sci.*, **31**, 2189 (1974).

[31] K. S. GAGE and W. H. JASPERSON: *Mon. Weather Rev.*, **107**, 77 (1979).
[32] D. K. LILLY and E. L. PETERSEN: *Tellus A*, **35**, 379 (1983).
[33] E. J. HOPFINGER, K. BROWAND and Y. GAGNE: *J. Fluid Mech.*, **125**, 505 (1982).
[34] J. SOMMERIA and R. MOREAU: *J. Fluid Mech.*, **118**, 507 (1982).
[35] C. BASDEVANT, B. LEGRAS, R. SADOURNY and M. BELAND: *J. Atmos. Sci.*, **38**, 2305 (1981).
[36] C. BASDEVANT and R. SADOURNY: *J. Méc. Théor. Appl.*, Numéro Special, 243 (1983).
[37] A. N. KOLMOGOROV: *J. Fluid Mech.*, **13**, 82 (1962).
[38] E. A. NOVIKOV and R. W. STEWART: *Izv. Akad. Nauk SSSR, Ser. Geofyz.*, **3** (1964).
[39] U. FRISCH, P. L. SULEM and M. NELKIN: *J. Fluid Mech.*, **87**, 719 (1978).
[40] J. C. MCWILLIAMS: in *Predictability of Fluid Motions*, edited by G. HOLLOWAY and B. WEST (American Institute of Physics, New York, N. Y., 1984), p. 205.
[41] W. BLUMEN: *J. Atmos. Sci.*, **35**, 774 (1978).
[42] J. R. HERRING: *J. Atmos. Sci.*, **37**, 969 (1980).
[43] R. SALMON: *Geophys. Astrophys. Fluid Dyn.*, **10**, 25 (1978).
[44] R. SALMON: *Geophys. Astrophys. Fluid Dyn.*, **15**, 167 (1980).
[45] J. M. HOYER and R. SADOURNY: *J. Atmos. Sci.*, **39**, 707 (1982).
[46] R. SALMON, G. HOLLOWAY and M. C. HENDERSHOTT: *J. Fluid Mech.*, **75**, 691 (1976).
[47] J. G. CHARNEY: *J. Meteorol.*, **4**, 135 (1947).
[48] E. T. EADY: *Tellus*, **1**, 33 (1949).
[49] R. SADOURNY and J. M. HOYER: *J. Atmos. Sci.*, **39**, 2138 (1982).
[50] B. J. HOSKINS and F. P. BRETHERTON: *J. Atmos. Sci.*, **29**, 11 (1972).
[51] B. J. HOSKINS: *J. Atmos. Sci.*, **32**, 233 (1975).
[52] D. G. ANDREWS and B. J. HOSKINS: *J. Atmos. Sci.*, **35**, 509 (1978).
[53] J. G. CHARNEY and G. R. FLIERL: *Evolution of Physical Oceanography* (Cambridge, Mass., 1981), p. 504.

Thermal Convection in a Rotating Fluid Subject to a Horizontal Temperature Gradient.

R. HIDE

Geophysical Fluid Dynamics Laboratory, Meteorological Office (Met 0 21)
Bracknell, Berkshire RG12 2SZ, England, U. K.

1. – Introduction.

The problem of central interest to the participants in this summer course on turbulence and predictability in fluid dynamics and in climate dynamics is that of predicting the motions of the Earth's atmosphere from the laws of fluid dynamics and thermodynamics. Research towards a satisfactory solution to this problem must of necessity involve systematic quantitative studies of a hierarchy of fluid-dynamical systems, of which the atmosphere is but one, very complex, example. What is evidently required is a thorough understanding of thermal convection due to an impressed temperature gradient with horizontal components in a rotating fluid of low viscosity and thermal conductivity. Were not the formulation and analysis of mathematical or numerical models fraught with the severe technical difficulties encountered in most realistic theoretical studies in fluid dynamics, it would be unnecessary to consider alternatives to a direct mathematical approach. Fortunately and not entirely fortuitously, research on rapidly rotating fluids carried out by fluid dynamicists during the past thirty years includes laboratory experiments and related analytical and numerical work which are relevant not only to the theory of the global circulation of the Earth's atmosphere, but also to that of motions in the atmospheres of other planets such as Jupiter and Saturn which, in contrast to the Earth's atmosphere, exhibit certain highly predictable long-lived dynamical features. We should not be too disappointed that mathematics alone is far from sufficient. After all, the study of basic chemistry would not be very advanced if it had to rely on mathematical solutions of the equations of quantum mechanics!

Of these laboratory studies, the most relevant in the first instance are controlled and reproducible experiments on thermal convection due to axisymmetric differential heating and cooling of a fluid annulus that rotates steadily about its vertical axis of symmetry, when the impressed temperature field associated with the applied differential heating and cooling is independent of the vertical (axial) co-ordinate. The findings of these experiments concerning what has come to be called « sloping » or « slantwise » convection and other basic fluid dynamic processes provide a framework for the study of the predictability and other aspects of large-scale motions in planetary atmospheres.

2. – Geostrophy.

Before outlining the results of these experiments it is useful to consider certain general properties of the motion of a fluid of low viscosity that departs but little from solid-body rotation with steady angular velocity Ω when typical time scales of the relative motion greatly exceed $(2\Omega)^{-1}$. Such motion is « geostrophic » nearly everywhere, satisfying

$$(2.1) \qquad 2\varrho\Omega \times \boldsymbol{u} + \nabla p - \varrho \nabla V \doteq 0 \,.$$

Here $\boldsymbol{u}$ is the Eulerian relative flow velocity, ϱ denotes density, p pressure, and ∇V is the acceleration due to gravity and centripetal effects. Equation (2.1) is the leading approximation to the full equation of motion

$$(2.2) \qquad 2\varrho\Omega \times \boldsymbol{u} + \nabla p - \varrho \nabla V = \boldsymbol{A} \,,$$

where

$$(2.3) \qquad \boldsymbol{A} \equiv -\varrho \, \mathrm{D}\boldsymbol{u}/\mathrm{D}t + \varrho \boldsymbol{r} \times \mathrm{d}\Omega/\mathrm{d}t + \boldsymbol{F},$$

the « ageostrophic » contribution. It is valid in regions where the Coriolis term $2\varrho\Omega \times \boldsymbol{u}$ greatly exceeds the relative acceleration term

$$\varrho \, \mathrm{D}\boldsymbol{u}/\mathrm{D}t \equiv \varrho [\partial \boldsymbol{u}/\partial t + (\boldsymbol{u} \cdot \nabla)\boldsymbol{u}]$$

(where t denotes time), the « precessional » term $\varrho \boldsymbol{r} \times \mathrm{d}\Omega/\mathrm{d}t$ (where $\boldsymbol{r}$ is the position vector of a general point in a frame which rotates with angular velocity $\Omega = \Omega(t)$ relative to an inertial frame), and the term $\boldsymbol{F}$ which represents the viscous and other forces (*e.g.* Lorentz forces in magnetohydrodynamic systems) per unit volume.

Now eq. (2.1) is mathematically degenerate. Being of lower order than the full equation of motion, (2.2), it cannot be solved under the complete set of boundary conditions. For this to be possible it is necessary to include ageostrophic terms in the analysis. It follows that the flow cannot be geostrophic everywhere, so that:

(2.4) *Regions of highly ageostrophic flow occurring not only on the boundaries of the system but also in localized regions (detached shear layers, jet-streams, etc.) of the main body of the fluid are necessary concomitants of geostrophic motion.*

Within these highly ageostrophic regions, A is comparable in magnitude with $2\varrho\mathbf{\Omega}\times\mathbf{u}$ and the corresponding relative vorticity $\boldsymbol{\xi} \equiv \nabla\times\mathbf{u}$ can be comparable with or even exceed 2Ω in magnitude. Many examples of such vorticity concentrations are found in Nature and in laboratory systems, such as those considered below. They are often associated with steep gradients of temperature (« thermal fronts »), as in the case of jet-streams and western boundary currents found in atmospheres and oceans.

Equations (2.1) and (2.2) lead directly to another important finding, namely that:

(2.5) *The motion of a fluid of low viscosity that departs only slightly from steady rapid rigid-body rotation will not in general be symmetric about the rotation axis, even when the boundary conditions are axisymmetric.*

This simple result elucidates of the occurrence of large-scale nonaxisymmetric disturbances in the Earth's atmosphere and other natural systems and receives direct verification from the experiments outlined below. It can be deduced as follows. In cylindrical co-ordinates (r, φ, z) where $\mathbf{\Omega} = (0, 0, \Omega)$, the φ-component of eq. (2.2) gives

$$(2.6) \qquad \varrho u_r = (2\Omega)^{-1}\{-r^{-1}\partial p/\partial\varphi + A_\varphi\}$$

if $\mathbf{u} = (u_r, u_\varphi, u_z)$ and $\mathbf{A} = (A_r, A_\varphi, A_z)$, since $\partial V/\partial\varphi = 0$ by the assumption of axial symmetry in the boundary conditions. Now, over any cylindrical surface of radius r the rate of advective transport $M(r, t; Q)$ of any quantity Q (per unit mass), such as heat, angular momentum, etc., is given by

$$(2.7) \qquad M(r, t; Q) \equiv \int_{z_1}^{z_2}\int_0^{2\pi} \varrho r u_r Q \, d\varphi \, dz = \frac{1}{2\Omega}\int_{z_1}^{z_2}\int_0^{2\pi}\left\{-\frac{\partial p}{\partial\varphi} + rA_\varphi\right\} Q \, d\varphi \, dz.$$

Since the average contribution A_φ to eq. (2.6) decreases rapidly with increasing Ω, advective transport perpendicular to the axis of rotation, as measured by $M(r, t; Q)$, will be negligible unless the flow pattern departs significantly from the axis of symmetry. In the axisymmetric case we have $\partial p/\partial\varphi = 0$, and $M(r, t; Q)$ is consequently of the order of the small ageostrophic contribution.

This argument is the basis of (2.5). There may be singular cases when the flow remains axisymmetric and in consequence any advective transport perpendicular to the rotation axis is negligible. Indeed, such cases can be realized in the laboratory by taking certain special precautions, but the general conclusion to be drawn from the laboratory studies outlined below is that (2.5) is a correct inference from the geostrophic equation.

3. – Regimes of thermal convection in a rotating fluid annulus.

The earliest laboratory experiments on thermal convection in a rotating fluid cylindrical annulus which rotates with angular speed Ω about its vertical axis of symmetry and is subject to axisymmetric applied differential heating showed that the general character of the flow is largely determined by two dimensionless parameters, namely

(3.1) $$\Theta \equiv gd\,\Delta\varrho/\bar{\varrho}\Omega^2(b-a)^2$$

and

(3.2) $$\mathcal{T} \equiv 4\Omega^2 L^4/\nu^2$$

(see fig. 1). Here g denotes the acceleration of gravity, d the fluid depth, b and a are the radii of the outer and inner side-walls, $\bar{\varrho}$ is the mean density and ν the kinematic viscosity of the fluid, L depends on b, a and d and has the dimensions of length (and is equal to $(b-a)^5/d$ over a wide range of conditions), and $\Delta\varrho$ is a measure of the impressed horizontal temperature contrast associated with the applied differential heating and cooling, which can be taken as the difference between the maximum and the minimum density associated with the impressed temperature field $T_0(r)$.

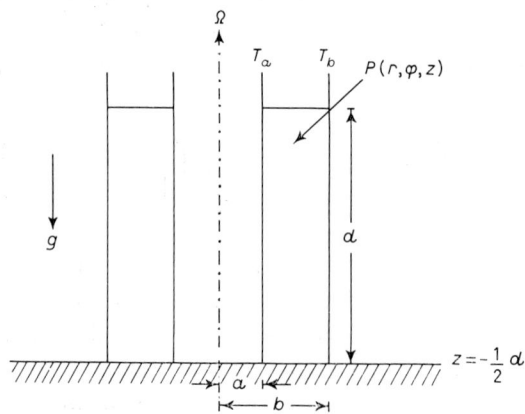

Fig. 1. – Schematic diagram of a rotating fluid annulus subject to a horizontal temperature gradient.

The side-walls are held at temperatures T_b and T_a, respectively, and the fluid is subject to diabatic heating per unit volume at a rate $q\bar{\varrho}$ times the specific heat c. The radial variation of the impressed temperature $T_0(r)$ takes the simple form

(3.3) $\quad T_0(r) = (b-a)^{-1}[(r-a)T_b + (b-r)T_a] - q(r-a)(r-b)/2\varkappa$

(where $\varkappa$ is the thermometric conductivity of the fluid) when $b-a \ll \frac{1}{2}(b+a)$ and q is constant. When there are no heat sources or sinks within the fluid, we have the most extensively studied case of all, often referred to as the « wall heated » case. When $q = 0$, eq. (3.3) becomes

(3.4) $\quad T_0(r) = (b-a)^{-1}[(r-a)T_b + (b-r)T_a]$, « wall heated »,

and $\Delta\varrho$ is $|\varrho(T_b) - \varrho(T_a)|$.

When $q \neq 0$, we have « internally heated » systems (which for convenience here include cases when $q < 0$ corresponding to « internally cooled » systems). There are three particularly interesting limiting cases of internally heated systems, namely the « inner wall cooled » case when the outer wall is a thermal insulator, so that $dT_0/dr = 0$ at $r = b$ and eq. (3.3) becomes

(3.5) $\quad T_0 = T_a + q(r-a)(2b-a-r)/2\varkappa \quad$ {internal heating; inner wall cooled};

the « outer wall cooled » case, when the inner wall is a thermal insulator so that $dT_0/dr = 0$ at $r = a$ and eq. (3.3) becomes

(3.6) $\quad T_0 = T_a - q(r-a)^2/2\varkappa \quad$ {internal heating; outer wall cooled};

and the « both walls cooled » case, when heat is removed at the same rate via both outer and inner side-walls simultaneously, so that dT_0/dr at $r = a$ is equal to $-dT_0/dr$ at $r = b$ and eq. (3.3) becomes

(3.7) $\quad T_0 = T_a - q(r-a)(r-b)/2\varkappa \quad$ {internal heating; both walls cooled}.

In the first two of these internally heated cases, $\Delta\varrho$ is the same as in the wall-heated case, $|\varrho(T_b) - \varrho(T_a)|$, but in the « both walls cooled » case we have

$$\Delta\varrho \equiv |\varrho(T_0(r = \tfrac{1}{2}(a+b))) - \varrho(T_a)|.$$

Careful studies of the principal spatial and temporal characteristics of flows over a wide range of precisely specified and carefully controlled experimental conditions led to the discovery of several fundamentally different free types of flow, only one of which is symmetrical about the axis of rotation (see fig. 2 and 3). When $\mathscr{T}$ is less than a certain critical value (see fig. 4) of about $2 \cdot 10^5$, viscosity ensures that the motion is essentially axisymmetric for all values of Θ.

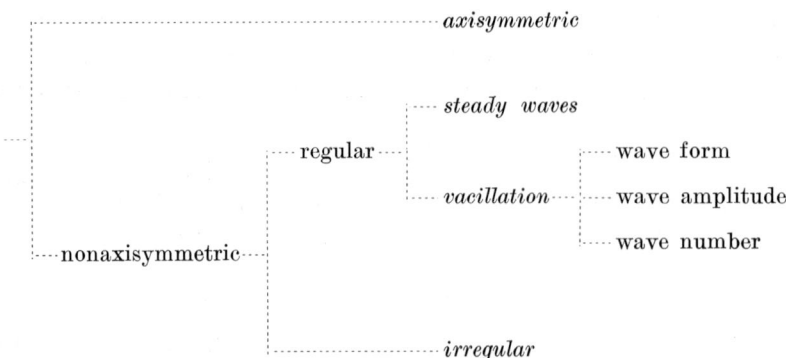

Fig. 2. – Broad classification of modes of free thermal convection in a rotating fluid annulus under axisymmetric boundary conditions.

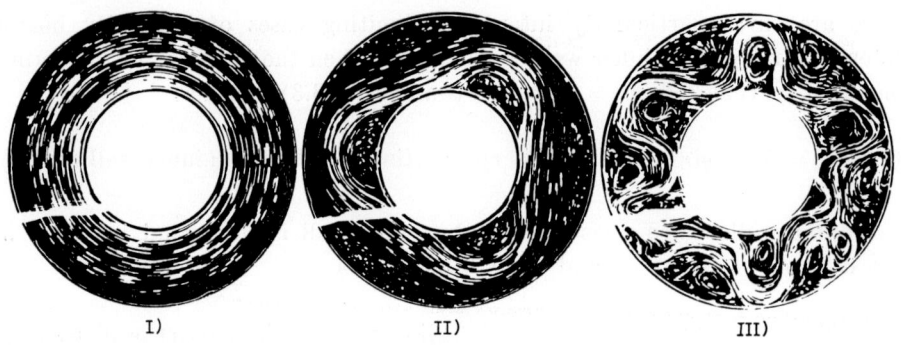

Fig. 3. – Streak photographs giving one example of each of the main modes of thermal convection in a rotating fluid annulus subject to a radial temperature field of the form given by eq. (3.4), namely I) axisymmetric flow ($\Omega = 0.341$ rad/s), II) regular (steady) nonaxisymmetric flow ($\Omega = 1.19$ rad/s) and III) irregular nonaxisymmetric flow ($\Omega = 5.02$ rad/s).

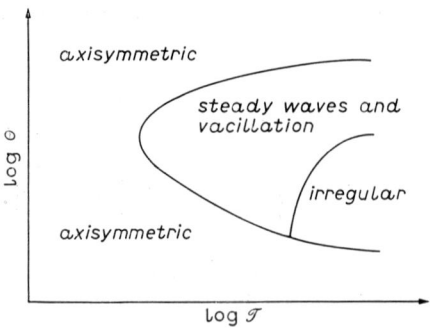

Fig. 4. – Schematic diagram illustrating the dependence of the mode of free thermal convection in a rotating fluid annulus under axisymmetric boundary conditions on the two principal dimensionless parameters found to specify the system, Θ and $\mathcal{T}$ (see eqs. (3.1) and (3.2)).

However, when $\mathcal{T}$ exceeds this critical value, there exists a range of Θ, namely $\Theta_R > \Theta > \Theta_L$ (where Θ_R and Θ_L depend on $\mathcal{T}$), within which highly nonaxisymmetric sloping convection occurs (see fig. 5). These nonaxisymmetric motions are either « regular » or « irregular » depending on the values of Θ

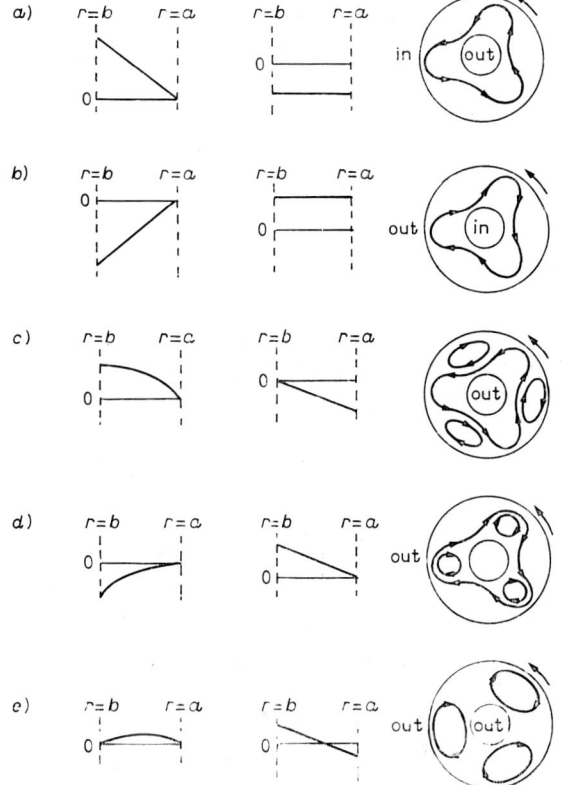

Fig. 5. – Schematic illustrations of the radial variation of the impressed temperature (left column), of the radial variation of the impressed radial temperature gradient (middle column) and of the corresponding upper-surface relative flow pattern in the regular regime of thermal convection in a rotating fluid annulus (right column), as predicted on the basis of eq. (4.8): wall heating: a) inner wall cooled, b) outer wall cooled; internal heating: c) inner wall cooled, d) outer wall cooled, e) both walls cooled.

and $\mathcal{T}$. The regular flows, which occur when $\Theta_R > \Theta > \Theta_I$, are spatially periodic and often exhibit periodic temporal « vacillation » in amplitude, shape or even wave number, but under certain conditions these periodic variations are so slight that, apart from a steady azimuthal drift of the flow pattern relative to the walls of the apparatus, the flow is virtually steady. In sharp contrast to this behaviour, irregular flows (which occur within the range $\Theta_I > \Theta > \Theta_L$) exhibit complicated a periodic fluctuation in both space and time. Axisymmetric flow occurs when $\Theta > \Theta_R$ or $\Theta < \Theta_L$.

An experiment in which all the impressed conditions are kept fixed except Ω, which is increased in steps from low values to high values, can be represented by a series of points on a straight line inclined at 45° to the Θ and $\mathcal{T}$ axes in the regime diagram of fig. 4, moving from the upper left part of the diagram to the lower right. The critical value Ω_R of Ω at which the so-called « upper symmetric » flow gives way to regular nonaxisymmetric flow on increasing the value of Ω corresponds to the point where $\Theta = \Theta_R(\mathcal{T})$; the critical value Ω_I at which regular nonaxisymmetric flow gives way to irregular axisymmetric flow corresponds to the point where $\Theta = \Theta_I(\mathcal{T})$; and the critical value Ω_L at which nonaxisymmetric flow gives way to axisymmetric flow on increasing the value of Ω corresponds to the point where $\Theta = \Theta_L(\mathcal{T})$.

4. – Patterns of regular nonaxisymmetric flow.

The mathematical equations governing the flows we are considering are the equation of motion, eq. (2.2), together with the equations of continuity and state for a liquid, namely

(4.1) $$\nabla \cdot \boldsymbol{u} = 0$$

and

(4.2) $$\varrho = \hat{\varrho}\,[1 - \alpha(T - \hat{T})]$$

(where T denotes temperature, α thermal coefficient of cubical expansion (for convenience here taken as constant), and $\hat{\varrho}$ is the density at the reference temperature $\hat{T}$), and the equation of heat transfer

(4.3) $$\partial T/\partial t + \boldsymbol{u}\cdot\nabla T = \varkappa \nabla^2 T + q\,,$$

where $\varkappa$ is the thermal diffusivity (equal to the thermal conductivity divided by the product of ϱ and the specific heat c and here assumed constant) and $q\varrho c$ is the rate of diabatic heating per unit volume. Equations (3.3) to (3.7) are, of course, solutions of eq. (4.3) when $\partial T/\partial t = \boldsymbol{u} = 0$. Across any cylindrical vertical surface $r = \text{const}$, the rate of heat transfer (made up of a conductive component and a convective (or « advective ») component when radiative effects are negligible) is given by

(4.4) $$H(r, t) = \int_{-\frac{1}{2}d}^{\frac{1}{2}d} \int_0^{2\pi} \varrho c \left[\varkappa \frac{\partial T}{\partial r} + u_r T\right] r\, d\varphi\, dz$$

if the fluid extends in the axial direction from $z = -\frac{1}{2}d$ to $z = \frac{1}{2}d$. As anticipated in sect. 2 (see eq. (2.7)), the geostrophic contribution to the advective heat flow term on the right-hand side of eq. (4.4) would vanish if the flow were axisymmetric, since the geostrophic part of u_r is proportional to $\partial p/\partial \varphi$. As we have already indicated in sect. **2**, this result points to the *raison d'être* of the nonaxisymmetric regimes of flow found when Ω is sufficiently large; geostrophic flow cannot convey heat perpendicularly to the axis of rotation unless it is nonaxisymmetric!

The boundary conditions on $\boldsymbol{u}$ under which the governing equations must be satisfied are that $\boldsymbol{u} = 0$ at a rigid bounding surface and that the stress should vanish at a free surface. The thermal boundary conditions require continuity of heat flow which, at a bounding surface, is purely conductive and proportional to $\varkappa \nabla T$. When the boundary conditions on the side-walls at $r = r^*$, where $r^* = a$ or b, are combined with the geostrophic relationship given by eq. (2.1) and used in conjunction with the standard relationship for the radial flow in the Ekman boundary layers on $z = -\frac{1}{2}d$ and $z = \frac{1}{2}d$ to evaluate the radial heat flow at $r = r^*$ (see eq. (4.4)), it is found that

$$(4.5) \quad H(r^*, t) \doteq -\left(\frac{\nu}{\Omega}\right)^{\frac{1}{2}} \left[\overline{T}\left(r^*, \frac{1}{2}d, t\right) - \overline{T}\left(r^*, -\frac{1}{2}d, t\right)\right] \cdot \left[\Gamma\left(r^*, \frac{1}{2}d, t\right) - \Gamma\left(r^*, -\frac{1}{2}d, t\right)\right],$$

where

$$(4.6) \quad \overline{T}(r^*, z, t) \equiv \frac{1}{2\pi} \int_0^{2\pi} T(r^*, \varphi, z, t)\, d\varphi$$

and

$$(4.7) \quad \Gamma(r^*, z, t) \equiv \int_0^{2\pi} U_\varphi(r^*, \varphi, z, t)\, r\, d\varphi,$$

$U_\varphi(r^*, \varphi, z, t)$ being the value of u_φ evaluated just outside the viscous boundary layer on $r = a$ or $r = b$, as the case may be. Now it may be shown that $\overline{T}(r^*, \frac{1}{2}d, t) > \overline{T}(r^*, -\frac{1}{2}d, t)$ (when $\alpha > 0$) and that either $\Gamma(r^*, \frac{1}{2}d, t) = -\Gamma(r^*, -\frac{1}{2}d, t)$ or $|\Gamma(r^*, \frac{1}{2}d, t)| \gg |\Gamma(r^*, -\frac{1}{2}d, t)|$ according as the upper surface is in contact with a rigid lid or is free. Hence

$$(4.8) \quad H(r^*, t) = \text{negative definite quantity} \times \Gamma(r^*, \tfrac{1}{2}d, t).$$

This relationship between the heat flow at a side-wall and the line integral of the tangential velocity near the side-wall embodies the arguments that have been used to provide a general interpretation of the upper-level flow pattern in the case when $q = 0$ everywhere (see eq. (4.4)), heat being introduced into the system *via* one of the side-walls and removed *via* the other side-wall. The corresponding impressed radial temperature gradient has the same sign at all values of r and the upper-level pattern of motion in the regular flow regime consists of a single jet-stream meandering in a wavy pattern between the bounding cylinders, with a positive (*i.e.* « westerly ») azimuthal component when heat enters *via* the outer side-wall and leaves *via* the inner side-wall (so that the

Fig. 6. – Streak photographs illustrating top-surface flow patterns of thermal convection in a rotating fluid annulus subject to internal heating in the axisymmetric regime (when $\Omega < \Omega_R$, see left column) and regular nonaxisymmetric regime (when $\Omega_R < \Omega < \Omega_I$, see right column): *a*) internally heated, inner wall cooled; *b*) internally heated, outer wall cooled; *c*) internally heated, both walls cooled. They show the dependence of the general characteristics of the flow pattern on the way heat is removed from the system and confirm predictions based in eq. (4.8). The cases *a*), *b*) and *c*) correspond to cases *c*), *d*) and *e*), respectively, of fig. 5. In the most striking case of all (see fig. 5*e*) and 6*c*)), sloping convection takes the form of closed anticyclonic eddies with the main motion concentrated in a jet-stream at the periphery of each eddy.

impressed radial temperature gradient is positive), and a negative one (« easterly ») when the overall radial heat transfer is in the opposite direction, from the inner to the outer cylinder (see fig. 5a), b)).

As a further test of eq. (4.8), experiments have been carried out using internal heating, so that the term q in eq. (4.3) is not equal to zero. This was done by passing an alternating electric current through the fluid. Heat could be removed *via* the inner side-wall, the outer side-wall, or both side-walls. The observed upper-surface flow patterns (see fig. 6) were found to be in good agreement with predictions for these three cases made on the basis of eq. (4.8) (see fig. 5c), d), e)). In the cases when heat is removed *via* one side-wall only, $H(r^*, t)$ vanishes at the other side-wall and, by eq. (4.8), the quantity $\Gamma(r^*, \frac{1}{2}d, t)$ must also vanish. For this to happen, $U_\varphi(r^*, \varphi, \frac{1}{2}d, t)$ will be positive at some values of φ and negative at others (or zero at all values of φ, as in the axisymmetric regime that occurs when $\Omega < \Omega_R$). Figure 5c), d) show how this requirement can be satisfied by adding closed eddies to the wavy pattern found in the cases when $q = 0$ (cf. fig. 5a), b)).

The most striking case of all studied in the experiments is the one illustrated by fig. 5e). Then heat is removed *via* both side-walls, implying that the impressed radial temperature gradient changes sign near mid-radius. The corresponding upper-level flow consists of several separate closed eddies, each circulating anticyclonically, in accordance with eq. (4.8), with the horizontal flow confined to a narrow jet-stream at the periphery of each eddy (see fig. 6c)).

5. – Atmospheric flows.

The meandering jet-streams within which the upper-level tropospheric air flow is mainly concentrated in the Earth's atmosphere are manifestations of sloping convection produced by differential solar heating, which maintains a systematic temperature contrast between tropical and polar regions in each hemisphere. These atmospheric motions are much less regular than those depicted in fig. 5a), c) and 2a) (cf. fig. 3), presumably because the Earth's angular speed of rotation exceeds the critical value Ω_I, although horizontal variations of surface conditions introduce complications which are not yet fully understood.

The atmosphere of the planet Jupiter is heated from below at about the same rate as its upper reaches are heated by solar radiation. Unlike the terrestrial case (where nonsolar atmospheric heating is utterly negligible), north-south temperature gradients in Jupiter's atmosphere change sign several times between equator and pole, and there is no evidence of any significant systematic temperature contrast between equatorial and polar regions. There are abundant observations of Jovian atmospheric motions at the upper cloud level, some of which go back many decades and even longer, and the Pioneer and Voyager space probes have added further details. Our knowledge of what goes on below the cloud level is meagre and this produces difficulties with the interpretation of observations of upper-level atmospheric motions. Indeed it has been argued that the main task of the « Jovian meteorologist » should perhaps be to use these observations to improve our knowledge of the vertical structure of the planet! But here is not the place to discuss these observations and review the many interesting though largely controversial issues being debated by those of us who take an interest in the interpretation of the observations in terms of basic dynamical processes. There is, however, one striking phenomenon upon which laboratory experiments on sloping convection in a fluid annulus subject to internal heating might have some bearing. The highly stable closed anticyclonic eddies with the main motion concentrated in a jet-stream at the periphery of each eddy that are depicted in fig. 5e) and 6c) are remarkably similar dynamically to the long-lived eddies to be seen in Jupiter's atmosphere in the southern hemisphere. The largest and most durable and conspicuous of these is the Great Red Spot in the South Tropical Zone, which may be at least three hundred years old. Next in size and age are the three White Ovals that formed in 1939 at the boundary between the South Temperate Belt and the South Temperate Zone, apparently as the residue of the highly irregular South Tropical Disturbance that was first seen in 1901. The smallest of the long-lived eddies are clearly seen in the magnificent Voyager picture as about a dozen oval markings somewhat closer to the pole. The motion in each of these Jovian eddies is anticyclonic and largely confined to a narrow region at the edge of the eddy. It is tempting, therefore, to suppose that the eddies are manifestations of sloping convection in Jupiter's atmosphere, implying that they derive their kinetic energy directly from the potential energy due to the action of gravity acting on density variations produced by internal and solar heating, and that they transport heat from the interior to the edges of the latitudinal bands in which they arise (see fig. 7).

Preliminary calculations indicate that there is nothing unreasonable about this hypothesis so far as its implications for the vertical structure and other properties of Jupiter's atmosphere are concerned, but a detailed examination of these implications and a critical comparison of the hypothesis with other proposals as to the nature of the long-lived anticyclonic eddies will have to be considered elsewhere. The hypothesis raises a number of fluid-dynamical questions which are now being resolved by further laboratory work. A particularly striking result of laboratory studies is the occurrence of a single isolated eddy (reminiscent of Jupiter's Great Red Spot) when $\mathcal{T}$ is high enough and Θ is very slightly less than Θ_R. The instability of the strongly sheared flow in the jet-stream itself also requires further study. Experiments with a wall-heated annulus provide some evidence that when viscous effects are sufficiently small the jet-stream develops local instabilities on one side but not on the other. Pictures of Jupiter show highly irregular flow on a comparatively small scale just outside the Great Red Spot (and the other long-lived eddies),

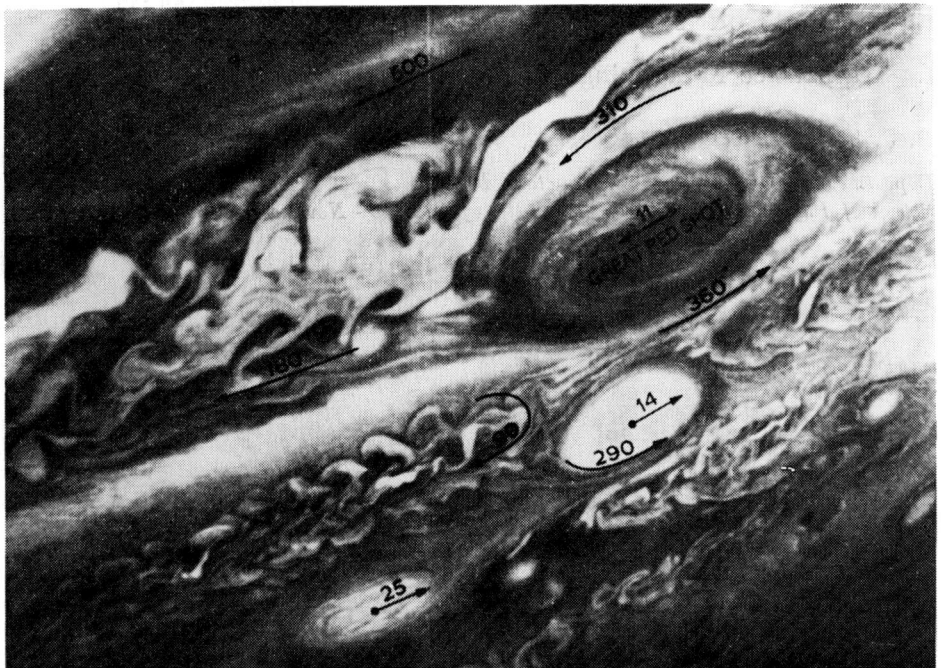

Fig. 7. – Long-lived anticyclonic eddies in Jupiter's atmosphere, namely the Great Red Spot (which is about 20 000 km long) and one of the three White Ovals. The arrows indicate the sense of relative motion. The speeds (as determined by Dr. R. F. BEEBE of the New Mexico State University) are given in kilometres per hour; for comparison note that points of Jupiter's equator rotate at about 35 000 km per hour.

but not on the inside. It will be important to establish by experiment and theory whether this highly irregular flow arises as a result of a « one-sided » instability of the jet-stream at the edge of the main eddy.

BYBLIOGRAPHY

For extensive lists on references see:

R. HIDE: *Meteorol. Mag.*, **110**, 335 (1981).
R. HIDE: *Philos. Trans. R. Soc. London, Ser. A*, **306**, 223 (1982).
R. HIDE: *On the dynamics of rotating fluids and planetary atmospheres*, in *Predictability of Fluid Motion*, edited by G. HOLLOWAY and B. WEST (American Institute of Physics, New York, N. Y., 1984), p. 79.
R. L. PFEFFER, G. BUZYNA and R. KUNG: *J. Atmos. Sci.*, **37**, 2129 (1980).
P. L. READ and R. HIDE: *Nature (London)*, **308**, 45 (1984).
D. J. TRITTON and P. A. DAVIES: *Instabilities in geophysical fluid dynamics*, in *Hydrodynamic Instabilities and the Transition to Turbulence*, edited by H. L. SWINNEY and J. P. GOLLUB (Springer-Verlag, New York, N.Y., 1981), p. 229.

Experiments on Turbulence in Geophysical Fluid Dynamics.

I. - Turbulence in Rotating Fluids.

D. J. TRITTON

Department of Geophysics and Planetary Physics
School of Physics, University of Newcastle-upon-Tyne - Newcastle-upon-Tyne NE1 7RU,
U. K.

1. – Introduction.

Laboratory experiments have played a key role in the development of our understanding of turbulence in « classical » or « nongeophysical » fluid dynamics [1-3]. Do they have a similar role in geophysical fluid dynamics? An answer to this question in an article of the present length must be illustrative rather than full, and I have chosen to discuss turbulence in rotating, homogeneous fluids. I gave a similar review [4] in 1978, and the present survey, whilst intended to be self-contained, emphasizes developments since then.

Even within this limited topic, some selection of material has been necessary. (One important subject not discussed is the turbulent Ekman layer [5, 6].) Some quantitative matters are discussed in a perhaps over-qualitative way because precise definitions (of, for example, the Rossby number of turbulence) would involve lengthy explanation of the scales involved. I urge readers to turn to the original papers to gain a more complete appreciation of the significance of the results and their relationship to theory.

For two reasons, I doubt whether laboratory work will become quite as central to « rotating » turbulence as it has been to « nonrotating ». Firstly, the experiments are more difficult. For example, the problem of the effect of distant walls will be a recurring point in this article. Secondly, at the time of the major work contained, for example, in ref. [2, 3], facilities did not exist for complex numerical modelling. Nevertheless, as I hope this article will show, experiments have much to tell us. And, in my opinion, descriptions of turbulence dynamics based on numerical modelling still require evidence from observation, in Nature or in the laboratory, if they are to inspire confidence. The experiments discussed below all aimed at elucidating the basic dynamics of turbulent motion, not at modelling some complex natural situation.

2. – Homogeneous turbulence.

The theoretical concept of homogeneous turbulence and the fact that an approximation to it is generated by flow through a grid have been major sources of ideas about turbulence. One naturally asks whether similar experiments can be done in a rotating frame. A particular question that arises concerns whether the rotation increases or decreases the rate of decay. Arguments can be advanced for either possibility [7]: rotation is generally stabilizing and so might be expected to suppress the turbulence, but it may also make the turbulence more two-dimensional and so inhibit the cascade of energy to dissipating length scales.

TRAUGOTT [8] and WIGELAND and NAGIB [9] developed wind tunnels in which fluid rotation could be generated. One cannot achieve Rossby numbers lower than around 1 in this way, and the observed effects of rotation are small. Such trends as can be discerned are towards increased turbulence intensity, perhaps indicating that the second process above is beginning to act.

IBBETSON and TRITTON [4, 7] developed a special apparatus in which a single traverse of two low open-area ratio grids generated turbulence of which the subsequent development and decay were observed. Very low values of the Rossby number were achieved in the later stages of decay. Rotation produced a markedly more rapid decay of the turbulence. The most plausible interpretation, based on quantitative details of the decay and of the spectral evolution, is that energy is transported by an inertial-wave mechanism to the boundaries and partially dissipated during reflection. Although this provides insight into the effects of rotation on the turbulence, it limits the inferences that can be made about the processes occurring in the interior of the fluid.

3. – Vibrating-grid turbulence.

The limitations to the parameter ranges achievable in the above ways lend special interest to a different arrangement employed by HOPFINGER and various co-authors [10-13] and by DICKINSON and LONG [14]. Turbulence is generated locally by a horizontal grid which vibrates vertically in a tank rotating about a vertical axis. The turbulence diffuses to other parts of the tank, with the intensity decreasing and the length scale increasing with increasing distance from the grid. It can thus be arranged that the Rossby number is high in the vicinity of the grid, but falls to low values some distance from it. The turbulence generation process should thus be little affected by the rotation, but differences between results with and without rotation appear as one traverses away from the grid.

This configuration shares with the experiments mentioned in sect. **2** the feature that there is no mean motion (or only a uniform mean motion) and thus (unlike the shear flows to be considered in sect. **5** and **6**) no turbulence energy generation throughout the turbulent region. It differs, however, from the previous experiments in that, instead of decaying, the turbulence is sustained by transport (*i.e.* by terms of the form $\partial(\frac{1}{2}\overline{wq^2} + \overline{wp})/\partial z$ in the turbulence energy equation, ref. [1], p. 247). Observed changes with rotation rate of the way in which the turbulence intensity falls off with distance from the grid may thus be due to the rotation affecting either or both of the transport process and the dissipation. It may not be easy to identify the effect on one of these. However, this complication is offset by the advantages of having a statistically steady low-Rossby-number region and of smaller wall effects.

The direction of propagation of turbulence energy is parallel to the rotation axis. It is highly likely, both on general principle and in the light of interpretations to be outlined below, that the results would be different for a different direction of propagation. DICKINSON and LONG [14] included a few experiments with radial energy propagation, *i.e.* the grid was replaced by a rough rod vibrating on the axis. It proved more difficult to get repeatable results with this arrangement, but it did show some significant differences from the main experiment as will be considered below.

Although Hopfinger's experiment and Dickinson and Long's main one used similar arrangements, their observations were complementary rather than overlapping. HOPFINGER and his colleagues observed primarily the structure of the motion when a statistically steady state had been established sufficiently long after the initiation of an experiment. DICKINSON and LONG observed primarily the propagation of the turbulence front into nonturbulent fluid immediately after the grid oscillations were initiated.

The two procedures gave rise to an interesting common result. The demarcation between the region closer to the grid where the behaviour was little different from that in a nonrotating fluid and the region further from it with evident strong rotational effects was remarkably sharp. This is based partly on flow visualization and partly on quantitative measurements. Figure 1 shows variations of the turbulence intensity with distance from the grid in a steady-state experiment. Figure 2 shows the distance advanced by the turbulence front as a function of time after the start of grid oscillation. The line on each graph represents the trend in the absence of rotation. (In fig. 2, both co-ordinates involve Ω, the rotation rate, but, since the line has slope $\frac{1}{2}$ on logarithmic scales, it is independent of Ω and can represent the $\Omega = 0$ case.) The differences of the rotating turbulence from the nonrotating start, in each case, at a Rossby number based on local turbulence length and velocity scales of around 0.2. In fig. 2, the points for Ωt greater than about 7 fit to a line of slope 1, corresponding to a constant velocity of advance of the front. This velocity may plausibly be identified with the group velocity parallel to the rotation axis of inertial waves with wave number related to the length scale of the turbulence near the transition.

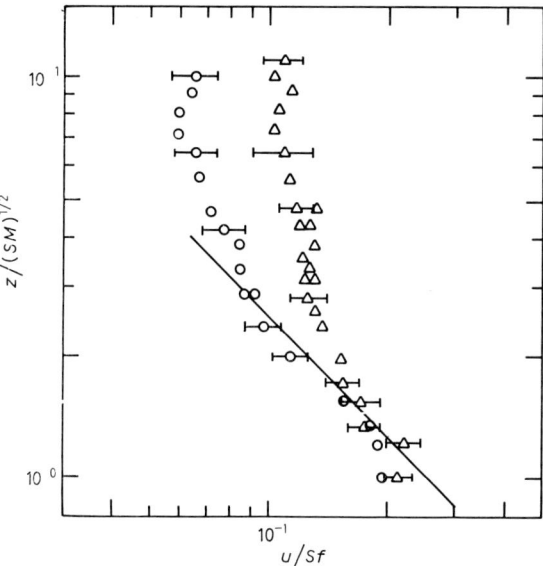

Fig. 1. – Variation of r.m.s. azimuthal turbulent velocity u with distance z from vibrating grid: ○ $f/2\Omega = 6.6$, △ $f/2\Omega = 3.3$. (M, S and f are the mesh length, stroke and frequency of the grid.) From E. J. HOPFINGER, R. W. GRIFFITHS and M. MORY: *The structure of turbulence in homogeneous and stratified rotating fluids*, J. Mec. Théor. Appl., Numéro Special, 21 (1983), © Gauthier-Villars (ref. [13]).

The quantitative agreement between the positions at which rotation first affects each of the propagation of the front and the structure of the turbulence in the steady state strongly suggests that the two are regulated by similar processes, *i.e.* that the change in the steady-state structure is dominantly a

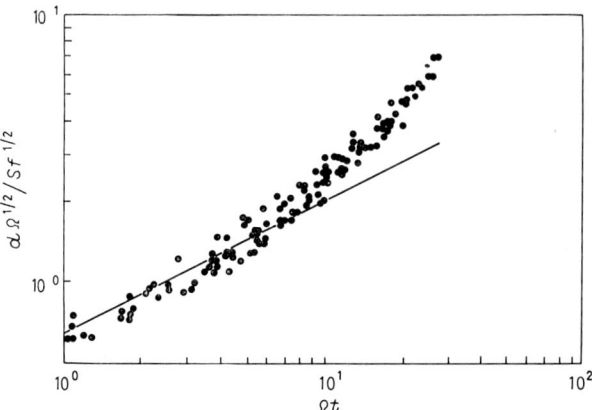

Fig. 2. – Distance d from vibrating grid reached by turbulence front in time t. (Other symbols as for fig. 1.) Data from S. C. DICKINSON and R. R. LONG: *Oscillating-grid turbulence including effects of rotation*, J. Fluid Mech., **126**, 315 (1983), © Cambridge University Press (ref. [14]).

change in the energy transport mechanism. The fact that, up to the transition, the mechanism is quantitatively, not just qualitatively, similar to that in a nonrotating fluid is perhaps surprising.

Figure 3 shows the (dimensional) counterpart of fig. 2 for Dickinson and Long's experiments with a radially propagating turbulence front. There is some reduction of the propagation rate even close to the vibrating rod. More strikingly, further out the rate drops to zero, so that the turbulence is always confined to a finite region. The latter feature can be related to the above discussion by noting that the group velocity of inertial waves perpendicular to the rotation axis is zero.

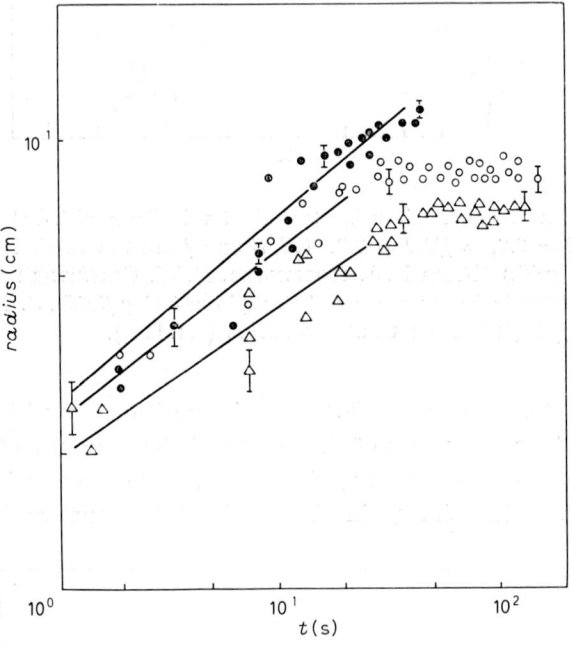

Fig. 3. – Radial distance from vibrating rod reached by turbulence front: ● $\Omega = 0$, ○ $\Omega = 0.38$ rad/s, ▵ $\Omega = 0.97$ rad/s. From S. C. DICKINSON and R. R. LONG: *Oscillating-grid turbulence including effects of rotation*, J. Fluid Mech., **126**, 315 (1983), © Cambridge University Press (ref. [14]).

Returning to the main experiments, the existence of a region in which there are distinctive differences from the nonrotating case obviously leads to the question of the detailed structure of this turbulence. We do not have space for a complete discussion and so look at one predominant feature. (There are several other interesting aspects considered particularly in ref. [10, 13].) Figure 4, from Hopfinger, Browand and Gagne's experiments, shows the motion in a plane perpendicular to the rotation axis. (A corresponding visualization in the absence of rotation is, by comparison, rather featureless.) The rotation has led to the generation of distinctive vortices. These are both cyclonic and anticyclonic, but the former are both more numerous and stronger. The region over which a given vortex dominates the velocity field has a length scale comparable with the turbulence integral length scale in the nonrotating fluid, but the vortex core is an order of magnitude smaller. In the direction parallel

Fig. 4. – Horizontal cross-section of region strongly affected by rotation in vibrating-grid turbulence. From E. J. HOPFINGER, F. K. BROWAND and Y. GAGNE: *Turbulence and waves in a rotating tank*, J. Fluid Mech., **125**, 505 (1982), © Cambridge University Press (ref. [10]).

to the rotation axis, the vortices extend throughout the rotationally dominated region, although they are not everywhere aligned exactly parallel to that axis (fig. 5).

It is worth mentioning, even without detail, that travelling disturbances on the vortex cores have been observed [10, 12]. They have been interpreted as solitons generated by the turbulent fluctuations and may be important as a mechanism of turbulence energy transport.

The origin of the vortices will be discussed in sect. **4**.

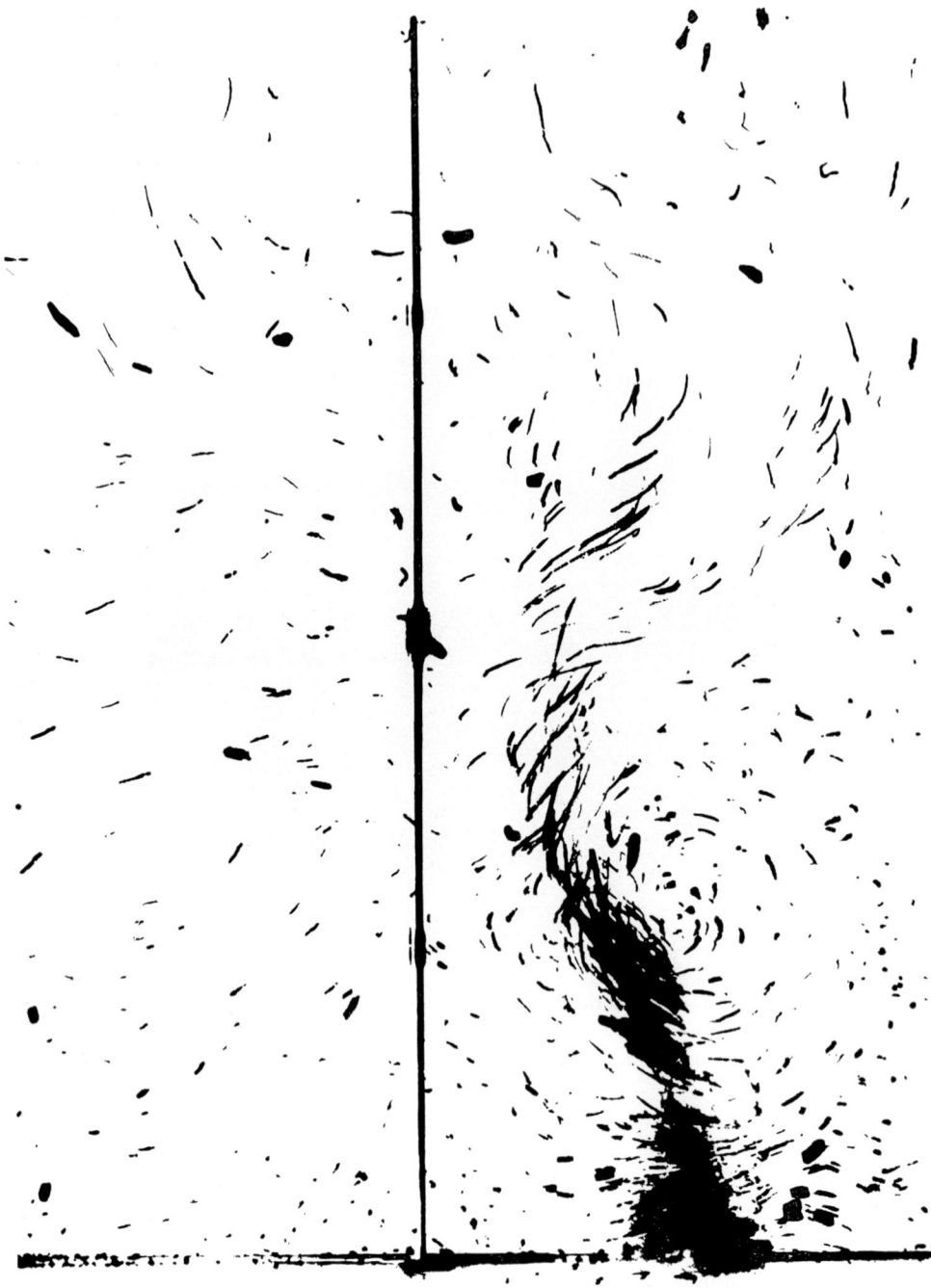

Fig. 5. – Vortex in vibrating-grid turbulence. The grid is beyond the top of the picture, and the change from slight to strong effect of rotation may be seen about a quarter of the way down. From S. C. DICKINSON and R. R. LONG: *Oscillating-grid turbulence including effects of rotation*, J. Fluid Mech., **126**, 315 (1983), © Cambridge University Press (ref. [14]).

4. – Large vortex-generating mechanisms.

Two experiments that I have not yet mentioned are important for an understanding of vortex formation by turbulence, those of McEwan [15] and Colin de Verdiere [16]. Superficially these experiments look similar; both consist of a rotating tank in which the flow is driven by an array of sources and sinks in the base. However, the details of operation and the Rossby number regime are quite different. In McEwan's experiment the turbulence arises from instability of the jets from the sources, and the Rossby number is in the high to moderately low range. Colin de Verdiere's apparatus is relatively shallow and operates at low Rossby number, so that the motion is two-dimensional; the « turbulence » is artificially introduced by randomizing the sources and sinks and switching them with various phases.

Pictures that give a general impression very similar to that given by fig. 4 are to be found in both McEwan's and Colin de Verdiere's papers, though detailed examination and quantitative measurements show that significant differences may lie behind the superficial similarity.

As usual in turbulence studies, in the absence of a full theory, one attempts to understand what is occurring in more general, descriptive terms. Two processes may be envisaged. The first is simply that the formation of the vortices is a manifestation of the « anticascade » of two-dimensional turbulence. The general fact that such turbulence involves the transfer of energy to larger scales is, of course, theoretically well founded, but its precise consequences in any particular flow are more speculative.

The second process is the one variously known as « vorticity expulsion » or « angular-momentum mixing ». It has been suggested, in various contexts, that the effect of small-scale turbulent motions on large-scale vorticity (such as that associated with frame-of-reference rotation) is to redistribute the latter into boundary layers or a concentrated core or cores (depending on the detailed geometry); vorticity is « expelled » from the remaining regions. If a single core is formed coincident with the rotation axis, and if the flow is viewed from a nonrotating reference frame, this may alternatively be seen as a process of making the angular momentum uniform. There is no rigorous proof that such a process must occur, but several lines of argument suggest that it may [4, 17]. Also some other experiments by McEwan [18], focussed particularly on this point, provide evidence for it. Thus neither the detailed mechanism nor its consequences are very specific in the proposal. However, one point is necessarily common to all ways of viewing the process: the turbulence must be three-dimensional, because nonconservation of the vorticity of a fluid particle in an inviscid fluid is implicit. This makes the process distinct from the « anticascade » process.

From the parameter ranges involved and from other evidence about the flow structure, it is likely that Colin de Verdiere's observations corresponded to anticascading and McEwan's to vorticity expulsion. In the former, the large vortices were cyclonic and anticyclonic in equal numbers (so long as the layer was of constant depth; depth variations to simulate the β-effect could produce differences). In McEwan's experiment, all the vortices were cyclonic. This is probably what one would expect from each of the two mechanisms.

This discussion is presumably relevant also to the vibrating-grid experiments; the mechanism of forcing is probably much less important than the frequencies (in relation to the rotation rate) involved in the turbulence generation in its effect on the flow structure. The fact that, in these experiments, both cyclonic and anticyclonic vortices appear but in unequal numbers suggests that both anticascading and vorticity expulsion are operative—or some more complex mechanism of which these two processes may be considered opposite limits. A further point of contrast with the consequences of anticascading alone has been pointed out to me by HOPFINGER: the vortices in Colin de Verdiere's experiment do not appear to have the small (intense) cores mentioned in sect. 3.

This interpretation implies that, altough the larger-scale motions seem to be predominantly two-dimensional [13], three-dimensionality of the turbulence as a whole is essentially involved. The only experiments to date that involve sustained « two-dimensional turbulence » are those (Colin de Verdiere's) in which the randomness was artificially introduced, not spontaneously arising. Decaying two-dimensional turbulence—or an approximation to it—was observed by HOPFINGER, GRIFFITHS and MORY [13] after the grid oscillation had been switched off.

Two further points may be made about the above interpretation of the vibrating-grid experiments. Firstly, what prevents the formation of cyclonic vortices in the region close to the grid? MCEWAN [18] suggests that their generating process occurs in principle in even the most weakly rotating fluid. However, in a fluid of finite depth, there is an upper limit to the Rossby number range over which vortices arise due to the process being opposed by spin-down by Ekman layer interaction. This implies that one of the changes occurring at the transition is quantitatively influenced by the overall size of the system, even though in most respects this type of experiment avoids wall effects more successfully than other types. Further work on this aspect would be useful.

Secondly, both the anticascading and the vorticity expulsion processes involve energy transfer from smaller-scale turbulence to the scale of the vortices. It may be queried whether the vortices observed in the vibrating-grid experiment are really large enough for such an interpretation. The correspondence (see sect. 3) of their size to the length scale of nonrotating turbulence looks significant (though the energy spectrum will, of course, extend to smaller scales). Speculatively, the processes under discussion are operative, but their consequences are affected by the continuous transfer of energy in physical space—from the region close to the grid.

It is not surprising that the discussion should become complex and speculative. The vortices are in a sense the « large eddies » of this kind of turbulence; and the large eddies of nonrotating turbulence are much more fully documented than understood!

5. – Turbulent shear flows.

We turn to the effect of rotation on shear flows, confining attention to the important special case in which the shear vorticity is parallel or antiparallel to the rotation axis of the whole system. We consider some general properties in this section and a particular experiment in sect. **6**.

The effect of rotation may be either stabilizing or destabilizing. Initially, as one increases the rotation rate from zero, it is stabilizing when the system vorticity has the same sign as the shear vorticity and destabilizing when it has the opposite sign. However, in the latter case, the trend is reversed for larger rotation rates, and rapidly rotating systems are stabilized for either sense of rotation.

These statements may be expressed more quantitatively in terms of the parameter [19, 20]

$$B = S(1+S),$$

where

$$S = -\frac{2\Omega}{\partial U/\partial y},$$

the basic flow having velocity U in the x-direction varying only in the y-direction, and the system having angular Ω velocity about an axis in the z-direction. The flow is destabilized relative to the case $\Omega = 0$ when B is negative (*i.e.* when $0 > S > -1$) and stabilized otherwise. In most flows, of course, $\partial U/\partial y$ itself varies with y and the above is a local criterion; the effect of rotation on the flow as a whole may be the net effect of regions of various sizes (and possibly signs) of B. S is a form of reciprocal Rossby number. We are thus concerned with weak and moderately strong, but not very strong, rotation. This range shows more significant differences from the nonrotating flow than it does for the configurations considered previously.

Justification for the above assertions about stabilization and destabilization and derivation of the role of B may be provided in various ways, which I have discussed in more detail elsewhere [4, 21]. A displaced-particle argument [21] can show how the interaction of Coriolis effects with the velocity variations can lead to amplification of a perturbation when $B < 0$. Such arguments apply most directly to the first instability of a laminar flow rather than to the structure of a turbulent one. However, it may have implications for the latter.

The argument relates to the growth of longitudinal vortices (similar to Görtler vortices in curved flow) uniform in the x-direction but involving all three components of velocity fluctuation. When $B < 0$, this type of disturbance will interact with the disturbance arising from Kelvin-Helmholtz or Tollmien-Schlichting instability. One may expect fully three-dimensional turbulent motion to be established quickly. Conversely, when $B > 0$, the stabilization may be expected to suppress the three-dimensional development from Kelvin-Helmholtz or Tollmien-Schlichting instability to full turbulence, rather than to suppress the onset of that instability.

An alternative approach (essentially equivalent, but quite differently formulated) to understanding the promotion or suppression of the turbulence by Coriolis effects is that originally due to JOHNSTON, HALLEEN and LEZIUS [20]. This focusses attention on the production terms in the turbulence energy component and Reynolds stress equations. We will use this approach to interpret some of the results in sect. **6**.

These theories are far from rigorous; experiments are required to verify their predictions and to discover the consequences of the stabilization or destabilization. Previous experiments include a major investigation of channel flow by JOHNSTON, HALLEEN and LEZIUS [20] and one on reattaching free shear layers by ROTHE and JOHNSTON [22]. Here I want to discuss some more recent experiments on free shear layers. These are being carried out by BIDOKHTI at Newcastle and were planned with the above ideas rather specifically in mind.

6. – An experiment on free shear layers.

The arrangement is shown schematically in fig. 6. Two parallel water streams with speeds U_1 and U_2 ($\simeq 0.5 U_1$) form a shear layer between them downstream of the splitter plate. The water is recirculated between A and B and between

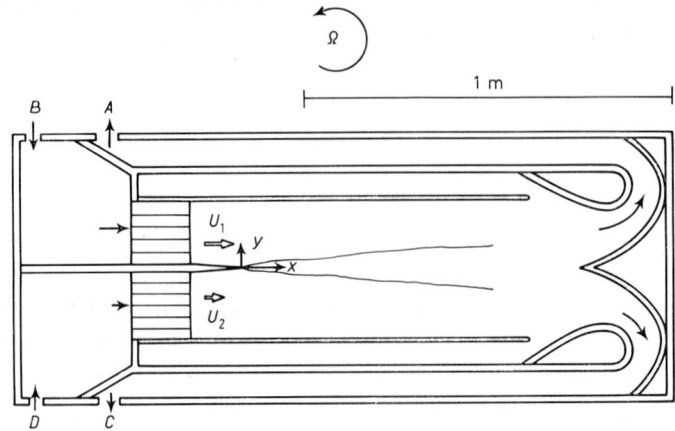

Fig. 6. – Schematic horizontal section of free-shear-layer apparatus.

C and D by pumps of controllable speed and passes through a pair of flow-smoothing sections before entering the working section. The whole system is mounted on a rotating table with its axis in the z-direction; it can rotate in either sense. The tank depth in the z-direction is 0.3 m; it is intended that, so far as possible, the mean flow should be two-dimensional.

The width δ of the shear layer increases with distance downstream. The Rossby number $(U_1 - U_2)/2|\Omega|\delta$ decreases—or, equivalently, the magnitude of S_0, the value of S based on the maximum value of $\partial U/\partial y$, increases. Hence, rotational effects become stronger with the distance downstream; the flow itself scans through the effects of varying S. For the most rapid rotation rates for which a satisfactory mean flow can be maintained $|S_0|$ reaches about 2 towards the downstream end.

There are, of course, interactions of the shear layer with the Ekman layers on the top and bottom of the channel, and these must to some extent cause departures from the assumed conditions throughout the flow. It is difficult to assess these effects quantitatively, but we do not think they are greatly altering our results. This is, however, a further reason why it might be difficult to extend to much higher $|S_0|$ with an apparatus of this type.

The following presents, in a somewhat simplified form, a few of the results, selected to show the main effects discussed in sect. **5**. We shall be preparing a full paper (authors: A. BIDOKHTI and D. J. TRITTON) when a few more results have been obtained. In addition to results summarized below, correlations and spectra are being measured.

Here we look at the results of hydrogen bubble flow visualization and of quantitative measurement with hot-film anemometers. For the former, the bubble source wire was close behind the edge of the splitter plate. The pictures thus correspond to the upstream part of the flow where the effects of rotation are relativaly weak.

Quantitative measurements have been made over the whole range of x, but we shall look at those further downstream, where a self-similar behaviour seems to be established. Results can be reduced to common trends by using the scaled distance

$$X = -2\Omega x/(U_1 - U_2).$$

The sign is chosen so that X has the same sign as S. In the self-similar region, S_0 is about $0.10X$.

The pictures and measurements thus do not relate to just the same part of the flow. It might be said that the pictures show transition to turbulence and the measurements are for fully turbulent flow, but this is a matter over which there has been controversy in the corresponding nonrotating case [23]. The pictures and measurements illustrate the same trends in a complementary way.

It is likely that the features that are apparent in the flow visualizations are also present as large eddy structures in the region for which measurements are presented but not so distinct from smaller-scale motions.

Figure 7 shows nonrotating, stabilized and destabilized flows. The first may be compared with previous studies of shear layers and shows the large « roller eddies » which have been extensively observed and discussed [24, 25]. They

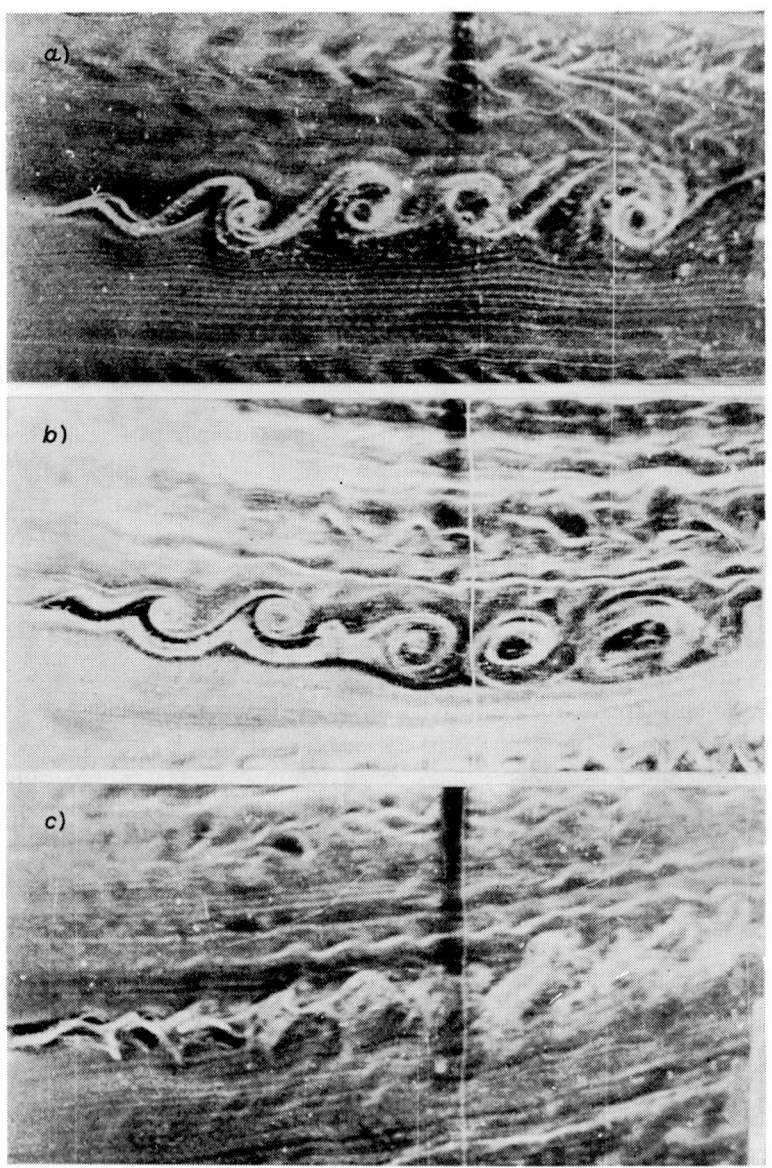

Fig. 7. – Free shear layer in nonrotating, stabilized (S) and destabilized (D) flows: a) $\Omega = 0$, b) $\Omega = 1.5$ rad/s (S), c) $\Omega = 0.26$ rad/s (D).

may be considered as a rather direct consequence of Kelvin-Helmholtz instability. They are again to be seen in the stabilized flow. Their onset appears little affected by the rotation, but there are some changes in their size and shape. Most importantly they persist as well-defined structures much further downstream. The stabilization delays the development of smaller-scale three-dimensional motions (as expected from the discussion in sect. 5). In contrast, fig. 7 shows the opposite rotation sense leading to rapid disruption of the roller eddies; no clear-cut structure is evident. The flow is becoming a fully three-dimensional turbulent one very quickly. (This description is an over-simplified but basically accurate summary of observations for a whole range of Ω of both signs.)

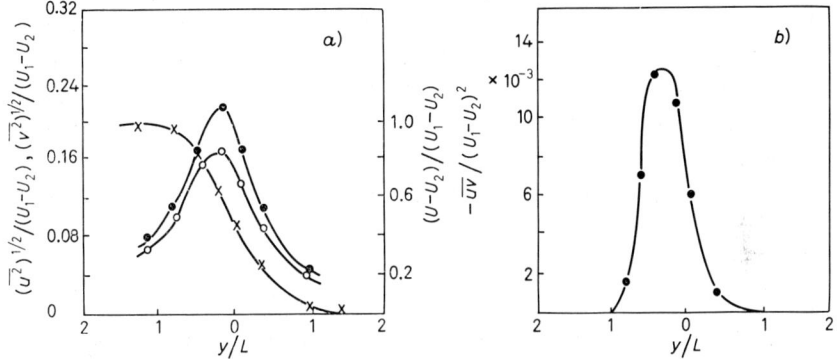

Fig. 8. – a) Mean-velocity profile and longitudinal and transverse intensity distributions for $X = -5.5$ ($\Omega = 0.31$ rad/s (D), $x = 42$ cm), × mean velocity, ○ u, ● v. b) Corresponding Reynolds stress distribution, L is distance over which $U - U_2$ changes from $0.1(U_1 - U_2)$ to $0.9(U_1 - U_2)$.

To illustrate the general nature of the quantitative data, fig. 8a) shows typical distributions across the shear layer of the mean velocity and turbulent intensities in the x and y directions, $(\overline{u^2})^{\frac{1}{2}}$ and $(\overline{v^2})^{\frac{1}{2}}$. (The z-intensity, $(\overline{w^2})^{\frac{1}{2}}$, has also been measured.) Figure 8b) shows the corresponding Reynolds stress distribution ($-\overline{uv}$).

A useful way of seeing the effects of rotation is to consider just the maxima in the intensity and Reynolds stress profiles. Figure 9 illustrates this by showing normalized values of $(\overline{v^2})^{\frac{1}{2}}_{\max}$ plotted against X. Despite considerable scatter the trends are clear, and in fig. 10 these are represented by a continuous curve, along with the corresponding curves for the other two intensity components and the Reynolds stress. Obviously, these curves are approximate and the details may change as further data are obtained, but we think they show the correct trends.

The broad features are immediately apparent: reduced turbulence for stabilizing rotation (positive X), initially increased turbulence for the opposite rotation sense, but reversal of the latter trend as $-X$ becomes large. All these confirm the expectations discussed earlier.

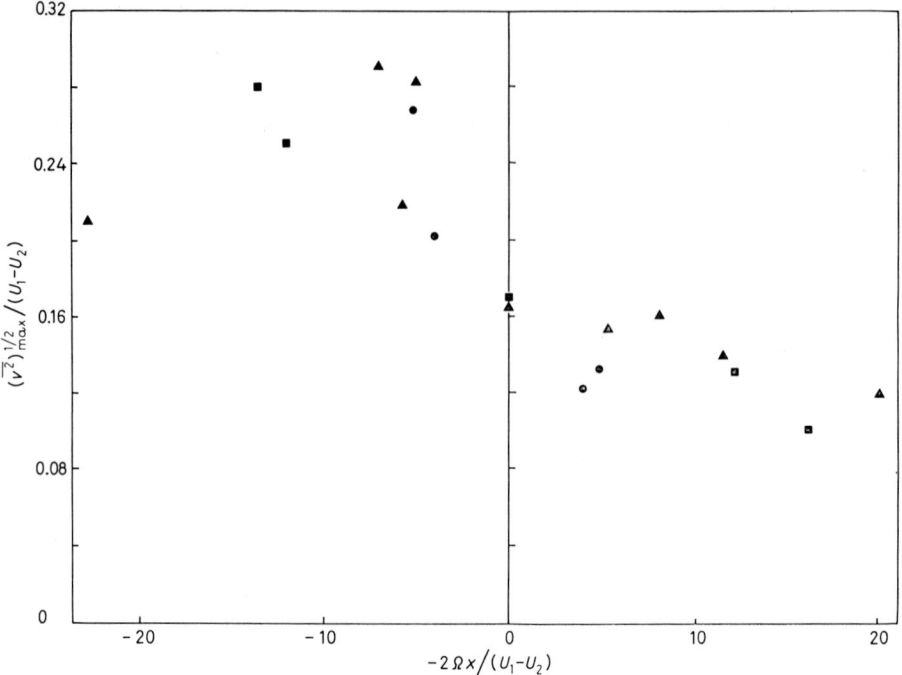

Fig. 9. – Variation of transverse intensity with X: ● $x = 30$ cm, ▲ $x = 42$ cm, ■ $x = 66$ cm.

There are also obvious differences in the way in which the rotation affects the different components. Some of these can be understood through the energy and Reynolds stress equations mentioned in sect. **5** [4, 20]. These may be written

(1) $\qquad \mathrm{D}(\tfrac{1}{2}\overline{u^2})/\mathrm{D}t = -\overline{uv}(1+S)\partial U/\partial y + [\mathrm{O.T.}]$,

(2) $\qquad \mathrm{D}(\tfrac{1}{2}\overline{v^2})/\mathrm{D}t = -\overline{uv}(-S)\partial U/\partial y + [\mathrm{O.T.}]$,

(3) $\qquad \mathrm{D}(\tfrac{1}{2}\overline{w^2})/\mathrm{D}t = [\mathrm{O.T.}]$,

(4) $\qquad \mathrm{D}(-\overline{uv})/\mathrm{D}t = [\overline{v^2} - (\overline{u^2} - \overline{v^2})S]\partial U/\partial y + [\mathrm{O.T.}]$.

([O.T.] stands for « other terms » which are algebraically identical to terms for the nonrotating case, although they can be and probably are indirectly changed by rotation.) The stabilizing or destabilizing effect of rotation is contained essentially in the term involving S in eq. (4). So long as $\overline{u^2} > \overline{v^2}$ (see fig. 10) positive S implies inhibition of Reynolds stress production and negative S promotion. The trend reversal, as $-S$ is increased, is associated with change in sign of $\overline{u^2} - \overline{v^2}$. The differences in the $\overline{u^2}$ and $\overline{v^2}$ curves themselves are related to the terms involving S in eqs. (1) and (2). Since $-\overline{uv}\partial U/\partial y$ must be positive

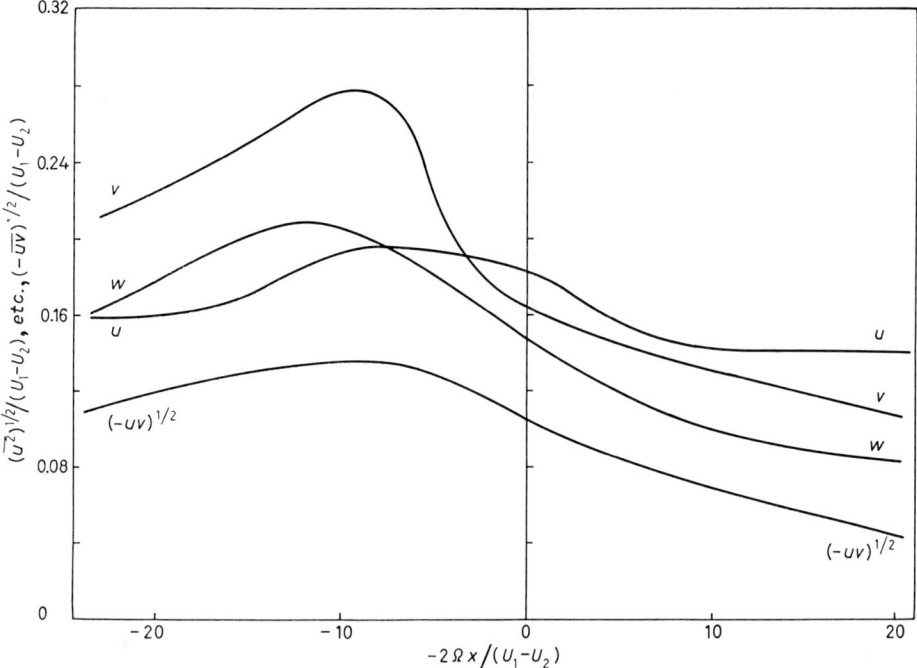

Fig. 10. – Smoothed variations of all intensity components and the square root of the Reynolds stress with X.

for overall energy supply to the turbulence, these terms represent transfer of energy from (or to) $\overline{v^2}$ to (or from) $\overline{u^2}$ when S is positive (or negative); the former will, therefore, be decreased (increased) relative to the latter as the overall intensity decreases (increases). This implies larger variation of $\overline{v^2}$ than of $\overline{v^2}$ on either side of $X = 0$, as seen in fig. 10.

$\overline{w^2}$ always receives its energy entirely by turbulent transfer from $\overline{u^2}$ and $\overline{v^2}$, and for negative X shows a trend intermediate to the other two. For positive X it falls off more rapidly than the other two, and this might be associated with the persistence of two-dimensional motion discussed in the context of the flow visualization.

Quantitative comparison between the theory and the values of X involved is difficult because S varies from place to place. At a given distance downstream $|S_0|$ is the lowest value of $|S|$. Counteracting this, the structure at any x will be affected by the dynamics further upstream where rotational effects are weaker. The fact that, averaging roughly over the various curves in fig. 10, strongest destabilization corresponds approximately to $X = -8$ (whilst the maximum value of $-B$ is at $S = -\frac{1}{2}$) suggests that the upstream influence is the more significant of these two effects.

Figure 11 shows the normalized Reynolds stress. For negative and mildly positive X, variations are slight, but, as X is increased to larger positive values,

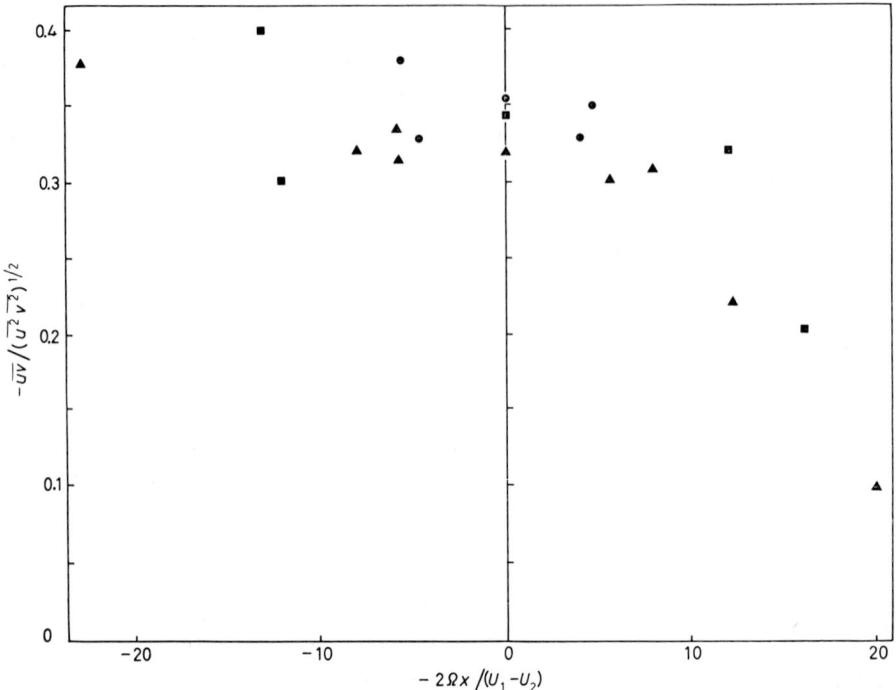

Fig. 11. – Variation of normalized Reynolds stress with X: ● $x = 30$ cm, ▲ $x = 42$ cm, ■ $x = 66$ cm.

there is a marked drop. The mean flow/turbulence interaction is getting weaker not just because the turbulence intensity is falling, but also because the turbulence structure is changing in a way that reduces the correlation of u and v. Extrapolation of fig. 11 suggests that, at still larger X, one might find a region of energy transfer from the turbulence to the mean flow. Indeed there is more positive evidence of this, although based at present on rather few measurements. Figure 12 shows the Reynolds stress distribution across the layer for $X = 20$ (cf. fig. 8b)). There is a region, on the low-mean-velocity side of the layer,

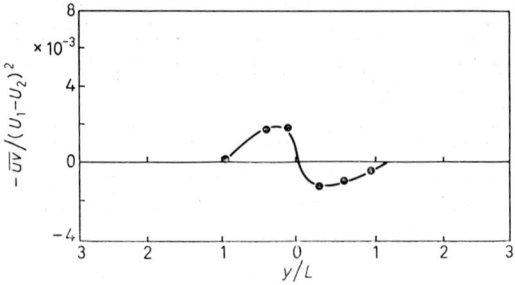

Fig. 12. – Reynolds stress distribution for $X = 20$ ($\Omega = -1.04$ rad/s (S), $x = 42$ cm).

where the sign is reversed (but, because of the larger positive region on the other side, this profile is represented by a positive point in fig. 11). We hope to extend measurements to larger X to see whether the whole profile reverses. Equation (4) may be used to understand how such a reversal may occur; the term $-\overline{u^2}S\partial U/\partial y$ may become the dominant one (cf. fig. 10). This behaviour is possible, of course, only because there is a region of different structure upstream, and it must lead ultimately to the shear layer becoming laminar again.

7. – Stewartson layers.

In all the topics so far we have been looking at the effects of rotation on a configuration for which there is a well-documented nonrotating counterpart. An alternative approach is to take a flow that is specifically characteristic of rotating fluids and that becomes turbulent. There has been much less work along these lines, but the approach should not go without mention. As an example of a flow in which turbulent motion has recently been observed and which might repay further investigation, we look briefly at the experiments by POTTER and TITMAN [26] on the instability of Stewartson layers.

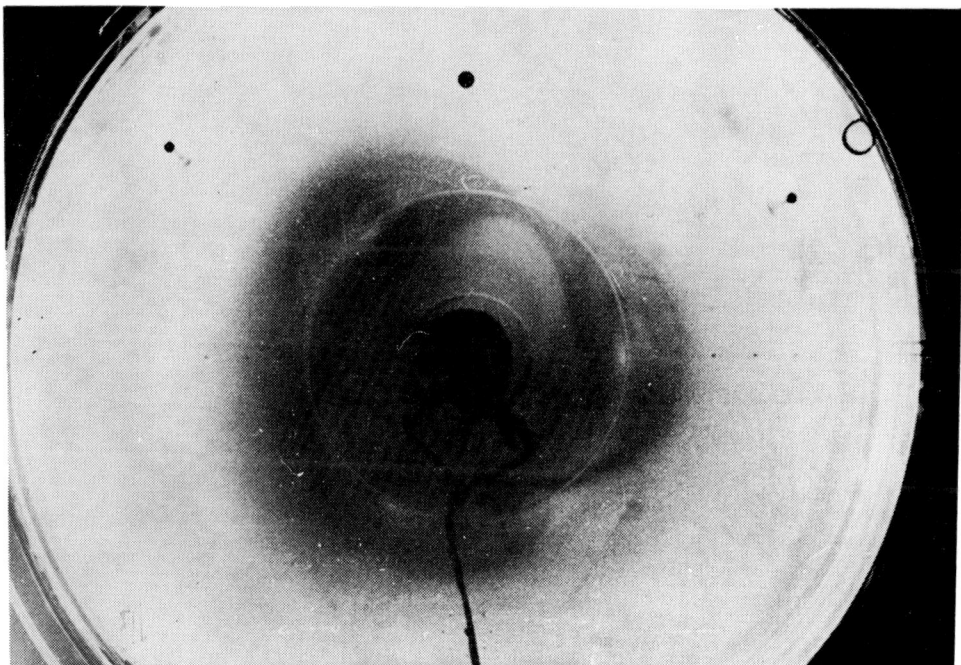

Fig. 13. – Stewartson layer vortices (Rossby number = 0.140, Ekman number = $1.34 \cdot 10^{-4}$).

A tank rotating about a vertical axis contains a pair of similar horizontal discs separated in the axial direction. The discs themselves rotate, with a common angular velocity, relative to the tank. The resulting basic flow [27] consists of a region between the discs and rotating with them, separated from fluid not rotating relative to the tank by an annular shear layer.

As was known from earlier work by HIDE and TITMAN [28] using a single disc, this layer can become unstable with the production of an array of vortices; an example from the two-disc experiment is shown in fig. 13. The major part of Potter and Titman's work has been to provide much additional information about this vortex flow [29]. However, they have also observed the vortex pattern becoming unsteady; figure 14 shows an instantaneous pattern during a vacillation between modes of wave number 1 and 2. Further transitions lead to irregular unsteadiness and to a regime in which flow visualization shows little apparent structure and which can probably be labelled fully turbulent.

The sequence of axisymmetric flow, steady vortex pattern, vacillation and turbulence may be observed by increasing the Rossby number (defined as the ratio of the relative angular velocity of the discs to the angular velocity of the tank), although a full regime diagram also involves the Ekman number. The Rossby numbers for which the turbulent flow has been observed are not very small compared with unity; the basic flow may have become somewhat different

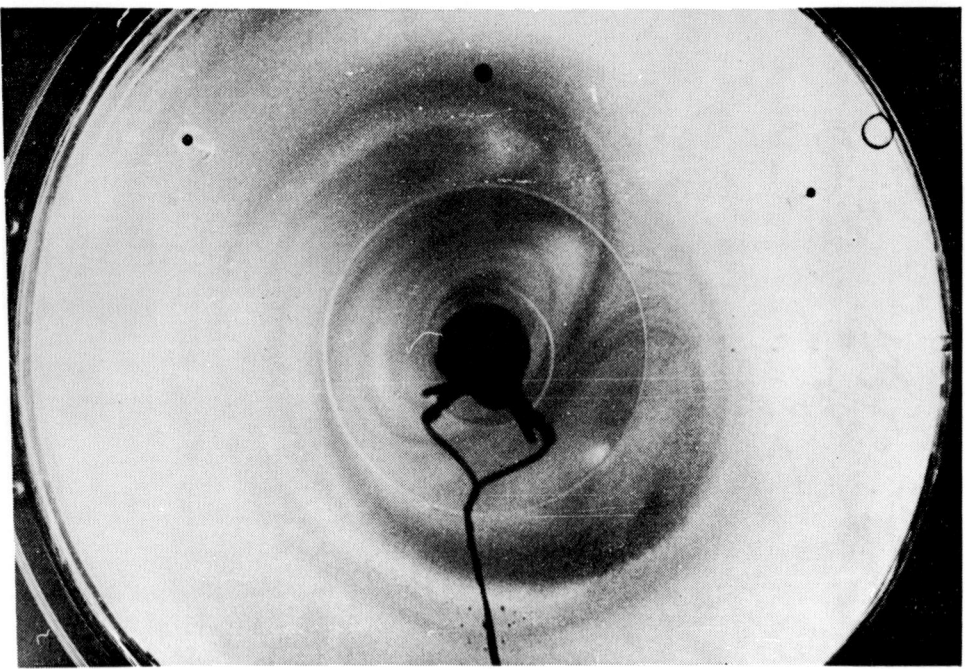

Fig. 14. – Instantaneous pattern during vacillation of Stewartson layer vortices (Rossby number = 0.232, Ekman number = $1.27 \cdot 10^{-4}$).

from the pattern described above. It remains, however, essentially a rotating-fluid situation; if the tank were not rotating, no shear layer would arise in the region where the turbulence is observed.

8. – Concluding remarks.

If, as seems likely, this article leaves the reader with the feeling that there are many interesting observations and ideas but no real unification of these, then that is, I think, an accurate reflection of the present state of the subject. I concluded my previous review [4] with a list of a few ideas which do recur in various contexts, and these probably hold good. But let me round off this article with a mildly provocative paragraph relevant to the aims of future experiments.

Twenty years ago any attempt to explan what one meant by calling a flow « turbulent » usually included the statement that turbulence is necessarily three-dimensional. More recently, there has been much discussion of « two-dimensional turbulence ». Sometimes the production of this is seen as a major purpose of turbulence experiments in rotating fluids. If one uses the phrase turbulent flow to mean a flow that should be described statistically because a sequence of instabilities has led to a low level of predictability (as seems to me the best approximation to a definition of turbulence [1]), then perhaps the older view stands. That is not to deny that experiments on two-dimensional flows in which randomness is introduced artificially may tell us much of relevance to the atmosphere and oceans. But I think it is more than just semanticism to distinguish them from experiments on turbulence.

REFERENCES

[1] D. J. TRITTON: *Physical Fluid Dynamics* (Wokingham, 1977).
[2] J. O. HINZE: *Turbulence*, 2nd edition (New York, N. Y., 1975).
[3] A. A. TOWNSEND: *The Structure of Turbulent Shear Flow*, 2nd edition (Cambridge, 1976).
[4] D. J. TRITTON: in *Rotating Fluids in Geophysics*, edited by P. H. ROBERTS and A. M. SOWARD (London, 1978), p. 105.
[5] D. R. CALDWELL, C. W. VAN ATTA and K. N. HELLAND: *Geophys. Fluid Dyn.*, **3**, 125 (1972).
[6] G. C. HOWROYD and P. R. SLAWSON: *Boundary-Layer Meteorol.*, **8**, 201 (1975).
[7] A. IBBETSON and D. J. TRITTON: *J. Fluid Mech.*, **68**, 639 (1975).
[8] S. C. TRAUGOTT: *Nat. Advis. Comm. Aeronaut., Tech. Note* 4135 (1958).
[9] R. A. WIGELAND and H. M. NAGIB: Illinois Institute of Technology, Fluids and Heat Transfer Reports, R78-1 (1978).
[10] E. J. HOPFINGER, F. K. BROWAND and Y. GAGNE: *J. Fluid Mech.*, **125**, 505 (1982).

[11] E. J. Hopfinger and F. K. Browand: in *Intense Atmospheric Vortices*, edited by L. Bengtsson and J. Lighthill (Berlin, 1982), p. 285.
[12] E. J. Hopfinger and F. K. Browand: *Nature (London)*, **295**, 393 (1982).
[13] E. J. Hopfinger, R. W. Griffiths and M. Mory: *J. Méc. Théor. Appl.*, Numéro Special (1983), p. 21.
[14] S. C. Dickinson and R. R. Long: *J. Fluid Mech.*, **126**, 315 (1983).
[15] A. D. McEwan: *Nature (London)*, **260**, 126 (1976).
[16] A. Colin de Verdiere: *Geophys. Astrophys. Fluid Dyn.*, **15**, 213 (1980).
[17] D. O. Gough and D. Lynden-Bell: *J. Fluid Mech.*, **32**, 437 (1968).
[18] A. D. McEwan: in *Intense Atmospheric Vortices*, edited by L. Bengtsson and J. Lighthill (Berlin, 1982), p. 271.
[19] P. Bradshaw: *J. Fluid Mech.*, **36**, 177 (1969).
[20] J. P. Johnston, R. M. Halleen and D. K. Lezius: *J. Fluid Mech.*, **56**, 533 (1972).
[21] D. J. Tritton and P. A. Davies: in *Hydrodynamic Instabilities and the Transition to Turbulence*, edited by H. L. Swinney and J. P. Gollub (Berlin, 1981), p. 229.
[22] P. H. Rothe and J. P. Johnston: *J. Fluids Eng.*, **101**, 117 (1979).
[23] C. Chandrsuda, R. D. Mehta, A. D. Weir and P. Bradshaw: *J. Fluid Mech.*, **85**, 693 (1978).
[24] G. L. Brown and A. Roshko: *J. Fluid Mech.*, **64**, 775 (1974).
[25] B. J. Cantwell: *Annu. Rev. Fluid Mech.*, **13**, 457 (1981).
[26] D. Potter and C. W. Titman: in preparation.
[27] H. P. Greenspan: *The Theory of Rotating Fluids* (Cambridge, 1968), sect. $2\cdot 18$.
[28] R. Hide and C. W. Titman: *J. Fluid Mech.*, **29**, 39 (1967).
[29] D. Potter: M.Sc. dissertation, University of Newcastle-upon-Tyne (1980).

Experiments on Turbulence in Geophysical Fluid Dynamics.

II. - Convection of a Very Viscous Fluid.

D. J. TRITTON

Department of Geophysics and Planetary Physics
School of Physics, University of Newcastle-upon-Tyne - Newcastle-upon-Tyne, NE1 7RU,
U. K.

1. – Introduction.

This short article describes some work on free convection of high-Prandtl-number fluids. This was motivated by ideas about convection in the Earth's mantle (see sect. **4**). It is being presented here, however, not so much for its geophysical applications as for its relevance to the topic of predictability, *i.e.* to the question of when do flows show stochastic behaviour. In essence the point is that such a behaviour can arise in fluids convecting at a very low Reynolds number so that dynamical inertia is negligible; nonlinearity enters entirely through thermal inertia (the $\boldsymbol{u} \cdot \nabla T$ term in conventional notation). The motion is very different from most familiar turbulent flows, but could properly be called turbulence on the definition mentioned in sect. **8** of part I [1].

Another experiment has recently been reported which points to the same conclusion about stochastic behaviour of viscosity-dominated systems. KOSTER and MÜLLER [2] studied convection in a Hele-Shaw cell and obseved the onset of time-dependent flow and the subsequent appearance of stochastic features—indicated by a change in the form of the spectrum.

2. – The experiment.

Our experiment is described in full elsewhere [3, 4], including more detailed justification for the conclusions mentioned here. The apparatus is basically simple—a large tank of fluid with its base maintained at a temperature higher than the ambient temperature. Conduction of heat from the base into the fluid produces a thermal boundary layer, which becomes unstable, thus generating a free convective flow in the main body of the fluid. Ideally, this process is little affected by the top surface and side walls, and the convection resembles that in an extremely large expanse of fluid heated from below. This type of experiment has been carried out in fluids of lower Prandtl number by several workers; Townsend's [5] experiments in air used an arrangement particularly similar to ours.

We have worked with two grades of silicone oil with Prandtl numbers of $2.6 \cdot 10^3$ and $8 \cdot 10^4$. The corresponding Reynolds numbers, based on observed length and velocity scales, are of the order of 1 and 10^{-2}. It is, of course, the second that is specially relevant to the main point of this article, but the observations are generally similar for the two oils.

With these very viscous fluids, in an apparatus of practicable size, the top and sides do in general influence the whole flow. However, after the base heating is initiated, there is a period (of several hours in the case of the more viscous oil) during which the unwanted influences on the thermal-boundary-layer instability and on its consequences are slight. One can do a statistically quasi-steady experiment.

3. – Observations.

As will be illustrated below, the consequence of the instability is the eruption from the boundary layer of « thermals »—columns of hot rising fluid of horizontal extent comparable with the boundary layer thickness and much larger vertical extent. Cold fluid replaces the hot advected away by the thermals, is then heated by conduction from the base and forms in turn a new generation of thermals. (The process is qualitatively but not wholly quantitatively similar to the heuristic model due to HOWARD [6].)

Thus far the description is essentially similar to that for much lower Prandtl number fluids [5]. The thermals themselves, however, are different. In air, for example, there are vigorous temperature fluctuations at any point within a thermal, arising from smaller-scale turbulence. In our oils, each thermal is essentially a laminar structure.

It is when one looks at the long time scale associated with the origin, existence as recognizable structures and fading-away of a substantial number of thermals that the stochastic aspect appears. When the base heating is first started, there is a quiescent period in which the boundary layer builds up. Thermals then arise almost simultaneously in various places. After these have died away, not so simultaneously, there is another quiescent period, but the second generation of thermals arises at different times in different places. The correlation between events in different places is quickly lost. The properties of a truly statistically steady state may be inferred from the subsequent behaviour. At any instant, over a large enough area, one would find thermals in all stages of development—just forming, fully developed and fading away. Over a small area (containing, say, an average of 2 thermals) the instantaneous number is very variable (from 0 to 5). The duration of thermals is also very variable, although some other properties, such as the typical width of a thermal and the speed with which its head rises, show less variation.

Figures 1 to 4 illustrate these statements. Figure 1 was obtained by dye injection close to the heated base plate and shows the development of a thermal (with an earlier thermal that has penetrated to the top of the tank also in the

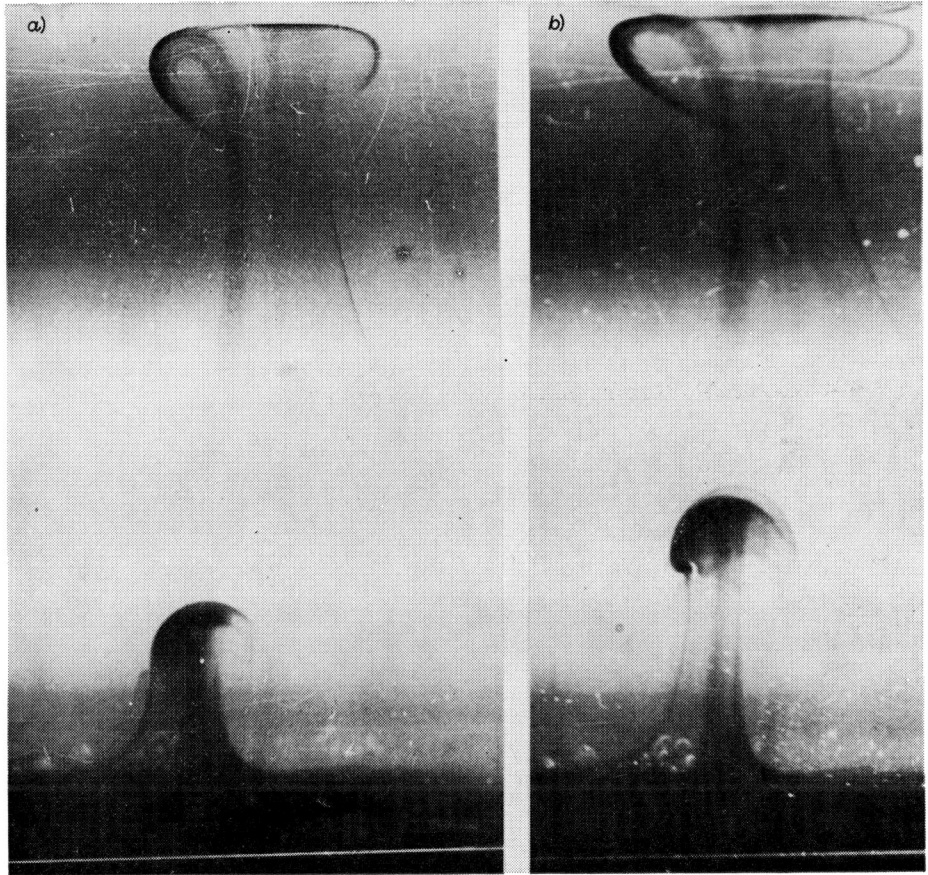

Fig. 1. – Two pictures, separated by a short time interval, of dye injection experiment ($Pr = 8 \cdot 10^4$).

field of view); the more viscous oil was being used. Figure 2 is a schlieren flow visualization of the temperature field in the less viscous oil; one sees the integrated effect across the tank along the line of sight. The picture shows, in fact, a late stage of development of the first generation of thermals and is already giving a somewhat chaotic appearance.

Figure 3 is the result of temperature measurements at a height of about 3 times the boundary layer thickness above the base plate (although the results

Fig. 2. – Schlieren flow visualization (Pr = $2.6 \cdot 10^3$).

would look essentially similar anywhere else except in the boundary layer). Fifteen thermistors were placed in a horizontal line and the outputs combined into this three-dimensional representation of temperature against position and time. The time scale starts when the base plate heating is initiated (but does not continue for the full duration of a typical run). One can see the initial quiescent period, the first onset of thermals (there are lines without peaks passing between those with) and subsequent developments as outlined above, superimposed on a gradual rise in temperature of the whole system.

A simple way of demonstrating the random nature of the phenomenon may be based on temperature observations such as those illustrated in fig. 3. The peaks on the traces can be counted with only occasional ambiguities. Figure 4 is a histogram of the number of peaks in a time of 200 Ψ, where

$$\Psi = [\nu^2/g^2\alpha^2(T_1-T_0)^2\varkappa]^{\frac{1}{3}}$$

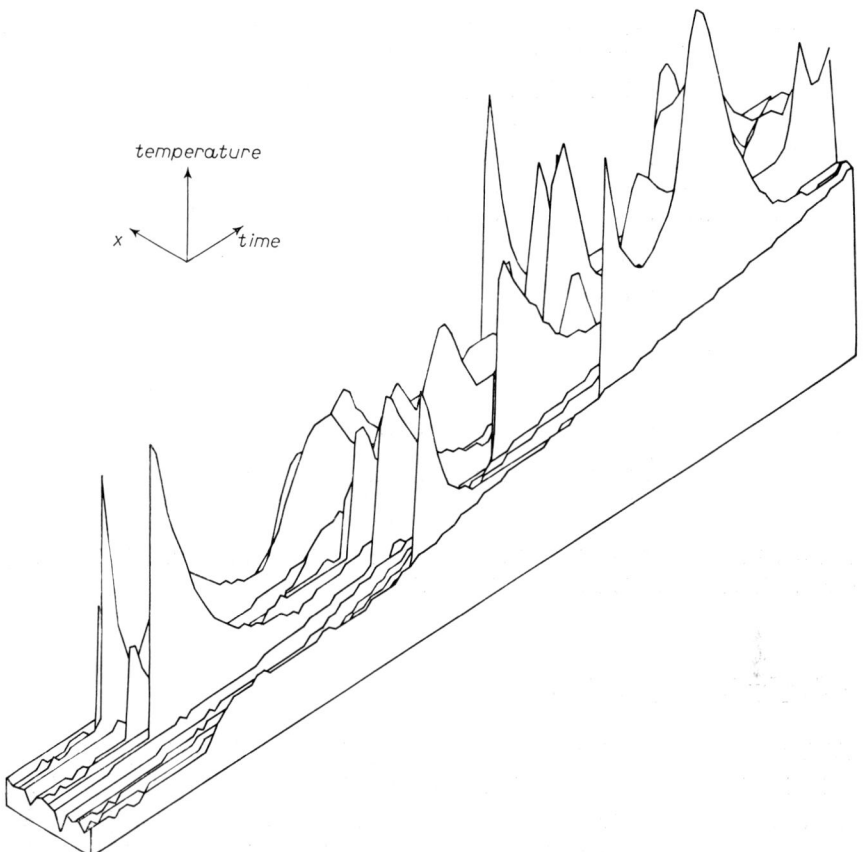

Fig. 3. – Development of temperature with time at horizontal row of points ($Pr = 8 \cdot 10^4$).

and is the basic time scale for convection over an infinite surface [3, 4] (T_1 and T_0 are the base plate and ambient temperatures, g is the acceleration due to gravity, and α, ν and $\varkappa$ are the expansion coefficient, the kinematic viscosity and the thermal diffusivity of the fluid). The observations are compared with the Poisson distribution for the same mean, and this appears to be a good representation.

This is, of course, a very simplified approach. Reference [4] contains a much fuller analysis of the data, including intensities, spectra and autocorrelations to show the statistical behaviour. The results are incorporated into the general picture given by experiments for other values of the Prandtl number.

All the quantitative observations so far relate to the temperature field. We are currently developing a method for measuring the velocity field.

In conclusion, we like to say (at least for dramatic effect) that we have the only laboratory turbulence in the world with a Reynolds number well below 1. It is necessary to include the word « laboratory »; there may have been such turbulence « in the world » for a long time (see sect. 4).

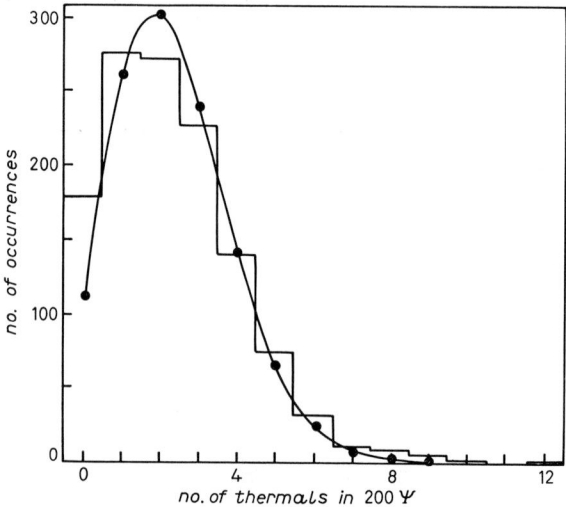

Fig. 4. – Histogram of number of thermals in 200 Ψ (Pr = $8 \cdot 10^4$). Blocks correspond to observations. Curve is Poisson distribution.

4. – Geophysical application.

Although it is peripheral to the main point of this article, some readers may be intrigued by the mention of convection in the Earth's mantle. This section, therefore, appends a few remarks about this. References [7-9] are suggested to readers wanting to know more about the subject as a whole, including the possibility that some of the ideas also apply to other terrestrial planets [7, 8].

The mantle is the solid region immediately below the Earth's crust. It is thought, however, that it can undergo solid-state creep and, on a geological time scale, this can resemble a fluid flow. In particular, the hypothesis that it is convecting thermally is the most widely—although by no means universally—accepted explanation of the now almost universally accepted «fact» of continental drift and associated processes such as mountain building, continental rifting and sea-floor spreading. Within this general hypothesis there are many different proposals about the extent and structure of the convection, outrunning the evidence available for discriminating between them. Although the present epoch of continental drift dates from less than 5% of the age of the Earth, there is increasing evidence that changing patterns of mantle convection have been a major factor in the evolution of the Earth over much of its history [10].

The experiments described above were not intended to model the mantle, but were motivated by the view that refinement of the hypothesis requires not only more geophysical data but also a fuller understanding of basic convection phenomena of very viscous fluids. One particular difference (in addition to the obvious geometrical ones) between the mantle and the laboratory system should be mentioned: the mantle, although plausibly Newtonian, almost certainly has very large temperature-induced variations of viscosity.

In my view, the main message of the laboratory observations for geophysical theories is a general point: changes in the convection pattern need not necessarily be associated with causal changes: instability can lead to a spontaneously changing pattern with steady imposed conditions. This is, of course, a commonplace of fluid dynamics. It has, however, sometimes been neglected by geophysicists or, if acknowledged, supposed not to apply to the extremely low Reynolds number flow of the mantle.

Any more specific application of the results to the mantle is, of course, open to the above comment about proposals outrunning the evidence. However, it was obviously worth scaling the results to the mantle [3]. The time scale implicit in fig. 4 corresponds to about 10^8 y, a time characteristic of changes in mantle behaviour. The length scales come out rather small and imply more detailed structure to the mantle than is often supposed, but not in conflict with the observational evidence. Convection strongly influenced by thermal-boundary-layer instability must be considered a significant possibility in the mantle [3, 11, 12]; it becomes quite probable if the convection extends throughout the depth of the mantle (rather than being confined to only part of it or being divided into several layers) and if the average viscosity is towards the lower end of the suggested range.

REFERENCES

[1] D. J. TRITTON: this volume, p. 172.
[2] J. N. KOSTER and U. MÜLLER: *J. Fluid Mech.*, **125**, 429 (1982).
[3] D. J. TRITTON, D. M. RAYBURN and M. A. FORREST: in *Mechanisms of Continental Drift and Plate Tectonics*, edited by P. A. DAVIES and S. K. RUNCORN (London, 1980), p. 267.
[4] D. J. TRITTON and L. J. RICKARDS: in preparation.
[5] A. A. TOWNSEND: *J. Fluid Mech.*, **5**, 209 (1959).
[6] L. N. HOWARD: in *Proceedings of the XI International Congress of Applied Mechanics* (Munich, 1964), p. 1109.
[7] G. SCHUBERT: *Annu. Rev. Earth Planet. Sci.*, **7**, 289 (1979).
[8] P. A. DAVIES and S. K. RUNCORN, Editors: *Mechanisms of Continental Drift and Plate Tectonics* (London, 1980).
[9] A. M. DZIEWONSKI and E. BOSCHI, Editors: *Proc. S.I.F., Course LXXVIII* (Amsterdam, 1980).
[10] D. H. TARLING: in *Evolution of the Earth's Crust*, edited by D. H. TARLING (London, 1978), p. 361.
[11] G. M. JONES: *J. Geophys. Res.*, **82**, 1703 (1977).
[12] D. A. YUEN and W. R. PELTIER: in *Proc. S.I.F., Course LXXVIII* (Amsterdam, 1980), p. 432.

An Introduction to Dynamo Theory.

S. Childress

Courant Institute of Mathematical Sciences, New York University - New York, NY 10012

Part I

Survey of the Principal Concerns of the Theory.

1. – Introductory remarks.

One intriguing property of the cosmos is the frequent association, in many objects, of rotation with a magnetic field. In the case of the Earth, it is now widely recognized that the reason for this association must lie in a dynamo operating somewhere within the fluid core, sustained by some continual source of energy. It is thought that the rotation of the planet plays an essential role in creating, by some very robust dynamical mechanism, a fluid flow with the right structure for maintaining magnetic energy in the core against natural losses (through Joule heating and viscous dissipation). Although there is optimism that the theory in its broad outlines is correct, there is as yet no fully acceptable physical model for the details of the process. The reason for optimism lies in the success of a number of restricted systems which seem likely to be an inevitable part of the complete dynamo mechanism, and it is the results obtained from these approximate models which we shall summarize in this paper. The reason for lack of complete success in securing a model must rest with the almost total lack of direct observations of the core dynamics (information obtained by seismic means being an exception). Indeed, probably the best probe of the core motion is the magnetic field itself! For this reason, the construction of a successful model of the geodynamo has many of the features of an inverse problem.

Our viewpoint will be somewhat special in that many interesting problems which will undoutedly bear on the success of the theory (such as wave propagation in the core, inferences concerning core dynamics from the detailed structure of the surface magnetic field, implications of the temporal-spatial structure of magnetic reversals, physics of the core and mantle material, etc.) will be omitted in favor of a few coarse but hopefully robust examples of the modeling. We also shall generally favor a deterministic approach to the fluid dynamics, although there is no reason to doubt that the motion of the core fluid is effectively turbulent. The student interested in probing deeper into the subject and reading a more balanced account of it should consult the recent books of Moffatt [1] and Parker [2]. For a thorough account of turbulent models see ref. [3]. There are also a large number of excellent review articles, discussing various stages of the development of the theory, for both Earth and solar fields [3-10]. In part the present paper represents a compression of chapters 7-9 from a joint research monograph with GHIL [11].

A few observational facts. For present purposes the Earth may be taken as consisting of three concentric, approximately spherical regions. The *mantle* occupies the annulus from the crust (at radius 6400 km) to the boundary with the fluid core (at 3500 km). A sphere of radius 1400 km makes up the inner core. Relative to the fluid core the mantle and inner core may be taken as approximately rigid, leaving the annulus between 1400 and 3500 km as the domain of the geodynamo. We shall take the electrical conductivity of the fluid core material to be a *constant* σ. In this sense the dynamo must be *homogeneous*; we are not able to invoke the engineering solution of varying the conductivity spatially through an arrangement of copper wires and must revise our intuition concerning how the process might work.

The most prominent and well-known feature of the observed magnetic field is its approximately dipole structure, with axis currently at about 11° off the rotation axis. Superimposed on this dipole is a complicated pattern of higher harmonics, the so-called *secular variation field*, with an amplitude of about 5% r.m.s. of the principal dipole. It should not be thought, however, that the field within the Earth remains a continuation of the dipole down to some small region of dynamo activity. The higher harmonics grow rapidly with depth, so we see a geometrically filtered field. In addition, there are components of the core field (the toroidal components, see below) which are essentially invisible at the surface. Despite the small (relative to σ) conductivity of the mantle, there is also temporal filtering, so that the typical magnetic map (in contrast to its atmospheric counterpart) needs to be revised only about every twenty years. (The small conductivity of the mantle can in addition exert a dynamical effect through electromagnetic coupling with the core, an effect we shall not deal with here.)

An interesting feature of the secular field is its resemblance to a fixed spatial pattern drifting from East to West at about 0.2° longitude per year. This implies, if extrapolated down to the fluid core as a rigid rotation, a linear speed of about $3 \cdot 10^{-4}$ m/s, which we will take here as a typical fluid velocity within the core, relative to the rotating frame. Such an estimate may well be misleading, however, if this *westward drift* were in fact a manifestation of an internal wave.

Turning now to the principal surface component of the field, namely the dipole field, variations of direction and amplitude of the moment are known to have varied with time scales in the range $(10^3 \div 10^6)$ y over a considerable fraction of the lifetime $(10^9$ y) of the planet; this evidence comes primarily from archeomagnetic and paleomagnetic data [12]. Included in this variation is pole wandering as well as complete reversals of polarity every $(10^3 \div 10^5)$ y [13]. The latter seem to occur in a random sequence.

We shall, therefore, take the time scale of $(10^4 \div 10^6)$ y as typical of our magnetic «climate». It is interesting to see how these figures match with a time scale T derived from an estimate of electrical conductivity. Defining $\eta = (\sigma\mu)^{-1}$ to be the *magnetic diffusivity* (where $\mu =$ magnetic permeability in vacuum $= 4\pi \cdot 10^{-7}$ mkq^{-2} in MKSQ units) and taking $\sigma = 3 \cdot 10^5$ m^{-3} k^{-1} sq^2 (see ref. [14]), we obtain $\eta = 3$ m^{2}s^{-1}, a convenient number to keep in mind when approaching the theory of the geodynamo. If a typical length is taken as $3 \cdot 10^6$ m, then a *diffusive time scale* $T = L^2/\eta$ has the value of roughly 10^5 y. This suggests that we have isolated the physical mechanism for the magnetic climate. It is easily seen (see, *e.g.*, ref. [15]) that T is a time scale typical of the *decay* of the magnetic field within a *rigid* core of dimension L surrounded by vacuum. Since this decay time is about 10^{-4} of the age of the Earth, we are forced to consider a dynamo mechanism, capable of sustaining the field against this natural decay.

2. – Equations of the hydromagnetic dynamo.

The material within the fluid core is thought to be an essentially incompressible iron alloy [16], so most models of the geodynamo take the velocity field to be divergence-free. Relative to the rotating frame the equations of motion are then

$$(2.1) \qquad \frac{d\boldsymbol{u}}{dt} + \varrho^{-1}\boldsymbol{\nabla} p + 2\boldsymbol{\Omega} \times \boldsymbol{u} + \varrho^{-1}\boldsymbol{B} \times \boldsymbol{J} - \nu\nabla^2\boldsymbol{u} = c\boldsymbol{g} \,,$$

$$(2.2) \qquad \boldsymbol{\nabla} \cdot \boldsymbol{u} = 0 \,, \qquad \frac{d}{dt} = \frac{\partial}{\partial t} + \boldsymbol{u} \cdot \boldsymbol{\nabla} \,.$$

In the momentum equation (2.1), $(\boldsymbol{u}, p, \boldsymbol{B}, \boldsymbol{J}) =$ (velocity, pressure, magnetic field, current). We have also included a body force $c\boldsymbol{g}$ expressing a possible inhomogeneity of the fluid density, the latter having been expressed in the form ϱc, where ϱ is a constant reference value of density. The use of ϱ in place of $\varrho(1+c)$ in the other terms of (2.1) indicates that the Boussinesq approximation has been adopted [17].

Since time scales of interest to us are enormous compared to the transit time of light through the core, we may filter out electromagnetic radiation by adopting the « pre-Maxwell » system:

(2.3) $\qquad$ Ampere's law: $\quad \nabla \times \boldsymbol{B} = \mu \boldsymbol{J}$,

(2.4) $\qquad$ Faraday's law: $\quad \nabla \times \boldsymbol{E} = -\dfrac{\partial \boldsymbol{B}}{\partial t}$,

(2.5) $\qquad$ Ohm's law: $\quad \boldsymbol{J} = \sigma(\boldsymbol{E} + \boldsymbol{u} \times \boldsymbol{B})$,

(2.6) $\qquad\qquad\qquad\qquad \nabla \cdot \boldsymbol{B} = 0$,

(2.7) $\qquad\qquad\qquad\qquad \nabla \cdot \boldsymbol{E} = q/\varepsilon$.

In the last equation, q is the charge density and ε is the dielectric constant. $\boldsymbol{E}$ is the electric field.

To close the system it is necessary to provide an equation for the scalar field c. Actually the inclusion of this term has already prejudiced the model toward some kind of *convection* to provide the primitive driving force for the dynamo. Whether a convective mechanism, if correct, is thermal or compositional is an active and interesting issue at present [10, 16]. Any such convectively driven model can probably be adequately represented by an equation for c of the form

(2.8) $$\frac{\mathrm{d}c}{\mathrm{d}t} - \nabla \cdot D_c \nabla c = Q,$$

where D_c is a diffusion coefficient and Q is a source term.

The boundary conditions of the dynamo problem must reflect the fact that the mantle is a poor conductor, that the core-inner core and core-mantle interfaces are transitions from fluid to virtual solid and that the exterior field is a vacuum field decaying to zero at infinity (at least as fast as a dipole). Since we allow nonzero viscosity in (2.1), it is appropriate to impose a no-slip condition on $\boldsymbol{u}$ at the core-mantle interface and to determine the motion of the inner core by a torque balance. In the sequel we shall, however, disregard the inner core, except to mention, in part III, its possible role in models driven by compositional convection. We shall formulate boundary conditions when they are needed below and refer to ref. [17] for a detailed discussion of their derivation.

3. – Kinematic vs. hydromagnetic theory.

We shall summarize some of the recent developments in dynamo theory, to see how the problem has been divided into manageable pieces. It should perhaps be noted at the outset that full numerical solution of the system given in sect. **2** has not been possible, although a related simulation for the solar dynamo has recently been accomplished (see part III). To simplify our representation of the problem, we may eliminate the pressure p from (2.1) and the electric field from (2.5) by taking the curl, so the resulting primitive system (including conditions of matching between core and vacuum fields) can be abbreviated as

$$\dot{Z} = F(Z; Q).\tag{3.1}$$

Here $Z = (\boldsymbol{u}, \boldsymbol{B}, c)$ and in the functional F we have suppressed all parameters except the source density Q. Now it is obvious that one class of solutions of (3.1) will be a state of *pure convection*, which may be steady or time-dependent, but for which the magnetic field vanishes everywhere. Suppose, however, that this convective state is, in some range of Q, *unstable* to the electromagnetic field, in the sense that a small « seed » magnetic field is always amplified. If $\boldsymbol{u}_0$, c_0 denotes the purely convective state, then the above instability will lead to dynamo action, since the system, once started, can never settle into the subspace $Z_0 = (\boldsymbol{u}_0, 0, c_0)$ of the phase space Z. On the other hand, in the presence of viscous and electrical dissipation, thermodynamic constraints should and do restrict solutions of the system to a bounded region of the phase space (see part III). Since it is very clear that F is a nonlinear functional of Z, the stage is set for complicated dynamic behavior including the possibility of the observed, irregular pattern of geomagnetic reversals.

The amplification of small seed field which initiates this behavior suggests a series of models in which $\boldsymbol{u}$ and $\boldsymbol{B}$ decouple. But, then, we might as well simplify matters further and disgregard the dynamical process which gave rise to $\boldsymbol{u}$ in the first place. In other words, a highly manageable piece of the problem simply takes $\boldsymbol{u}$ as a *given* function of position $\boldsymbol{r}$ and time t. This approximation, which carries with it some associated boundary and matching conditions, defines the *kinematic*-dynamo problem. Mathematically, the key equation for this theory is obtained by eliminating $\boldsymbol{E}$ from eqs. (2.3)-(2.6), using the vector identities

$$\nabla \times (\nabla \times \boldsymbol{B}) = \nabla(\nabla \cdot \boldsymbol{B}) - \nabla^2 \boldsymbol{B},\tag{3.2}$$

$$\nabla \times (\boldsymbol{u} \times \boldsymbol{B}) = \boldsymbol{u}(\nabla \cdot \boldsymbol{B}) + \boldsymbol{B} \cdot \nabla \boldsymbol{u} - \boldsymbol{B}(\nabla \cdot \boldsymbol{u}) - \boldsymbol{u} \cdot \nabla \boldsymbol{B},\tag{3.3}$$

to obtain (recall $\boldsymbol{u}$ is divergence-free)

$$\frac{\mathrm{d}\boldsymbol{B}}{\mathrm{d}t} - \eta \nabla^2 \boldsymbol{B} = \boldsymbol{B} \cdot \nabla \boldsymbol{u}.\tag{3.4}$$

Thus the key question of kinematic-dynamo theory is: given our freedom to pick $\boldsymbol{u}$, what choices will give dynamo action? Since (3.4) is, in fact, a *linear* parabolic equation for $\boldsymbol{B}$, the problem seems straightforward until it is realized that it is the coefficients of the equation, supplied by u, which are unknown! A « solution » of this problem will then be a u or a class of u which ensure the existence of complex growth rate with positive real part (*i.e.* an unstable mode). We note that a satisfactory solution u should not accomplish this by extraordinary means, such as the continual and rapid growth of kinetic energy with time. Note also that the kinematic version of the theory narrows and focuses the « inverse » nature of the problem, by connecting a spectrum to coefficients, a situation analogous to the celebrated problem of « hearing the shape of a drum » as well as to other inverse problems in geophysics [18].

Since the kinematic problem as formulated above initiated the modern studies of the geodynamo, we digress for a very abbreviated summary of some key steps in these developments. The proposal of Larmor [19] that a dynamo process might explain the Sun's magnetic field was made in 1919. In 1934 COWLING [20] showed that models of the kind contemplated here cannot succeed if both the magnetic field and the velocity field are symmetric with respect to the same axis. This celebrated theorem, which we discuss in sect. 4, was a disappointment to the dynamo theorists since the axis of rotation is a natural axis of symmetry. Following studies by ELSASSER [20], BULLARD and GELL-MAN [21] obtained numerical evidence for dynamo action in a sphere in a truncated version of the eigenvalue problem (cf. part II). Only recently has it been shown [22] that, in fact, the velocity field adopted in ref. [21] fails to be a kinematic dynamo. In retrospect, however, these tentative results had the positive effect of stimulating the optimists to study other mechanisms. In 1955 PARKER [23] introduced a new line of attack, by taking the velocity field to have a much smaller spatial scale than the core. He also introduced the idea of *averaging* the effects of the small-scale velocity field, to obtain an effective dynamo operating on the scale of the core.

Other examples of kinematic dynamos, using various different mechanisms, followed soon after [1]. In 1964, an important step was taken by BRAGINSKY[24]. He analysed the kinematic problem under a dual limit of small diffusivity η and near axial symmetry of u and B, a procedure which observation suggests to be highly reasonable. We end this brief tour with the discovery in 1966 of the formalism of mean-field electrodynamics (see ref. [25]) by STEENBECK, KRAUSE and RÄDLER, an approach which also utilizes averaging methods. We shall describe the smoothing methods and the technique of Braginsky in part II.

The most difficult problem in dynamo theory is, of course, to determine the fate of a growing mode of magnetic activity. Regardless of how the kinematic problem was solved (the most desirable situation being that u is a state of pure convection), it is likely that the magnetic field, through the $B \times J$ force in (2.1), will react on u, which will change the form of eq. (3.4) and rearrange the scalar field c, etc. For reason which we take up in part III, as yet no completely satisfactory theory of this nonlinear problem, which we call the *hydromagnetic-* dynamo problem, has emerged.

4. – A disc model.

Simple models based on laboratory dynamos have played a prominent role in developing approaches to this subject [26-28]. In the present section we describe one such model, whose relevance to contemporary dynamo theory was emphasized by MALKUS [29]. A detailed analysis was carried out by ROBBINS [30]. It is a *shunted disc dynamo*, shown in fig. 1. The equations for the

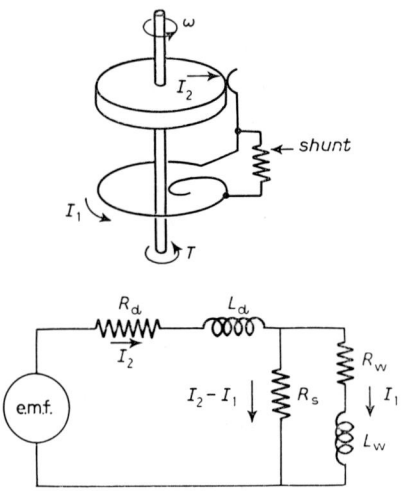

Fig. 1. – The shunted disc model. The subscripts w, s and d refer to the coiled wire, the shunt and the disc.

currents $I_{1,2}$ in the winding and through the shunt and the angular velocity ω are

$$M\omega I_1 = L_d \dot{I}_2 + R_d I_2 + R_s(I_2 - I_1), \tag{4.1}$$

$$(I_2 - I_1)R_s = L_w \dot{I}_1 + I_1 R_w, \tag{4.2}$$

$$C\dot{\omega} = T + MI_1 I_2 - \nu\omega, \tag{4.3}$$

where M is a mutual inductance, $M\omega I_1$ is the e.m.f., T is the torque applied to the shaft, ν is a friction coefficient, and C is a moment of inertia. All parameters of the model (including T) are taken to be constant. By means of the transformation [30]

$$\begin{cases} [t, \omega, I_1, I_2] \to [\tau t, (x - R_s/M)\delta, \beta y, \alpha z], \\ \tau = L_d/R_d, \quad \delta = (R_w + R_s)(R_d + R_s)/MR_s, \\ \alpha^2 = (R_w + R_s)/MR_s \nu \delta, \quad \beta = R_s \alpha/(R_w + R_s), \end{cases} \tag{4.4}$$

the equations may be brought into the form

(4.5) $$\dot{x} = R - yz - vx,$$

(4.6) $$\dot{z} = xy - z,$$

(4.7) $$\dot{y} = \sigma(x - y).$$

It is interesting that a second transformation $(x, y, z) \to (R/v - z, x, y)$ converts (4.5)-(4.7) into the Lorentz system! As a result, it is possible to translate the complicated dynamical behavior known to occur for this system into a model for irregular « magnetic reversals », as determined by the sign of the currents.

To formulate the analog of the kinematic-dynamo problem in this setting, we consider the special case $R_s = \infty$, so that $I_1 = I_2 = I$, and take ω as a given quantity. Then, from (4.1)-(4.3) we have

(4.8) $$L\dot{I} = (M\omega - R)I, \quad L = L_w + L_d, \quad R = R_w + R_d.$$

We thus see that current will grow or decay exponentially depending upon whether ω is greater than or less than R/M. In other words, when a velocity field of given *structure* (here determined by the disc geometry) is to be tested as a kinematic dynamo, the *speed* at which the system is driven is an eigenparameter in the problem. The critical eigenvalue R/M corresponds to the onset of dynamo activity [31]. The simplicity of this argument and eq. (4.8) compared to the possible complexity of solutions of the Lorentz systems is perhaps suggestive of the relationship of the actual kinematic and hydromagnetic dynamo problems!

5. – Some remarks on the underlying topology.

In addition to Cowling's theorem (see sect. **6**), there have been number of « antidynamo » theorems discovered in the last 50 years [1]. These have been useful for setting limits to the *u* which should be considered as potential kinematic dynamos, but, in fact, they isolate an extremely small set of motions within the family of possible flow fields in three dimensions. As we shall see in part II, whenever the question has been posed precisely, it has turned out that « most » of the motions are dynamos, provided only that the admitted class was sufficiently wide to begin with. The reason that rather complicated motions in three dimensions seem to be relevant is suggested by some elementary arguments based on eq. (3.4). According to this equation, when $\eta = 0$, corresponding to a fluid of infinite electrical conductivity, magnetic lines of force are material lines [17], and flux tubes can be stretched, compressed, flattened and folded by the flow. If this process brings lines with different orientations into close proximity, there will be enhanced diffusion of flux once η is made finite. In fig. 2a) we show a side view of a sheet of flux being folded by an essentially two-dimensional flow in the plane. In order to continue the folding, the sheet must be stretched and (by the material property of the lines of force)

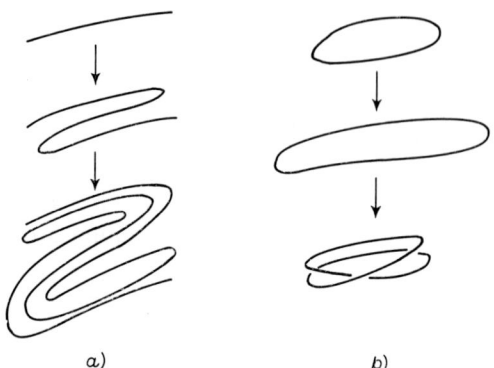

Fig. 2. – a) A thin sheet of magnetic flux is continuously stretched and folded by a two-dimensional motion of the fluid. b) A thin tube of magnetic flux is stretched and folded in three dimensions.

the magnetic field increases in strength. It is easily shown [11], however, that with finite η the diffusion of flux between folds ultimately defeats the field enhancement due to stretching. This is a basic obstacle to a dynamo process in two dimensions, and there are possibly related arguments for the other cases covered by antidynamo theorems. In fig. 2b) we indicate one step of a stretching and folding operation which can be accomplished likely to lead to dynamo action with finite η; this is an observation made by ALFVÉN.

6. – Cowling's theorem in a spherical core.

In order to acquaint the reader with the notation frequently used in treating homogeneous dynamos in a spherical core, we outline now the proof of Cowling's theorem given by BRAGINSKY [24].

Consider a spherical core V surrounded by an insulator V_0. If, as COWLING supposed, both $\boldsymbol{u}$ and $\boldsymbol{B}$ are axisymmetric, it is then possible to represent both divergence-free fields in the form [15]

(6.1) $$\boldsymbol{u} = \boldsymbol{u}_\mathrm{T} + \boldsymbol{u}_\mathrm{P} = U\hat{\varphi} + \boldsymbol{\nabla} \times \psi\hat{\varphi},$$

(6.2) $$\boldsymbol{B} = \boldsymbol{B}_\mathrm{T} + \boldsymbol{B}_\mathrm{P} = B\hat{\varphi} + \boldsymbol{\nabla} \times A\hat{\varphi},$$

involving scalar fields A, B, U, ψ. The subscripts denote *toroidal* and *poloidal*, and $\hat{\varphi}$ is a unit vector defined from a polar angle. Given $\boldsymbol{u}$, the equations for

A, B within the core follow from (3.4) and can be expressed in cylindrical polar (z, s, φ) co-ordinates as follows:

$$\frac{\partial A}{\partial t} + s^{-1} \boldsymbol{u}_\mathrm{P} \cdot \boldsymbol{\nabla}(sA) = \eta(\nabla^2 - s^{-2})A \,, \tag{6.3}$$

$$\frac{\partial B}{\partial t} + s\boldsymbol{u}_\mathrm{P} \cdot \boldsymbol{\nabla}(B/s) - s\boldsymbol{B}_\mathrm{P} \cdot \boldsymbol{\nabla}(U/s) = \eta(\boldsymbol{\nabla}^2 - s^{-2})B \,. \tag{6.4}$$

In V_0, electric currents must vanish and we have

$$(\nabla^2 - s^{-2})A = 0\,, \qquad B = 0\,. \tag{6.5}$$

Thus the toroidal component of the field does not, in this approximation of zero mantle conductivity, penetrate the core. Continuity of the magnetic field at the core ($r = r_0$ say) requires

$$A, B, \frac{\partial A}{\partial r} \text{ continuous on } r = r_0\,. \tag{6.6}$$

Finally, to exclude sources of magnetic energy at infinity, we require that

$$A = O(r^{-2})\,, \qquad r \to \infty, \tag{6.7}$$

since this enforces decay at least as fast as a dipole. We indicate the general

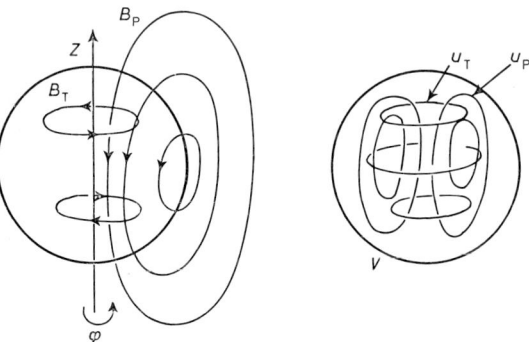

Fig. 3. – The poloidal-toroidal geometry in a spherical core V.

structure of these fields in fig. 3, where we indicate the symmetry which seems most realistic [1].

Cowling's theorem may now be established by multiplying (6.3) by $s^2 A$ and (6.4) by $s^{-2}B$ and integrating over V. The result of this calculation, after several applications of the divergence theorem, is

(6.8) $$\frac{d}{dt}\int_V (sA)^2 \, dV = -\eta \int_{V+V_0} (\nabla sA)^2 \, dV \, ,$$

(6.9) $$\frac{d}{dt}\int_V (B/s)^2 \, dV = \int_V (B/s)\boldsymbol{B}_P \cdot \boldsymbol{\nabla}(U/s) \, dV - \eta \int_V [\nabla(B/s)]^2 \, dV \, .$$

In the first equation a surface integral which arises upon integration has been expressed, using (6.5), as an integration over V_0. We have assumed here that $\boldsymbol{u}_P$ has vanishing normal component on the boundary. From (6.8) we see that the poloidal component of the field decays to zero, so the source term in (6.9) cannot be sustained and B decays as well. Thus the dynamo fails. We have also assumed that $\max_V |\boldsymbol{\nabla}(U/s)| \leqslant$ some constant for all time.

Note that it is really the absence of any coupling with B in (6.8) which prevents dynamo action. The « source » term in (6.9) is easily interpreted when U is taken to have the general structure indicated in fig. 3. It represents the possibility of distortion of a field line as shown in fig. 4. Because of the twisting,

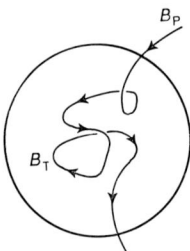

Fig. 4. – Twisting and stretching and poloidal-field line to create toroidal field.

the poloidal field becomes a source of toroidal field. The difficult part of kinematic-dynamo theory is finding a step which closes the loop, producing poloidal field from toroidal field. Cowling's theorem asserts that such a step is not present under conditions of strict axial symmetry.

7. – A necessary condition.

A number of authors have discovered conditions on $\boldsymbol{u}$ which are necessary for dynamo action [15, 32-34]. Among the simplest to state and derive for a spherical core of radius r_0 is the condition [32]

(7.1) $$U_m r_0/\eta > \pi \, , \qquad u_m = \max_V |\boldsymbol{u}| \, .$$

The proof utilizes the energy balance equation [15]

$$(7.2) \qquad \frac{\mathrm{d}}{\mathrm{d}t} \int_{V+V_0} (|\boldsymbol{B}|^2/2) \, \mathrm{d}V = \int_V \boldsymbol{u} \cdot (\boldsymbol{B} \times \boldsymbol{\nabla} \times \boldsymbol{B}) \, \mathrm{d}V - \eta \int_V |\boldsymbol{\nabla} \times \boldsymbol{B}|^2 \mathrm{d}V \, .$$

The first term on the right of (7.2) is the contribution from a dynamo effect. The second term is the rate of loss of magnetic energy due to Joule heating. For the Cauchy-Schwarz inequality it follows from (7.2) that

$$(7.3) \qquad \frac{\mathrm{d}}{\mathrm{d}t} \int_{V+V_0} (|\boldsymbol{B}|^2/2) \, \mathrm{d}V \leqslant u_m \left[\int_{V+V_0} |\boldsymbol{B}|^2 \mathrm{d}V \int_V |\boldsymbol{\nabla} \times \boldsymbol{B}|^2 \, \mathrm{d}V \right]^{\frac{1}{2}} - \eta \int_V |\boldsymbol{\nabla} \times \boldsymbol{B}|^2 \mathrm{d}V \, .$$

On the other hand, it is known that [15]

$$(7.4) \qquad \int_V |\boldsymbol{\nabla} \times \boldsymbol{B}|^2 \mathrm{d}V \geqslant (\pi^2/r_0^2) \int_{V+V_0} |\boldsymbol{B}|^2 \mathrm{d}V \, .$$

Thus

$$(7.5) \qquad \frac{\mathrm{d}}{\mathrm{d}t} \int_{V+V_0} (|\boldsymbol{B}|^2/2) \, \mathrm{d}V \leqslant [u_m r_0/\pi - \eta] \int_V |\boldsymbol{\nabla} \times \boldsymbol{B}|^2 \, \mathrm{d}V$$

and the conclusion follows easily. Note that we have in no way restricted the co-ordinate system, so that (7.1) must be satisfied relative to *any* rotating frame.

PART II

Projection Methods for the Kinematic Problem.

8. – General remarks.

The mathematically inclined reader will probably find the various techniques devised to solve the kinematic-dynamo problem to be the most interesting part of the subject. Certainly it is the most firmly established. In a broad sense, these analytic methods avoid a direct attack using asymmetric three-dimensional fields and the exact partial differential equation. The earlier work [20, 21] focused on expansion in the natural orthogonal functions in the representation of $\boldsymbol{B}$. By projecting (3.4) onto a finite number of these modes (having first selected the structure of $\boldsymbol{u}$) there results a truncated system of ordinary differential equations (in t) for the coefficients which determine $\boldsymbol{B}$ [35]. However, without carefully adjusting the structure of $\boldsymbol{u}$, the convergence of the procedure with number of modes may be too slow to allow firm deductions concerning the eigenvalues, a difficulty which again emphasizes the inverse nature of the problem. In retrospect, the most successful methods have all depended on an ingenious selection of a *class* of admissible u, chosen to be sufficiently « wide », yet simple enough to allow analysis of (3.4). In this way the dynamos could be *deduced* from the admissible class. The most easily visualized examples of this general scheme utilize discrete regions of simple motion within an otherwise rigid core [1], the first example of this type being the dynamo of Herzenberg [36]. His model utilized two spheres in solid-body rotation within a larger spherical core. BACKUS [15] chose instead a temporal sequence of special flows. These procedures and the methods to be discussed in more detail below all share to some degree a *projection* of the exact eigenvalue problem, visualized as formulated in a rather large function space, onto a restricted subspace with some simplifyng features. Another way to think of this is to regard B as split into a principal part B_1 whose evolution can be followed rather easily and a « noise » B_2 generated as B interacts with u in (3.4). In the method of Backus [15], for example, B_1 consisted of magnetic modes whose amplitudes remained dominant after each cycle of u. We emphasize this point because of its similarity to « slow-manifold » projection in the computation of atmospheric dynamics. Also, the method of Braginsky (sect. **10** below) projects onto the axisymmetric fields, a technique which is quite close to the one-dimensional modeling of climate [11]. It is important to realize, however, that one cannot just disgregard the complement of the projected field; it will be essential to dynamo action within the subspace of analysis. For example, Cowling's theorem ensures that Braginsky's method fails if the asymmetric components are neglected.

9. – Methods based on smoothing.

The general procedure outlined above is perhaps most easily illustrated when B_2 is a field with smaller spatial scale than B_1, so that the projection can be realized as a smoothing. There is a considerable literature concerning variants of this idea, involving temporal as well as spatial smoothing, turbulent velocity fields, waves, etc. (See ref. [25] and chapt. **7** of ref. [1].) The roots of the idea are to be found in the classic paper of Parker [23]. Mathematically, the methods are perhaps most easily developed for u which are periodic in space and time, for which the eigenvalue problem becomes similar to Bloch wave analysis [37, 38]. The simplest special case when u is independent of time and periodic on a scale L much smaller than the core radius r_0 and has magnitude u_0. We shall outline how the smoothing method proceeds for this case.

We tentatively take B_1 in the above decomposition to vary on the core scale r_0, and we also suppose that the component B_2, which presumably varies on both spatial scales, to have magnitude small compared to B_1. The dominant *small-scale* term on the right of (3.4) is thus $\boldsymbol{B}_1 \cdot \nabla \boldsymbol{u}$, so that $\boldsymbol{B}_2$ will satisfy

$$\text{(9.1)} \qquad \frac{\partial \boldsymbol{B}_2}{\partial t} - \eta \nabla^2 \boldsymbol{B}_2 \simeq \boldsymbol{B}_1 \cdot \nabla \boldsymbol{u} \, .$$

From (9.1) we may estimate that $\boldsymbol{B}_2$ has magnitude of order $B_1 L u_0 / \eta$, and so our guess concerning the relative magnitude of $\boldsymbol{B}_2$ will be correct provided that

$$\text{(9.2)} \qquad R \equiv L u_0 / \eta \ll 1 \, .$$

The dimensionless quantity R is the *magnetic Reynolds number* of the velocity field. Since it must be small, there is relatively little distortion of the basic magnetic field by the flow [17]. For $\boldsymbol{u}$ which are also time-dependent with time scale T, (9.2) may be replaced by min $(R, R^*) \ll 1$, $R^* = u_0 T/L$.

To obtain dynamo action on the projected (smooth) fields $\boldsymbol{B}_1$, we average (3.4) over a domain of size intermediate between L and r_0. This gives

$$\text{(9.3)} \qquad \frac{\partial \boldsymbol{B}_1}{\partial t} - \eta \nabla^2 \boldsymbol{B}_1 = \nabla \times [\text{average } (\boldsymbol{u} \times \boldsymbol{B}_2)] \equiv \nabla \times \boldsymbol{\varepsilon} \sim B_1 L u_0^2 / \eta r_0 \, .$$

Assuming that $\boldsymbol{B}_1$ varies on a time scale r_0^2/η, we see from (9.3) that the right-hand side is significant if

$$\text{(9.4)} \qquad r_0 L u_0^2 / \eta^2 \sim 1 \, .$$

From (9.2) and (9.4) we see that this balance can be achieved provided that $R_0 = u_0 r_0 / \eta$ is *large*. This last condition does not have a physical interpretation in terms of line distortion, since r_0 is not the scale of the velocity field.

The vector $\boldsymbol{\varepsilon}$ on the right of (9.3) can be computed for a large class of spatially periodic fields and there results

$$\text{(9.5)} \qquad \boldsymbol{\varepsilon} \simeq \boldsymbol{A} \boldsymbol{B}_1 \, ,$$

$\boldsymbol{A}$ a symmetric pseudotensor. For example, if $\boldsymbol{u} = (0, \cos kx, \sin kx)$, then one computes from (9.1) the periodic solution $(L = 2\pi/k)$

$$\text{(9.6)} \qquad \boldsymbol{B}_2 \simeq (R/2\pi) \boldsymbol{B}_1 \cdot \boldsymbol{i} \, [0, -\sin kx, \cos kx] \, .$$

Note that we have neglected the slow spatial variation of $\boldsymbol{B}_1$ relative to $\boldsymbol{u}$ in solving for $\boldsymbol{B}_2$. This emphasizes the essentially recursive nature of the smoothing method. The result of its systematic implementation is an infinite series in a parameter representing the ratio of scales. From (9.6) we then compute that, for this simple $\boldsymbol{u}$, $\boldsymbol{A}$ is a diagonal matrix with $u_0^2/\eta k$ in the first entry and zeros elsewhere. This implise that $\boldsymbol{u}$ acts on the x-component of $\boldsymbol{B}_1$, perturbing the field lines in such a way that a *mean current* in the x-direction is produced. Since this source of current has emerged from one step in the recursive smoothing algorithm, the resulting equation for B_1 is said to be that of *first-order smoothing* [1].

The important point is that this calculation can be carried to the same stage for a large class of fields (indeed, for the periodic case the entire series of terms may be studied [37, 38]), so that the functional dependence of A upon u (which is here quadratic) can be investigated in detail, and the best u for dynamo action thereby deduced. For example, the choice

$$(9.7) \quad u = u_0(\sin ky + \cos kz, \sin kz + \cos kx, \sin kx + \cos ky)$$

produces $A = \alpha I$, I the identity tensor and $\alpha = u_0^2/\eta k$; here u_0 can be an arbitrary function on the spatial scale r_0. This symbol is used because this magnetohydrodynamic consequence of first-order smoothing is known as the *alpha effect* [25].

The many variants of the smoothing method yield any number of such *effects* as the series is contructed [25]. Nevertheless, it is fair to say that the production of mean current from mean field which is embodied in the alpha effect is the principal result of the smoothing theory of the kinematic dynamo. Interestingly enough, essentially the same mathematical expression is obtained by the averaging methods of Parker [23] and Braginsky [24], through quite different arguments. The alpha effect is thus generally invoked in one way or another to produce poloidal field from toroidal field; this step, in conjuction with the mechanism depicted in fig. 4, is thus a favorite candidate for the cycle sustaining the geomagnetic field.

In spite of this appealing state of affairs, the mathematical restrictions which allow smoothing to work should not be forgotten in the process of modeling a specific system. The principle « Nature abhors an asymptotic expansion » is not a bad working hypothesis for magnetohydrodynamical systems of this complexity, although it is not always true. In the specific case considered above, once the fluid is free to move according to the dynamics, will the magnetic field sustain the effective separation of scales (energy in the fluid on small scales, energy in the field on large ones) [39]? If one is tempted to answer « yes » to this, the question might then revert to why the fluid eddies should maintain a small local magnetic Reynolds number. Certainly this is not what is observed in the solar convection zone [8]. It seems that the crucial problem currently facing this aspect of the theory is just how smoothing can be carried out when the field is highly distorted on the small scale (in our working example above, this amounts to smoothing at large R). The computation of the alpha effect at large R was carried out by CHILDRESS for some special geometries [40], but there is as yet no means available for using such results in a global theory of the mean field.

10. – Braginsky's method.

Since a particularly readable account of this method may be found in ref. [1], we will confine our remarks here to a few essential points, which we hope will aid the reader in penetrating some rather formidable mathematical machinery.

The compelling motivation behind Braginsky's approach is the probable strong link between rotation of a planet and its magnetic field, as evidenced by the close alignment of the axis of symmetry of the field with the axis of rotation. This is true not only of the Earth, but also of Jupiter and Saturn [41]. Since exact symmetry is precluded by Cowling's theorem, the *slight* breaking of symmetry could be the basis of a perturbational algorithm. It turns out that such a position is closely allied with a highly conducting core, or more precisely with a large core magnetic Reynolds number $R_0 = u_0 r_0/\eta$, since the dominant symmetry could then be explained as a result of the mechanism of fig. 4. Note that the values of u_0, r_0 in the Earth's core given in sect. **2** yield a R_0 of 300 and so a value in the range $10 \div 100$ is not at all unreasonable. BRAGINSKY thus bases his analysis on the simultaneous limit of $R_0 \to \infty$ combined with fields which are small perturbations of axially symmetric ones. In the context of sect. **8**, the projection is onto the axially symmetric subspace.

In the original analysis [24] an expansion in powers of $R_0^{-\frac{1}{2}}$ was carried out for an appropriately chosen $\boldsymbol{u}$. Since $\boldsymbol{u}$ was taken as dominantly but not exactly axisymmetric, it was also represented by a (finite) series in this parameter. In particular, the axisymmetric poloidal part of $\boldsymbol{u}$ was chosen to be smaller by a factor $1/R_0$ than the toroidal part, while the asymmetric components were smaller by a factor $R_0^{-\frac{1}{2}}$. For a *steady* flow of this form, the particle paths deviate slightly from circles about the symmetry axis (at least over one revolution of the particle). The expansion thus obtained is complicated by the singular nature of the limit (small η in (3.4)), and it was necessary to carry the series further than absolutely necessary to obtain equations for the dominant fields. More astonishing was the equivalence of Braginsky's final equations to Parker's result [23], provided certain « effective variables » emerging from Braginsky's expansion were identified with the dependent variables of Parker's theory. In an attempt to understand these similarities, SOWARD considered an alternative approach and, in a series of papers culminating in ref. [42], was able to greatly clarify the mathematical basis of Braginsky's results.

SOWARD focused on a invariance of eqs. (3.4) with η identically zero, under volume-preserving transformations of space. It can be checked that, if

(10.1) $$\frac{\partial \boldsymbol{B}}{\partial t} + \boldsymbol{u} \cdot \nabla \boldsymbol{B} - \boldsymbol{B} \cdot \nabla \boldsymbol{u} = 0$$

and we are given a function $\tilde{\boldsymbol{r}}\,(\boldsymbol{r}, t)$, then

(10.2) $$\frac{\partial \tilde{\boldsymbol{B}}}{\partial t} + \tilde{\boldsymbol{u}} \cdot \tilde{\nabla} \tilde{\boldsymbol{B}} - \tilde{\boldsymbol{B}} \cdot \tilde{\nabla} \tilde{\boldsymbol{u}} = 0$$

provided that we define

(10.3) $$\widetilde{\boldsymbol{B}}(\widetilde{\boldsymbol{r}}, t) = \boldsymbol{B}(\widetilde{\boldsymbol{r}}, t) \cdot \nabla \widetilde{\boldsymbol{r}},$$

(10.4) $$\widetilde{\boldsymbol{u}}(\widetilde{\boldsymbol{r}}, t) = \frac{\partial \widetilde{\boldsymbol{r}}}{\partial t} + \boldsymbol{u}(\boldsymbol{r}, t) \cdot \nabla \widetilde{\boldsymbol{r}}.$$

Recall that our object is to reduce the magnetic field to its axially symmetric projection, an operation that can be performad directly on a vector function, expressed in cylindrical polar co-ordinates, by averaging each component of the polar angle (longitude). SOWARD notes that this can be accomplished in the case of (3.4), without having to deal with quadratic contributions from asymmetric parts of $\boldsymbol{u}$ and $\boldsymbol{B}$, *provided we first perform a transformation of 3-space so that $\boldsymbol{u}$ is axisymmetric* (or nearly so). Since we actually want to allow for (small) diffusion in (3.4), the exact invariance as obtained in (10.1), (10.2) is not possible and the resulting changes give rise, upon projection in co-ordinates in which $\boldsymbol{u}$ is nearly axisymmetric, to a effective alpha-type mechanism.

To explain this a little more directly, let us suppose that $\boldsymbol{u}$ is to be described in Lagrangian co-ordinates. Let us label a particle by $\boldsymbol{r}_0$ and then write the dominant axisymmetric motion of the core as the Lagrangian function $\boldsymbol{r}(\boldsymbol{r}_0, t)$. (A natural and likely candidate for this flow is a geostrophic motion in a spherical core.) We denote this change of variables, $\boldsymbol{r}_0 \to \boldsymbol{r}$, by F and visualize F as mapping a point into a circle about the axis of symmetry, if contributions to the axisymmetric flow which are of order R_0^{-1} or higher are disregarded. However, we must deal with the asymmetric «wiggles» of order $R_0^{-\frac{1}{2}}$. We thus introduce a secondary transformation $\widetilde{F}$, $\boldsymbol{r} \to \widetilde{\boldsymbol{r}}$, yielding a Lagrangian description of the physical velocity field (through terms of order $R_0^{-\frac{1}{2}}$ inclusive). Note that $\widetilde{F}$ is near to the identity. We indicate these steps in fig. 5.

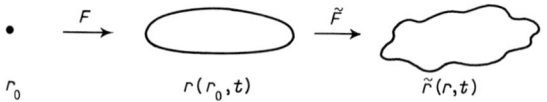

Fig. 5. – The transformations F and $\widetilde{F}$.

SOWARD interprets Braginsky's construction as initiated in physical or «tilde» variables, since an asymmetric velocity field is considered. (Put tildes everywhere in (3.4).) Now apply the transformation $\widetilde{F}^{-1}$ to this equation, utilizing expressions (10.3), (10.4). The fortunate effect of this «pullback» to «untilded» co-ordinates is to make the $\boldsymbol{u}$ axisymmetric, so that projection onto axisymmetric fields goes through without quadratic coupling. However, since (3.4) rather than (10.1) is being solved, the pullback does not leave the equation invariant but generates subsidiary terms, and an identification with the alpha effect is immediate [42]. It is interesting that it is *diffusive* effects which are, therefore, responsible for the alpha effect. By the way, exactly the same conclusion can be drawn from the smoothing technique described in the previous section.

In suitable dimensionless variables (which effectively replace η by R_0 in (3.4)), the equations thus obtained by BRAGINSKY and SOWARD have the form (cf. (6.3), (6.4))

$$\frac{\partial A_e}{\partial \tau} + s^{-1} \boldsymbol{u}_{P_e} \cdot \boldsymbol{\nabla}(sA_e) = \eta(\nabla^2 - s^{-2})A_e + \alpha B , \tag{10.5}$$

$$\frac{\partial B}{\partial \tau} + s\boldsymbol{u}_{P_e} \cdot \boldsymbol{\nabla}(B/s) - s\boldsymbol{B}_{P_e} \cdot \boldsymbol{\nabla}(U/s) = \eta(\nabla^2 - s^{-2})B , \tag{10.6}$$

where now (6.2) is replaced by

$$\boldsymbol{B} = B\hat{\varphi} + R_0^{-1} \boldsymbol{\nabla} \times A\hat{\varphi} \tag{10.7}$$

and $t = R_0 \tau$. In these results the subscript e indicates an « effective » variable which can be explicitly related to physical variables and the velocity field. The crucial variable α in (10.5) is similarly a function of the velocity field. We point out that the α in (10.5) represents a small alpha effect, smaller by a factor R_2^{-2} than $O(1)$ terms in (10.6). If the subscript e is dropped, (10.5), (10.6) are identical to the equations proposed by PARKER [23].

11. – Other results.

The method of Braginsky and Soward makes no special assumption regarding magnitudes of spatial scales of velocity and magnetic fields, since its success is instead linked to the near axial symmetry and the zero-diffusivity limit. If surface observations of other geophysical phenomena are any indication, however, we expect core motions to involve components of many different scales, with perhaps dominance by the scale of the core itself [43, 44]. KHRAICHNAN [45] has hypothesized that such a multiplicity of scales could give rise to modulation of the alpha effect and that such spatial variations could, after a secondary averaging, contribute possible negative terms to the expression for the turbulent effective duffusivity η_T. It seems possible that this might lead locally to *negative* values of η_T [1].

In a related analysis, we have examined [46] the interaction of isolated domains of movement, any one of which is capable of an alpha effect, but unable to act independently as a dynamo. Using an approximate renormalization technique, it was found that a situation could arise in which smoothing failed to yield an alpha effect. The reason was simply that, when these critical conditions prevailed, large numbers of domain pairs could co-operate to excite small-scale components (« microdynamos »), rendering the global smoothing meaningless. There appears to be a close analogy with other changes of phase on spatial scales. A systematic group-theoretic renormalization analysis, within the framework of mean-field electrodynamics, has recently been reported by MOFFATT [47].

PART III

Hydromagnetic Models.

12. – Implications of linear and near-linear analysis.

As we indicated in part I, current models tend to favor some kind of convectively driven motion as the source of geomagnetism, and our purpose in this final part is to describe some of the elements which are likely to be important in such hydromagnetic models of the dynamo.

Some time ago, ELTAYEB and ROBERTS pointed out that convection in a rapidly rotating sphere of viscous, electrically conducting fluid could be dramatically destabilized by a magnetic field [48]. In terms of the Taylor number T and the Hartmann number M, defined by

$$(12.1) \qquad T = \Omega^2 r_0^4/\nu^2, \qquad M^2 = B_0^2 r_0^2/\mu\varrho\nu\eta,$$

it was argued that a critical Rayleigh number Ra_c, known to be $O(T^{\frac{2}{3}})$ as $T \to \infty$ with $M = 0$ (see, e.g., ref. [49]), would be lowered to a value $\sim T^{\frac{1}{2}}$ once $M \sim T^{\frac{1}{4}}$. Although it is difficult to estimate T for the Earth's core because of the uncertain value of viscosity [43], it is certainly enormous and the reduction in Ra_c is, therefore, significant. Fortunately, viscosity cancels out of the ratio $M^2/T^{\frac{1}{2}}$ and for the Earth one estimates values of the order of $1 \div 5$, corresponding to a core field of about 100 G.

It happens that, in some systems, convectively driven dynamos can operate stably with M small and $\mathrm{Ra}_c \sim T^{\frac{2}{3}}$, as we indicate below. Nevertheless, the Eltayeb-Roberts scenario would go as follows: sufficiently large seed fields would destabilize the core motions, resulting in greater speed and, therefore, through kinematic-dynamo action, would lead to increases in field energy. The process would ultimately be arrested at a field energy at which magnetic structures with spatial scales comparable to the core size could provide the needed dissipation. This balance is now customarily referred to as the *strong-field* regime. It is a matter of debate to what extent this balance determines the strength of a planet's magnetic field [7], but it would appear to be an inescapable part of many convectively driven models. We indicate the general bifurcation structure suggested by the Eltayeb-Roberts ordering in fig. 6.

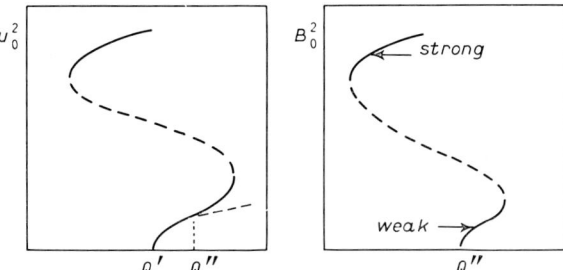

Fig. 6. – Stable (solid) and unstable (dashed) branches of a hydromagnetic dynamo according to the Eltayeb-Roberts scenario. The abscissa is the strength of some energy source density, Q. The point Q' corresponds to the onset of pure convection. The point Q'' marks the onset of dynamo action. The weak- and strong-field branches are indicated.

Nevertheless, for analytical reason it is tempting to study the *onset* of dynamo action (at the point Q'' in fig. 6) and make use of the methods of nonlinear stability and bifurcation theory near a point of criticality [50, 51]. For rapidly rotating systems it happens that the pure convection state (where we now refer solely to thermal convection [49]) usually involves motions on spatial scales which are small compared to the dimensions of the container, which allows the smoothing method of sect. **9** to be used used. These *weak-field* dynamos then have the property that the coupling between the velocity and magnetic fields is slight, but just sufficient to maintain a dynamo stably. For the Benard model [50] the instability of the system to increases of magnetic energy, as indicated in fig. 6, has been explicitly verified. It occurs once $M \sim T^{1/12}$ [52].

Before turning to the problems which beset analysis in the strong-field regime, we mention the interesting prospect that the convective engine driving the geodynamo is *compositional* rather than thermal. A compositional dynamo links the magnetic field to the continual formation of the solid inner core. It is hypothesized that light material is released during solidification at the boundary between the inner core and the fluid core, which then floats up and releases gravitational energy. (For reference to the literature see the recent paper by LOPER and ROBERTS [53].) Some process of this type may well be unavoidable, despite the existence of heat sources within the Earth, because of limits on the thermal efficiency of a heat engine [54, 55].

13. – Attempts to model the strong-field regime.

Although as yet there have apparently been no numerical computations of a spherical dynamo based directly on the primitive equations (2.1)-(2.8), GILMAN and MILLER [56] and GILMAN [57] have modeled the solar dynamo numerically from an analogous primitive system. The results obtained so far do not bode well for kinematic-dynamo models based upon smoothing, for the amplitude of the alpha effect, which has been adopted *a priori* to match the solar field, does not emerge from the numerical simulation. We mention this unfortunate circumstance since, in attempting to construct hydromagnetic geodynamos, theorists have adopted a *prescribed* alpha effect and focused their efforts upon the analysis of the dynamical balance between field and flow. These *macrodynamical* models, which we discuss in this section, may, therefore, be relying on a less than secure kinematic theory. In this connection we note that the alpha effect present in (10.5) will now be referred to as the *primary* effect. It is possible in principle to introduce a *secondary* alpha effect which produces poloidal current from poloidal field. This has the effect of adding a term $\nabla \times (\alpha^*(\nabla \times A_e \hat{\varphi}))$ to the right-hand side of (10.6), α^* being an additional function to prescribe.

An important dynamical constraint on these models, first noted by TAYLOR [58], can be indicated by writing (2.1) in the form

$$(13.1) \qquad \varrho^{-1} \nabla p + 2\Omega \times u + \varrho^{-1} B \times J = f.$$

Taylor's result is then that, if the normal component of u vanishes on the core boundary, and if $\hat{\varphi} \cdot f = 0$, then

$$(13.2) \qquad \int_{S_0} [B \times J] \cdot \hat{\varphi} \, dS = 0,$$

where S_0 is the intersection of a cylinder $s = s_0$ (in cylindrical polar co-ordinates (z, s, φ)) with the core. Physically, (13.2) states that the electromagnetic torque on any such cylinder must vanish. This result plays an important role in the proof of impossibility of dynamo action in a class of models. If i) the above condition on u is in force and ii) both α^* and f vanish, then the dynamo fails to be sustained regardless of what α is chosen to be [4]. BRAGINSKY [59] has generalized this result to include some effects not specifically accounted for above, namely the existence of an Ekman layer at the core-mantle interface and electrodynamic coupling of the core and the mantle. These two effects, by the way, accelerate the decay by introducing viscous dissipation as well as additional Joule losses.

The efforts to construct working hydromagnetic models have circumvented this nonexistence result in two ways. One possibility is retain $f = 0$ but admit a secondary alpha effect [60, 61]. Simulations suggested that fields were evolving so as to ultimately satisfy (13.2), but it is not clear that the equilibration process is actually this simple [62].

BRAGINSKY [59] has proposed a completely different interpretation of the Taylor constraint. He disgregards a secondary alpha effect, admits Ekman layers and core-mantle coupling and sets $\boldsymbol{f} = c\boldsymbol{g}$, with c prescribed rather than obtained from (2.8). Instead of (13.2), one then obtains a constraint in the

$$\tag{13.3} s_0^{-2} \frac{\mathrm{d}}{\mathrm{d}s_0} s_0^2 \int_{-z_0}^{+z_0} B \frac{\partial A_e}{\partial z} \mathrm{d}z = \delta ,$$

where $2z_0$ is the height of the cylinder of radius s_0 introduced with Taylor's constraint above. Estimates suggest that the function δ is small, so we are in a position to impose (13.2) to first order. Instead, BRAGINSKY argues that the left-hand side of (13.3) is *nominally* small, in that A_e is very nearly independent of z. This implies that the poloidal field in the core is very nearly aligned with the axis of rotation and requires the existence of intense current sheets at the core-mantle boundary in order to effect a match with the exterior dipole field.

It is puzzling that solutions with such different structure should be compatible with the same set of equations, but CHILDRESS [46] has found that a simple system of five ordinary differential equations, chosen to mimic the equations for the velocity field as well as values of A_e and B at two points lying on a straight line $s = s_0$, $\varphi = \varphi_0$ in the core, exhibits multiple equilibria. One equilibrium, in a limit analogous to that employed by BRAGINSKY, indeed yields « A_e independent of z »; it is, however, an unstable equilibrium. The other solutions are analogous to those suggested by MALKUS and PROCTOR [60] and PROCTOR [61].

It might be thought that strong-field dynamos should be sought which exploit to the fullest the advantages of the smoothing method, but this neglects the basic reason for strong-field equilibration, namely that both magnetic and velocity fields acquire a structure on the scale of the core.

Another strategy for dealing with strong fields is simply to write down plausible ordinary differential equations which reflect as closely as possible the underlying physics. A closely related approach utilizes highly truncated modal decomposition [63], but an unrealistic choice of modes can easily produce spurious results. In ref. [52] a five-equation model of a convective dynamo, containing some of the physics of destabilization represented in fig. 6, was put forward. This system has the form

$$\tag{13.4} \dot{A} + \mu A = \alpha B ,$$

$$\tag{13.5} \dot{B} + \eta B = \omega A + CA ,$$

$$\tag{13.6} \dot{C} + C = uc ,$$

$$\tag{13.7} \dot{c} + \chi c = Ku(1 - kC) ,$$

$$\tag{13.8} \varepsilon \dot{u} = u[\mathrm{Ra}\,(1 - kC) - f(E) - u^2] .$$

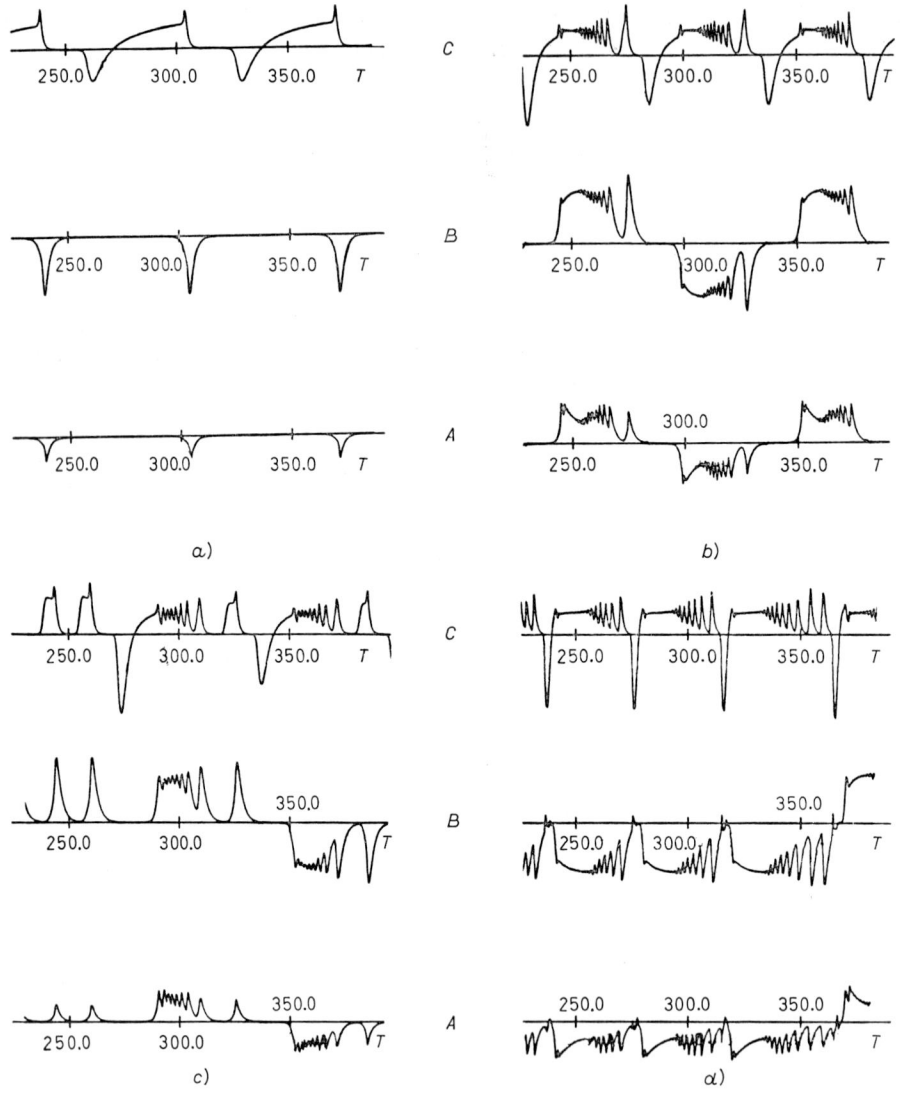

Fig. 7. – Calculated values of A, B and C for the model system (13.4)-(13.9): a) $E_2 = 3$, Ra = 1; b) $E_2 = 4$, Ra = 2; c) $E_2 = 3$, Ra = 2; d) $E_2 = 3$, Ra = 7.

With thermal convection in mind, c and C represent perturbation and mean temperatures, respectively. A and B have the same significance as in (10.5), (10.6), u is the microscale velocity, ω the large-scale (geostrophic?) component of velocity, and $E = A^2 + B^2$. It was assumed in ref. [52] that $\omega = C - AE$, but we shall instead take $\omega = -AAB$, to bring the system closer to the five-equation system mentioned previously [11, 46]. We also set

(13.9) $\qquad \alpha = u^2(1 + \lambda E)^{-1}, \quad f(E) = E(E - E_1)(E - E_2), \qquad 0 < E_1 < E_2.$

Thus the right-hand side of (13.4) represents the primary alpha effect, the right-hand side of (13.5) allows for the field distortion of fig. 4. According to (13.7) microscale temperature fluctuations are created by microscale advection of the *local* mean temperature profile, the latter being determined by turbulent advection through (13.6). The destabilization of the convection field by the magnetic field, consistent with fig. 6, is evident from (13.8) together with the form of f in (13.9). The dependence of ω on A and B is suggestive of (13.3), and we also incorporate, in (13.9), a quenching of the alpha effect at high magnetic energy.

Although the analysis of this system has only just begun, the solution structure appears to be very rich. When ε is small in (13.8), highly intermittent behavior is possible and we have encountered extreme sensitivity to the computational algorithm in certain runs. We show, in fig. 7, some typical results with $\eta = \mu = 0.5$, $k = K = 0.2$, $\varepsilon = \lambda = \Lambda = 1$, $E_1 = 0.1$ and $\chi = 10^{-4}$, for various E_2 and « Rayleigh number » Ra. Note that there are several different « signatures » which occur during reversals in some runs. Run c) appeared to be chaotic.

14. – Concluding remarks.

It should be clear from this brief review that there remain many unresolved problems in modeling of the geomagnetic dynamo, due to the complexity of the magnetohydrodynamics and the difficult parameter range. In spite of these obstacles, one can be hopeful that the construction of a satisfactory « climatic » model will be possible in the not too distant future. It would appear that there is a need for additional fundamental analysis of the kinematic-dynamo problem when the eddy magnetic Reynolds number is *large* and the flow fully three-dimensional. It is not clear how to exploit smoothing in such a setting.

The hydromagnetic models suggest that the global dynamic balance, in the context of macrodynamic postulates, is considerably more involved than one might suspect from the form of the Taylor constraint. But it must also be realized that one cannot yet be confident that a presumed alpha effect can actually be realized in the hydromagnetic case. These are all fundamental issues which will very likely repay the efforts of the interested researcher.

* * *

This paper was prepared under sponsorship of the National Science Foundation under Grant MCS-8301809 at New York University.

REFERENCES

[1] H. K. MOFFATT: *Magnetic Field Generation in Conducting Liquids* (Cambridge, 1978).
[2] E. N. PARKER: *Cosmical Magnetic Fields* (Oxford, 1979).
[3] R. HIDE and P. H. ROBERTS: in *Physics and Chemistry of the Earth*, edited by L. H. AHRENS, K. RANKAMA and S. K. RUNCORN (London, 1961), p. 25.
[4] P. H. ROBERTS: in *Mathematical Problems in the Geophysical Sciences*, edited by W. H. REID, Vol. **2** (Providence, R.I., 1971), p. 129.
[5] N. O. WEISS: *Q. J. R. Astron. Soc.*, **12**, 432 (1971).
[6] D. J. GUBBINS: *J. Geophys.*, **43**, 453 (1977).
[7] F. H. BUSSE: *Annu. Rev. Fluid Mech.*, **10**, 435 (1978).
[8] M. STIX: *Solar Phys.*, **74**, 79 (1981).
[9] T. G. COWLING: *Annu. Rev. Astron. Astrophys.*, **19**, 115 (1981).
[10] F. H. BUSSE: *Annu. Rev. Earth Planet Sci.*, **11**, 241 (1983).
[11] M. GHIL and S. CHILDRESS: *Topics in Geophysical Fluid Dynamics: Atmospheric Dynamics, Dynamo Theory and Climate Dynamics* (New York, N. Y., 1985), in press.
[12] D. W. STRANGWAY: *History of the Earth's Magnetic Field* (New York, N. Y., 1970).
[13] A. COX: *Science*, **163**, 237 (1969).
[14] P. H. ROBERTS and A. M. SOWARD: *Annu. Rev. Fluid Mech.*, **4**, 117 (1972).
[15] G. E. BACKUS: *Ann. Phys. (N. Y.)*, **4**, 372 (1958).
[16] D. R. FEARN and D. E. LOPER: *Science*, **289**, 393 (1981).
[17] P. H. ROBERTS: *An Introduction to Magnetohydrodynamics* (New York, N. Y., 1967), p. 8.
[18] F. GILBERT: in *Mathematical Problems in the Geophysical Sciences*, edited by W. H. REID, Vol. **2** (Providence, R.I., 1971), p. 107.
[19] J. LARMOR: *Rep. Br. Assoc. Adv. Sci.*, **159** (1919).
[20] W. M. ELSASSER: *Phys. Rev.*, **69**, 106 (1946).
[21] E. C. BULLARD and H. GELLMAN: *Philos. Trans. R. Soc. London, Ser. A*, **247**, 213 (1954).
[22] R. D. GIBSON and P. H. ROBERTS: in *The Application of Modern Physics to the Earth and Planetary Sciences*, edited by S. K. RUNCORN (New York, N. Y., 1969), p. 577.
[23] E. N. PARKER: *Astrophys. J.*, **122**, 293 (1955).
[24] S. I. BRAGINSKY: *Sov. Phys. JETP*, **20**, 726 (1964).
[25] F. KRAUSE and F.-H. RÄDLER: *Mean-Field Magnetohydrodynamics and Dynamo Theory* (Oxford, 1980).
[26] E. C. BULLARD: *Proc. Cambridge Philos. Soc.*, **51**, 744 (1955).
[27] D. W. ALLAN: *Proc. Cambridge Philos. Soc.*, **58**, 671 (1962).
[28] A. E. COOK and P. H. ROBERTS: *Proc. Cambridge Philos. Soc.*, **68**, 547 (1970).
[29] W. V. R. MALKUS: *Trans. Am. Geophys. Union*, **53**, 617 (1972).
[30] K. A. ROBBINS: *Proc. Cambridge Philos. Soc.*, **82**, 309 (1977).
[31] The exponential growth of current is misleading when compared with a more precise treatment of the geometry. See H. K. MOFFATT: *Geophys. Astrophys. Fluid Dyn.*, **14**, 147 (1979).
[32] S. CHILDRESS: *Lecture Notes*, Institute Poincaré, University of Paris (1969).
[33] F. H. BUSSE: *J. Geophys. Res.*, **80**, 278 (1975).
[34] M. R. E. PROCTOR: *Geophys. Astrophys. Fluid Dyn.*, **9**, 89 (1977).

[35] T. G. Cowling: *Magnetohydrodynamics* (New York, N. Y., 1957), Chapt. 5.
[36] A. Herzenberg: *Philos. Trans. R. Soc. London, Ser. A*, **250**, 543 (1958).
[37] S. Childress: in *The Application of Modern Physics to the Earth and Planetary Interiors*, edited by S. K. Runcorn (New York, N. Y., 1969), p. 629.
[38] G. O. Roberts: *Philos. Trans. R. Soc. London, Ser. A*, **266**, 535 (1970). In this paper it is shown that a periodic motion will act as a dynamo if a certain scalar functional, quadratic in the Fourier coefficients defining the velocity field, does not vanish. The nondynamos thus lie on a lower-dimensional manifold, and almost all periodic motions excite a magnetic field.
[39] It seems unlikely that this happens in the geodynamo, cf. Part III.
[40] S. Childress: *Phys. Earth Planet. Inter.*, **20**, 172 (1979).
[41] F. H. Busse: *Phys. Earth Planet. Inter.*, **12**, 350 (1976).
[42] A. M. Soward: *Philos. Trans. R. Soc. London, Ser. A*, **272**, 431 (1972).
[43] P. H. Roberts and A. M. Soward: *Annu. Rev. Fluid Mech.*, **4**, 117 (1972).
[44] F. H. Busse: *Geophys. J. R. Astron. Soc.*, **42**, 437 (1975).
[45] R. H. Kraichnan: *J. Fluid Mech.*, **59**, 745 (1973).
[46] S. Childress: in *Stellar and Planetary Magnetism*, edited by A. M. Soward (New York, N.Y., 1983), p. 81 and 245.
[47] H. K. Moffatt: in *Stellar and Planetary Magnetism*, edited by A. M. Soward (New York, N.Y., 1983), p. 3.
[48] I. A. Eltayeb and P. H. Roberts: *Astrophys. J.*, **162**, 699 (1970).
[49] S. Chandrasekhar: *Hydrodynamic and Hydromagnetic Stability* (Oxford, 1961).
[50] A. M. Soward: *Philos. Trans. R. Soc. London, Ser. A*, **275**, 611 (1974).
[51] Busse, in ref. [44], utilizes a geometry much closer to that which might be expected to prevail in the Earth's core, when compared with the Bernard layer used by Soward.
[52] Y. Fautrelle and S. Childress: *Geophys. Astrophys. Fluid Dyn.*, **22**, 235 (1982).
[53] D. E. Loper and P. H. Roberts: *Geophys. Astrophys. Fluid Dyn.*, **16**, 83 (1980).
[54] J. M. Hewitt, D. D. McKenzie and N. O. Weiss: *J. Fluid Mech.*, **68**, 721 (1975).
[55] G. E. Backus: *Proc. Natl. Acad. Sci. USA*, **72**, 1555 (1975).
[56] P. A. Gilman and J. Miller: *Astrophys. J. Suppl.*, **46**, 211 (1981).
[57] P. A. Gilman: *Astrophys. J. Suppl.*, **53**, 243 (1983).
[58] J. B. Taylor: *Proc. R. Soc. London, Ser. A*, **274**, 274 (1963).
[59] S. I. Braginsky: *Phys. Earth Planet. Inter.*, **11**, 191 (1976).
[60] W. V. R. Malkus and M. R. E. Proctor: *J. Fluid Mech.*, **67**, 417 (1975).
[61] M. R. E. Proctor: *J. Fluid Mech.*, **80**, 769 (1977).
[62] G. R. Ierley: *Bull. Am. Phys. Soc.*, **27**, 549 (abstract, 1982).
[63] S. Childress: in *Problems of Stellar Convection*, edited by E. A. Spiegel and J. P Zahn (Berlin, 1976), p. 195.

Large-Scale Turbulence in the Jovian Atmosphere.

JIM L. MITCHELL (*)

Jet Propulsion Laboratory, California Institute of Technology
4800 Oak Grove Drive, Pasadena, CA 91109

T. MAXWORTHY

Departments of Mechanical and Aerospace Engineering, University of Southern California
University Park, Los Angeles, CA 90007

1. – Introduction.

We present evidence, based upon a summary of the results of a more comprehensive report by MITCHELL [1], that a large inertial subrange exists within the atmosphere of Jupiter whose dynamics and energetics are dominated by a strong reverse cascade of energy from smaller scales. This regime is quite different from that of geostrophic turbulence which plays such an important role in the dynamics of the terrestrial atmosphere. Such a reverse energy cascade is capable of supporting planetary-scale vortices, such as the Great Red Spot, whose dynamic balance appears to be consistent with the solitary-Rossby-wave hypothesis.

2. – Turbulent planetary cascades.

KRAICHNAN [2] suggested the simultaneous existence of both energy and enstrophy cascading regimes within two-dimensional turbulent systems. Provided that the spatial domain of these systems is large enough to allow for the inertial growth of turbulent structures, he demonstrated the presence of a direct enstrophy cascade on spatial scales smaller than some monochromatic forcing (external or internal) at length scale L_f and the presence of a reverse energy cascade on spatial scales larger than L_f.

(*) Present affiliation: Naval Ocean Research and Development Activity, Remote Sensing Branch, NSTL, Mississippi 39529.

The existence of each of these regimes, albeit not necessarily simultaneous, has been suggested in geophysical contexts by CHARNEY [3], whose geostrophic-turbulence regime was characterized by direct cascades of enstrophy to smaller spatial scales, and by RHINES [4], whose reverse-energy-cascade regime was characterized by « reverse » cascades of kinetic energy to larger spatial scales. The former regime appears to dominate within the terrestrial troposphere (though reverse cascades are important on the largest atmospheric scales), while the latter regime may well be manifest in oceanic western boundary current regions.

Several identifying characteristics of these two regimes are:

a) In the reverse energy cascade turbulent stresses and ambient mean vorticity gradients are positively correlated, with momentum and kinetic-energy cascades *from* the turbulent field *into* the mean field (defined either spatially or temporally); such positive correlations are not observed in the geostrophic-turbulence regime.

b) The distribution of turbulent kinetic energy as a function of planetary zonal wave index (k) in the reverse-cascade regime obeys a $k^{-5/3}$ power law [2, 5]; the power law obeyed by kinetic energy in the geostrophic-turbulence regime takes a k^{-3} form [2, 3].

c) The spatial dispersion (L) of Lagrangian tracers in the reverse-energy-cascade regime takes the following time (t) dependence:

$$L(t) \sim t^{1/2} \tag{1}$$

being consistent with a constant viscosity coefficient; in the enstrophy cascade regime the dispersion takes a form

$$L(t) \sim \exp[t] \tag{2}$$

as suggested by LIN [6].

The co-existence of both regimes in the Earth's atmosphere has been a subject of much interest and debate over the past several decades. Investigators such as SALTZMAN and TEWELES [7], OORT [8] and STARR [9] have presented evidence for reverse cascades at low wave numbers in the terrestrial atmosphere, while JULIAN et al. [10] have performed spectral analyses of upper tropospheric winds which suggest a k^{-3} law at somewhat higher zonal wave indices, presumably indicative of an enstrophy cascade in the band of zonal wave indices $7 < k < 20$. In the terrestrial troposphere baroclinicity excites large-scale turbulence at an effective baroclinic deformation wave index $k_\mathrm{D} \simeq 7$ (deformation radius L_D) where

$$L_\mathrm{D} \sim \frac{NH}{f_0} \tag{3}$$

for Brunt-Väisälä frequency N, atmospheric depth scale H and Coriolis frequency f_0. Thus the observed k^{-3} power law on scales smaller than k_D is characteristic of Charney's geostrophic-turbulence regime.

The technical difficulty of performing dispersion studies with Lagrangian tracers has greatly limited the use of this discriminating approach to distinguishing instances of the two cascade regimes or their simultaneous existence. Although this approach has received more attention from oceanographers than from atmospheric dynamicists, its most interesting application has been in the deployment and study of the motion of constant-volume ballons in the ter-

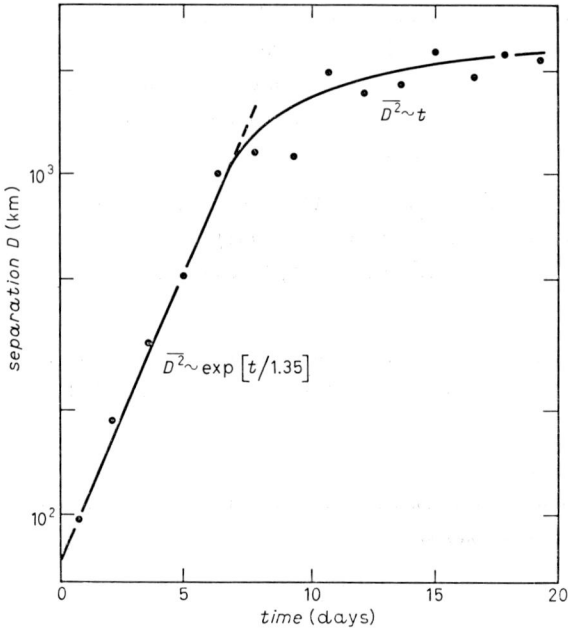

Fig. 1. – Separation distance (D) as a function of time (t) in the constant-volume balloon experiment of Morel and Larcheveque [11]. Note the exponential nature of $D(t)$ for separation distances shorter than approximately 1500 km which is characteristic of an enstrophy cascading regime. The $t^{\frac{1}{2}}$ dependence of D for longer separation distances suggests a reverse energy cascade at these scales. The location of the break in the form of $D(t)$ near the deformation scale is anticipated for baroclinically forced large-scale turbulence.

restrial stratosphere by MOREL and LARCHEVEQUE [11]. Referring to their fig. 8 (reproduced here as fig. 1), it is tantalizing to contemplate the break at approximately the baroclinic radius of deformation (~ 1500 km) from balloon dispersions governed by eq. (2) to those governed by eq. (1). This break clearly suggests the simultaneous existence of an enstrophy cascade for $L < L_D$ and a reverse energy cascade for $L > L_D$. The very small forcing (deformation) scales relative to the scale of the spatial domain within active oceanic regions have to date precluded the performance of a similarly relevant experiment in the oceans.

What other evidence is there in Nature for the existence of such turbulent inertial regimes? STARR [9] originally suggested that the large-scale eddies observed in Jupiter's atmosphere might be manifestations of a reverse energy cascade. In this paper we present an overview of several pieces of evidence that indeed large-scale turbulence in the Jovian atmosphere does manifest a reverse energy cascade. In turn, this regime provides the context for the existence of the largest Rossby wave scale features like the Great Red Spot and the long-lived White Ovals. To a large extent this paper briefly summarizes some of the preliminary results reported by MITCHELL [1], though this paper is not meant to be a comprehensive review of all of the evidence in this extensive preliminary report.

3. – Jovistrophic turbulence.

Best evidence from retrieved atmospheric temperature profiles using the Voyager Infrared Imaging Spectrometer (IRIS) indicates that the baroclinic deformation scale (estimated according to eq. (3)) in the Jovian upper troposphere is of order 10^3 km or roughly the same as in the terrestrial troposphere [12]. RHINES [4] demonstrated the growth of turbulent structures on a β-plane up to the scales of planetary Rossby waves given by

$$(4) \qquad k_\beta \sim \sqrt{\beta/2U}$$

for Coriolis gradient β and velocity scale U. At these k_β scales, resonant wave interactions begin to dominate the less restricted turbulent interactions with the result that the reverse cascade of kinetic energy is truncated. Typical estimates of these β-length scales within the zonal jets of Jupiter's upper troposphere are the order of 10^4 km. Under the assumption that the observed large-scale turbulence in the Jovian atmosphere is forced on smaller scales (which would be the case for baroclinic forcing), one might anticipate that the key differences between the Jovian and terrestrial atmospheres are governed by two factors:

a) the diminished value of β within the Jovian atmosphere due to the order-of-magnitude larger radius of the planet and

b) the higher mean wind speeds associated with the Jovian zonal jets.

Both of these factors tend to increase the wavelength at which the β-restoring force becomes effective, and, hence, give rise to a wider wave band within the reverse-energy-cascade regime. These considerations lead WILLIAMS [13] to suggest that large reverse cascades of kinetic energy occur in the Jovian atmosphere with k_β scale features, such as the strong zonal jets, being the ultimate repository of this energy. In this summary we will briefly examine some of the mounting evidence that this « Jovistrophic » turbulent regime (to contrast with Charney's geostrophic regime) is indeed manifest in the dynamics of Jupiter's upper troposphere.

4. – Global turbulent interactions and energetics.

BEEBE et al. [14] and INGERSOLL et al. [15] reported positive correlation between zonally averaged eddy stresses and the meridional gradient in the zonal jets. Their analyses were based upon large data sets of cloud-tracked winds in pairs of Voyager photographs with typical spatial resolution of the order of 10^2 km. Their estimates of mixing lengths were roughly of the order of $(10^2 \div 10^3)$ km, perhaps consistent with forcing on IRIS-deduced deformation scales. The cloud-tracked wind vectors used in their analyses are somewhat sparse and are nonuniform in their Jovigraphic distribution. SROMOVSKY et al. [16] have argued that the distribution of selected cloud-tracers can bias the deduced stress-strain rate correlations. They have performed analyses similar to those of Beebe et al. and Ingersoll et al. on substantially lower-spatial-resolution photographs. Their cloud-tracers are purposely selected so as to be distributed uniformly. They report the lack of any meaningful stress-strain rate correlations. Unfortunately, their analyses are severely weakened by the fact that the spatial resolution of the photographs used in their study (over 500 km) is probably not adequate to resolve the important mixing length scales as deduced by BEEBE et al. and INGERSOLL et al.

We have approximately doubled the number of cloud-tracers in the Voyager 2 set of global photographs used by INGERSOLL et al. and added to the Jovigraphic coverage with previously unused pairs of photographs (all at approximately 10^2 km spatial resolution) in an attempt to provide as dense and as uniform a set of global wind vectors at as high a spatial resolution as is possible. Analysis of the stress-strain rate correlation in this extensive Voyager 2 data set indicates a global average transport of kinetic energy from turbulent scales of order 10^3 km into the zonal mean flow at a rate of $2.24 \cdot 10^{-4}$ W kg^{-1}. This is very nearly an average of the $3.0 \cdot 10^{-4}$ W kg^{-1} reported from Voyager 1 photographs by BEEBE et al. and the $1.5 \cdot 10^{-4}$ W kg^{-1} reported from Voyager 2 photographs by INGERSOLL et al.

CONRATH, GIERASCH and NATH [17] applied the baroclinic-instability model of Green [18] to Jupiter and argued that baroclinic release of energy is possible only if

$$\frac{\partial \bar{u}}{\partial z} < 0,$$

where $\bar{u}$ is the zonally averaged jet speed and z is the upward positive local vertical co-ordinate. Based upon assumed boundary conditions that the vertical velocity vanishes as $z \to \infty$ and that $\bar{u} = 0$ at $z = 0$ (i.e. the bottom of the atmosphere), they pointed out that the release of baroclinic energy, which may drive the Jovistrophic regime, can occur only in westward jets. We have examined the stress-strain rate correlations in eastward vs. westward jets and find stronger positive correlations in the eastward jets. Thus, if Green mode baroclinicity provides the ultimate excitation for Jovistrophic turbulence, the appropriate upper boundary condition would appear to be $\bar{u} = 0$ as $z \to \infty$. This condition is clearly consistent with the deeper circulating atmosphere suggested by BUSSE [19].

Our attempts to estimate kinetic-energy transports by zonally symmetric stresses (as suggested by STONE [20]) and simple Hadley overturning (as suggested by HESS and PANOFSKY [21]), while far from conclusive, indicate the potential importance of these processes in any considerations of the global momentum and energy budgets. Unfortunately, our estimates of the transports due to zonally symmetric fields are very dependent upon measurement of the zonally averaged meridional velocity ($\bar{v}$) which lies near the noise level of the cloud-tracking process (of order 1 m s^{-1}).

MITCHELL and ALLISON (see [1]) formulate the first energy budget for Jupiter's general circulation based upon kinetic energies estimated from cloud-tracked winds and potential energies as estimated from the spatial distribution of IRIS-deduced vertical temperature profiles. One of the several difficulties with their analysis is their inability to demonstrate appropriate reference of their potential energies to an isentropic surface. Another problem seems to be the marginal spatial resolution of the IRIS system compared with the defor-

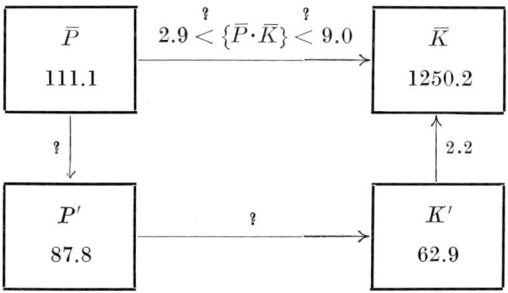

Fig. 2. – Oort energy budget diagram of the Jovian upper troposphere. Potential energies ($\bar{P}$ and P') are evaluated using IRIS observed temperatures at the 196 mb level. Kinetic energies ($\bar{K}$ and K') are evaluated using observed cloud top motions in the Voyager 2 world map data set. Uncertainties in measured values of $\bar{v}$ result in the indicated uncertainties in the symmetric conversion of $\bar{P}$ into $\bar{K}$, while the observed stress-strain rate correlation in global sets of cloud-tracked winds gives the estimate for conversion from K' into $\bar{K}$. Quantities within boxes have units of J kg^{-1}; transport quantities have units of 10^{-4} W kg^{-1}.

mation scale within the Jovian upper troposphere. Nevertheless, their energy budget (summarized as fig. 2) suggests the importance of baroclinic processes, with inferred $\overline{P} \to P'$ conversion leading up to the $K' \to \overline{K}$ conversion, in maintaining Jovian large-scale turbulence. The vast amount of energy stored within the zonal mean kinetic-energy reservoir is consistent with deep, geostrophic-mode circulation and slow radiative-relaxation time scales. Also, they note a correlation in the observed meridional distributions of both eddy kinetic and eddy potential energies as would be anticipated in a baroclinically forced atmosphere. Figure 3 (reproduced from [1]) shows this correlation.

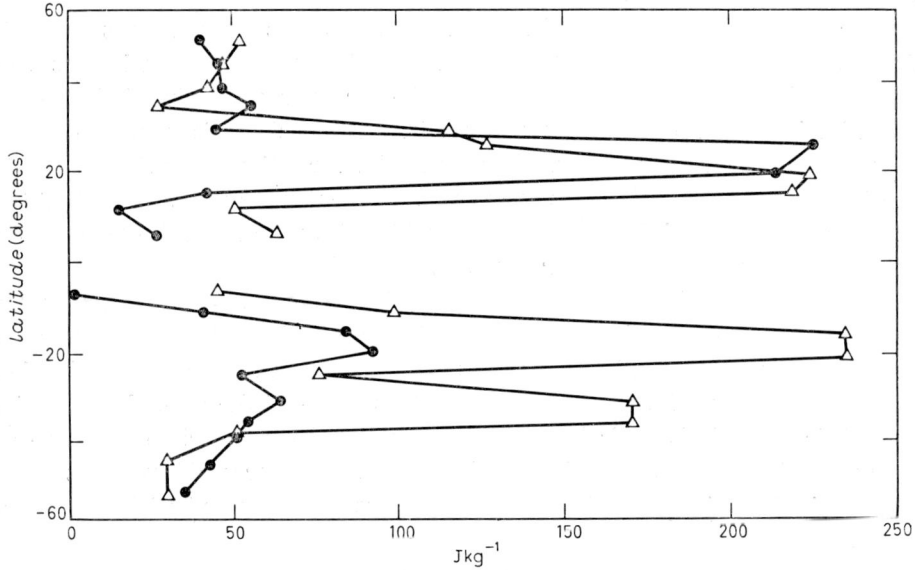

Fig. 3. – Meridional profile of observed cloud top eddy kinetic energy (K') from the Voyager 2 world map (•) and observed IRIS distributions of eddy available potential energy (P') at the 196 mb level from the Voyager 1 so-called high-resolution data sets (△).

We have performed spectral analyses of the global cloud-tracked winds to look for evidence of the reverse energy cascade suggested by the positive correlations in the stress-strain rate fields. Our preliminary approach has been to smooth the original global velocity field to a uniform grid with 2.8° zonally spaced grid points (a Nyquist wave index of 64) and then perform a fast-Fourier-transform analysis. While this naive approach to spectral analysis may well be deficient, the basic results of a composite of the analyses of 13 major zonal jets (both eastward and westward) are quite interesting. As seen in fig. 4, a very good broken linear fit with a breakpoint at $2\pi L = 10\,000$ km can be made to the data. The fact that the slope of this broken linear fit suggests a $k^{-5/3}$ power law for longer wavelengths and a k^{-3} power law for shorter wavelengths is clearly consistent with forcing near scales $L \sim 1500$ km (*i.e.* near the first-baroclinic-mode deformation radius) and a simultaneous enstrophy cascade toward smaller scales with an energy cascade toward larger scales. Clearly,

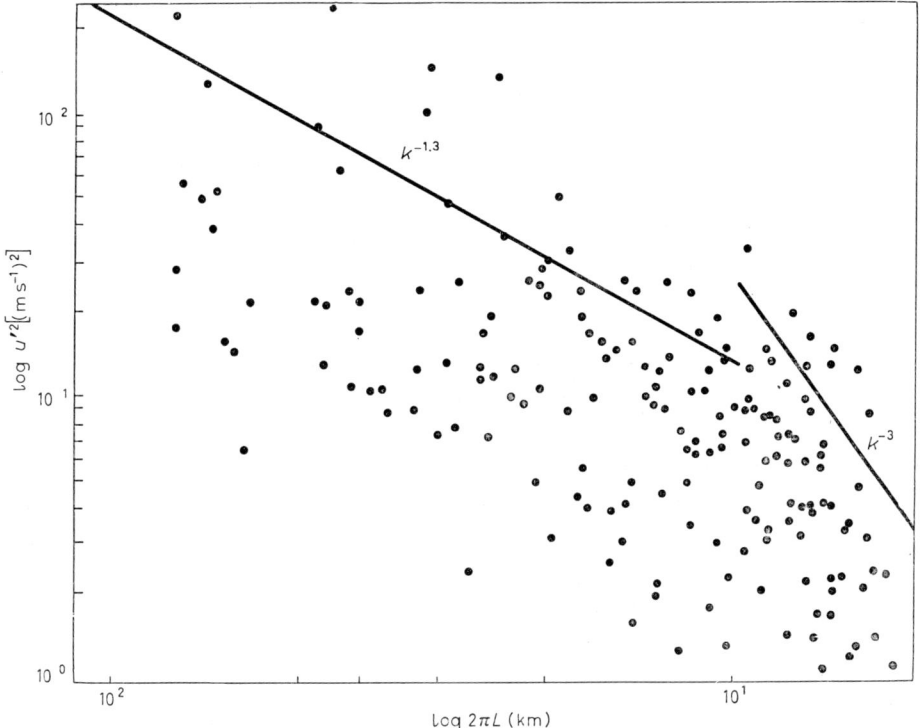

Fig. 4. – Two-segment linear fit to a composite of the fast-Fourier-transform (FFT) spectra of u' velocity component in 13 major zonal jets as deduced from cloud-tracked winds. Plotted on a double logarithmic scale are u'^2 ($\propto K'$) and $2\pi L$ for wavelength L. Also shown are linear least-squares fits to the data. A breakpoint in the slopes of these fits near $2\pi L = 10\,000$ km seems to give the best overall fit to the data. Interestingly, the resulting slopes are roughly -3.0 and -1.3 on the short- and long-wavelength sides of this breakpoint, respectively. Data points represent 4-point running averages of the 13 individual FFTs, each with Nyquist wave indices of 64. Since the data are plotted on a double logarithmic scale, the actual least-squares fits must be performed as fits weighted by $1/\sigma^2$, where $\sigma = \log e/u'^2$. The reader is cautioned not to attempt a visual fit of the data when plotted in this fashion. The apparent discontinuity at $2\pi L = 10\,000$ km is within the uncertainty of the shorter-wavelength fit. On the other hand, the difference in slopes on either of this breakpoint is much greater than the uncertainty in either fit.

a more careful spectral analysis is called for before too many conclusions are made. These preliminary results are, nevertheless, very enticing; whether they are truly meaningful remains to be seen.

5. – Rossby-wave energetics.

The largest Jovian eddies, such as the Great Red Spot (GRS) and White Ovals, have attracted much historical interest. Cloud-tracked Voyager winds indicate that these largest-scale vortices reside within regions of maximum meridional shear in the zonal jets and typically have cross-shear widths roughly equal to the meridional spacing between the westward and eastward jets themselves. As suggested by WILLIAMS [13] and estimated by MITCHELL [1], these cross-shear widths scale approximately as the appropriate β scale lengths. Thus the largest Jovian vortices, as well as the zonal jets themselves, might be expected to represent the ultimate repository of kinetic energy and momentum within the Jovistrophic regime. We briefly review the evidence that such is indeed the case and suggest that the dynamic balances associated with the GRS support the solitary-Rossby-wave hypothesis of Maxworthy and Redekopp [22].

Using four independent sets of high-resolution (roughly 40 to 100 km) Voyager images of the GRS, we have been able to isolate the eddy stress field associated with the GRS. Zonal correlations of this stress field with the ambient meridional shear, as measured in global wind maps, are each found to be consistently negative, indicating a net flow of momentum and kinetic energy from the zonal mean field into the GRS field. An average net energy exchange rate of approximately $-23.7 \cdot 10^{-4}$ W kg^{-1} is deduced from the two Voyager 1 and two Voyager 2 data sets. Based upon an average cloud top GRS kinetic-energy field containing $6.7 \cdot 10^{21}$ J, this suggests e-folding spin-up periods as brief as perhaps one to two rotations of the GRS vortex (period of roughly 6 terrestrial days). Comparison of the meridional profile of the westward jet which overrides the GRS with the shape of a neutrally stable barotropic jet (a parabola on which the meridional gradients of planetary vorticity and shear vorticity are equal) suggests the barotropic-instability mechanism as the continuous supplier of GRS momentum with dissipation counterbalanced by barotropic momentum exchange. One may speculate that the net result is a mean vorticity field of near neutral stability as observed in fig. 5.

Ultimately, the meridional profile of the zonal wind shear through the axis of the GRS should be compared with this ambient profile as a check on the validity of Maxworthy and Redekopp's solitary-wave hypothesis. The work of Beaumont [23] in this respect is not entirely satisfactory as he does not include a critical layer (at which the ambient flow velocity and the linear phase speed of the wave are equal) within his domain for solution of the barotropic-stability equation. An independent check on the validity of the solitary-wave hypothesis may be performed by comparing the IRIS-deduced vertical structure of the GRS with the hypothetical structure of a soliton solution as in the following section. At this point our discussion will, by necessity, become somewhat more mathematical. This is necessary for an explicit understanding of the dynamic balances which define a solitary Rossby wave.

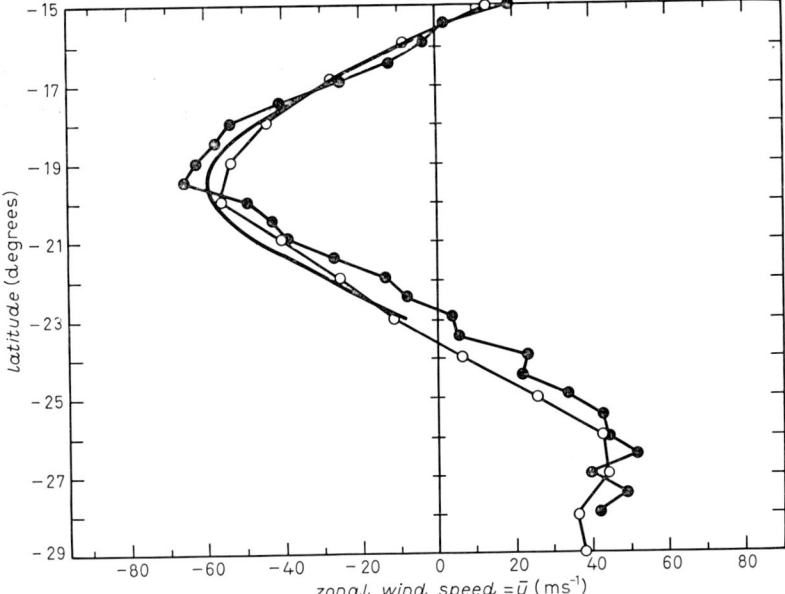

Fig. 5. – Global mean profiles of $\bar{u}(y)$ through the westward jet which runs over the northern edge of the GRS and the region of mean anticyclonic shear in which the GRS is embedded as computed from the world map data sets: • Voyager 1 global average, ○ Voyager 2 global average. The solid parabola represents a locus for which the gradient of the shear vorticity and the gradient of the planetary vorticity are equal for a central latitude of 19.5° (*i.e.* $d^2\bar{u}/dy^2 = \beta$).

6. – Vertical structure of the Great Red Spot.

The solitary Rossby wave represents a first-order perturbation solution to the quasi-geostrophic potential vorticity equation which may be written in logarithmic pressure co-ordinates as

$$(5) \quad \left(\frac{\partial}{\partial t} + \Psi_y \frac{\partial}{\partial x} - \Psi_x \frac{\partial}{\partial y}\right)\left(\frac{\partial^2}{\partial x^2} + \frac{\partial^2}{\partial y^2} + \frac{1}{\varrho_0}\frac{\partial}{\partial z}\left(\frac{\varrho_0 f_0^2}{N^2}\frac{\partial}{\partial z}\right)\right)\Psi + \beta\frac{\partial \Psi}{\partial x} = 0,$$

where

$z = -H \ln(p/p_0)$ for pressure scale height H, pressure p and reference pressure p_0;

Ψ is the total streamfunction;

ϱ_0 is the Boussinesq density $(\varrho_0 = \varrho_0(z))$;

f_0 is the mean Coriolis frequency;

N is a representative Brunt-Väisälä frequency;

and x and y are, respectively, the zonal and meridional local Cartesian coordinates.

Notice that eq. (5) is homogeneous, which is strictly consistent with the observed large momentum fluxes only on longer time scales over which the net momentum flux is presumably balanced by dissipation. Equation (5) simply represents a balance between changes in potential vorticity due to local advection, vortex tube stretching and planetary advection. The nondimensional form of eq. (5) is appropriate only for perturbation length scales (L) of the same order as the first-mode baroclinic-deformation radius (L_D). While it may upon cursory examination appear that the spatial scale of the GRS greatly exceeds this internal deformation scale (indeed, from an *energetics* viewpoint, we have argued that this is the case), we feel that the appropriate length scale governing the *dynamics* of this shear-fed perturbation is the cross-shear length scale. For the GRS this scale might be argued to be of order L_D. Nevertheless, the seeming inconsistency of scaling the GRS as L_β in the context of its energy balance and as L_D in the context of its dynamical balance remains an unresolved issue and a major source of objection to the solitary-wave hypothesis.

Following MAXWORTHY and REDEKOPP [22], nondimensionalization of eq. (5), introduction of a perturbation streamfunction ψ and a multiple scale expansion in the nondimensional amplitude of the perturbation leads to a separable solution for the perturbation of the form

(6) $$\psi = \sum_n A_n(\xi, \tau)\varphi_n(y)Z_n(z),$$

where to first order in perturbation amplitude $A_n(\xi, \tau)$ satisfies a Korteweg-de Vries equation in ξ and τ (the rescaled version of x and t), $\varphi_n(y)$ satisfies the barotropic-stability equation given by

(7) $$\varphi_n'' - k_n^2 \varphi_n + \frac{\beta - U''}{U - C_{0n}}\varphi_n = 0,$$

where C_{0n} is the linear long-wave phase speed for a Rossby wave in the observed shear $\bar{u}(y)$, and $P_n(z)$ satisfies the vertical-structure equation given by

(8) $$P_n''(z) + \left[\frac{N^{2\prime}(z)}{N^2(z)} + \frac{T_0'(z)}{T_0(z)} + 1\right]P_n'(z) - \frac{N^2(z)H^2}{f_0^2 L^2}k_n^2 P_n(z) = 0,$$

where primes denote derivatives with respect to z, and $T_0(z)$ is the ambient, unperturbed thermal profile. H and L are, respectively, the depth and length scale for nondimensionalization, while $N^2(z)$ is once again the Brunt-Väisälä frequency which may be written as

(9) $$N^2(z) = \frac{g^2}{RT_0^2}T_0'(z) + \frac{g^2}{c_v T_0}$$

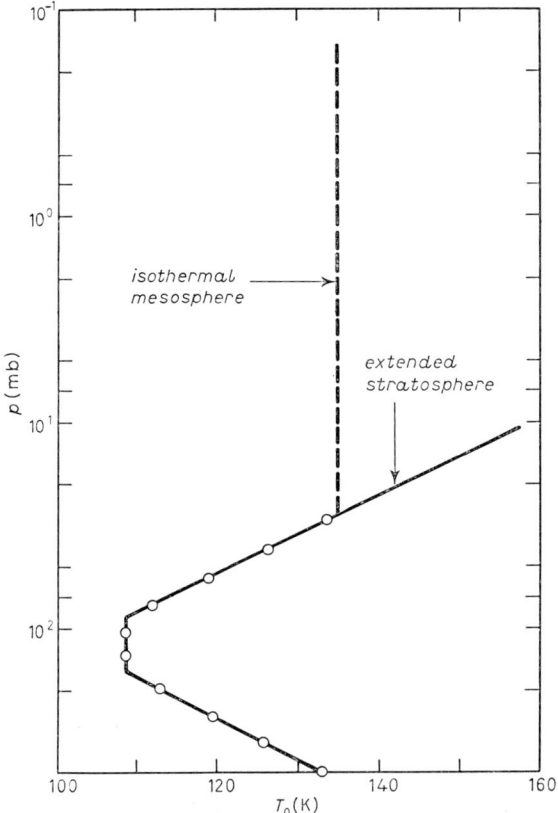

Fig. 6. – Profile of zonal mean brightness temperature over the South Equatorial Belt/South Temperate Zone taken from the Voyager 1 North/South map data set. Center of IRIS weighting functions are indicated by open dots. This profile is used to define the ambient stratification of the atmosphere. It is not clear to what level the stratosphere extends.

for gravitational acceleration g, specific heat at constant pressure c_p and gas constant R.

Use of eq. (9) allows the vertical-structure equation to be written in the form of a second-order eigenvalue problem whose coefficients depend only upon the ambient thermal profile $T_0(z)$ and the depth and length scales H and L. It is important to note that to first order in perturbation amplitude the meridional-structure equation (eq. (7)) and the vertical-structure equation (eq. (8)) remain coupled through the common nondimensional wave number k_n. As we will see, this coupling provides a consistency check on the solitary-wave hypothesis.

Using zonally averaged IRIS temperatures over Jupiter's South Equatorial Belt to define the mean thermal profile (as illustrated according to a simple data fit in fig. 6), MITCHELL [1] uses a fourth-order Runge-Kutta scheme to

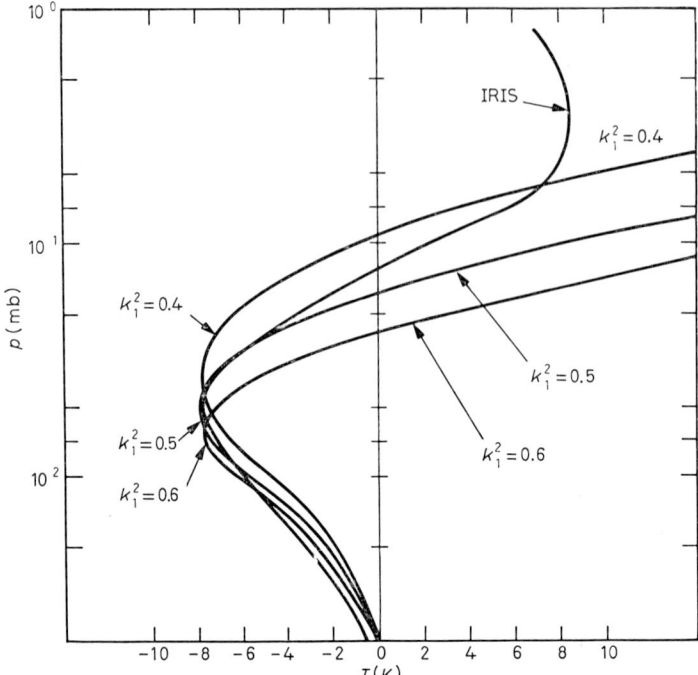

Fig. 7. – Observed perturbation temperature profile above the Great Red Spot as deduced from high-resolution IRIS data sets (see [17]) and computed perturbation temperature profiles based upon numerical solution of the compressible-vertical-structure equation for a range of eigenvalues (k_1). Note that $k_1 \simeq 0.45$ yields the most realistic result. The computed perturbation temperature profiles have been normalized to the observed minimum in the IRIS profile occurring in the lower stratosphere. Vertical propagation of the solitary Rossby wave may readily explain the observed temperature excess above 10 mb.

integrate upward and obtain the set of eigenfunctions for $n = 1$ (*i.e.* the gravest baroclinic mode) whose vertical derivatives are illustrated in fig. 7. Notice that these eigenfunctions satisfy the questionable boundary condition that $P'_n(z) = T(z) = 0$ at $z = 0$, where $T(z)$ is the perturbation (*i.e.* the GRS) thermal profile and $z = 0$ corresponds to the cloud top level. One must also bear in mind that the magnitude of the eigenvalue k_1 is scaled by one's choice of the ratio L^2/L_D^2. As already discussed, the solitary-wave solution given by eq. (6) is valid only for $L^2/L_D^2 \sim O(1)$, *i.e.* strongly stratified cases. Within this strongly stratified regime fig. 7 suggests that the most realistic vertical structure is displayed for $k_1^2 \simeq 0.45$ ($k_1 \simeq 0.7$). Interestingly, this corresponds to the value of $k_1 = 0.7$ selected by MAXWORTHY and REDEKOPP on the basis of Howard and Drazin's [24] and Lipps' [25] barotropically neutral solutions for ambient shears of the form

$$\bar{u}(y) = \operatorname{tgh} y,$$

which is a reasonable fit to the velocity profile in the neighborhood of the GRS. Again, it is obvious that the choice of appropriate length scale (L) is crucial to an understanding of GRS dynamics. Clearly this issue needs further examination.

7. – Conclusion.

In this brief summary of the work of Mitchell [1] we have presented evidence that most of the Jovian cloud top turbulence observed by the Voyager spacecraft occurs within a reverse energy cascading regime. Further, there is evidence to suggest that baroclinicity may play an important role by providing mixing on relatively smaller scales which gives rise to such a regime. On the other hand, large vortices, such as the Great Red Spot, may represent β scale sinks of this up-gradient flux of momentum and kinetic energy. This context may be consistent with the solitary-Rossby-wave hypothesis for the GRS, and, indeed, the IRIS-deduced temperature profile within the GRS suggests the validity of solitary-wave dynamics for describing this feature.

REFERENCES

[1] J. L. MITCHELL: *The nature of large-scale turbulence in the Jovian atmosphere*, Ph.D. Thesis, University of Southern California and JPL publication 82-34 (1982).
[2] R. H. KRAICHNAN: *Phys. Fluids*, **10**, 1417 (1967).
[3] J. G. CHARNEY: *J. Atmos. Sci.*, **28**, 1087 (1971).
[4] P. B. RHINES: *J. Fluid Mech.*, **69**, 417 (1975).
[5] C. E. LEITH: *Phys. Fluids*, **11**, 671 (1968).
[6] J. T. LIN: *Relative dispersion in the enstrophy-cascading inertial range of homogeneous two-dimensional turbulence*, NCAR manuscript MS 71-107 (1971).
[7] B. SALTZMAN and S. TEWELES: *Tellus*, **16**, 432 (1964).
[8] A. H. OORT: *Mon. Weather Rev.*, **92**, 483 (1964).
[9] V. P. STARR: *Physics of Negative Viscosity Phenomena* (New York, N. Y., 1968).
[10] P. R. JULIAN, W. M. WASHINGTON, L. HEMBREE and C. RIDLEY: *J. Atmos. Sci.*, **27**, 376 (1970).
[11] P. MOREL and M. LARCHEVEQUE: *J. Atmos. Sci.*, **31**, 2189 (1974).
[12] R. HANEL, B. CONRATH, M. FLASAR, V. KURDE, P. LOWMAN, W. MAGUIRE, J. PEARL, J. PIRRAGLIA, R. SAMUELSON, D. GAUTIER, P. GIERASCH, S. KUMAR and C. PONNAMPERUMA: *Science*, **204**, 972 (1979).
[13] G. P. WILLIAMS: *J. Atmos. Sci.*, **35**, 1399 (1979).
[14] R. F. BEEBE, A. P. INGERSOLL, G. E. HUNT, J. L. MITCHELL and J. P. MULLER: *Geophys. Res. Lett.*, **7**, 1 (1980).
[15] A. P. INGERSOLL, R. F. BEEBE, J. L. MITCHELL, G. W. GARNEAU, G. M. YAGI and J. P. MULLER: *J. Geophys. Res.*, **86**, 8733 (1980).
[16] L. A. SROMOVSKY, H. E. REVERCOMB, V. E. SUOMI, S. S. LIMAYE and R. J. KRAUSS: *J. Atmos. Sci.*, **39**, 1433 (1982).

[17] B. J. CONRATH, P. J. GIERASCH and N. NATH: *Icarus*, **48**, 256 (1981).
[18] J. S. A. GREEN: *Q. J. R. Meteorol. Soc.*, **86**, 237 (1960).
[19] F. H. BUSSE: *Icarus*, **29**, 255 (1976).
[20] P. H. STONE: *J. Atmos. Sci.*, **24**, 642 (1967).
[21] S. L. HESS and H. A. PANOFSKY: *The atmospheres of the other planets*, in *Compendium of Meteorology*, Americal Meteorological Society (1951), p. 391.
[22] T. MAXWORTHY and L. G. REDEKOPP: *Icarus*, **29**, 261 (1976).
[23] D. N. BEAUMONT: *Icarus*, **41**, 400 (1980).
[24] L. N. HOWARD and P. G. DRAZIN: *J. Math. Phys. (N.Y.)*, **43**, 83 (1964).
[25] F. B. LIPPS: *J. Fluid Mech.*, **21**, 225 (1965).

Part IV

PREDICTABILITY OF GEOPHYSICAL FLOWS

The Growth of Errors in Prediction.

E. N. LORENZ

Massachusetts Institute of Technology - Cambridge, MA

PART I

General Aspects of Error Growth.

1. – Introductory remarks.

Among the innumerable systems which exist in Nature, in the laboratory, or as mathematical abstractions, some are convergent, or stable, while others are divergent, or unstable. By a stable system we mean one whose future succession of states, if the present state should be slightly disturbed, will converge toward the succession of states which would have occurred if there had been no disturbance. By an unstable system we mean just the opposite—a system whose future following a slight disturbance will diverge from what its future would have been without a disturbance.

Stable and unstable systems may be very simple. Consider, for example, a smooth plain in which there is a single bowl-shaped depression. Consider the motion of a ball, which is placed somewhere in the depression and allowed to roll until friction stops it. It will ultimately come to rest at the bottom of the depression. If it had been placed in a slightly different location in the depression, it would still have come to rest at the bottom. Equivalently, if two identical balls had been placed in slightly different locations, they would have come to rest at the same place. The system is stable.

Consider next a smooth plain from which there rises a single dome-shaped hill. If a ball is placed somewhere on the hill and is allowed to roll, it will come to rest somewhere on the plain. If a second ball is placed in a slightly different location on the hill, it may come to rest at a considerably different location on the plain, particularly if the initial locations are near the top of the hill. The system is unstable.

In the former example we do not need a detailed knowledge of the laws of motion to predict the final location of the ball. It is sufficient to know that the ball will seek the lowest point. Moreover, we do not need to know the initial location. It is sufficient to know that it is in the depresssion rather than on the plain. The accuracy of our prediction will be limited only by the degree of precision with which we can locate the bottom of the depression.

In the latter example our prediction of the final location depends upon the details of the laws of motion, including the manner in which friction acts. This does not mean that we must know the laws; we could, for example, have previously placed many balls in different locations on the hill and have learned by experience what final location corresponds to what initial location. We must, however, know the initial location of the ball. If the precision of our measurements allows us to say only that the ball is initially in some small region on the hill, we can say only that the final location will be somewhere in a much larger region on the plain. It is thus apparent that stability favors predictability, while instability opposes it.

Most systems are much more complicated than the two which we have described. Many of them are somewhat analogous to a ball rolling on an undulating surface with numerous depressions and hills of different sizes. Here there are additional possibilities for instability; for example, two balls rolling down a hill along slightly different paths may subsequently encounter another hill and be deflected by its curvature into widely diverging paths. This increases the likelihood that they will ultimately come to rest in the bottoms of different depressions.

2. – The concept of error growth.

In the above examples we may define the state of the system at any particular time as the combined position and velocity of the ball at that time. We may represent the state by a set of four numbers—two position components and two velocity components. Likewise, we may represent the laws of motion, applied to the system, by a set of four first-order ordinary differential equations. Strictly speaking, our systems are more complicated; in addition to rolling, the ball may be spinning about an axis perpendicular to the ground, and its spin may influence its future path. More general systems often require hundreds or thousands of numbers to represent their states, even approximately, and an equal number of ordinary differential equations to represent the governing laws. Alternatively, the laws may sometimes be represented by a relatively small number of partial differential equations.

We shall define an error as the difference between two possible states of a system. In our examples an error may be represented by four numbers, obtained by subtracting the numbers representing one state from those representing the other. The logic of this definition becomes apparent when we consider the case where one state is the true state, and the other is the state which has been observed to exist or is believed to exist, with the inevitable lack of perfect precision in the observations. If each state subsequently varies according to the physical laws, the future error becomes the error in prediction which we would make, using an optimal prediction procedure. The decay or growth of errors, therefore, influences the possible accuracy of predictions. In our first example the ultimate error is merely the error in determining the lowest point in the depression; in the second example it is the difference between two points in a possibly extensive region of the plain.

It is sometimes useful to extend the definition of an error to apply to the difference between states of two different systems. The systems must, of course, be enough alike for the state of one to be subtractable from the state of the other. In our examples the second system could be one where a ball of a different mass or radius is allowed to roll. For practical purposes two systems are different only if the equations governing their states are different. The usefulness of the extended definition becomes evident when one initial state is the true state of a real system, and its equations are the true equations, while the other is the assumed state, and its equations are approximations to the true equations.

To study the predictability of a system, we may investigate the growth or decay of errors. We need not know in advance whether the most feasible method of predicting the behavior of the system actually involves starting out from an initial state. If our investigation indicates that errors will decay, it will tell us that some simpler method is probably available. If it indicates that errors will grow, it will imply that we cannot predict the future without knowing the present or some recent past state.

The systems in our examples are somewhat specialized in that the motion of the ball is a strictly transient phenomenon. In each example the system eventually acquires a steady state. Geophysical fluid systems where prediction is of interest, such as the atmosphere, undergo fluctuations which never cease. It is systems of this sort whose predictability will be our primary concern. Despite the differences, the concepts of stability and instability, and decay and growth of errors, are still relevant.

3. – Simple numerical examples.

Before presenting a general treatment of error growth in systems whose states continue to vary, we shall consider some simple numerical examples. In each of these the state is defined by a single variable $x(t)$, and its variation with time t is given by the quadratic difference equation

(1) $$x_{n+1} = x_n^2 - c$$

rather than by a differential equation, where $x_n = x(t_n)$ and $t_0, t_1, t_2, \ldots$ is a sequence of times. Once the constant c has been specified, an initial value x_0 determines a sequence $x_0, x_1, x_2, \ldots$. If c lies between 0 and 2, and x_0 lies between $-c$ and $+c$, x_n will lie between $-c$ and $+c$. Because a state is defined by a single number, the example cannot illustrate all aspects of error growth. Equation (1) has received much recent attention from mathematicians because of the many types of solutions which it exhibits [1, 2].

TABLE I. – *Particular solutions x_n and y_n of quadratic difference equation $x_{n+1} = x_n^2 - c$, with $c = 1.2$, difference $y_n - x_n$, particular solution z_n with $c = 1.201$, and difference $z_n - x_n$. All values have been multiplied by 10 000.*

n	x_n	y_n	$y_n - x_n$	z_n	$z_n - x_n$
0	5 000	5 010	10	5 000	0
1	− 9 500	− 9 490	10	− 9 510	− 10
2	− 2 975	− 2 994	− 19	− 2 966	9
3	− 11 115	− 11 104	11	− 11 130	− 15
4	354	329	− 25	378	24
5	− 11 987	− 11 989	− 2	− 11 996	− 9
6	2 370	2 374	4	2 380	10
7	− 11 438	− 11 436	2	− 11 444	− 6
8	1 084	1 079	− 5	1 086	2
9	− 11 883	− 11 884	− 1	− 11 892	− 9
10	2 120	2 122	2	2 132	12
11	− 11 551	− 11 550	1	− 11 555	− 4
12	1 342	1 340	− 2	1 343	1
13	− 11 820	− 11 821	− 1	− 11 830	− 10
14	1 971	1 972	1	1 984	13
15	− 11 612	− 11 611	1	− 11 616	− 4
16	1 483	1 481	− 2	1 484	1
77	− 11 708	− 11 708	0	− 11 716	− 8
78	1 708	1 708	0	1 716	8
79	− 11 708	− 11 708	0	− 11 716	− 8
80	1 708	1 708	0	1 716	8

In our first example $c = 1.2$. Table I compares a basic or « true » solution x_n where $x_0 = 0.5$ with a perturbed or « predicted » solution y_n where y_0 has been « observed » to be 0.501. The error $y_n - x_n$ is also shown. If the example represented a real physical system whose state could be measured, the « observational » error 0.001 might not be unreasonably large.

We see that the error amplifies several fold in the first few steps, but subsequently undergoes damped oscillations. Before step 80 the solutions are alike to four decimal places. Evidently the system is stable.

Table I also compares x_n with a solution z_n of (1), with $c = 1.201$. In a real system involving physical constants the error 0.001 in determining one constant c might not be unreasonable. Again, the initial error growth soon ceases, and the error ultimately oscillates between -0.0008 and $+0.0008$.

It is apparent that with this stable system we do not need to know the governing law exactly to make reasonably good predictions far in advance. Neither do we need to know the initial state very closely; it is sufficient to know that x_0 lies between -0.6 and $+0.6$.

In our second example $c = 1.8$. Table II is similar in format to table I. We see that the initially small error $y_n - x_n$ proceeds to grow rather irregularly, generally gaining an order of magnitude in about five steps, until it becomes comparable to x_n itself. At this point the prediction y_n for x_n has become

TABLE II. – Same as table I, with $c = 1.8$ for x_n and y_n, and $c = 1.801$ for z_n.

n	x_n	y_n	$y_n - x_n$	z_n	$z_n - x_n$
0	5 000	5 010	10	5 000	0
1	$-15 500$	$-15 490$	10	$-15 510$	-10
2	6 025	5 994	-31	6 046	21
3	$-14 370$	$-14 407$	-37	$-14 355$	15
4	2 650	2 757	107	2 595	-55
5	$-17 298$	$-17 240$	58	$-17 336$	-38
6	11 922	11 722	-200	12 045	123
7	$-3 786$	$-4 260$	-474	$-3 502$	284
8	$-16 566$	$-16 185$	381	$-16 784$	-218
9	9 444	8 196	$-1 248$	10 160	716
10	$-9 080$	$-11 282$	$-2 202$	$-7 688$	1 392
11	$-9 755$	$-5 271$	4 484	$-12 100$	$-2 345$
12	$-8 484$	$-15 222$	$-6 738$	$-3 370$	5 114
13	$-10 802$	5 170	15 972	$-16 874$	$-6 072$
14	$-6 331$	$-15 328$	$-8 997$	10 465	16 796
15	$-13 992$	5 493	19 485	$-7 059$	6 933
16	1 578	$-14 982$	$-16 560$	$-13 227$	$-14 805$
17	$-17 751$	4 447	22 198	$-1 041$	16 710
18	13 510	$-16 023$	$-29 533$	$-17 902$	$-31 412$
19	253	7 672	7 419	14 037	13 785
20	$-17 994$	$-12 114$	5 880	1 693	19 687

worthless. The system is patently unstable. Of course, the error cannot amplify forever, since both y_n and x_n remain bounded.

If $z_0 = x_0$, but c is taken to be 1.801 instead of 1.8, the situation is similar. The error amplifies at roughly the same rate and levels off similarly. Evidently it makes little difference whether the original uncertainty is in the observed state or in the governing law.

It is evident that the temporary growth rate of $y_n - x_n$ or $z_n - x_n$ is highly dependent on the true state x_n. In a large system with many components, some components might undergo their more rapid growth, while others undergo their slower growth, and the overall growth might be smoother. With the present system we can produce smooth growth by averaging a large ensemble of solutions.

To obtain a representative ensemble, we have extended the solution x_n in table II to 10 000 steps, subsequently using each of the 10 000 values x_n as an

TABLE III. – *Geometric mean ε_n of absolute values of differences $y_n - x_n$ between 10 000 particular solutions x_n and corresponding solutions y_n, with $y_0 = x_0 + 0.001$, of quadratic difference equation $x_{n+1} = x_n^2 - 1.8$. All values have been multiplied by 10 000.*

n	ε_n	n	ε_n
0	10	20	5 534
1	15	21	5 845
2	23	22	6 130
3	34	23	6 380
4	51	24	6 639
5	76	25	6 752
6	114	26	6 901
7	172	27	6 816
8	258	28	6 850
9	385	29	7 034
10	578	30	7 101
11	853	31	7 134
12	1 273	32	7 244
13	1 794	33	7 258
14	2 358	34	7 277
15	2 904	35	7 226
16	3 430	36	7 282
17	4 138	37	7 407
18	4 918	38	7 540
19	5 406	39	7 526

initial value in a new solution. Each new initial value is perturbed by adding an error 0.001. Table III shows the behavior of the average error. At first the growth is almost perfectly exponential, with an amplification factor of about 1.5 per step. Later the growth subsides, and by step 40 it has nearly ceased. The average is actually a geometric mean; the arithmetic mean would grow somewhat less smoothly.

A noteworthy feature of table I is that each solution asymptotically approaches a periodic sequence. This behavior is, in fact, demanded by the stability of the system. Since x_n must be between $-c$ and $+c$, a value x_M closely approximating a previous value x_L must occur in due time. Because

TABLE IV. – *Same as table I, with $c = 1.5$ for x_n and y_n, and $c = 1.501$ for z_n.*

n	x_n	y_n	$y_n - x_n$	z_n	$z_n - x_n$
0	5 000	5 010	10	5 000	0
1	− 12 500	− 12 490	10	− 12 510	− 10
2	625	600	− 25	640	15
3	− 14 961	− 14 964	− 3	− 14 969	− 8
4	7 383	7 392	9	7 397	14
5	− 9 549	− 9 536	13	− 9 538	11
6	− 5 881	− 5 907	− 26	− 5 912	− 31
7	− 11 541	− 11 511	30	− 11 514	27
8	− 1 680	− 1 751	− 71	− 1 752	− 72
9	− 14 718	− 14 693	25	− 14 703	15
10	6 661	6 590	− 71	6 608	− 53
11	− 10 563	− 10 637	− 74	− 10 643	− 80
12	− 3 841	− 3 642	199	− 3 682	159
13	− 13 524	− 13 674	− 150	− 13 655	− 131
14	3 291	3 697	406	3 635	344
15	− 13 917	− 13 633	284	− 13 689	228
16	4 368	3 587	− 781	3 728	− 640
17	− 13 092	− 13 713	− 621	− 13 620	− 528
18	2 139	3 806	1 667	3 540	1 401
19	− 14 542	− 13 552	990	− 13 757	785
20	6 148	3 365	− 2 783	3 915	− 2 233
21	− 11 220	− 13 868	− 2 648	− 13 477	− 2 257
22	− 2 411	4 231	6 642	3 154	5 565
23	− 14 419	− 13 210	1 209	− 14 016	403
24	5 790	2 449	− 3 341	4 633	− 1 157

of the stability the solutions u_n with $u_0 = x_L$ and v_n with $v_0 = x_M$ will approach one another, i.e. the sequence x_n will eventually be almost unchanged by replacing n by $n + M - L$, and the solution will be periodic, with period $M - L$. In our example $M - L = 2$. In the unstable solutions in table II there is no evidence of periodicity.

The most general behavior may be a superposition of periodicity and aperiodicity. To illustrate this possibility, we present a third example, with $c = 1.5$. Table IV shows basic and perturbed solutions x_n and y_n and an additional solution z_n with $c = 1.501$. Again we observe an irregular but unmistakable amplification, implying that the system is unstable, but the amplification ceases well before $y_n - x_n$ or $z_n - x_n$ is comparable to x_n. There is a continual oscillation between small negative or positive values of x_n when n is even and large negative values when n is odd. To someone unaware of this periodicity, y_n and z_n might seem to be moderately good predictions for x_n, for all values of n. If, however, the periodicity is removed from x_n, y_n and z_n by subtracting the average value for all the even, or odd, numbered steps from each value at an even, or odd, step, there remain three aperiodic sequences, the second and third of which do not constitute good predictions for the first in the distant future.

4. – More general systems.

We now consider error growth in more general systems. First of all, the result that stability creates periodicity still holds, provided that the system is one in which analogues, i.e. close approaches to previous states, must occur [3, 4]. A sufficient but not necessary condition for the occurrence of analogues is that the system may be represented by a finite number of numbers, each having a finite range. An equivalent result is that any system which is observed to very aperiodically must be unstable.

In general an unstable system will possess some periodic solutions, but, if these are even slightly disturbed, the periodicity will disappear. The unstable system defined by eq. (1) with $c = 1.8$ possesses many periodic solutions; one of these is an oscillation between $-0.5 + \sqrt{1.05}$ and $-0.5 - \sqrt{1.05}$.

The mere presence of aperiodicity does not reveal the rate at which small errors grow. For a quantitative treatment, let the state of the system at time t be given by M numbers $X_1(t), ..., X_M(t)$, which may be treated as elements of a matrix X with M rows and one column, or a M-dimensional vector. Let the equation governing the system be

(2) $$dX/dt = F(X),$$

and let the elements of F be $F_1, ..., F_M$: Let a basic solution and a perturbed solution be given by X and $Y = X + x$, so that x is the error. If x is sufficiently small, it is approximately governed by the homogeneous linear equation

(3) $$dx/dt = Gx,$$

where the elements G_{ij} of the square matrix G are the partial derivatives $\partial F_i/\partial X_j$. Between a time t_0 and a later time t_1 eq. (3) may be integrated to yield the solution

(4) $$x(t_1) = Ax(t_0),$$

where $A = A(t_1, t_0)$ is a square matrix which depends upon the behavior of X between t_0 and t_1.

If we define the magnitude of the error as the magnitude of the vector x, small errors of a given magnitude at time t_0 satisfy the equation

(5) $$\tilde{x}x = \varepsilon^2 I,$$

where I is the identity of order one.

At time t_1 these errors, therefore, satisfy the equation

(6) $$\tilde{x}(A\tilde{A})^{-1}x = \varepsilon^2 I.$$

Equations (5) and (6) define a sphere and an ellipsoid in M-dimensional space. Whether or not any small errors grow between t_0 and t_1 depends upon whether any semi-axis of the ellipsoid exceeds the radius ε of the sphere. The semi-axes of the ellipsoid are the quantities $\varepsilon\lambda_i$, where $\lambda_1, \ldots, \lambda_M$ are the singular values of A, i.e. $\lambda_1^2, \ldots, \lambda_M^2$ are the eigenvalues of $A\tilde{A}$. These may be numbered in order of decreasing magnitude. Error growth, therefore, depends upon whether the singular value λ_1, or the eigenvalue λ_1^2, exceeds unity.

If t is a later time than t_1, it is evident from (4) that

(7) $$A(t, t_0) = A(t, t_1)A(t_1, t_0).$$

This does not mean that the singular values of $A(t, t_0)$ are products of singular values of $A(t, t_1)$ and $A(t_1, t_0)$. It is even possible that each of the latter two matrices possesses a singular value exceeding unity, while the product matrix does not. To investigate the ultimate growth or decay of small errors, as opposed to temporary growth or decay, we should make $t - t_0$ large. The limiting values

(8) $$l_i = \lim_{t \to \infty} \lambda_i^{1/(t-t_0)}$$

are called the Liapunov numbers of the system, while their logarithms

(9) $$a_i = \lim_{t \to \infty} \log \lambda_i/(t - t_0)$$

are called the characteristic exponents. For many well-behaved systems these are independent of the choice of the state at t_0. For eq. (1), with $c = 1.8$, the single Liapunov number is 1.5; with $c = 1.2$ it is 0.89. (There are other definitions of Liapunov numbers and characteristic exponents, in which the quantities λ_i are eigenvalues of A instead of square roots of eigenvalues of $A\tilde{A}$.)

To evaluate $A(t_1, t_0)$, we may first choose the initial state $x(t_0)$ and then integrate (2) numerically from t_0 to t_1, obtaining $x(t_1)$. We then perturb $x(t_0)$ with M separate one-column matrices $y_1, ..., y_M$, where the j-th component y_{ij} of y_i is $\varepsilon\delta_{ij}$ and ε is very small, and integrate numerically M times from t_0 to t_1: We obtain $x(t_1) + z_i$ for each i, from which we may subtract $x(t_1)$. The M columns of $A(t_1, t_0)$ are the M one-column matrices z_i/ε.

5. – Approximate formulae.

Unless a_2 and a_1 are nearly equal, a randomly chosen small error will eventually behave as if the only characteristic exponent were a_1. The magnitude E of the error will then be governed approximately by the equation

(10) $$dE/dt = a_1 E .$$

A popular measure of the growth of an error is the doubling time, which in this case is given by $\log 2/a_1$.

We have noted that errors do not grow forever. The processes which limit the error growth must be represented by nonlinearities in (2), which have been purposely omitted from (3). A simple assumption, which often gives fairly realistic results, is that these processes are quadratic in E, so that (10) may be replaced by

(11) $$dE/dt = - c_1 E^2 + a_1 E ,$$

where c_1 is chosen so that the limiting value of E is a_1/c_1. Equation (11) cannot be rigorously justified even in the frequent cases in which eq. (2) is quadratic in X, but it often affords a useful means for dealing approximately with the life history of an error. The numbers in table III fit (11) rather well, with $a_1 = 1.5$ and $c_1 = 2$.

If the governing equation (2) is not perfectly known, the prediction $X + x$ for X will obey an equation

(12) $$d(X + x)/dt = F(X + x) + f(X + x) ,$$

where f as well as x is small. In this event eq. (3) will be replaced by

(13) $$dx/dt = Gx + f(X) .$$

The term Gx, depending on x, implies a possible exponential amplification, while the term $f(X)$, which is independent of x, implies a possible additional linear accumulation. Integrating (13) between t_0 and t_1, we obtain

$$z = Ay + B, \tag{14}$$

where A is the same as in (4), and B is a one-column matrix depending only upon the behavior of X between t_0 and t_1. Equation (10), when the approximations leading to it are justified, is then replaced by

$$dE/dt = a_1 E + b_1, \tag{15}$$

where b_1 is a constant. Integration of (15) shows that the eventual exponential growth rate is independent of b_1, i.e. independent of the fact that the wrong governing equation is used. This result is consistent with the behavior observed in table II. Likewise, when E becomes large, (11) is replaced by

$$dE/dt = - c_1 E^2 + a_1 E + b_1. \tag{16}$$

Equation (16) includes the case in which b_1 is large, whence the solutions of (11) and (16) are not alike. This indicates that the proper exponential growth will not be revealed with a completely erroneous governing equation.

The case in which (11) and (16) fail occurs when the most rapidly amplifying mode grows very rapidly to a small limiting value, after which another mode continues to grow more slowly to a larger limiting value. A better approximation than (11) might then be the pair of equations

$$dE_i/dt = - c_i E_i^2 + a_i E_i, \tag{17}$$

for $i = 1$ and 2, where $a_1 > a_2$ and $a_1/c_1 < a_2/c_2$ and $E_1 + E_2 = E$. More generally there may be many modes $E_1, E_2, ...$ which grow at different rates while small and level off at different amplitudes. This is sometimes the case in geophysical fluid systems which possess a wide variety of scales of motion.

Part II

Error Growth in the Atmosphere.

6. – The general atmospheric problem.

Investigations of error growth acquire special significance when the system being studied is one whose behavior has been the object of numerous attempts at prediction. Probably no system fits this description better than the atmosphere.

There is no question but what the atmosphere is an unstable system. Evidence is its lack of perfect or nearly perfect periodicity [3, 4]. The prominent periods in temperature, wind, moisture and other weather elements are the diurnal and annual periods and their principal overtones (semidiurnal, semiannual, etc.). Other periodic variations such as a lunar tide are detectable with careful measurements. Nevertheless, when all known periods are subtracted out, the remaining variations are still large. These include the weather changes associated with the passage of migratory cyclones and anticyclones across the continents and oceans.

The periodic variations can easily be predicted without a detailed knowledge of the governing laws or the present weather pattern; it takes little skill, for example, to predict that during the next century the winters will continue to be colder than the summers. On the other hand, there is a range beyond which the locations of the migratory storms cannot be predicted with detectable accuracy; if this were not the case, these storms would also occur periodically.

As we have noted, the fact that a system is varying aperiodically does not indicate the rate at which errors will grow. The most direct way to determine this rate would be to disturb the system, *i.e.* to introduce an error at some « initial » time, and observe what happens. For some simple systems this procedure might work, but, if our system is the atmosphere, we would never know what would have happened if we had not disturbed it. We might attempt to predict what would have happened and subtract this prediction from what did happen, but, unless we have somehow introduced a disturbance which is as large as our uncertainty in observing the initial state, the uncertainty in the prediction, and hence in evaluating the error, will continue to be as large as the error itself.

As an alternative to introducing an error, we can search through the many years of past atmospheric states which have been recorded and archived. If we find two states which are analogues, *i.e.* which are very much alike, we may regard the second state as equal to the first plus a reasonably small error. We can then determine the growth of the error by subtracting states following the first state from states following the second. We shall presently examine what we have learned about error growth from studying analogue pairs; for the moment we simply mention that, among all states which are presently archived, there appear to be no pairs which would qualify as good analogues [5].

There remains the possibility of using the equations governing the atmosphere, and this is just what has been done in most investigations of atmospheric predictability. Since we cannot formulate the equations exactly, and since, even if we could do so, we could not solve them exactly, we must introduce some simplifying approximations. Systems of equations which to some degree approximate the true atmospheric equations are generally called « models ». In most instances the solutions have been obtained numerically.

It is to be expected that the least simplified models can yield the most realistic results. However, they also require the greatest computational effort. Since in general the growth rate of an error depends upon both the nature of the error and the state on which it is superposed, the labor involved in performing comprehensive investigations with the best available models is prohibitive. Very simple models had to be used some years ago, when computers were slower, or else the scope of the investigations had to be limited.

7. – A simple model.

The first systematic study of error growth was made with a model in which the state of the atmosphere was represented by only 28 numbers, and its behavior was governed by 28 ordinary differential equations [4, 6]. The study was concerned mainly with the growth of very small errors. The model was derived from the familiar two-level quasi-geostrophic equations, with thermal and mechanical damping and thermal forcing, by expanding the streamfunction at each level in a double Fourier series, and then retaining only 14 terms in each series. Two of these terms represented a zonally symmetric flow with a variable profile, while the remaining twelve represented three interacting waves with variable shapes and longitudinal phases.

The simplicity of the model permitted a special treatment of random errors. We may denote the dependent variables of the model by $X_1, ..., X_M$, where $M = 28$, and let the governing equation be eq. (2). Small errors then obey (3) and (4), and small errors of a given magnitude at time t_0 satisfy (5). The squared magnitude of an error at time t_1 is then

$$(18) \qquad \tilde{x}(t_1) x(t_1) = \tilde{x}(t_0) \tilde{A} A x(t_0) .$$

If an ensemble of errors is random at time t_0, aside from the existence of a common magnitude ε,

$$(19) \qquad \langle x_i(t_0) x_j(t_0) \rangle = \varepsilon^2 \delta_{ij}/M ,$$

where the pointed brackets denote an ensemble average. If follows, since the eigenvalues of $\tilde{A} A$ equal those of $A \tilde{A}$, that

$$(20) \qquad \langle \tilde{x}(t_1) x(t_1) \rangle = \varepsilon^2 \sum_{i=1}^{M} \lambda_i^2/M .$$

That is, the amplification of a random error is simply the root-mean-square singular value of A.

We first performed a basic run of 64 simulated days, using three-hour time steps in the integration, and retained the values of x at times $t_0, t_2, ..., t_{62}$, the subscript denoting the number of days since t_0. At each time t_j for $j = 0, 2, ..., 62$ we then determined the matrix $A(t_{j+2}, t_j)$; this required 28 two-day runs for each value of j. Finally we determined the singular values λ_i of $A(t_{j+2}, t_j)$ and, using (20), found the average two-day amplification factor $\alpha(t_{j+2}, t_j)$ of small errors which were random at time t_j.

We found that α varied greatly with X. On four of the 32 occasions $\alpha(t_{j+2}, t_j)$ was actually less than unity; on three occasions it exceeded 3.0.

We then determined the matrices $A(t_{j+4}, t_j)$ for $j = 0, 4, ..., 60$, $A(t_{j+8}, t_j)$ for $j = 0, 8, ..., 56$, etc., using (7). Three of the 16 amplifications $\alpha(t_{j+4}, t_j)$ were less than 2.0; two exceeded 10.0. Both 32-day amplifications were greater than 100-fold, while the single amplification $\alpha(t_{64}, t_0)$ was $2 \cdot 10^4$; this implied an average doubling time of 4.5 days. From these results, and from estimates of the accuracy with which real atmospheric states could be determined, we concluded that, pending further results from larger models more closely resembling the real atmosphere, the prospect was rather favorable for forecasting a week in advance and unfavorable for one-month forecasts.

8. – Early experiments with global circulation models.

At the time when the results of the 28-variable study appeared, the meteorological community was in the process of planning the Global Atmospheric Research Program (GARP). This was to be an international program involving many related observational and theoretical investigations. One of its stated aims was to extend the range at which useful predictions could be achieved [7].

As the 4.5-day doubling time indicated by the 28-variable model, and also the general result that an aperiodically varying system could not be predicted at sufficiently long range, became generally known, the question arose as to whether the goal of extended-range prediction might be unattainable. It soon became obvious that error growth studies ought to be performed with the most realistic models possible.

At that time the suitable models in existence were those which had been developed by SMAGORINSKY [8], LEITH [9] and MINTZ [10], who, at a special meeting devoted to the planning of GARP, consented to performing some error growth experiments with their respective models. To the extent possible, the experiments were to have a common format.

Each model had its individuality. Smagorinsky's model, which was the oldest, was a two-layer model covering most of the northern hemisphere. Heating was Newtonian in form, and surface friction and lateral diffusion provided the mechanical damping. The underlying surface was uniform, and the atmosphere was dry. Leith's model had six layers and was global in extent. The underlying surface was again flat, but the model included moisture and an accompanying release of latent heat. Radiation and small-scale convective heating were present, and a somewhat unrealistically large horizontal eddy diffusion coefficient was used to control the computational stability. Mintz's model had only two layers, but was also global. Moisture was not present explicitly, but the underlying surface possessed oceans and continents, and the continents had mountains. The model contained radiative and convective heating and a reasonably small eddy diffusion coefficient; computational stability was controlled by a differencing scheme designed by ARAKAWA [11]. It should be noted that none of these model was constructed for the purpose of investigating predictability.

With each model a basic run was performed and was followed by one or more perturbed runs in which the initial error was confined to the temperature field. In Leith's model, the error decayed rapidly during the first few days and then partially recovered, but leveled off while still small. Once the result was obtained, it became apparent that the large diffusion coefficient which prevented computational instability also suppressed much of the real atmospheric instability and that the migrating disturbances were passing by in essentially periodic sequence.

In Smagorinsky's models the errors grew rather slowly, but after two months acquired a reasonable amplitude before leveling off. The more rapid growth did not commence until the second month. Examination after the fact indicated that during the first month the variations were mainly periodic, subsequently becoming more irregular. During the second month the doubling times was about eight days.

In Mintz's model the error initially decayed, but then grew regularly, with a five-day doubling time. It leveled off with a reasonable amplitude. The visible outcome of these experiments was a report on the feasibility of a global meteorological experiment [7], which concluded that only Mintz's model displayed a reasonable absence of periodicity and that a five-day doubling time represented the best available estimate.

There were some interesting additional results. In the experiments common to all models the initial error field was sinusoidal. SMAGORINSKY and MINTZ also performed experiments with random and localized initial errors and found that the form of the error had little effect on the growth rate. MINTZ also compared the northern and southern hemispheres and found that the errors tended to be smaller in the southern hemisphere (where it was summer), particularly when the initial error was confined to the northern hemisphere, but that the growth rates in the two hemispheres were similar. We have already noted that SMAGORINSKY obtained different growth rates during different months, when the weather patterns were different.

9. – Later studies with global circulation models.

A significant advance in predictability studies with large models came with a new study by SMAGORINSKY [12]. His new model was a nine-level primitive-equation model covering the northern hemisphere. It contained all the advanced physical features of Leith's and Mintz's models; in addition, the underlying surface contained sea ice and land ice and snow, and the absorbers of radiation included water vapor, carbon dioxide and ozone. Actually the model had been in existence when the earlier experiments were performed, but it could not be used then because it would have required too much computation.

In format Smagorinsky's experiment was like the earlier ones. He found that after the first day small errors were doubling in about three days. As they became larger, the growth rate subsided, but the errors had not reached their ultimate size by three weeks, when the computations were terminated. Interesting additional results were that the temperature errors grew most rapidly in the lower troposphere, while, when a spectral analysis was performed, the smaller scales were found to grow most rapidly.

Following these studies, error growth experiments using newly developed or improved global circulation models made frequent appearances [13, 14]. We shall confine our description to one of the most recent ones, which we performed with the operational model of the European Centre for Medium Range Weather Forecasts (ECMWF) [15]. This is a 15-level primitive-equation model which contains most of the refinements developed in the decade or more since Smagorinsky's experiment, including a nearly complete hydrological cycle.

As with some of the earlier studies, the amount of computation required for a comprehensive study would have been prohibitive. However, it turned out that nearly all of the computations needed for a more limited study had already been performed in the course of preparing the operational forecasts.

Forecasts from one to ten days in advance are issued daily at ECMWF. The feature which makes it possible to use these forecasts in an error growth study is the rather high quality of the one-day forecasts; thus the one-day forecast for today's weather pattern may be regarded as equal to today's pattern plus a moderately small error. To see how much this error grows in one day, when both patterns are governed by the equations of the model, it is sufficient to compare the two-day forecast for tomorrow with the one-day forecast for tomorrow. Likewise, we can determine the error growth during the next day by comparing the three-day with the two-day forecast for the day after tomorrow, and we may, in fact, continue the process for nine days. Additional estimates of the growth rate of somewhat larger errors may be obtained by comparing K-day with J-day forecasts, for various values of J and K.

As a measure of the error we have used the root-mean-square difference of the 500 mb height fields; this choice effectively gives greater weight to middle and higher latitudes. We have used the analyses and forecasts for each day of the 100-day period beginning 1 December 1980. Table V shows the average error when J-day and K-day forecasts are compared. A zero-day forecast is simply an analysis.

TABLE V. – *Average global root-mean-square 500 mb height differences E_{JK}, in meters, between J-day and K-day forecasts for the same day, during the period 1 December 1980 - 10 March 1981, made by the ECMWF operational model.*

J	E_{J1}	E_{J2}	E_{J3}	E_{J4}	E_{J5}	E_{J6}	E_{J7}	E_{J8}	E_{J9}	E_{J10}
0	24	38	51	63	75	85	93	99	104	108
1		29	45	59	71	82	90	97	102	106
2			36	53	67	78	88	95	100	104
3				44	62	74	85	93	99	103
4					52	70	81	91	97	102
5						61	77	86	95	100
6							68	82	91	97
7								74	87	94
8									79	90
9										84

The top row, comparing forecasts with analyses, reveals the average rate of error growth when two states are governed by two different systems of equations—the true atmospheric equations and the model. It, therefore, indicates the average performance of the model. The diagonal rows, where $K - J$ is constant, reveal the rate of growth when both states are governed by the model equations; this is the rate usually sought in predictability studies. Comparison of the diagonals for different values of $K - J$ indicates that, to a reasonable approximation, the amplification rate dE/dt is a function of the error E, so that the numbers in the lower diagonal may be extrapolated for several more days.

The smallest error, 24 m, doubles in about 3.5 days. The usually quoted doubling time is, however, the doubling time for very small errors. We have extrapolated the numbers on the lowest diagonal to smaller errors, using eq. (11). We find that very small errors double in 2.4 days.

Comparing the results of predictability studies performed with hemispheric or global circulation models, we find that the estimates of the doubling time have continually decreased. The first and simplest model, Smagorinsky's original model, suggested about eight days, when it was behaving aperiodically [7]. Mintz's model, with such features as oceans and continents, indicated five days [7]. Smagorinsky's more recent model, with nine levels, reduced the time to three days [12], while the most refined model, the ECMWF model, gave 2.4 days [15].

It seems unlikely that this continual decrease is coincidental. We suspect that the most appropriate doubling time is rather short, perhaps two days. The earlier models were necessarily the simplest, and one result of the simplifications appears to have been an underestimate of the atmosphere's instability. The ECMWF model, whether because of additional physical features, higher spatial resolution, or superior numerical techniques, appears to give the closest estimate.

Global circulation models are still undergoing development, and we have attempted to estimate the doubling time which some future model would yield by noting that the ECMWF model makes some systematic errors. Further refinements in the physics or mathematics will presumably remove these errors, but, in the mean time, we can subtract them out. We have done this and have repeated the study which led to table V. The new doubling time is 2.1 days [15]. This is consistent with our hypothesis that continual refinement will continue to shorten the doubling time.

10. – An analogue study.

To pursue the matter further, we turn to one of the few atmospheric predictability studies which involves no models [5]. It is based on the occurrence of analogues. As we have noted, if we can discover two weather patterns which closely resemble each other, their difference will constitute an error whose growth can be studied. If the patterns occur at the same time of year, they will be governed by effectively the same equations; at different times of year the different diabatic heating fields might cause them to behave rather differently. Ideally the values of each weather element should be nearly alike at all points of the globe.

Our study was based upon twice-daily observations for the five years 1963-1967. At that time there were no data which would reliably indicate whether two patterns were similar in the southern hemisphere, and even in the northern hemisphere we lacked large-scale cloudiness and moisture fields. We were ultimately led to using the height fields at 850, 500 and 200 mb in the northern hemisphere and defining the error as a weighted root-mean-square height difference. Patterns which are sufficiently alike in these fields should also be somewhat alike in temperature, according to the hydrostatic relation, and wind, according to the geostrophic relation. Today we could probably evaluate errors defined in terms of global fields of cloudiness, as measured by satellite.

We compared only those patterns which occurred within one month of the same time of year, but in different years (or different winters, if they occurred in December and January). This yielded a total of about 400 000 error values. These values were normalized so that two randomly chosen weather patterns at the same time of year would yield an error $E = 1$.

We had hoped to find some moderately small errors, say 0.2 or 0.3, but the smallest error encoutered was 0.62. We were thus forced to base our study on mediocre analogues. We first observed that the value of dE/dt, averaged over all cases in which E fell within a given narrow range, was a smoothly varying function of E, having the same sign as $1 - E$. Upon invoking the quadratic hypothesis as expressed by eq. (11), we were able to extrapolate to small values of E, and we found a doubling time of 2.5 days.

It should be noted that this study was performed before SMAGORINSKY had obtained the three-day doubling with his newer model and that the generally accepted value, among those who accepted the reality of error growth, was Mintz's five days [7]. In seeking to explain the discrepancy, we considered the possibility that the analogue results were invalid, but decided that models were probably more likely than observations to go astray. In particular, it appeared that the Arakawa differencing scheme [11] which ensured computational stability could also render the simulated atmosphere more stable [16]. The more recent model studies agree with the analogue study to within 20 percent one way or the other; closer agreement is unlikely in view of the differences in the dates and locations of the data and the quantities chosen to define the error.

11. – The influence of smaller scales.

Error growth studies based on global circulation models tell us only about the growth of errors in the spatial scales which the models resolve. Likewise, the study based on analogues tells us only about the scales appearing in the analyses. By omitting the smaller scales, we effectively make the initial errors in these scales large enough so that they undergo no further growth. Even if the smaller scales were resolved, the error growth in these scales would constitute only a minor part of the total growth, since these scales account for only a small part of the total variability.

The influence of errors in the smaller scales upon errors in the larger scales can nevertheless be large. In most models which retain only the larger scales explicitly, the smaller scales are treated as turbulent eddies, and their effect on the larger scales is represented in terms of coefficients of eddy viscosity and eddy conductivity. In reality the small-scale features which are present at any one time constitute no more than a statistical sample, and their effect is subject to sampling fluctuations. Thus the fact that the details of the smaller scales are uncertain introduces some uncertainty into the behavior of the larger scales. To predict the larger scales perfectly, we would also have to predict the smaller scales perfectly [17].

It appears that errors in very small scales produce an almost undetectably small direct accumulation of errors in the very large scales. Nevertheless, they may cause errors to accumulate fairly rapidly in slightly larger scales. Once these have become appreciable, they will produce an accumulation in still larger scales, etc., and the eventual result will be errors in the largest scales.

Whether errors in the small scales are of practical importance depends on how rapidly the accumulation in the large scales can occur. Suppose that a weather pattern contains an incipient storm and that the initial error consists of overlooking the storm. The error will then grow just as rapidly as the storm itself. If the storm is a migratory cyclone, the error may double in two days or less. If it is a thunderstorm, it may double in less than an hour.

It thus appears that errors in the very small scales require very little time before they have attained the size at which they can appreciably affect somewhat larger scales. These scales will require somewhat longer, but not too long, to affect still larger scales, etc. The initial amplitudes of the smallest-scale errors are, therefore, of little concern, provided that they are not zero. Even an uncertainty of a factor of ten in the magnitude of the error in the thunderstorm scale will cause an uncertainty of only a few hours in the time required for the larger scales to be affected. Equivalently, if we could somehow make perfect measurements of all scales larger than thunderstorms—a task which, incidentally, would be far more expensive than any observational program so far undertaken—, we would subsequently add only a few hours to the range of useful prediction by accomplishing the equally expensive task of improving our thunderstorm observations by a factor of ten.

Let us consider the possible procedures for determining how soon a small error in a small scale will significantly affect the large scales. We can certainly accomplish nothing by physically perturbing the small scales and observing what happens, because again we would have nothing with which to compare the perturbed state. We can construct global models with one-kilometer or even finer horizontal resolution, but to solve them numerically would vastly overburden even today's fastest computers. The studies which have provided tentative answers have taken existing atmospheric models—generally rather simple ones—and derived new systems of equations whose dependent variables represent the amplitudes of the errors, in the various scales [17-19]. The number of equations is minimized by introducing such simplifications as spatial homogeneity and assuming that the spectral amplitude is a smooth function of scale, so that fairly coarse spectral resolution is allowable.

In the first study of this sort [17], our atmospheric model was the simple barotropic-vorticity equation, with neither damping nor forcing. The new variables were the spectral amplitudes in 21 consecutive bands, extending from the 40 m scale to the 40 000 km scale. Growth rates of errors depend very much upon the properties of the basic state on which they are superposed, and the variance spectrum of the basic state was specified in advance. As is generally the case when the original equations are quadratic, the derived equations for means contain covariances, the derived equations for covariances contain mean triple products, etc., and at some point an auxiliary assumption must be introduced to close the system. Our assumption was that quadratic functions of the errors and quadratic functions of the basic state were statistically independent. Subsequent studies with more realistic closure assumptions have yielded rather similar results [18, 19].

The derived equations were formally linear. The nonlinear effects, which, as in eq. (1), should prevent the errors from growing indefinitely, were incorporated by replacing each variable by a constant as soon as that variable, representing the amplitude of the error in one scale, reached the prespecified amplitude of the basic state in that scale.

TABLE VI. – Times $t_n(10)$ and $t_n(90)$ required for mean square error in spectral band n, with average wave length L_n, to attain 10 percent and 90 percent, respectively, of its limiting value, in theoretical model with no spectral gap, and similar times $t'_n(10)$ and $t'_n(90)$ in theoretical models with strong spectral gap centered near band 10, when initial errors are confined to band 20.

n	L_n (km)	$t_n(10)$	$t_n(90)$	$t'_n(10)$	$t'_n(90)$
12	12	0.8 h	1.3 h	1.0 h	1.5 h
11	25	1.3	2.1	1.8	2.9
10	50	2.2	3.4	3.6	7.6
9	100	3.5	5.6	7.8	3.5 d
8	200	5.8	9.2	2.3 d	4.6
7	400	9.6	15.2	3.7	5.5
6	800	15.9	1.1 d	4.7	6.3
5	1 600	1.1 d	1.8	5.2	6.7
4	3 200	1.9	3.1	6.0	7.6
3	6 400	3.3	5.4	7.6	9.8
2	12 800	5.9	9.8	10.2	14.2
1	25 600	9.4	16.3	13.8	20.6

In the numerical integrations the initial error was confined to the smallest scales. The middle columns in table VI show the times required for the squares of various amplitudes to attain 10 percent, and 90 percent, of their limiting values. Predictions with a 10 percent error would generally be considered good, while those with a 90 percent error would be almost worthless.

We see that the error progresses up the scale rapidly at first and then more slowly, until, after about half a day, 10 percent errors have reached the scales commonly resolved by global circulation models. The largest scales remain predictable for a week or two.

The final columns in table VI were produced by similar computations, in which the basic state was assumed to possess a «spectral gap» centered in the mesoscale [20]. The errors progress through the smallest scales more or less as before, but they encounter considerable difficulty in crossing the gap and are delayed by several days in reaching the larger scales. Whether or not a spectral gap exists is still a topic for debate. In any event, it is apparent that a definite determination of the range at which the larger scales are predictable will require a more precise knowledge of normal atmospheric conditions than we presently possess.

12. – Concluding remarks.

We have hypothesized that the errors presently made in one-day prediction by the «improved» ECMWF operational model are similar in magnitude and spectral distribution to the errors which would be present in J-day prediction as a result of the inevitable errors in the very small scales, if the largest scales possessed no initial errors at all [20]. Assuming a somewhat weaker spectral gap than the one leading to the final columns in table VI, we have estimated that $J = 4$. It, therefore, appears reasonable to add about three days to estimates, based upon table V, of the range at which predictions of a given quality are possible.

Having seen that there is a limit to atmospheric predictability, it is relevant to ask why there should be a limit, $i.e.$ why the atmosphere should be unstable. In the 28-variable atmospheric model the cause is easy to identify; it is advection, which is the only nonlinear process. Advection appears in the model as a displacement of the temperature and vorticity fields by the wind field; since the wind which does the advecting is not uniform, it produces distortion as well as displacement. This increases the variety of temperature and vorticity patterns which can occur and reduces the likelihood of periodic repetition.

In the real atmosphere and also the global circulation models the cause of instability is probably advection also. There are other important nonlinear processes, including evaporation and condensation of water, and radiation. We suspect, however, that the latter processes by themselves would not produce instability, while advection by itself would. Very likely these processes are important modifiers to the instability which advection produces.

* * *

This work has been supported by the GARP Program of the Atmospheric Sciences Section of the National Science Foundation, Grant No. 82-14582 ATM.

REFERENCES

[1] J. GUCKENHEIMER: *Invent. Math.*, **39**, 165 (1977).
[2] M. FEIGENBAUM: *J. Stat. Phys.*, **19**, 25 (1978).
[3] E. LORENZ: *J. Atmos. Sci.*, **20**, 130 (1963).
[4] E. LORENZ: *Trans. N. Y. Acad. Sci.*, II, **25**, 409 (1963).
[5] E. LORENZ: *J. Atmos. Sci.*, **26**, 636 (1969).
[6] E. LORENZ: *Tellus*, **17**, 321 (1965).
[7] J. CHARNEY, R. FLEAGLE, V. LALLY, H. RIEHL and D. WARK: *Bull. Am. Meteorol. Soc.*, **47**, 200 (1966).
[8] J. SMAGORINSKY: *Mon. Weather Rev.*, **91**, 99 (1963).
[9] C. LEITH: *Methods Comput. Phys.*, **4**, 1 (1965).
[10] Y. MINTZ: *W. M. O. Tech. Notes*, **66**, 141 (1964).
[11] A. ARAKAWA: *J. Comput. Phys.*, **1**, 119 (1966).
[12] J. SMAGORINSKY: *Bull. Am. Meteorol. Soc.*, **50**, 286 (1969).
[13] R. JASTROW and M. HALEM: *Bull. Am. Meteorol. Soc.*, **51**, 490 (1970).
[14] D. WILLIAMSON and A. KASAHARA: *J. Atmos. Sci.*, **28**, 1313 (1971).
[15] E. LORENZ: *Tellus*, **34**, 505 (1982).
[16] E. LORENZ: *Bull. Am. Meteorol. Soc.*, **50**, 345 (1969).
[17] E. LORENZ: *Tellus*, **21**, 289 (1969).
[18] C. LEITH: *J. Atmos. Sci.*, **28**, 148 (1971).
[19] C. LEITH and R. KRAICHNAN: *J. Atmos. Sci.*, **29**, 1041 (1972).
[20] E. LORENZ: *Predictability of Fluid Motion*, Am. Inst. Phys., Conf. Proc. (1983).

Two-Dimensional Coherent Structures.

C. E. Leith

National Center for Atmospheric Research (*) - *Boulder, CO 80307*

1. – Enstrophy cascade in 2D turbulence.

Turbulent flows in two dimensions differ qualitatively from those in three dimensions owing to the existence of new inviscid integrals related to the conservation of vorticity in addition to the energy integral common to both geometries. Of these the enstrophy (half-squared vorticity) plays a dominant role.

In three-dimensional turbulence, energy in the larger scales of motion can cascade by means of nonlinear processes toward smaller scales, where, at some sufficiently small scale, viscosity becomes important and energy is removed. The Kolmogorov-Obukhov scaling arguments for the inertial range of scales through which only cascading of energy takes place leads in wave number space to the well-known spectral law

$$E = \alpha \varepsilon^{2/3} k^{-5/3} . \tag{1}$$

For the dimensions of the spectral energy density E [$L^3 T^{-2}$], the energy cascade rate ε [$L^2 T^{-3}$] and the wave number k [L^{-1}], eq. (1) is the only dimensionally consistent combination. The dimensionless coefficient has been determined by observations to be about 1.5.

In two-dimensional turbulence a new possibility arises of an enstrophy cascading inertial range. Dimensional scaling arguments now lead to the energy spectrum

$$E = \beta \eta^{2/3} k^{-3} \tag{2}$$

characterized by the constant enstrophy cascade rate η [T^{-3}], whose dimensions, differing from those of ε, change the spectral dependence in eq. (2) from that in eq. (1).

(*) The National Center of Atmospheric Research is sponsored by the National Science Foundation.

Observations, numerical simulations and stochastic turbulence models show that this is more than a theoretical possibility. Two-dimensional spectra are characterized, for intermediate wave numbers, by eq. (2); it is enstrophy that is predominantly cascaded toward small scales to be removed by viscosity or by a breakdown of two-dimensional constraints.

What then happens to the energy? The arguments leading to eq. (1) would appear to remain valid in two dimensions, and, in fact, a two-dimensional energy cascading inertial range can exist but with two differences: the coefficient α is believed to be larger and the cascade direction is reversed. The tendency for energy to be transferred by nonlinear processes to larger scales is the major qualitative difference of two-dimensional turbulence.

One imagines then the possibility of energy and enstrophy being introduced into a two-dimensional turbulent flow at some input wave number k_i, at rates ε and $\eta = k_i^2 \varepsilon$, respectively. The enstrophy then flows toward higher wave number at a rate η and is finally dissipated at such a high wave number $k_d \gg k_i$ that the associated energy dissipation $k_d^{-2} \eta$ is negligible compared to $\varepsilon = k_i^{-2} \eta$. The energy meanwhile is flowing toward low wave numbers to be removed perhaps by scale-independent drag effects at wave numbers $k_e \ll k_i$ and thus with negligible enstrophy dissipation $k_e^2 \varepsilon$ compared to $\eta = k_i^2 \varepsilon$.

The simple back cascade of energy just described has been harder to observe than the forward cascade of enstrophy. A more realistic general observation would seem to be that enstrophy cascades toward small scales to be dissipated, but that energy is trapped in large scales. Numerical simulations of decaying two-dimensional turbulence show just this. Enstrophy decreases much more rapidly than does energy.

2. – Selective decay.

We deal here with an example of a more general process that has been called selective decay [1]. Although inviscid integrals are in general subject to dissipation by viscous effects, some may be more so than others owing to turbulent cascade processes that take them into small scales where viscosity is more effective. This distinction becomes increasingly relevant as the Reynolds number increases and molecular viscosity becomes of less direct importance than eddy viscosity.

Integrals that are not readily dissipated through turbulent cascade processes are called rugged; those that are may be called dissipated. Thus for two-dimensional turbulence energy is rugged and enstrophy is dissipated.

The hypothesis that selective decay of turbulent motions may occur, that is, that some integrals are rugged and others are dissipated, poses immediately an interesting variational problem. For one may expect that the final state of turbulent decay would be one with the minimum value of a dissipated integral for fixed values of rugged integrals.

BRETHERTON and HAIDVOGEL [2] applied the selective-decay principle to the quasi-two-dimensional large-scale flow of the ocean over bottom topography. There was some tendency in their numerical simulations for the flow to settle down to that one determined by variational analysis for which the enstrophy was a minimum subject to fixed energy and topographic constraints.

MATTHAEUS and MONTGOMERY [1] examined the consequences of selective decay for two-dimensional flow in a periodic domain and found in numerical simulations a tendency for the flow to decay to the gravest mode for their domain, thus the one with the least enstrophy for a given energy.

They also examined the selective decay of magnetohydrodynamic turbulence. Here energy is dissipated but magnetic or cross-helicity may be rugged integrals. The minimum-energy reversed-field pinch configuration described by TAYLOR [3] can be considered to be a consequence of selective decay with magnetic helicity fixed.

Another quite different example of selective decay has been given by HASEGAWA, KODAMA and WATANABE [4] for turbulent solutions of a slightly damped version of the Korteweg-deVries (KdV) equation. The KdV equation has an infinite number of integrals and no cascades, but the introduction of a small damping makes all integrals dissipated except for a momentum and an energy integral which remain rugged. When they minimized the lowest-order dissipated integral keeping momentum and energy fixed, they found the classical simple solutions, namely, a soliton in an infinite domain and a single cnoidal wave in a periodic domain. This is a clear example of the selective decay of chaotic motion to a coherent stable structure.

The stability of such final flows is a consequence of thir minimal properties. After the turbulent motion has decayed away, the dissipated integrals can again play their inviscid role. Since these are at a minimum, they provide through Liapunov functional arguments sufficient conditions for stability.

3. – Method of anticipated vorticity.

In numerical simulation of two-dimensional turbulent flows at large Reynolds numbers, it is not feasible to resolve explicitly all scales of motion down to those at which dissipation is occurring. It becomes then necessary to introduce an artificial mechanism to simulate the influence of unresolved scales on those that are resolved. This cannot, of course, be done in a completely satisfactory manner simply because the details of this influence are not available from knowledge only of resolved scales.

The most commonly used prescriptions introduce an eddy viscosity of a magnitude sufficient to remove any cascading energy or enstrophy at scales that are small but still resolved. Efforts have been made to localize dissipation as much as possible on barely resolved scales by use of higher powers of the Laplacian in the dissipation laws.

A quite different approach has been taken by SADOURNY and BASDEVANT [5] based on the selective-decay hypothesis. They point out that for resolution truncation of a numerical simulation in the enstrophy cascading inertial range, as is usually the case, what is desired is a mechanism for removing enstrophy while conserving energy. Such a scheme would then be completely compatible with the hypothesis that enstrophy is dissipated but that energy is rugged.

Two-dimensional flow is governed by the vorticity equation

$$\frac{dq}{dt} = \frac{\partial q}{\partial t} + \mathbf{V} \cdot \mathbf{grad}\, q = 0 , \tag{3}$$

where q is the vorticity and $\mathbf{V}$ is the nondivergent two-dimensional velocity. The enstrophy in a periodic domain

$$G = \tfrac{1}{2} \int q^2 \, d\sigma \tag{4}$$

is conserved by eq. (3), but, if eq. (3) is modified to

$$\frac{\partial q}{\partial t} + \mathbf{V} \cdot \mathbf{grad}\, (q + \Delta q) = 0 , \tag{5}$$

then

$$\frac{dG}{dt} = - \int \Delta q \, (\mathbf{V} \cdot \mathbf{grad}\, q) \, d\sigma . \tag{6}$$

Here $\int d\sigma$ indicates integration over the periodic domain.

If now one chooses

$$\Delta q = \theta \mathscr{L}(\mathbf{V} \cdot \mathbf{grad}\, q) , \tag{7}$$

where $\mathscr{L}$ is any nondimensional positive definite linear operator and θ is some characteristic time, then eq. (6) leads to enstrophy dissipation.

No matter what Δq is in this formulation, energy is conserved. To see this, note that the vorticity equation (5) may also be written as

$$\frac{\partial q}{\partial t} + J(\psi, q + \Delta q) = 0 \tag{8}$$

in terms of the stream function ψ and the Jacobian $J(A, B) = A_x B_y - A_y B_x$. The energy integral may be expressed by partial integration as

$$E = - \tfrac{1}{2} \int \psi q \, d\sigma \tag{9}$$

and, since $q = \nabla^2 \psi$, its rate of change, as

(10) $$\frac{dE}{dt} = -\int \psi \frac{\partial q}{\partial t} d\sigma = \int \psi J(\psi, q + \Delta q) d\sigma = \frac{1}{2} \int J(\psi^2, q + \Delta q) d\sigma = 0$$

with the last step a consequence of the vanishing of the integral of any Jacobian over a periodic domain.

The positive definite operator $\mathscr{L}$ is most naturally chosen as a filter with the form, in the spectral domain where it is diagonal,

(11) $$\mathscr{L} = (k/k^*)^{2p}.$$

Here p is a positive integer, say 4, and k^* is the truncation wave number.

This is called the method of anticipated vorticity since it effectively evaluates the vorticity used in the dynamics equation not at the time t but at a time $t + \tau(k)$, where the anticipation time

(12) $$\tau(k) = (k/k^*)^{2p}\theta$$

is a function of scale. The implementation of this method is particularly straightforward in spectral-transform models in which the spectral form of the advection term $\boldsymbol{V} \cdot \boldsymbol{grad}\, q$ is readily available.

Note that at $k = k^*$, that is, at the limit of resolution, eq. (12) results in $\tau = \theta$. It is then natural to choose the constant time θ as the characteristic eddy time of the smallest resolved eddies, but this is also the basis for choosing the time step Δt in the calculation, thus $\theta = \Delta t$ turns out to be a suitable choice.

We see then that the method of anticipated vorticity can be interpreted as a frequency-damping mechanism, and this has a particularly valuable consequence. For stationary flows, or in regions of stationary flow, there is no change in vorticity, the anticipated value does not differ from the current value, and the correction vanishes. The method of anticipated vorticity is, therefore, particularly well suited to the numerical study of the selective decay of two-dimensional flow toward a stationary final state, since in that final state the prescribed enstrophy dissipation will cease.

Unfortunately, the method of anticipated vorticity is not invariant for Galilean transformations. Thus a structure that is stationary except for a uniform translation will continue to be damped by the method.

4. – Minimum-enstrophy modon.

A major discovery in the search for coherent structures in two-dimensional flows was Stern's modon [6]. In beta-plane dynamics for geophysical flows, in which the Coriolis coefficient is approximated as $f = f_0 + \beta y$ with y the co-ordinate in the northward direction, the vorticity equation (3) is replaced by an absolute vorticity equation

$$\frac{d}{dt}(q + \beta y) = \frac{\partial}{\partial t}(q + \beta y) + J(\psi, q + \beta y) = 0. \tag{13}$$

Although any circularly symmetric vortex is a stationary solution of eq. (3), that is not the case for eq. (13). However STERN found a dipole solution to eq. (13), called a modon, with a stream function of the form in polar co-ordinates

$$\psi = -\beta \gamma^{-2} \sin\theta \, [r - R J_1(\gamma r) J_1^{-1}(\gamma R)] \tag{14}$$

for $r < R$ and vanishing for $r > R$. Here γ is a constant such that $J_2(\gamma R) = 0$; the smallest possible γ is such that $\gamma R = 5.136$. Note that $\psi(R, \theta) = 0$ and that the modon amplitude is determined by the choice of R and the root γR.

The most general stationary solution of eq. (13) requires that some functional relation exists

$$q + \beta y = \mathscr{F}(\psi) \tag{15}$$

in order that $J(q + \beta y, \psi)$ vanish. For the modon of eq. (14), eq. (15) is satisfied for $\mathscr{F}(\psi) = -\gamma^2 \psi$, and the vorticity is given by

$$q = -\beta R \sin\theta J_1(\gamma r) J_1^{-1}(\gamma R). \tag{16}$$

Note that $q(R, \theta) = -\beta Y \neq 0$, where $Y = R \sin\theta$, but that the absolute vorticity anomaly $q + \beta y$ vanishes for $r = R$.

STERN has shown by a variational analysis that the modon of eqs. (14) and (16) is the solution of eq. (13), satisfying appropriate boundary conditions at $r = R$, for which the enstrophy is a minimum. It follows that a modon is a natural final flow for selective decay of enstrophy and should be stable relative to small local disturbances.

There have been a number of extensions of Stern's modon summarized by FLIERL et al. [7]. Although these are observed to be remarkably stable in numerical experiments by MCWILLIAMS et al. [8], neither their minimal properties nor their stability have yet been examined analytically.

5. – Minimum-enstrophy vortices.

For strictly nondivergent flow in two dimensions without a β term, thus governed by eq. (3), there also exist minimum-enstrophy structures that may serve as final flows of a selective-decay process. These are circularly symmetric vortices. The remainder of these lectures will deal with their properties as found by variational analysis.

The search for such vortices was stimulated by the numerical studies of McWilliams [9] that showed, as did earlier studies by FORNBERG [10], circularly symmetric vortices arising apparently spontaneously from decaying two-dimensional turbulence. Coherent vortex structures have also been observed in laboratory experiments with two-dimensional turbulence in a rotating cylindrical tank by HOPFINGER et al. [11], and, of course, large-scale quasi-two-dimensional flows in the atmosphere clearly exhibit vortex structure in cyclones and anticyclones.

Two minimum-enstrophy vortices have been found. In each, energy E is a rugged integral and enstrophy G dissipated. They differ in that in one, labeled MEV M, angular momentum M is treated as a rugged integral, and the total circulation C vanishes. In the other, labeled MEV C, the angular momentum M becomes infinite, but the circulation is finite nonzero and is taken as the second rugged integral. In either case, the vorticity is confined to a disk of radius R that is determined by the specified values of E and M or C, respectively.

Since, for given E and R, the enstrophy G is a minimum, we are interested in the gravest, that is, the largest-scale, structure that will fit. If this is a circularly symmetric vortex, its detailed structure in the two cases is derived in the next two sections.

6. – MEV M.

For a circularly symmetric vortex all dependent variables are a function of radius r alone. We let $u(r)$ be the tangential velocity and $q(r) = \mathrm{d}u/\mathrm{d}r + u/r$ be the vorticity. Then the key integrals of interest are

$$(17) \qquad M = 2\pi \int_0^R u r^2 \, \mathrm{d}r \,,$$

$$(18) \qquad E = \pi \int_0^R u^2 r \, \mathrm{d}r \,,$$

$$(19) \qquad G = \pi \int_0^R q^2 r \, \mathrm{d}r \,.$$

It is convenient to rescale r with R by letting $r = Rs$, $u(r) = U(s)$ and $q(r) = Q(s)/R$, where $Q(s) = \mathrm{d}U/\mathrm{d}s + U/s$. Then the key integrals become

$$(20) \qquad M = 2\pi R^3 \int_0^1 U s^2 \, \mathrm{d}s \,,$$

(21) $$E = \pi R^2 \int_0^1 U^2 s \, ds \,,$$

(22) $$G = \pi \int_0^1 Q^2 s \, ds \,.$$

To avoid a singular contribution to G, we must have $U(0) = U(1) = 0$.

We are interested in variations in these integrals with respect to variations δR in radius as well as δU and δQ in shape. The first variations are

(23) $$\delta M = 2\pi R^3 \int_0^1 \delta U \, s^2 \, ds + 3 M R^{-1} \, \delta R \,,$$

(24) $$\delta E = 2\pi R^2 \int_0^1 U \, \delta U \, s \, ds + 2 E R^{-1} \, \delta R \,,$$

(25) $$\delta G = 2\pi \int_0^1 Q \, \delta Q \, s \, ds = -2\pi \int_0^1 [dQ/ds] \, \delta U \, s \, ds$$

with the last step the result of partial integration.

To find stationary values of G for fixed E and M, we introduce Lagrange multipliers λ and μ and write the constrained first-variation equation as

(26) $$0 = \delta G + \lambda \, \delta E + \mu \, \delta M =$$
$$= 2\pi \int_0^1 [-dQ/ds + \lambda R^2 U + \mu R^3 s] \, \delta U \, s \, ds + (2\lambda E + 3\mu M) R^{-1} \, \delta R$$

to be valid for arbitrary variations δR and δU with $\delta U(1) = 0$.

From the δR requirement we find

(27) $$\mu = -\frac{2}{3} \frac{E}{M} \lambda$$

and from the δU requirement

(28) $$dQ/ds - \lambda R^2 U + \frac{2}{3} \lambda \frac{E}{M} R^3 s = 0 \,.$$

A particular solution of this inhomogeneous equation is a solid-body rotation

(29) $$U(s) = \frac{2}{3} \frac{E}{M} R s \,,$$

(30) $$Q(s) = \frac{4}{3} \frac{E}{M} R \,.$$

The general solution satisfying $U(0) = U(1) = 0$ is

(31) $$U(s) = -\frac{2}{3}\frac{E}{M}\frac{R}{J_1(\gamma R)}[J_1(\gamma Rs) - J_1(\gamma R)s].$$

For this general solution we may evaluate the integrals

(32) $$M = \frac{1}{3}\pi\frac{E}{M}R^4\left\{1 - 4(\gamma R)^{-1}\frac{J_2(\gamma R)}{J_1(\gamma R)}\right\},$$

(33) $$E = \frac{1}{3}\pi\left(\frac{E}{M}\right)^2 R^4\left\{1 - 4(\gamma R)^{-1}\frac{J_2(\gamma R)}{J_1(\gamma R)} + \frac{2}{3}\frac{J_2^2(\gamma R)}{J_1^2(\gamma R)}\right\},$$

whence by division we must have

(34) $$J_2(\gamma R) = 0,$$

that is, γR must be a positive root of $J_2(x)$.

It then follows that for the vorticity we may write

(35) $$Q(s) = -\frac{4}{3}\frac{E}{M}\frac{R}{J_0(\gamma R)}[J_0(\gamma Rs) - J_0(\gamma R)].$$

Note that $Q(1) = 0$ and that vorticity is continuous as is velocity.

To select among possible stationary solutions that one for which G is a true minimum requires an analysis of second variations. It is convenient first to transform to a spectral representation of the problem by setting

(36) $$U(s) = \sum_{n=0}^{\infty} U_n g_n(s),$$

where

(37) $$g_0(s) = 2s,$$

(38) $$g_n(s) = \sqrt{2}\, J_1(\alpha_n s)/J_1(\alpha_n), \qquad n > 0,$$

with $J_2(\alpha_n) = 0$ form an orthonormal set of basis functions. The U_n are expansion coefficients satisfying, since $U(1) = 0$, the constraint

(39) $$2U_0 + \sqrt{2}\sum_{n=1}^{\infty} U_n = 0.$$

In the spectral representation the key integrals become

(40) $$M = \pi R^3 U_0,$$

$$\text{(41)} \qquad E = \pi R^2 \sum_{n=0}^{\infty} U_n^2,$$

$$\text{(42)} \qquad G = \pi \sum_{n=1}^{\infty} \alpha_n^2 U_n^2.$$

By rescaling with $U_n = U_0 V_n$ and using relation (40) between M and U_0, we may write

$$\text{(43)} \qquad E = \pi^{-1} M^2 R^{-4} \left[1 + \sum_{n=1}^{\infty} V_n^2\right],$$

$$\text{(44)} \qquad G = \pi^{-1} M^2 R^{-6} \left[\sum_{n=1}^{\infty} \alpha_n^2 V_n^2\right]$$

with condition (39) becoming

$$\text{(45)} \qquad \sum_{n=1}^{\infty} V_n = -\sqrt{2}.$$

The stationary solutions already found correspond, with $\gamma = \alpha_N R^{-1}$, to

$$\text{(46)} \qquad V_N = -\sqrt{2}, \qquad V_n = 0 \qquad \text{for } n \neq N,$$

whence we find

$$\text{(47)} \qquad E = 3\pi^{-1} M^2 R^{-4},$$

$$\text{(48)} \qquad G = 2\pi^{-1} \alpha_N^2 M^2 R^{-6} = \tfrac{2}{3} \alpha_N^2 R^{-2} E.$$

For given E and M, eq. (47) determines R. For given E and R, eq. (48) shows that G is the smallest for the stationary solution with $N = 1$. This is the gravest mode which we then take as the MEV M solution.

At the MEV M solution the constrained second variation may be written as

$$\text{(49)} \qquad \delta^2 G + \lambda \delta^2 E = 2\pi^{-1} M^2 R^{-6} \left[\sum_{n=1}^{\infty} (\alpha_n^2 - \alpha_1^2)(\delta V_n)^2 - \tfrac{2}{3}\alpha_1^2 (\delta V_1)^2\right]$$

with

$$\text{(50)} \qquad \sum_{n=1}^{\infty} \delta V_n = 0.$$

It is not yet clear that this second variation may not be negative. A further rescaling with

$$\text{(51)} \qquad \beta_n = -\delta V_n / \delta V_1,$$

$$\text{(52)} \qquad \sum_{n=2}^{\infty} \beta_n = 1$$

leads to the second-variation expression

(53) $$\delta^2 G + \lambda \delta^2 E = 2\pi^{-1} M^2 R^{-6} p (\delta V_1)^2,$$

whose positivity depends on the factor

(54) $$p = \sum_{n=2}^{\infty} (\alpha_n^2 - \alpha_1^2) \beta_n^2 - \tfrac{2}{3} \alpha_1^2.$$

We now carry out a variational analysis of eq. (54), using eq. (52) as a constraint, to find a stationary point for p at

(55) $$\beta_n = (\alpha_n^2 - \alpha_1^2)^{-1} \Big/ \sum_{n=2}^{\infty} (\alpha_n^2 - \alpha_1^2)^{-1},$$

which is, in fact, a minimum, and where we find

(56) $$p_{\min} = \left[\sum_{n=2}^{\infty} (\alpha_n^2 - \alpha_1^2)^{-1} \right]^{-1} - \tfrac{2}{3} \alpha_1^2.$$

A mathematical property of the roots α_i of the Bessel functions $J_n(x)$ is that

(57) $$\alpha_1^2 \sum_{i=2}^{\infty} (\alpha_i^2 - \alpha_1^2)^{-1} = \frac{n+1}{2}.$$

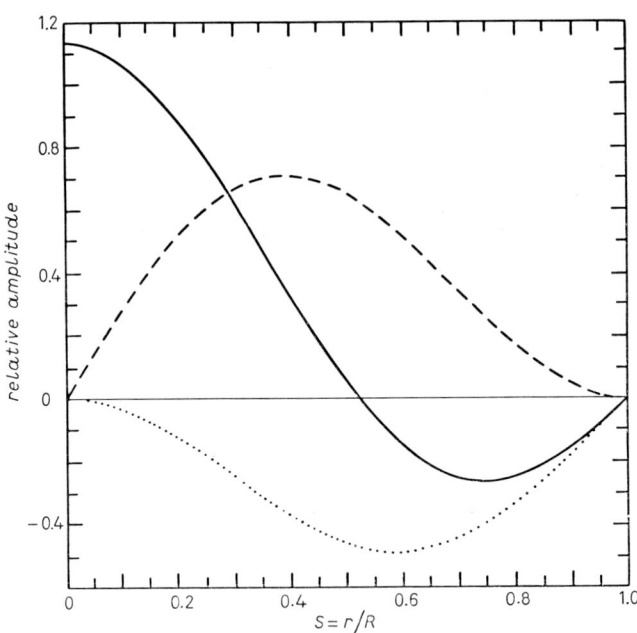

Fig. 1. – Velocity (dashed), vorticity (solid) and strain rate (dotted) as a function of r/R for MEV M.

If we use this property, in this case for $n = 2$, we find that

$$p_{\min} = 0 \,. \tag{58}$$

We conclude then that the MEV M given by eqs. (31) and (35) has indeed a minimum enstrophy. The functions $U(s)$ and $Q(s)$ are shown in fig. 1 together with the corresponding strain rate

$$Q(s) - 2U(s)/s \tag{59}$$

which is of interest for the dynamics of the eddy enstrophy cascade.

7. – MEV C.

For MEV C we treat as a rugged integral the energy within a disk of radius R^* which is finite but much larger than the radius R within which the vorticity will eventually be found. Since the circulation C is constant for $r > R$, we have

$$u = (2\pi)^{-1} C r^{-1} \tag{60}$$

and the contribution of the annulus $R < r < R^*$ to the energy integral is

$$\Delta E = (4\pi)^{-1} C^2 \ln (R^*/R) \,. \tag{61}$$

The key integrals for MEV C are, for fixed R^*,

$$C = 2\pi \int_0^R q r \, dr = 2\pi R u(R) \,, \tag{62}$$

$$E = \pi \int_0^R u^2 r \, dr + (4\pi)^{-1} C^2 \ln (R^*/R) \,, \tag{63}$$

$$G = \pi \int_0^R q^2 r \, dr \,. \tag{64}$$

With the scaling introduced for MEV M, we may rewrite these as

$$C = 2\pi R U(1) \,, \tag{65}$$

$$E = \pi R^2 \int_0^1 U^2 s \, ds + (4\pi)^{-1} C^2 \ln (R^*/R) \,, \tag{66}$$

$$G = \pi \int_0^1 Q^2 s \, ds \,. \tag{67}$$

We introduce a stream function ψ such that

(68) $$U(s) = d\psi/ds,$$

(69) $$\psi = U(1) \ln s \qquad \text{for } s > 1,\ \psi(1) = 0,$$

(70) $$Q = \nabla^2 \psi,$$

and rescale in terms of $U_1 = U(1)$ by letting $\tilde\psi = \psi/U_1$, $\tilde U = U/U_1$ and $\tilde Q = Q/U_1$: Then, using relation (65) between C and U_1, we may write the key integrals as

(71) $$E = (4\pi)^{-1} C^2 \left[\int_0^1 \tilde U^2 s\, ds + \ln(R^*/R) \right],$$

(72) $$G = (4\pi)^{-1} C^2 R^{-2} \int_0^1 \tilde Q^2 s\, ds.$$

The variational analysis for minimum G with C, E and R^* fixed is analogous to that for MEV M. The resulting functions defining MEV C are, for $s \leqslant 1$,

(73) $$\tilde\psi = -\gamma^{-1} J_0(\gamma s)/J_1(\gamma),$$

(74) $$\tilde U = J_1(\gamma s)/J_1(\gamma),$$

(75) $$\tilde Q = \gamma\, J_0(\gamma s)/J_1(\gamma)$$

with $J_0(\gamma) = 0$, γ being the smallest positive root, and, for $s > 1$,

(76) $$\tilde\psi = \ln s,$$

(77) $$\tilde U = s^{-1},$$

(78) $$\tilde Q = 0.$$

These functions and the strain rate are shown in fig. 2.
The energy integral (71) reduces for this solution to

(79) $$E = (4\pi)^{-1} C^2 [\tfrac{1}{2} + \ln(R^*/R)],$$

which, for given C, E and R^*, suffices to determine R. The associated enstrophy integral is

(80) $$G = (8\pi)^{-1} \gamma^2 R^{-2} C^2$$

and a second-variation analysis similar to that for MEV M shows that this is a true minimum.

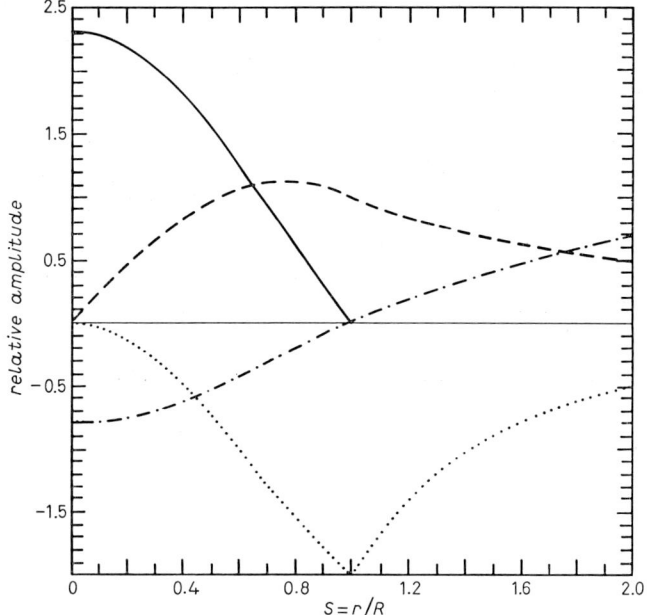

Fig. 2. – Stream function (dash-dotted), velocity (dashed), half-vorticity (solid) and strain rate (dotted) as a function of r/R for MEV C.

8. – Conclusion.

It has already been pointed out that the minimal properties of a structure like a MEV provide a Liapunov functional to assure stability relative to small local perturbations. Consider, for example, such a perturbation of a MEV that changes E to $E + \Delta E$, and either M to $M + \Delta M$ or C to $C + \Delta C$. Then for the new values of E and M or C there exists a MEV with minimum enstrophy G_0. The perturbation in general, however, gives an enstrophy $G = G_0 + \delta G > G_0$. Treat G now as an integral of the motion and a Liapunov functional to see that the perturbed motion must be trapped in a small region of dynamical phase space about the MEV point with enstrophy G_0.

The foregoing analysis only examines circularly symmetric vortices and perturbations, but one may wonder, especially for MEV M, whether there is an asymmetric state with lower enstrophy. Some numerical experiments, using spectral-transform methods with anticipated vorticity, have provided preliminary answers to this question. When a circular MEV M was subjected to a spatially white perturbing vorticity field with resolved amplitudes of order 50% of that of the MEV M itself, a relatively rapid removal of perturbation enstrophy took place in the region of the MEV as expected from shear-induced selective decay. On a longer time scale, however, the MEV broke up into a rotating

$(-1, 2, 1)$ triad of circulation elements with the same integral values of E and M but with G lower by about 20%. Further analytic and numerical work is needed to show whether the circular MEV M is metastable or unstable relative to asymmetric perturbations.

Similar experiments have not been done with MEV C. These are harder to set up owing to long-range influences. A lower-enstrophy asymmetric state is, however, less likely in this case owing to the single sign of the vorticity in the solution of sect. 7.

The major unanswered dynamical question about MEV's is whether a turbulent decay goes all the way to a MEV or stops at some other stationary circularly symmetric vortex having exhausted the eddy enstrophy needed for radial mixing and adjustment. It may be that MEV's are ideal structures toward which flows tend but which they never reach. In any case they are of interest as further examples of two-dimensional coherent structures.

REFERENCES

[1] W. Matthaeus and D. Montgomery: *Ann. N. Y. Acad. Sci.*, **357**, 203 (1980).
[2] F. Bretherton and D. Haidvogel: *J. Fluid Mech.*, **78**, 129 (1976).
[3] J. B. Taylor: *Phys. Rev. Lett.*, **33**, 1139 (1974).
[4] A. Hasegawa, Y. Kodama and K. Watanabe: *Phys. Rev. Lett.*, **47**, 1525 (1981).
[5] R. Sadourny and C. Basdevant: *C. R. Acad. Sci. II*, **292**, 1061 (1981).
[6] M. Stern: *J. Mar. Res.*, **33**, 1 (1975).
[7] G. Flierl, V. Larichev, J. McWilliams and G. Reznik: *Dyn. Atmos. Oceans*, **5**, 1 (1980).
[8] J. McWilliams, G. Flierl, V. Larichev and G. Reznik: *Dyn. Atmos. Oceans*, **5**, 219 (1981).
[9] J. McWilliams: in *Predictability of Fluid Motions*, edited by G. Holloway and B. West (American Institute of Physics, New York, N. Y., 1984), p. 205.
[10] B. Fornberg: *J. Comput. Phys.*, **25**, 1 (1977).
[11] E. Hopfinger, F. Browand and Y. Gagne: *J. Fluid Mech.*, **125**, 505 (1982).

Theoretical Predictability of Small-Scale Motions.

D. K. LILLY

University of Oklahoma - Norman, OK

1. – Summary.

Recently summarized data on the kinetic-energy spectra of the troposphere in the small and mesoscale domains allow extension to these scales of the estimates of predictability due to LORENZ [1] and LEITH and KRAICHNAN [2]. Several factors limit the validity of these estimates, however, the most important of which may be the highly intermittent nature of small-scale weather events. Some evaluations of this intermittency are summarized. As an example of the effects of intermittency on predictability, a two-dimensional flow consisting of widely spaced Rankine vortices is analyzed. The intermittency of velocity predictability is found to be greater than that of the velocity field itself. Aspects of the strategy of observation and prediction of highly intermittent flows are discussed.

2. – Introduction and scope.

Consideration of the problem of prediction of the smaller scales of atmospheric activity suggests existence of a number of important factors which do not exist or are less important for the scales defined by the global observing net. I will first outline the straightforward application of a conventional large-scale approach to the smaller scales, then discuss the circumstances which limit the validity of that approach. Most attention is given to the effects of highly non-Gaussian probability distributions, *i.e.* intermittency.

3. – Conventional estimates of small-scale predictability.

The problem of predictability of atmospheric evolution encompasses, in principle, all the complexities of understanding of the atmosphere, since prediction is classically the true test of scientific understanding. Some success in evaluating the feasibility of prediction has been attained by application of some simple assumptions that do not require direct knowledge of the physics of atmospheric evolution. It is assumed that the atmosphere is a uniformly turbulent fluid, that it is everywhere unstable to small perturbations and that errors in the initial state grow until they are as large as the difference between two randomly chosen atmospheric states. The growth of error is not uniform with spatial scale, however, so that the smaller scales of motion lose their predictability first. Figure 1, from [1], shows a calculation of the loss

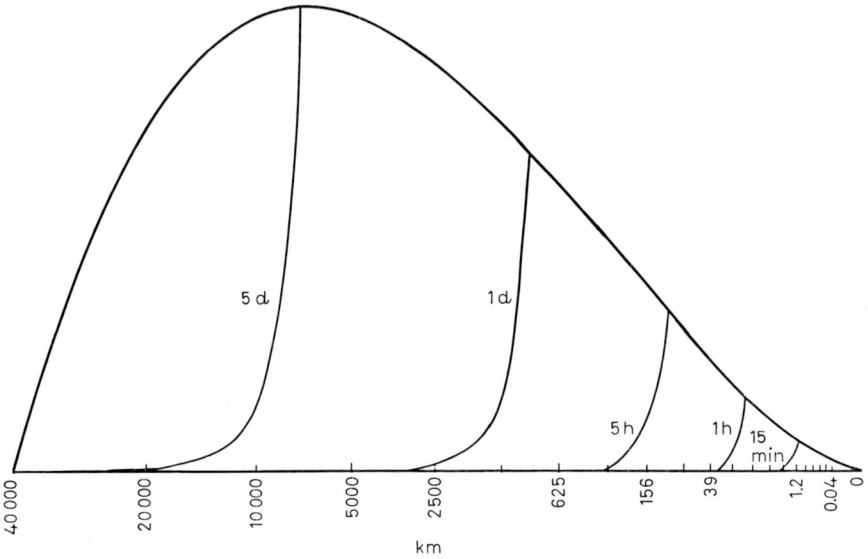

Fig. 1. – Results of a closure model calculation of the rate of loss of predictability of two-dimensional flow where the kinetic-energy spectrum is given by the upper curve. The short parabolic arcs show the left edge of the unpredicted spectrum as it proceeds to larger scales. From [1].

of predictability for an idealized distribution of atmospheric energy, corresponding to a $k^{-5/3}$ spectrum at the high wave numbers. The calculation was made from a linearized quasi-normal closure model. The initial state consisted of a small error at the highest available wave number, corresponding to a wavelength of 38 m, but it was shown that the results would be similar for any small initial error. LEITH and KRAICHNAN [2] carried out similar calculations using an improved closure theory (test field model) and obtained generally similar results. For an initial error at a very small wavelength progressing up a two-dimensional $k^{-5/3}$ spectrum, they found that predictability is lost for wave number k at a time τ proportioned to the eddy turn-over time, i.e. $\tau \sim 6/[k^3 E(k)]^{1/2}$. For the same spectral magnitudess, the Leith-Kraichnan predictability durations appear to be 3 to 4 times greater than Lorenz's, apparently due to the different closure model.

Lorenz's use of a $k^{-5/3}$ spectrum at subsynoptic scales was speculative and contradictory to the then-unfolding recognition of the existence of a k^{-3} inertial decay range for two-dimensional and geostrophic turbulence. More recent evidence tends to confirm Lorenz's assumption, however. Figure 2 depicts hori-

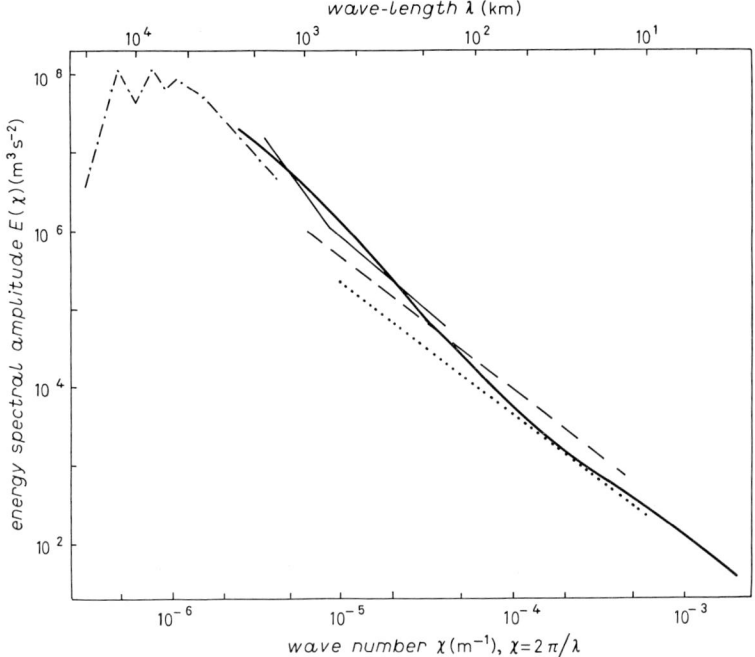

Fig. 2. – Spectra of two-dimensional energy obtained from various sources: —— NASTROM and GAGE, —— LILLY and PETERSEN, – – – VINNICHENKO, ··· BALSLEY and CARTER, –·–·– CHEN and WIIN-NIELSEN (from ref. [3]).

zontal velocity spectra over the scale range from planetary wavelengths to a few kilometers. The solid lines show upper tropospheric spectra obtained by LILLY and PETERSEN [3] and by NASTROM and GAGE [4], which tend to confirm the earlier time-space transformed estimates by VINNICHENKO [5], and spectra of Doppler radar velocities due to BALSLEY and CARTER [6]. The composite picture is of a power spectrum proportional to the $-5/3$ or -2 power of wave number at scales smaller than those of the short -3 range obtained from sonde data by CHEN and WIIN-NEILSEN [7] and others. GAGE [8] and LILLY [9] have theorized that this spectrum is produced by upscale transport of energy generated by thermal convection and shear at small scales and partially converted to two-dimensional turbulence by the effects of thermal stratification. Table I exhibits the eddy velocities $[kE(k)]^{1/2}$ and the maximum predictability time $6/[k^3 E(k)]^{1/2}$ for wavelengths from 10 to 1000 km, based on fig. 2. The eddy velocities are about a factor of two smaller than Lorenz's, mainly due to his neglect of the k^{-3} range between 1000 and 400 km wavelengths, and the predictability times are 5 to 10 times longer. Intuitively the small-scale velocities seem too weak and predictabilities too long to apply to periods of significant meteorological events.

TABLE I. – *Eddy velocities and predictability times.*

Wavelength (km)	$[kE(k)]^{\frac{1}{2}}$ (ms^{-1})	$6/[k^3 E(k)]^{\frac{1}{2}}$ (h)
1000	4	72
500	3	48
200	1.7	34
100	1.3	24
50	1.0	14
20	0.7	8
10	0.5	6

For a variety of reasons the above estimates of small-scale predictability should be regarded with caution or suspicion. ANTHES et al. [10] show evidence that some small-scale events can be predicted without knowledge of the initial conditions on that scale. The most clear-cut example is that of a stationary mountain wave, which develops in a few hours when a large-scale (and, therefore, predictable) flow field with favorable vertical structure passes over small-scale terrain. ANTHES also regards frontogenesis as being predictable from knowledge of the large-scale flow alone, though it seems likely that only the occurrence of the process, not its detailed temporal and spatial location, can be anticipated independently of the small-scale initial state. Some kinds of weather events, like ducted traveling waves and perhaps intense vortices like hurricanes, may maintain a stable identity much longer than their local eddy turn-over time, although their velocity and subsequent positioning accuracy are subject to the predictability of the environmental flow field. Elsewhere [11] I propose that rotating thunderstorms are similarly stable and predictable. The frequency and importance of this kind of event brings out the principal point of this lecture, which is to cast doubt on the usefulness of the Lorenz-Leith-Kraichnan analysis for small-scale flows because of its implicit Gaussianity requirements. For Gaussian variables, amplitudes greater than about 3 times the standard deviation are extremely rare, so that predictions made from the Leith-Kraichnan analysis are locally as well as globally valid. This is no longer true for flow with a non-Gaussian, highly intermittent character, which I believe characterizes small-scale meteorology, at least during the time periods when prediction is of interest.

4. – Atmospheric intermittency and effects on predictability.

The probability distribution functions of standard atmospheric analysis variables like pressure, temperature, stream function and humidity at a given location and season are superficially Gaussian in form. Rainfall is not, of course, being a one-signed and relatively rare quantity at most locations. Some studies of definable atmospheric disturbances, such as clouds, show log-

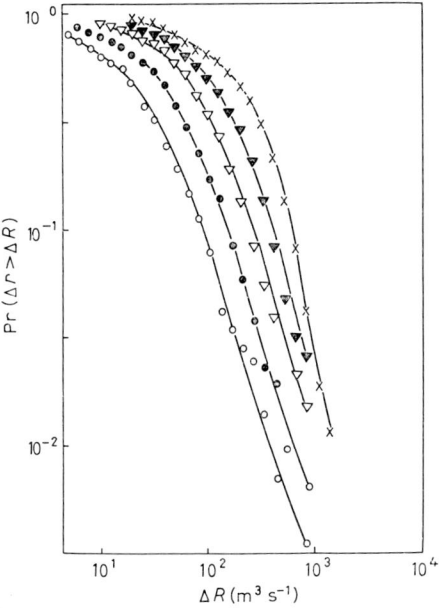

Fig. 3. – Probability of the rainfall rate Δr exceeding values ΔR, averaged over specified time intervals: ○ 5 min increments, ● 10 min increments, ▽ 20 min increments, ▼ 40 min increments, × 80 min increments. The rainfall rate is obtained from radar data, integrated over an isolated storm. From [14].

normal distribution functions, i.e. Gaussian in the logarithm of the variable [12]. The critical area is the frequency of extreme values, as estimated, for example, by the root mean cubes and higher powers. For a log-normal distribution all these moments are higher than they are for a Gaussian distribution with the same variance. MANDELBROT [13] suggested the existence of power law, or

« hyperbolic » distributions, such that the probability of some variable x exceeding a value x' is proportional to $x'^{-\alpha}$, where α is some positive number.

This suggestion has been confirmed for rain rates [14] and for squared vertical differences of velocity and potential temperature between levels of balloon soundings [15], as shown in fig. 3 and 4. Such distribution functions imply that moments like $\overline{x^\alpha}$ do not exist, $i.e.$ their estimated value continues to grow with the number of samples. I am unaware of comparable statistics of horizontal differences having been presented, though they are in principle accessible from records of high-resolution moving instrument platforms, such as aircraft, and from scanning remote sensors like satellites and Doppler radar.

Attempts to predict variables whose moments diverge or become very large would seem unverifiable and, therefore, virtually futile, since the difference between the variable and its prediction will probably also have divergent statistics. We may be in that position regarding rainfall prediction, since the

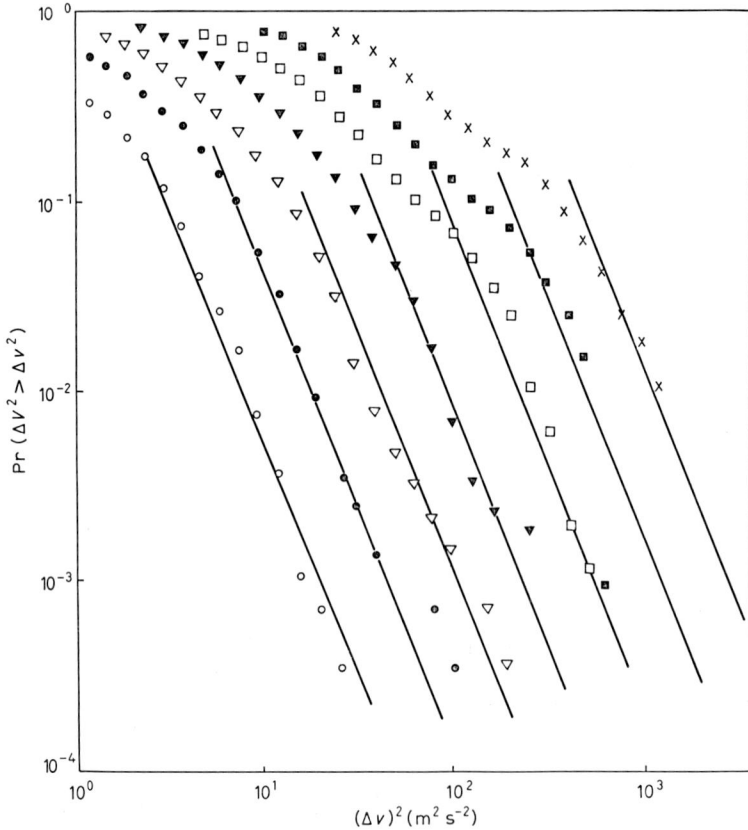

Fig. 4. – Probability of the square of the velocity difference $(\Delta V)^2$ exceeding the value $(\Delta v)^2$, where the difference is taken between levels of a balloon sounding record at spatial separations of 50, 100, 200, 400, 800, 1600 and 3200 m (left to right curves). From [15].

spatial and temporal derivatives of rainfall seem to be virtually unbounded as the distance between raingauges decreases. Perhaps it is as hopeless to try to predict rainfall at a point in space as it is to try to determine the length of the coast of Italy.

5. – Predictability of flow around widely spaced vortices.

As an example of the kind of predictability problem which seems to arise in small scale meteorology, let us consider a flow consisting of widely spaced Rankine vortices. Such a flow is a crude but easily definable analog to the tropical atmosphere in a hurricane-prone area, or to an Oklahoma afternoon in the severe storm season, or perhaps to the flow found by McWilliams [16] to be typical of numerical simulations of decay of two-dimensional turbulence. For a single Rankine vortex located at the origin, the tangential velocity is given by

$$(5.1) \quad v = \begin{cases} \Gamma r/a^2, & r < a, \\ \Gamma/r, & r > a. \end{cases}$$

With $v = \partial \psi/\partial r$, the stream function ψ is then given by

$$(5.2) \quad \psi = \begin{cases} \Gamma r^2/2a^2, & r < a, \\ \Gamma/2 + \Gamma \ln(r/a), & r > a, \end{cases}$$

where a is the radius of the vorticity-containing region and Γ is the circulation. The stream function is arbitrarily taken to be zero at the origin. If the total flow is described by an ensemble of such vortices, of different strengths and sizes, then the stream function for the i-th vortex within and outside its vorticity-containing core is

$$(5.3) \quad \psi_i = \begin{cases} \Gamma_i r^2/2a_i^2 & \text{for } |\mathbf{r} - \mathbf{r}_i| < a_i, \\ \Gamma_i [\tfrac{1}{2} + \ln(|\mathbf{r} - \mathbf{r}_i|/a_i)] & \text{for } |\mathbf{r} - \mathbf{r}_i| > a_i, \end{cases}$$

and that for the ensemble of vortices outside all of their cores is

$$(5.4) \quad \psi = \sum_i \Gamma_i [\tfrac{1}{2} + \ln(|\mathbf{r} - \mathbf{r}_i|/a_i)].$$

If we assume that vortices of both signs are equally likely, so that the circulation around the total area disappears, then the maximum kinetic-energy values occur near the core of each vortex, i.e.

$$(5.4)' \quad E_{max} = \frac{\Gamma^2}{2a^2},$$

where $\Gamma^2 \equiv \overline{\Gamma_i^2}$ and $a^2 \equiv \overline{a_i^2}$. The most probable or modal kinetic energy, on the other hand, is that occurring midway between vortices, whose average spacing is taken to be L, and is

$$E_{\text{mod}} = \frac{\Gamma^2}{2L^2}. \tag{5.5}$$

The mean kinetic energy is determined essentially by integrating the energy of each vortex within the area where its circulation dominates, *i.e.* a box of width L, and is

$$\overline{E} = \frac{\Gamma^2}{L^2} \ln (L/2a). \tag{5.6}$$

If it is assumed that $L \gg a$, then the maximum, average and modal kinetic energies are greatly different from each other.

Since each vortex influences the motion of all others, errors in positions or intensities of each feed back nonlinearly, and the total flow will be unstable, excepting for degenerate cases when only a small number of vortices exist and they are in certain favored configurations. The doubling time of an error in vortex position can be expected to be roughly the time it takes a vortex to move the distance between them, *i.e.* $T = L^2/\Gamma$. The maximum, modal and mean kinetic-energy errors caused by a position error δ are then given as follows

$$(\Delta E)_{\max} = \frac{\Gamma^2 \delta^2}{a^4} = \frac{2\delta^2}{a^2} E_{\max},$$

$$(\Delta E)_{\text{mod}} = \frac{\Gamma^2 \delta^2}{L^4} = 2 \frac{\delta^2}{L^2} E_{\text{mod}},$$

$$\overline{\Delta E} = \frac{\Gamma^2}{L^2} \frac{\delta^2}{a^2} = \frac{\delta^2/a^2}{\ln (L/a)} \overline{E}.$$

Thus, for a given magnitude of vortex position error, not only the local energy errors but their ratios to local energy are much greater near the vortex cores than elsewhere. While the exponential growth rate of error may be similar everywhere, its influence takes effect much sooner near the vortex cores. In addition, a point on the apparent trajectory of a vortex moves from a low to a high error ratio region as the core comes closer, so that the effective error growth rate there is much faster than the nominal exponential rate. Thus it may be said that the velocity predictability is more intermittent than the velocity itself.

All the above considerations only put into a semi-analytic form the well-known problems of predicting such events as hurricanes and severe storms. These problems are intuitively accepted by forecasters and, to a considerable degree, the affected public. Prediction is much more difficult near the center of an intense meteorological event, but it is also much more important, and even modest success is rewarding. The evaluations of success can hardly be made on the conventional bases of predictability comparisons. As a hurricane approaches, the predicted wind at a point or small region will surely have an error that is much larger than the r.m.s. velocity for the season and location, but the prediction will also be far better than either climatology or persistence, provided the existence of the hurricane is recognized. In acknowledgment of these basic alterations of the prediction and verification problem, forecasters focus attention on the location and amplitude of the center of the storm, and in communications with the public state how far out from that center damaging winds, rain, etc. are expected to extend.

Operational meteorology has for decades depended most heavily on networks of fixed observing stations, reporting at specified uniform intervals. The high intermittency of smaller scales suggests a greater reliance on mobile and remote sensors, with some of them operating on an « on-call » basis when strong events are imminent. Again this need is intuitively recognized and fulfilled by the use of radar, satellites, hurricane reconnaissance aircraft, tornado « chase » teams (used for research purposes in the past, but now also carried out by radio and television stations in the Oklahoma City area) and airborne Doppler radar, which can locate, track down and accurately observe the internal characteristics of intense storms from a safe distance.

REFERENCES

[1] E. N. Lorenz: *Tellus*, **21**, 289 (1969).
[2] C. E. Leith and R. H. Kraichnan: *J. Atmos. Sci.*, **29**, 1041 (1972).
[3] D. K. Lilly and E. L. Petersen: *Tellus A*, **35**, 379 (1983).
[4] G. D. Nastrom and K. S. Gage: *Tellus A*, **35**, 383 (1983).
[5] N. K. Vinnichenko: *Tellus*, **22**, 158 (1970).
[6] B. B. Balsley and D. A. Carter: *Geophys. Res. Lett.*, **9**, 465 (1982).
[7] T.-C. Chen and A. Wiin-Nielsen: *Tellus*, **30**, 313 (1978).
[8] K. S. Gage: *J. Atmos. Sci.*, **36**, 1950 (1979).
[9] D. K. Lilly: *J. Atmos. Sci.*, **40**, 749 (1983).
[10] R. A. Anthes, Y.-H. Kuo, S. G. Benjamin and Y.-F. Li: *Mon. Weather Rev.*, **110**, 1187 (1982).
[11] D. K. Lilly: *Dynamics of rotating thunderstorms*, in *Mesoscale Meteorology, Theories, Observation and Models*, edited by D. K. Lilly and T. Gal-Chen (Dordrecht, 1983), p. 531.
[12] R. E. Lopez: *Mon. Weather Rev.*, **105**, 865 (1977).
[13] B. B. Mandelbrot: *J. Fluid Mech.*, **62**, 331 (1974).
[14] S. Lovejoy: *A statistical analysis of rain areas in terms of fractals*, in *XX Conference on Radar Meteorology* (Boston, Mass., 1981), p. 476.
[15] D. Schertzer and S. Lovejoy: *The dimension of atmospheric motions*, unpublished report, EERM/CRMO, Météorologie Nationale, Paris, France (1983).
[16] J. C. McWilliams: *Emergence of isolated coherent vortices in turbulent flow*, in *J. Fluid Mech.*, in press (1984).

Scale and Variance Analysis of Atmospheric Motion.

G. D. ROBINSON

The Center for the Environment and Man, Inc. - Hartford, CT 06120

1. – Introduction.

In 1969, LORENZ published a paper with the title « The predictability of a flow which possesses many scales of motion ». The purpose of this lecture is, in an elementary manner, to examine the atmosphere as an example of such a flow—how are the scales of motion defined, how are statistics concerning them derived and what is the order of magnitude of these statistics—in the context of predictability and the equations used for prediction. Sections **2** and **3** discuss the nature of the velocity variable and its spatial gradients in the equations of motion used by dynamical meteorologists and expose some of the constraints implied by the requirement for spatial averaging or truncated spectral decomposition of the velocity field. Section **4** contains the results of various determinations of the variance spectrum of atmospheric velocity; and sect. **5** is one example of an analysis of the inertial interaction between the scales. In these sections the distinction is drawn between actual observations and the output from, or « initialized » input to, atmospheric models. Section **6** touches on the magnitude of the production of kinetic energy in the atmosphere and some unresolved problems concerning the mechanism of its dissipation.

In view of the content of many of the presentations to this School it must be emphasized that the atmosphere is a dissipative system driven by thermodynamic energy conversion and the force of gravity. As a dynamical system, it is essentially three-dimensional and essentially non-Hamiltonian.

2. – The concept of velocity.

We begin with an examination of an equation of motion and of the nature of the variables which it interrelates. We are concerned with a predominantly gaseous system and from kinetic theory we may derive equations for mass, momentum and total-energy conservation in terms of, amongst other variables, a time- and position-dependent velocity c which is the average of molecular velocities on the smallest feasible scale, *i.e.* the scale on which it is plausible to neglect the Brownian fluctuations.

If we refer to axes rotating with angular velocity $\boldsymbol{\Omega}$, the momentum conservation equation takes the form

(1) $$\underbrace{\frac{\partial \boldsymbol{c}}{\partial t}}_{(1.1)} + \underbrace{\frac{\partial}{\partial \boldsymbol{r}} \cdot \boldsymbol{cc}}_{(1.2)} + \underbrace{2\boldsymbol{\Omega} \times \boldsymbol{c}}_{(1.3)} + \underbrace{\left[\underbrace{\frac{1}{\varrho}\boldsymbol{\nabla}p}_{(1.4a)} - \underbrace{\nu\nabla^2\boldsymbol{c}}_{(1.4b)}\right]}_{} + \underbrace{\boldsymbol{g}}_{(1.5)} = 0 \; ,$$

where the terms have been numbered for convenient reference. Term (1.4) is derived from the pressure tensor, p being the static pressure and ν the kinematic viscosity, both expressible in kinetic-theory variables. Equation (1) bears a resemblance (perhaps remarkably close in view of the way it is derived) to eq. (2)

(2) $$\underbrace{\frac{\partial \boldsymbol{c}}{\partial t}}_{(2.1)} + \underbrace{\boldsymbol{c} \cdot \frac{\partial \boldsymbol{c}}{\partial \boldsymbol{r}}}_{(2.2)} + \underbrace{2\boldsymbol{\Omega} \times \boldsymbol{c}}_{(2.3)} + \underbrace{\left[\underbrace{\frac{1}{\varrho}\boldsymbol{\nabla}p}_{(2.4a)} - \underbrace{\boldsymbol{A}_\mathrm{F}}_{(2.4b)}\right]}_{} + \underbrace{\boldsymbol{g}}_{(2.5)} = 0$$

(where the incompressibility condition ($\partial/\partial \boldsymbol{r} \cdot \boldsymbol{c} = 0$) has been applied in term (2.2)), which is to be found in texts on dynamical meteorology. For example (with notation changes only), it is eq. (2.8) of [1], where $\boldsymbol{c}$ is described as the « velocity vector » and $\boldsymbol{A}_\mathrm{F}$ as the « frictional force ». HOLTON remarks « it is this form of the momentum equation which is basic to most work in dynamical meteorology ». Whatever the « velocity vector » of the dynamical meteorologist ($\boldsymbol{c}$ of eq. (2)) might be, it is not the « mean velocity » of the kinetic-theory derivation of the Navier-Stokes momentum equation ($\boldsymbol{c}$ of eq. (1)). Consideration of what it is leads us directly into the question of scales of motion.

3. – The concept of scale.

An order-of-magnitude analysis accompanies eq. (2) in Holton's text. He takes the « synoptic scale » of 10^6 m, when the horizontal components of term (2.2) are of order 10^{-4} m s^{-2}; the components of terms (2.3) and (2.4a) involving horizontal velocity gradients are of order 10^{-3} m s^{-2} and approximately in balance; and the vertical components of term (2.4a) and term (2.5), of order 10 m s^{-2}, are very closely in balance. We have not yet defined $\boldsymbol{A}_\mathrm{F}$, but it and term (2.2) are of the order of the small imbalances. We shall also be concerned with a mechanical-energy equation derived by taking the scalar product of $\boldsymbol{c}$ and all terms of eq. (2).

Now consider how velocity is observed in the atmosphere. Away from regions which can be reached by surface-anchored structures, it is measured by tracking objects (balloons, powered aircraft) moving with the wind. The tracking can be very precise, but the high precision, even if sought and achieved, is routinely degraded in the reporting process. Atmospheric inhomogeneities, *e.g.*, of density or water content, which approximately move with the wind, can also be tracked, but most of the data come from the radiosonde network. The measured velocities may roughly be considered as averages over a few tens of seconds time and a space of linear dimension about 10^2 m, reported to a precision of 1 m s^{-1} in speed and $10°$ in direction. Term (2.2), however, contains the gradient tensor of c and term (2.4a) the gradient of p. The smallest scale on which we can apply eq. (2) to the atmosphere is set not by the individual observation but by the spacing of the observations. On this scale the term $\nu\nabla^2 c$ would be of order 10^{-15} m s^{-2}. THOMPSON [2] was probably the first to call attention to the quantitative relation between this scale and predictability. We are effectively applying eq. (2) to data smoothed or averaged on a scale of $5 \cdot 10^5$ m; the theoretical deduction of the closely related eq. (1) implied an averaging scale of 10^{-5} m. Can the use of eq. (2) be justified other than empirically?

There are two possible approaches. The first is by averaging on the required scale, *i.e.* by specifying an averaging operator and applying it rigorously to all terms of the Navier-Stokes equation, eq. (1). The second is by a Fourier or analogous transform of each term, replacing the partial differential equation, eq. (1), by a set of ordinary differential equations for the arguments of the characteristic modes of a defined domain. Both approaches are mathematical rather than physical in content. The scale on which c is defined by averaging the equation has no dynamical basis. In prognostic studies, it is usually the grid size in a finite-difference solution. In many diagnostic studies, the scale is that of the latitude circle, but there is no physical mechanism which will create and sustain stationary zonal motion at any latitude. The zonal mean wind is an arithmetical artifact. In the alternative approach, spectral dissociation is always possible mathematically, but characteristic modes are not necessarily physically excited, nor is the solution for any one mode or incomplete set of modes necessarily a solution of the equation for the whole flow. If we assign physical significance to any quantity derived in either approach, we must have plausible physical evidence.

Consider first the averaging alternative. We are concerned in many problems, including those of weather prediction and of the spread of contaminants in the atmosphere, with the time evolution of a single realization of the system governed by eq. (2). The average must be an occasional—time or space rather than ensemble—average. REYNOLDS was the first to produce an equation for a time average—what he termed the « mean mean motion », his mean motion being the kinetic-theory average velocity of eq. (1). His averaging operator was

$$\int_{-\Delta t/2}^{t+\Delta t/2} \mathrm{d}\tau\,[\ ]\,.$$

There are some disturbing features of his derivation which I am told are universally recognized, but which are so frequently ignored that it is useful to recall them (see, e.g., [3]). Writing the averaging operator as an overbar, we have $\boldsymbol{c} = \overline{\boldsymbol{c}}(\Delta t) + \boldsymbol{c}'(\Delta t)$; term (2.1) becomes $\partial \overline{\boldsymbol{c}}/\partial t + \partial \boldsymbol{c}'/\partial t$ and term (2.2) becomes $\overline{\boldsymbol{c}} \cdot \partial \overline{\boldsymbol{c}}/\partial \boldsymbol{r} + \overline{\boldsymbol{c}} \cdot \partial \boldsymbol{c}'/\partial \boldsymbol{r} + \boldsymbol{c}' \cdot \partial \overline{\boldsymbol{c}}/\partial \boldsymbol{r} + \boldsymbol{c}' \cdot \partial \boldsymbol{c}'/\partial \boldsymbol{r}$. REYNOLDS sought a time-dependent equation for $\overline{\boldsymbol{c}}$ by a second application of the averaging operator and the postulates

$$\overline{\overline{\boldsymbol{c}}(\Delta t)} = \overline{\boldsymbol{c}}(\Delta t) \quad \text{and} \quad \overline{\partial \boldsymbol{c}/\partial \boldsymbol{r}} = \partial \overline{\boldsymbol{c}}/\partial \boldsymbol{r}.$$

With Reynolds' definition of the averaging operator, the postulates imply that $\overline{\boldsymbol{c}}$ is invariant in time. Similar arguments apply to a spatial average $\overline{\boldsymbol{c}}(\Delta \boldsymbol{r})$, in which case the postulates imply $\partial \overline{\boldsymbol{c}}/\partial \boldsymbol{r} = \text{const}$. From the point of view of prediction of the evolution of a nonuniform field of averaged velocity, the equation is logically useless. Even as a diagnostic equation, Reynolds' equation contains the term $\boldsymbol{c}' \cdot \partial \boldsymbol{c}'/\partial \boldsymbol{r}$ which is an unknown, since the detailed field of $\boldsymbol{c}$ is unknown, but which does take the place of term (2.4b), $A_\text{F} = \overline{\boldsymbol{c}' \cdot \partial \boldsymbol{c}'/\partial \boldsymbol{r}}$, though it could comprise driving as well as dissipative frictional acceleration on the averaging scale $\Delta \boldsymbol{r}$. If we ignore the requirement of invariance and use eq. (2) as a prediction equation applicable to a mean velocity on the scale of the separation of observations, we must find some empirical expression of $\overline{\boldsymbol{c}' \cdot \partial \boldsymbol{c}'/\partial \boldsymbol{r}}$ in terms of observables. Several artifices have been employed and the resulting modification of eq. (2) used as a prognostic equation for single realizations with considerable success in practice. The simplest of these empirical devices is to write $A_\text{F} \propto |\Delta \boldsymbol{r}|^{4/3} \nabla^2 \overline{\boldsymbol{c}}$, a scale-dependent formula for « eddy viscosity » with some observational support. It is sometimes asserted that those difficulties of interpretation of eq. (2) applied to an observable scale which arise from implied invariance of occasional averages can be avoided by the use of ensemble averages in cases in which the time development of statistics, rather than of a single realization, is sought. I have advanced what I consider a refutation of this assertion [3]. The question is relevant, in principle, to possibilities of climate prediction.

Now consider the alternative development of eq. (1)—spectral dissociation—for a simple one-dimensional case. $\boldsymbol{c}$ and p of eq. (1) are specified as functions of position x in a domain $0 \leqslant x \leqslant L$, $\boldsymbol{c} = f(x)$, $\partial \boldsymbol{c}/\partial x = f'(x)$, etc. The general term of eq. (1) is $\varPhi(x)$. We take the Fourier transform

$$\mathscr{F}(\varPhi(x)) = F(k) = (1/L) \int_{-\frac{1}{2}L}^{+\frac{1}{2}L} dx\, \varPhi(x) \exp[-ikx]$$

of each term of eq. (1), so that

(3) $$\varPhi(x) = \sum_{n=-\infty}^{n=+\infty} F(k_n) \exp[ik_n x].$$

The k are the wave numbers $k_n = 2\pi n/L$ of the characteristic modes of domain L (strictly the infinite limits should be replaced by $L/2\Delta x$, where Δx is the linear dimension of the «grain» on which the molecular mean velocity is defined). Equation (1) is transformed into an effectively infinite set of equations for $F(k_n)$. Term (1.2) is the transform of the product $f(x) f'(x)$—the convolution $\mathscr{F}(f(x)) * \mathscr{F}(f'(x))$—so that, in the equation for $F(k_n)$, the second term has the form

$$\sum_{n=n_1 \pm n_2} F(k_{n_1}) \cdot ik_{n_2} F(k_{n_2})$$

and the equation for the argument of each mode contains the arguments of all the other modes. In the situation in which our information is restricted to samples equally spaced at distance l within L, we can extract no information for $k_n > \pi/l$. The truncated spectral dissociation is always mathematically possible, but, if the deleted modes are associated with physical processes, the truncated equation must be empirically adjusted to simulate these processes, most elegantly by assuming a physically plausible analytical expression for $F(k)$ in the deleted spectrum and computing the convolutions, less arduously by associating some empirical fixed dissipative or forcing acceleration with each resolved mode (see, e.g., [4]).

It is in this context that the question of «inertial ranges» in spectra related to atmospheric motion has received particular attention. Hitherto we have considered the momentum equation and the spectral components $F(k)$ of an analysis in phase and amplitude of the velocity field. Many spectral analyses are made in connection with the energy equation and concern the quadratic forms $F \cdot F^*(k)$ and $k^2 F \cdot F^*(k)$, the spectral dissociations of kinetic energy and enstrophy, respectively. Here we will use the term «variance spectrum», $V(k)$, rather than the «kinetic-energy spectrum» to emphasize the mathematical rather than the physical basis for the spectral dissociation. The «inertial range» is, however, a physically based concept. If there is a range of scales in which a physical quantity, dimensions ($l^a t^b$), on which the variance density, dimensions ($l^3 t^{-2}$), depends is being exchanged between scales because of the inertial acceleration, but is not being produced or destroyed, $V(k)$ will be stationary and depend only on $k(l^{-1})$ and the rate ($l^a t^{b-1}$) at which the quantity passes through the range of scales so that $(l^3 t^{-2}) \equiv (l^a t^{b-1})^m (1)^{-n}$, $m = 2/(1-b)$, $n = (2a + 3b - 3)/(1-b)$. If the physical quantity is kinetic energy ($a = 2$, $b = -2$), then $n = -5/3$; if it is enstrophy ($a = 0$, $b = -2$), then $n = -3$. The dimensional argument does not depend on the direction of transfer through the scales: the physical consequences of the transfer are manifest outside rather than within the inertial range. Nothing in the argument mandates such a range.

Summarizing the position we have reached, we note that, particularly for economic reasons, we can observe atmospheric motions and manipulate the observations only as data effectively averaged on time and space scales which are large compared with those of details we know to be present. We develop equations of change, for motions specified on those large scales, which contain many terms involving the unspecified detail. We reduce the number of these terms, but do not completely remove them, by assumptions which involve invariance in time and space of the averages we know to change. We ignore the remaining detailed terms, or replace them by empirical functions of the averages, and use them as though time and space invariance of the main variable were not implied. We find that actual changes can be predicted, not precisely but very usefully. We know that the scales we have treated by approximation or neglect are responsible for all frictional dissipation of kinetic energy and enstrophy, and for the forcing associated with convective precipitation. Does an examination of the results of scale and variance analyses offer any explanation of the utility of the equation, or allow an estimate of the growth of error of prediction when it is used?

4. – Scale analysis of atmospheric motion.

In this section we will examine some of the results of scale analysis, choosing examples which illustrate differing approaches. The section is not intended as an exhaustive survey.

4'1. *Raw observations and manipulated data.* – We begin with a warning. Much of the material which has been used is not the result of direct observation, but of the adjustment of observations to fit theories, or rather approximations to theories. This apparent negation of the scientific method arises because the immediate, and economically justifying, use of the data is in forecasting models, and in the use of any model, from a child's toy upwards, to attempt to force against the built-in constraints is to court frustration or even disaster. The models also call for uniform separation of data points in space and time; so do the standard algorithms for harmonic analysis and numerical filtering. When attempting scale analysis, it is very convenient, if not imperative, to carry it out on the initialized input to, or even the output from, prognostic models. This must be kept in mind when using or interpreting the results of scale analysis, particularly if we are concerned with processes deleted from, or modified in, the model equations.

4'2. *Examples of data analysis.* – The first example we will consider is the exhaustive investigation of northern-hemisphere data for 1950-51 by SALTZMAN and collaborators, published between 1957 and 1962 (see, *e.g.*, [5]). The material they used was taken mainly from daily 500 mb contour charts for a full year. At this date the chart analysis was presumably by subjective interpolation and the « winds » used were derived from the geostrophic balance (effectively that among terms (2.3), (2.4a) and (2.5)). The nominal constraints are thus that the analyzed motion is that of an unforced inviscid fluid and the integrated effect of the inertial interactions between scales (term (2.2)) is zero. Standard harmonic-analysis methods were used on data read every 10° of longitude and amplitude and phase for modes up to $n = 15$ computed. The analysis was made each day for both zonal and meridional components. Annual and seasonal

TABLE I. – *Examples of variance spectra: variance in mode* ($m^2 s^{-2}$).

Mode	Spectrum						
	1	2	3	4	5	6	7
0	360	165	—	—	175	185	—
1	32	28	24	18	30	20	44
2	26	30	24	13	15	8	6
3	28	26	26	18	5	15	30
4	20	22	24	22	10	20	12
5	18	20	21	25	15	24	16
6	18	16	22	18	17.5	16	16
7	14	14	17	12	16	15	18
8	12	11	12	11	17.5	15	22
9	8	8	8	10	10	10	18
10	6	7	6	6	10	10	14
11	4	5	5	5	10	10	11
12	4	4	3	5	7.5	10	9
13	3	3	2	3	5	10	8
14	2.5	2.5	1.5	2	10	10	6
15	2.5	2	1.5	1	5	3.5	5
20	—	—	—	—	1.3	1.5	2.8
30	—	—	—	—	0.6	0.4	0.8
Note					(a)	(b)	

Note a. Spectrum 5 has continuum $V(n) = 1.84 \cdot 10^2 n^{-5/3}$ for $n > 15$ and error variance 25 $m^2 s^{-2}$.
Note b. Spectrum 6 has continuum $V(n) = 1.17 \cdot 10^4 n^{-3}$ for $15 < n < 40$ and is truncated at $n = 40$.

Spectrum

1 SALTZMAN et al. [5], geostrophic wind, 500 mb, $47\frac{1}{2}°$ N winter.
2 WIIN-NIELSEN et al. [6], balanced wind, all levels, northern hemisphere, year.
3 JULIAN et al. [7], balanced winds, 500 mb, 50° N, winter.
4 JULIAN et al. [7], raw observations, 500 mb, 50° N, winter.
5 BROWN and ROBINSON [8], raw observations, 500 mb, Europe, January-February.
6 BROWN and ROBINSON [8], NMC grid winds, 500 mb, Europe, January-February.
7 BOER and SHEPHERD [9], NMC grid streamfunction, 500 mb, global, January.

means and standard deviations of the daily values were tabulated. One example from this mass of data, expressed as a variance rather than a kinetic-energy spectrum, is included in our table I. The source is table II of [5].

Our second example is from another exhaustive investigation, that by WIIN-NIELSEN with collaborators BROWN and DRAKE, published between 1962 and 1967. The data used in these studies were subjected to considerable preliminary manipulation, but originated in charts of geopotential on standard pressure surfaces produced by the NMC (U.S. National Meteorological Center). Quoting WIIN-NIELSEN and DRAKE [6], « ... a streamfunction, ψ, was computed for each level at each observation time (on the grid used by NMC) by solving the balance equation

$$(4) \qquad \nabla^2 \psi = (1/f)\nabla^2 \Phi - (1/f^2)\nabla f \cdot \nabla \Phi ,$$

where Φ is the geopotential and $f = 2\Omega \sin \varphi$, φ being the latitude ». Φ is, of course, the result of the NMC smoothing and interpolation. Equation (4) is a linear equation implying nondivergent frictionless flow in which the integrated effect of the nonlinear inertial accelerations is zero. In Wiin-Nielsen's work the streamfunction so computed on the NMC grid was transferred by further interpolation (which presumably introduced no physical constraints, but which brings us three moves away from the measurements) to a spherical grid with a 2.5° separation in longitude, nominally allowing four times the resolution of Saltzman's analysis. The actual information content of the global meteorological observing network did not notably increase in the decade of the 1950's and WIIN-NIELSEN was content to limit the harmonic analysis to the first 15 zonal modes. The example in our table I is taken from fig. 21 of [10]. It is quoted as a variance spectrum and refers to all analyzed levels and latitudes in the northern hemisphere in the period February 1963 to January 1964.

JULIAN et al. [7] attempted a scale analysis of actual observations, combined with a further analysis of « balanced winds » interpolated from NMC gridded data. They used the same techniques as SALTZMAN for the « initialized » data and we reproduce in table I their variance spectrum for data at 50° N, 500 mb, sampled over the years 1963-1968 (from their table IV).

The « raw » data they used came from 16 radiosonde stations located between longitudes 52° N and 54° N and latitudes 123° W and 50° E. The basis of their analysis is the relation between the spectrum $V(k)$ and the correlation-separation function $R(r)$ which, for a one-dimensional statistically homogeneous set, is the Fourier cosine transform

$$(5) \qquad V(k) = \overline{c^2}\int dx \, R(x) \cos kx .$$

To define a continuous spectrum $R(x)$ is required as a continuous function, for a line spectrum it must be specified at integral multiples of some minimum separation. The two-point products of all possible pairs of a quasi-random array of observations do not supply the necessary material—some form of interpolation is called for and fig. 1, which is fig. 8 of [7], gives some insight into the nature of the problem; the continuous line being a cubic spline fit to the points, which are the correlations for all possible pairs of the 16 observing

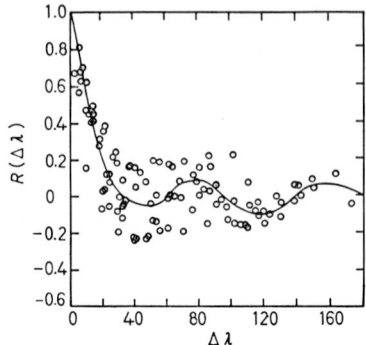

Fig. 1. – Correlation-separation function for the radiosonde stations analyzed; based on fig. 8, p. 384, of the paper by JULIAN et al. (from J. Atmos. Sci., **27**, 376 (1970), © Americal Meteorological Society).

stations. The investigators comment that the spread of points may be due to inadequate sampling or to inhomogeneity in the basic statistics. They settle for sampling fluctuations, but they have no real choice, since eq. (5) is valid only for homogeneous statistics. Our table I contains the example of a « raw data » spectrum, averaged for winter and summer, from table IV of [7], but the situation relating to the comparison of spectrum from « raw » and « initialized » data is perhaps better conveyed by their fig. 12, reproduced here as fig. 2. The two sets of data are not from the same geographical area and time period, so we cannot speculate on the significance of the differences at $n = 1$ to 3, but we certainly cannot conclude from this work that the initialization of the raw data does not significantly affect the spectral decomposition. It should be noted that computation of the correlation-separation function allows an estimate of the errors of observation and reporting of which JULIAN et al. did not take advantage. The two-point products contain this « error variance » as a factor independent of distance, which can, in principle, be extracted by extrapolation of the correlation to zero separation. Constraining the correlation to be unity at zero separation and zero at the maximum separation, in the manner of Julian et al., places physically unjustified constraints on the spectrum in the high-mode and low-mode regions, respectively.

BROWN and ROBINSON [8] also operated with raw radiosonde data. This presentation will concentrate on recent unpublished work, which used material from a two-dimensional array of 135 radiosonde stations over Europe and Western Asia, and analyzed in terms of the characteristic transverse oscillations of all diameters of a tangent plane of diameter equal to the length of the 50° latitude circle. They employed two methods. In the first method the variance $V(l)$ was taken within groups of increasing numbers of stations characterized

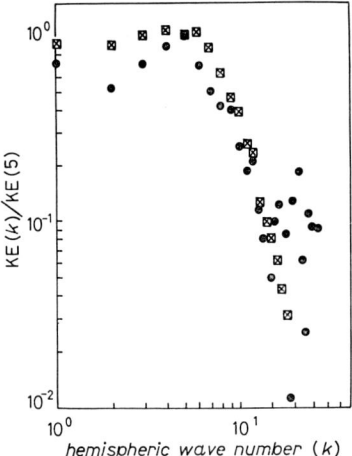

Fig. 2. – Kinetic-energy spectra of observed winds and gridded data, 500 mb, $(50 \div 55)°$ N; ⊠ objective analysis 360° winter and summer; • observed winds, average runs 4-6, winter and summer. Based on fig. 18, p. 383, of the paper by JULIAN et al. (from J. Atmos. Sci., **27**, 376 (1970), © American Meteorological Society).

by linear dimension l. This is related to the variance spectrum $V(k)$ by a filter; for the mode of analysis employed

(6) $$V(l) = \int dk\, V(k)[1 - \sin^2 \tfrac{1}{2} kl/(\tfrac{1}{2} kl)^2] .$$

The second method was an adaptation to the two-dimensional field of the Fourier transform of the correlation-separation function. For a one-dimensional analysis in which the wave vector $\boldsymbol{k}$ has a direction at angle θ to the separation vector $\boldsymbol{r}$ the correlation is

$$\overline{c^2} R(r, \theta) = \int dk\, V(k) \cos(kr \cos\theta) ,$$

which for a statistically isotropic field integrates to the Fourier-Bessel transform

(7) $$\overline{c^2} R(r) = \int dk\, V(k) J_0(kr) .$$

The procedure adopted by BROWN and ROBINSON was to estimate a spectrum $V(k)$ by trial-and-error inversion of eq. (6), transform this $V(k)$ by eq. (7) and compare with the empirically determined correlation, adjusting to produce the best fit to both $V(l)$ and $R(r)$.

We illustrate their results by an example covering a 20-day period in January-February 1979, using raw data from the 00Z ascents at 135 stations and the NMC gridded wind data for 00Z at 135 points covering the same area and period. It is not easy to discover details of the process by which the NMC grid values were computed in 1979. The account of McPherson et al. [11] is specific in the matter of objective smoothing of the readings from adjacent stations—which should minimize observing and reporting errors—but less so concerning the genesis of the initial estimate of the wind field, which appears related in some way to the output of a forecasting model, *i.e.* the specific solutions of model equations with physical constraints. The spectra computed from the raw data and the grid values are shown in table I; the differences at wave numbers 1 to 5 are probably larger than any that could arise from the subjective method of analysis of the observations. Between $n = 6$ and $n = 20$, the two spectra are effectively identical. The subjective analysis is not sufficiently sensitive to distinguish between the two forms of continuum specification set out in

TABLE II. – *Illustrating Brown and Robinson's « variance function » analysis for observations and NMC grid data (20 January through 14 February 1979, 00Z, 500 mb). p is the number of stations or grid points in the group. l is the linear dimension of the group. Spectrum 4.7 (line 5 of table I) is fitted to the observations, spectrum 12 to the grid data. Truncation of 4.7 is at mode 40. (In group $p = 81$, some data are used more than once.)*

a) FGGE observations.

p	l (m)·10^6	Vector variance (m² s⁻²)	
		empirical	spectrum 4.7
3	0.6	78	80
9	1.05	116	116
27	1.8	161	161
81	3.1	198	196
135	4.0	205	206

b) NMC grid winds.

p	l (m)·10^6	Vector variance (m² s⁻²)			
		empirical	spectrum 4.7	spectrum 4.7 (truncated)	spectrum 12
3	0.6	41.1	87	39	40.1
9	1.15	81.4	124	76	81.5
27	2.0	132.4	170	122	132.3
81	3.5	174.1	202	154	172.0
135	4.5	183.9	210	162	184.0

table I, but truncation at mode 40 of the best-fit spectrum to the observations accounts quantitatively for the difference in small-scale variance (~ 50 m^{2}s^{-2}) between the grid values and the observations. Only 25 m^{2}s^{-2} of this can be attributed to error of observing and reporting. The remainder is true subgrid-scale variance. The degree of precision sought in the trial-and-error inversion of the variance function is illustrated in table II. The Fourier-Bessel transform of the spectrum of observed winds is illustrated in fig. 3, together with the empirical correlations.

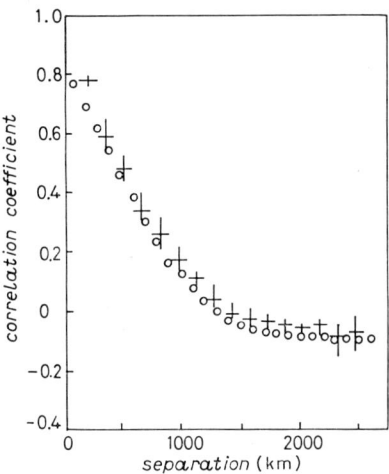

Fig. 3. – Correlation-separation function for the wind velocity vector; Europe-Western Asia, January-February 1979, 00Z 500 mb. Horizontal lines are the correlation coefficients averaged over all separation directions, vertical lines show the variation with direction, circles are the Fourier-Bessel transform of spectrum 4.7 (line 5, table I) (BROWN and ROBINSON).

BROWN and ROBINSON obtained similar results in an analysis of observations and NMC grid data at 500 mb over the same area for 20 days in July 1979. In this case, there was an increase in $V(l)$ at the smallest scale from about 50 m^{2}s^{-2} to 80 m^{2}s^{-2} between the 00Z and 12Z observations, with no corresponding change in the grid data. Again about 25 m^{2}s^{-2} was attributable to error variance in each observation set, so that the true small-scale variance increased by a factor of about 2 to become about 25 percent of the total variance on all scales. The NMC initialization removed all this small-scale variance, as did truncation at mode 40 of the spectrum fitted to the observations. There again appeared to be significant differences in the resolved modes between the « initialized » and the raw data.

A different type of spectral analysis for the period January 1979 has been published by BOER and SHEPHERD [9]. They perform a global analysis in terms of the normal modes on the sphere—the spherical harmonics, the analogue of eq. (3) being

(8) $$\Phi(\lambda, \varphi) = \sum_n \sum_{|m|\leqslant n} \Phi^m Y_n^m(\lambda, \varphi),$$

where m is the zonal wave number and $n - m$ the number of zeros between the poles. The total wave number n takes the place of the total wave number $|\boldsymbol{k}|$ on the plane: the Laplacian of Y_n^m is $-n(n+1)R^{-2}Y_n^m$ (where R is the radius of the sphere), and modes with the same n do not interact inertially. BAER [12] describes the method and its advantages.

BOER and SHEPHERD apply this analysis to a streamfunction derived from the NMC gridded winds. The analysis requires specification of a global field and, when applied to obtain variance spectra, implies global statistical homogeneity and isotropy. In view of these physically unreal stipulations, it seems particularly important to keep in mind the distinction between pattern analysis and physical analysis. HOSKINS [13] touches on this and also points to possible variants of the method, *e.g.* « one can now envisage any stationary wave field as being described as the sum of modes on different great circles » (which, with $|m| = n$, would be the spherical analogue of Brown and Robinson's tangent-plane analysis). To return to BOER and SHEPHERD, in addition to the constraints of the NMC analysis, their use of the streamfunction implies a nondivergent wind. The variance spectrum attributed to BOER and SHEPHERD in our table I is read by interpolation from the kinetic-energy spectra of their fig. 4. Their

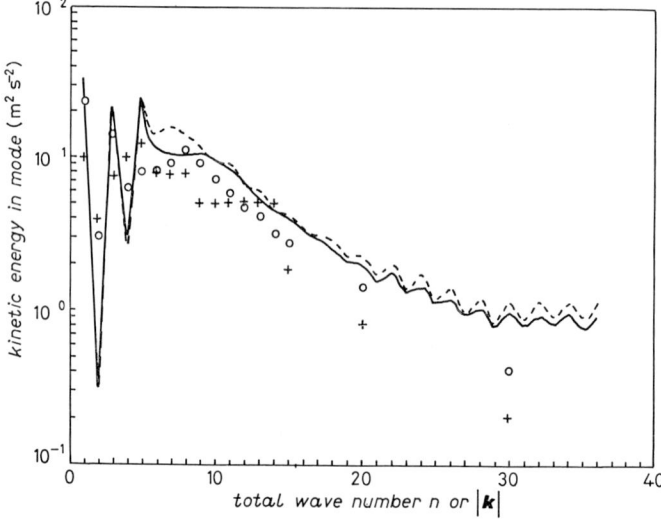

Fig. 4. – Various analyses of NMC gridded data (500 mb). Solid and dashed lines, respectively, nonblocking and blocking, BROWN; spherical, 2×northern hemisphere, streamfunction, 15 Januarys; ○ BOER and SHEPHERD, spherical 2×northern hemisphere, streamfunction, January 1979; + BROWN and ROBINSON, planar, Europe, vector wind, January-February 1979.

global analysis does not cover exactly the same period as Brown and Robinson's local analysis, but there are surprising similarities, for example the minimum at mode 2 and the general intensity level for modes higher than 10.

BROWN (unpublished) has analyzed NMC data for 15 Januarys in terms of spherical harmonics for a « double northern hemisphere » globe. He treated separately days with « blocking » and « nonblocking » synoptic patterns, defined objectively. Some of his results are illustrated in fig. 4, which shows also the Boer and Shepherd spherical and Brown and Robinson plane total wave number analyses of NMC data for January 1979.

We close this sample survey with references to two investigations in which the data used were time series of observations at a single location, and a transform from frequency ν to wave number k, $\nu = |\bar{c}|k$, was employed. This « Taylor » transformation was first used in laboratory studies. It has been referred to as « the frozen-turbulence hypothesis », since it would hold if a stationary pattern of motion were to move past the observing point with velocity $\bar{c}$. It might be expected to hold statistically for a statistically stationary and homogeneous pattern, but there is no convincing reason to expect it to be useful in an atmospheric investigation. It has been used because there seems at present to be no other economically feasible way of determining the detail of variance spectra in the range of scales less than about 1000 km. (Brown and Robinson's method determines the integrated true variance in these scales, but does not discriminate decisively between, for example, the $k^{-5/3}$ and k^{-3} spectral functions.)

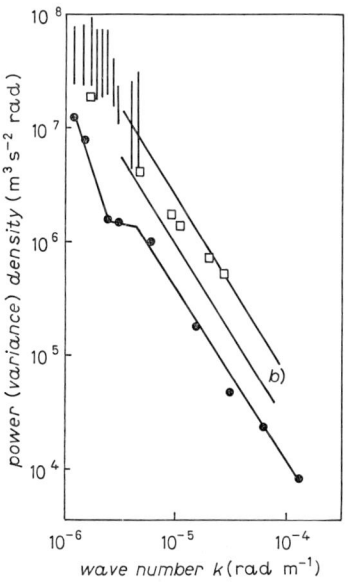

Fig. 5. – Variance spectra transformed from frequency spectra: □ VINNICHENKO et al., USSR, July 1966, 5 km; ● LARSEN et al., Alaska, spring, 6 km. The lines b) are the limits of estimates by BROWN and ROBINSON.

The first investigation we note is that of Vinnichenko et al. [14], who analyzed two sets of radiosonde ascents made at two-hourly intervals over a period of one month (from a site near Kharkov, USSR). The spectral analysis appears to have been made by numerical filter. In fig. 5 we plot their analysis at height 5 km for July 1966. (The actual data were read with some difficulty from a manuscript preprint made available at the Luzern meeting of UGGI in 1967.) The second example in fig. 5 is from LARSEN et al. [15], who observed, for a period of 40 days, the movement of atmospheric inhomogeneities by high-powered radar from a site in Alaska. In the diagram, their published frequency spectrum for height 5.99 km has been transformed by applying the mean wind (5.5 m s^{-1}) over their full 40-day period of observations (which, particularly since it covered the equinox, seems too long a period for methods implying statistical stationarity).

Figure 5 shows that both the investigations of frequency spectra, when transformed, suggest a spectrum of form $k^{-5/3}$ at scales < 1000 km and that

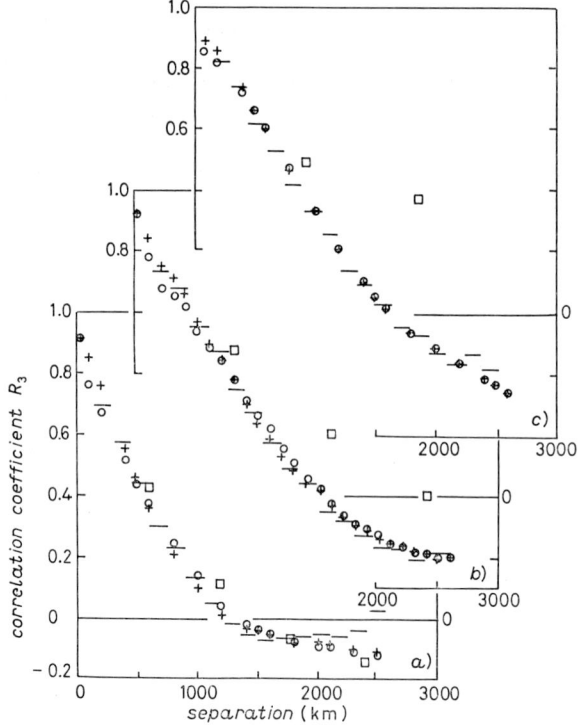

Fig. 6. – Correlation-separation function and Taylor transform of the time lagged correlation. Eastern Europe 500 mb, a) July 1966, b) January 1967, c) January 1971. Transforms of spectra fitted to the variance function are also shown; — empirical correlation $R(r)$, □ empirical Taylor-transformed time-lag correlation, ○ J_0 transform of $-\frac{5}{3}$ spectrum, + J_0 transform of -3 spectrum.

this is compatible with Brown and Robinson's integrated variances, though not established by them. It seems useful to look for evidence concerning the validity or otherwise of the Taylor transformation, and some is to be found in Brown and Robinson's work. In addition to the spatial correlation, they performed a time-lag correlation on all stations in their arrays, and transformed time-lag to spatial separation by the space-time mean wind. They found that in the two summer periods they investigated, July 1966 and July 1979, the Taylor transformation appeared valid up to separations of over 2000 km, but in the three Januarys (1967, 1971, 1979) it seemed marginally useful up to about 1000 km, but failed completely at larger scales. Figure 6 illustrates this. Note that the July 1966 and January 1967 example covered (by design) the same period as the work of Vinnichenko et al., whose observing point was central to the Brown and Robinson array.

5. – Analysis of the advective interactions.

We turn now from examination of the patterns of atmospheric wind fields to the interaction between scales which is implied by the patterns, i.e. the consequences of the advective accelerations (term (2.2)). We take as an example the work of Chen and Wiin-Nielsen [16]. Their basic data are the coefficients of a spherical harmonic analysis of streamfunction and temperature fields which was carried out by BAER on NMC gridded data for winter 1971-72 at levels 1000 mb to 100 mb, assuming a « double northern hemisphere » planet. The equations they employ are not eq. (2) and the corresponding total-energy equation, but « the vorticity and thermodynamic equations of nondivergent, adiabatic, frictionless atmospheric motion ». In considering the significance of their conclusion, we must keep in mind that atmospheric motion is neither nondivergent nor adiabatic nor frictionless, and that we are now considering not the almost balanced large terms of eq. (2) but the small residual terms, which would not exist without divergence, internal energy sources and friction, and without which there would be no turbulence. CHEN and WIIN-NIELSEN perform with these equations the equivalent of a computation of the convolutions in the spectral form of the energy equation analogous to eq. (2). Their harmonic coefficient series is truncated at $n = 31$, $n - |m| = 30$ (« triangular truncation »), and they note that because of the truncation « contributions beyond $n = 25$ are not representative of the observed data », i.e. of the NMC grid values. They compute the mean, over the full period and all levels, of the contributions to the kinetic energy and enstrophy of each mode by all other resolved modes. These are displayed in their table II, here reproduced with minor changes as our table III. They do not explicitly exhibit the variance spectrum which forms the basis of the computations. Note that the unit of 1 erg cm^{-2}s^{-1} for a column integration corresponds to 10^{-7} m^2 s^{-3} in the variance ($2 \times$ specific kinetic energy) units employed here, so that the time constant of the exchange process in the resolved modes is of order 10^6 s.

The numbers in table III, column 1, are the rates of loss or gain of kinetic energy due to internal interaction in a stationary spectrum, which must be maintained by processes not included in the equations employed in the com-

TABLE III. – *Time mean and vertical integral of nonlinear exchange of kinetic energy, I_n (erg cm^{-2} s^{-1}) (Chen and Wiin-Nielsen).*

n	I_n		
1		-69	
2	1009	964	
3		114	
4		-52	830
5		-9	
6		-598	
7		480	
8		-268	
9		-60	
10		-141	
11	-1274	-101	-877
12		-35	
13		-139	
14		-133	
15		-56	
16		-90	
17		-30	
18		-42	
19		10	
20		-19	
21		18	
22		26	
23		23	47
24		30	
25	265	41	
26		27	
27		37	
28		16	
29		30	
30		13	
31		13	

putations. In our suggested modified interpretation, approximately 1.3 W m^{-2} is exported from modes 4 through 18, 1 W m^{-2} to modes 1 through 3, and 0.3 W m^{-2} to modes 19 through 31. The nondivergent frictionless adiabatic system is importing energy in the mid-range modes and exporting it in the low and high orders. Mode 0 is excluded by the method of analysis and modes higher than 31 by truncation and by absence of parameterization.

6. – Some aspects of the global kinetic-energy cycle.

Consider the magnitude of these power exchanges in relation to global processes. Averaged over surface and the year the planet absorbs about 250 W m^{-2} from the solar radiation, the corresponding average black-body reradiation temperature is about 258 K (the actual outgoing radiation is emitted at temperatures ranging between about 300 K and 190 K). The bulk of the absorption (about 200 W m^{-2}) occurs at the surface at temperatures around $(290 \div 295)$ K. Considered as a heat engine, the planet has a maximum thermodynamic efficiency of about 0.1, and, since the heat engine works simply to maintain its internal motions against frictional dissipation, we must formally allow maximum thermodynamic efficiency. About 20 W m^{-2} must be dissipated in the atmosphere and the oceans. The usual estimate of the atmosphere's contribution (including the boundary layer) is about 5 W m^{-2}; this could well be in error by a factor of 2.

This argument does not preclude adiabatic exchanges between kinetic and potential energy on any scale, or between scales, but it does set the magnitude of the frictional dissipation. It does not quantify any partition of kinetic-energy generation between the major processes—baroclinic instabilities in midlatitudes in modes around 10 or vertical convection in the unresolved higher modes in all, but particularly in near-equatorial latitudes. The diurnal modulation noted in the European July wind data at 500 mb by BROWN and ROBINSON, 25 m^{2}s^{-2}, corresponds to a power input and dissipation of about 1.5 W m^{-2}, restricted to modes higher than $n = 40$.

What features of the atmospheric spectrum have we explained? Analyses of initialized data on the whole seem to indicate a n^{-3} spectrum for $n >$ about 10, as does the raw-data analysis of Julian et al. The analyses of raw data by BROWN and ROBINSON also indicate a steep fall-off in variance for $n >$ about 15, but do not explicitly indicate the spectral form at higher modes, and in some cases contain more variance in modes $n > 40$ than can be explained by either a -3 or a $-5/3$ spectrum. The two analyses of time series indicate a $-5/3$ spectrum in modes $n > 15$ if the Taylor transformation holds, and Brown and Robinson's work suggests that it does hold for one of the cases investigated by VINNICHENKO et al. There is, however, little reason to expect to find an inertial range in the atmosphere in any resolvable range of modes. The physical reason for expecting an inertial range is that kinetic energy and enstrophy are injected at scales very large compared with those on which appreciable frictional dissipation can occur. One cannot expect to discover such a range by methods, such as those of Chen and Wiin-Neilsen, which specifically exclude the possibility of an energy or enstrophy flux to the dissipative modes. Additionally, both kinetic energy and enstrophy cannot be simultaneously conserved and cascaded to dissipative scales—either kinetic energy must be directly extracted from the very largest scales by anisentropic processes or enstrophy must be created at smaller scales; scales which in the atmosphere are still too large for the divergence and twisting terms of the three-dimensional vorticity equation to affect the process significantly. In theoretical and computational investigations, the dilemma is usually avoided by one of two devices. Either a Newtonian-type frictional drag is introduced, acting directly on the largest scales, as though a body of air several thousand kilometers in extent were moving as a solid over a uniform surface (see, *e.g.*, [17]), or a scale-dependent eddy viscosity replaces the advective transfers to smaller scales, usually estimated in a manner which implies the existence of an energy cascading inertial range extending from the highest resolved modes (see, *e.g.*, [18]). Both devices are empirically successful; neither has a satisfactory physical basis. Can we explain this, and in particular are we likely to explain it by equations which ignore the critical question of dissipation, or by using data initialized by particular forms of solutions of such equations?

* * *

Preparation of this lecture and the unpublished work by BROWN and BROWN and ROBINSON which is reported were made possible by National Science Foundation Grants ATM-8242149 and ATM-8111575 to the Center for the Environment and Man, Inc.

REFERENCES

[1] J. R. HOLTON: *An Introduction to Dynamic Meteorology*, 2nd edition (New York, N.Y., 1979).
[2] P. D. THOMPSON: *Tellus*, **9**, 275 (1957).
[3] G. D. ROBINSON: *Mon. Weather Rev.*, **106**, 448 (1978).
[4] R. H. KRAICHNAN: *J. Atmos. Sci.*, **33**, 1521 (1976).
[5] B. SALTZMAN and A. FLEISCHER: *J. Atmos. Sci.*, **19**, 195 (1962).
[6] A. WIIN-NIELSEN and M. DRAKE: *Mon. Weather Rev.*, **93**, 79 (1965).
[7] P. R. JULIAN, W. N. WASHINGTON, L. HEMBRE and C. RIDLEY: *J. Atmos. Sci.*, **27**, 376 (1970).
[8] P. S. BROWN jr. and G. D. ROBINSON: *J. Atmos. Sci.*, **36**, 270 (1979).
[9] G. J. BOER and T. G. SHEPHERD: *J. Atmos. Sci.*, **40**, 164 (1983).
[10] A. WIIN-NIELSEN: *Tellus*, **19**, 549 (1967).
[11] R. D. MCPHERSON, K. R. BERGMAN, R. E. KISTLER, G. S. RAUSCH and D. S. GORDON: *Mon. Weather Rev.*, **107**, 1445 (1979).
[12] F. BAER: *J. Atmos. Sci.*, **29**, 649 (1972).
[13] B. J. HOSKINS: *Q. J. R. Meteorol. Soc.*, **109**, 1 (1983).
[14] N. K. VINNICHENKO, N. Z. PINUS and G. N. SHUR: *Meteorol. Gidrol.*, No. 3, 16 (1968).
[15] M. F. LARSEN, M. C. KELLEY and K. S. GAGE: *J. Atmos. Sci.*, **39**, 1035 (1982).
[16] T.-C. CHEN and A. WIIN-NIELSEN: *Tellus*, **30**, 313 (1978).
[17] B. L. GAVRILIN, A. P. MIRABEL and A. S. MONIN: *Isvestiya Acad. Sci. USSR Atmospheric and Oceanic Physics* (English edition), **8**, 275 (1972).
[18] G. D. ROBINSON: *J. Atmos. Sci.*, **34**, 1810 (1977).

PART V

EXTENDED-RANGE PREDICTION
AND CLIMATE DYNAMICS

Part I

Weather Predictability beyond a Week: an Introductory Review.

E. KALNAY

NASA/Goddard Space Flight Center, Laboratory for Atmospheric Sciences
Greenbelt, MD 20771

R. LIVEZEY

NOAA/Climate Analysis Center - Washington, DC 20233

1. – The problem of weather predictability: an example.

As an introduction to the problem of weather predictability, we will consider two forecasts, each of them 20 days long. They were performed with the Fourth Order General Circulation Model of the Goddard Laboratory for Atmospheric Sciences (GLAS), which is representative of present-day forecast models [1]. Its resolution is somewhat coarse (4° latitude ×5° longitude × ×9 levels) compared to more typical resolutions of about 2°, but this is partly compensated by the use of higher-order finite differences. The two forecasts are started from initial conditions corresponding to 9 January 1979, 00 GMT, obtained by applying the GLAS objective analysis scheme [2] to all conventional and satellite data available at that time [3]. The only difference between the two forecasts is that they start from initial conditions which have rather small differences arising from modifications in the analysis scheme. The differences between the two initial conditions are smaller than the uncertainty with which we can determine the state of the atmosphere (fig. 1.1).

Each of fig. 1.2, 1.3 and 1.4 presents the 500 mb geopotential height maps for the first forecast, the 2nd forecast and the verification (panels a), b) and c), respectively) at 5, 10 and 20 days. Several observations stand out: After 5 days there is more similarity between the two forecasts than between any of them and the verification, indicating limited skill in the 5-day forecast. This skill is also regionally dependent: the development of blocking over Eurasia is well forecasted, whereas the amplification of several short waves over North America is poorly predicted, and the Southern Hemisphere forecast has almost no skill. After 10 days there is virtually no skill in either forecast, but the two forecasts still resemble each other. After 20 days not only is there no forecast

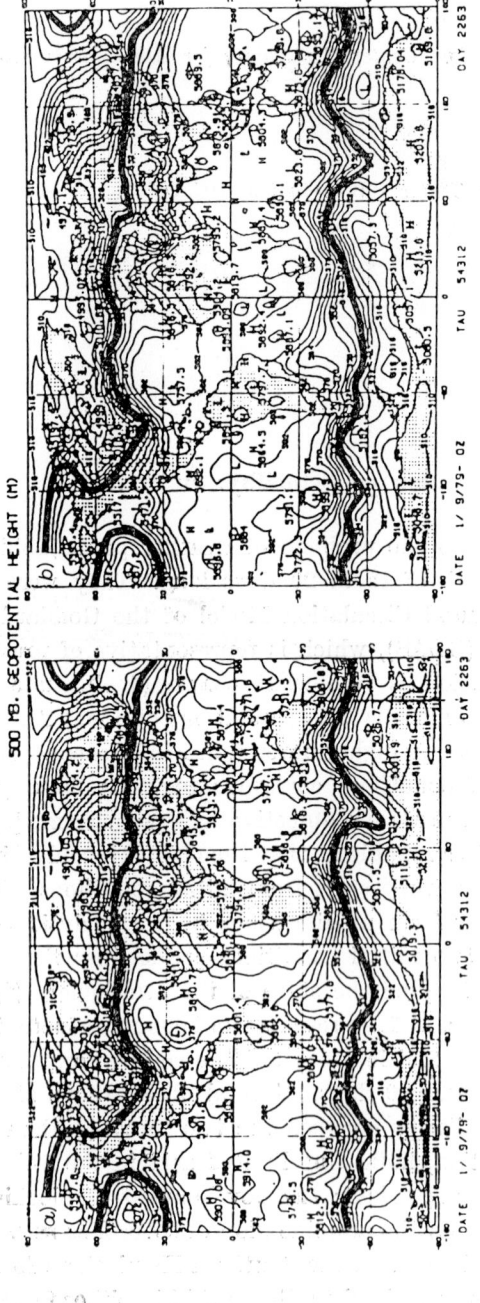

Fig. 1.1. – Initial conditions for 20-day forecasts: *a*) analysis *A*, *b*) analysis *B*, corresponding to 9 January 1979.

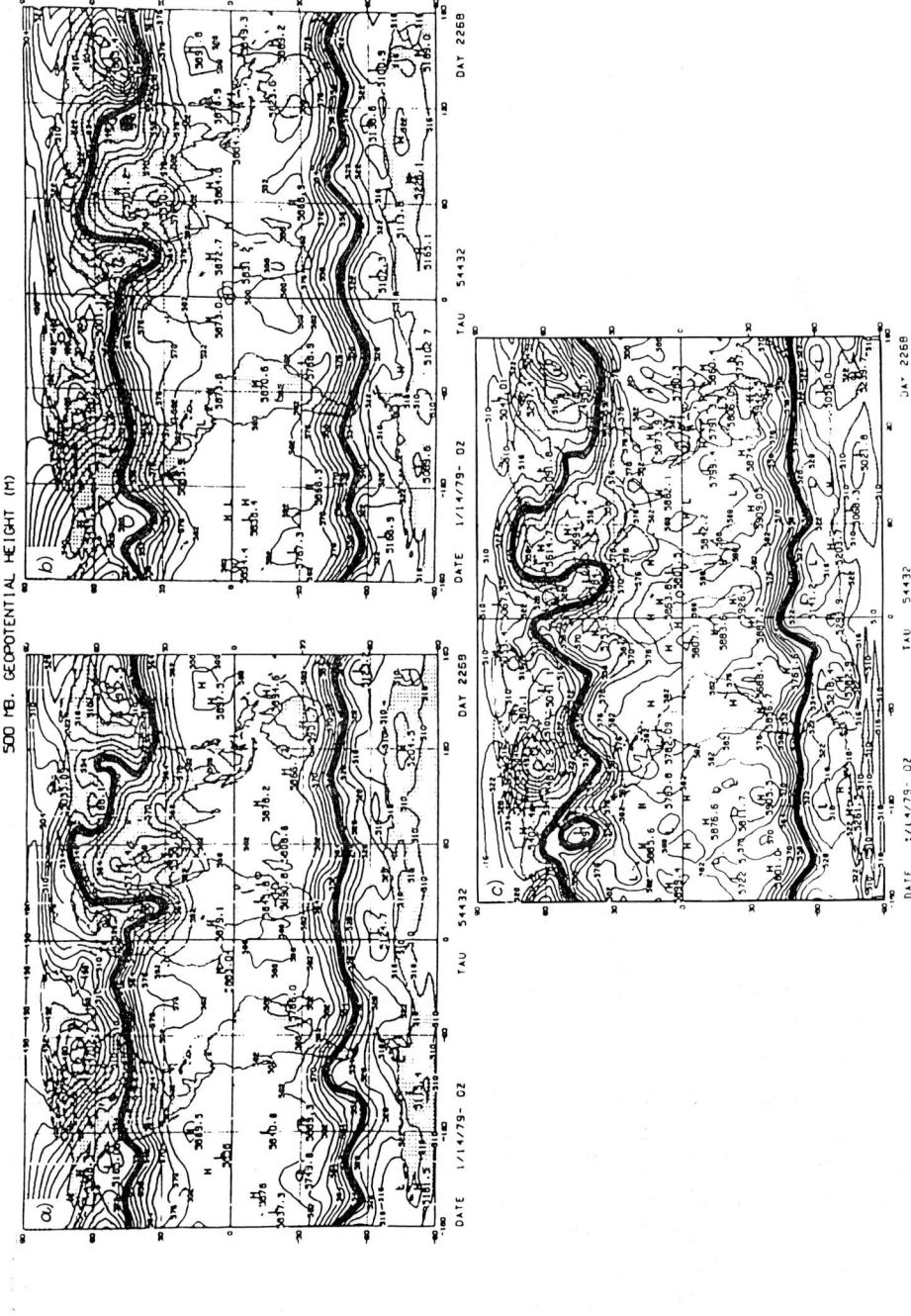

Fig. 1.2. – 5-day forecasts: a) from analysis A, b) from analysis B, c) verification corresponding to 14 January 1979.

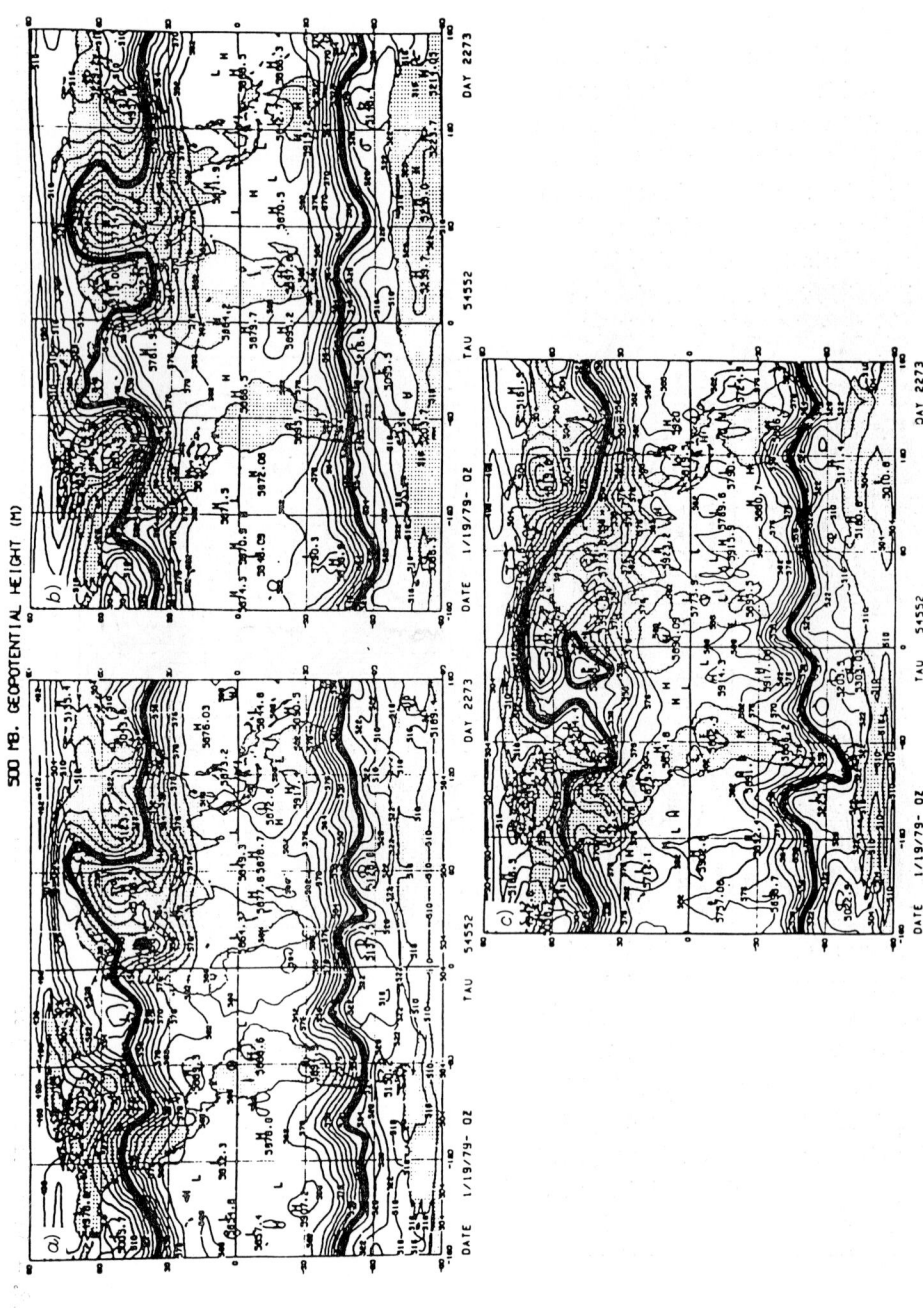

Fig. 1.3. – 10-day forecasts: a) from analysis A, b) from analysis B, c) verification corresponding to 19 January 1979.

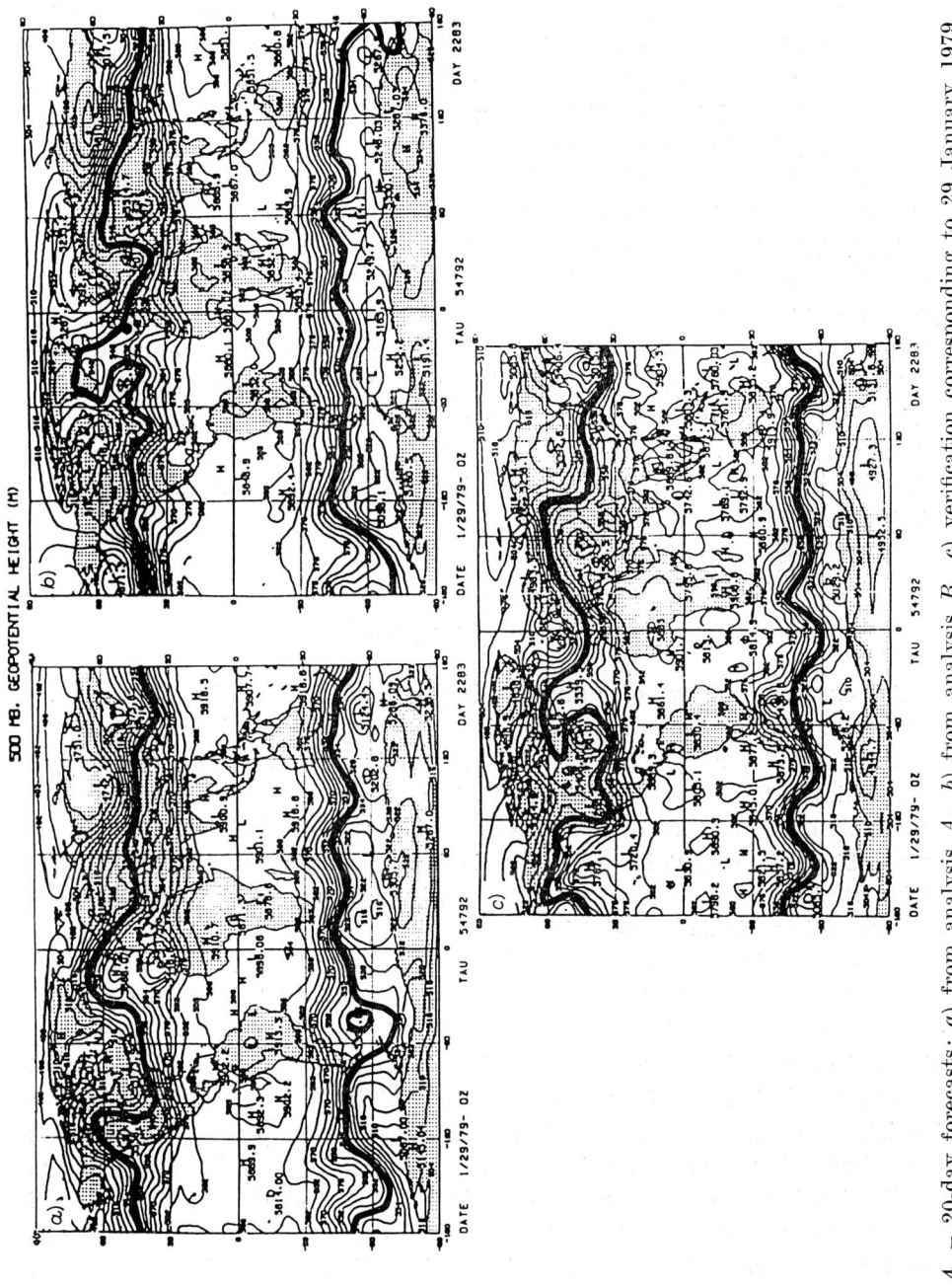

Fig. 1.4. – 20-day forecasts: a) from analysis A, b) from analysis B, c) verification corresponding to 29 January 1979.

skill in predicting the position of individual highs and lows, but the forecasts have also ceased resembling each other.

This example illustrates the fact that our present forecast skill is limited to 5 to 10 days. Furthermore, the divergence of the two forecasts made with the same model and from very similar initial conditions shows that, even if we had perfect models and very accurate observations, the growth of inevitable small errors in the initial conditions would result in the breakdown of the forecast of the *individual weather patterns* after two or three weeks.

In these lectures we briefly review the classic predictability studies which pointed out this intrinsic limitation of deterministic weather prediction (sect. **2**). We discuss other factors not included in the classical studies, such as the influence of slowly changing boundary conditions (sect. **3**) and identified climate oscillations (sect. **4**), which allow potential predictability of atmospheric anomalies on a longer time range. Finally we discuss the current state of the art in long-range weather prediction (sect. **5**). Since this is an introductory review, we have made no attempt to cover the subject exhaustively. The interested reader is directed to two recent reviews, where further extensive references are provided: *Proceedings of the Long-Range Forecasting Expert Study Meeting* [4] and NICHOLLS [5].

2. – Predictability and global forecast error growth.

An excellent review of atmospheric predictability studies is given by LORENZ in [6] (L from now on) and in this volume, to which the reader is referred. Briefly, as pointed out by LORENZ [7], the presence of dynamic instabilities in the atmosphere means that two states which are initially arbitrarily close will in time evolve into appreciably different states. Conversely, any stable system must have quasi-periodic solutions, since, once two trajectories in phase space become close, they will always remain close. The lack of completely periodic behavior in the atmosphere indicates that it is unstable.

Several studies have concentrated on determining the *error growth rate* by integrating an atmospheric model from two sets of initial conditions close to each other and measuring how fast the two integrations depart from each other. The models ranged from a 28-variable low-order model [8], several general circulation models of the 1960's [9, 10] to the European Center for Medium-range Weather Forecasting (ECMWF) model (L). By performing « identical twin » runs, *i.e.* using the same model in both integrations, the idealization is made that the atmospheric model is perfect, and, therefore, if the growth rate of small errors in such models is realistic, the experiments provide an *upper limit* of atmospheric predictability. The estimates of r.m.s. error growth rates have varied from a doubling time of 4 days [8], 5 days [9], 3 days [10] and (1.8÷2.4) days (L). The classic conclusion was that even with a perfect model it would not be possible to make predictions of individual weather patterns beyond about two weeks, because 500 mb geopotential heights, for example, are observed with uncorrelated r.m.s. errors of $(10 \div 20)$ m, and the r.m.s. differences between random states of the atmosphere range from 100 to 200 m.

LORENZ [11] pointed out that it is not possible to extend the forecast another $(2 \div 3)$ days by simply halving the observational errors. By means of a low-order model with widely different scale components, and making some reasonable statistical assumptions, he showed that the doubling time for error growth increases with wavelength. Therefore, inevitable errors in the smaller scales («the flap of the wings of a seagull») will quickly grow and propagate through nonlinear interactions into larger scales. ROBINSON [12] reached very similar conclusions by a simple scale analysis in which a diffusionlike term provided a time scale for the predictability of each horizontal scale. These results are strongly dependent on the assumed shape of the atmospheric energy spectrum. In 3-dimensional turbulence, with an inertial range such that the spectral energy $E \sim k^{-5/3}$, Lorenz' and Robinson's results are valid. In 2-dimensional turbulence, with $E \sim k^{-3}$, the predictability time is infinite if errors are assumed to be made at infinitesimally small scales (see SADOURNY, this volume). However, at very small scales, turbulence is 3-dimensional, and the $k^{-5/3}$ energy spectrum assumption becomes valid, implying a finite predictability time even for 2-dimensional turbulence. We should point out the very important question of what is the limiting factor of present forecasting skill, which falls short of the theoretical limit of predictability. Is it the lack of predictive skill in the smallest scales, or the insufficient accuracy in the observations of the large scales? This problem has not yet been thoroughly studied.

More recently LORENZ [6] has provided a lower bound as well as an upper bound for attainable numerical weather prediction (NWP) skill, the lower bound being given simply by the present ECMWF skill. He showed that the growth rate of the r.m.s. difference E *between two forecasts verifying on the same day* is well parameterized by the simple function

$$(2.1) \qquad \frac{dE}{dt} = aE(E^\infty - E),$$

where a is the growth rate corresponding to a doubling time of 2.4 days and E^∞ is the asymptotic value of the error at long times. This formula reflects the fact that infinitesimal errors grow exponentially in time, and that the average error growth must cease once the system reaches saturation, at which time the two solutions are as different as two random solutions.

Figures 2.1 and 2.2, adapted from L, show this behavior corresponding to the hyperbolic-tangent solution of (2.1) in the case of the thin lines, which represent identical twin comparisons. The actual r.m.s. forecast error growth

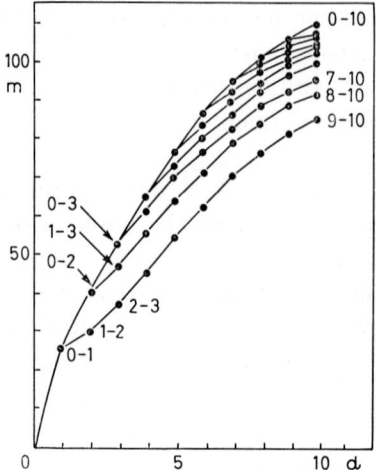

Fig. 2.1. – Global r.m.s. 500 mb height differences E_{jk}, in m, between j-day and k-day forecasts made by the ECMWF operational model for the same day for $j < k$, plotted against k. Values of (j, k) are shown beside some of the points. Left curve connects values of E_{0k}. Thin segments connect values of E_{jk} for constant $k - j$. (From [6].)

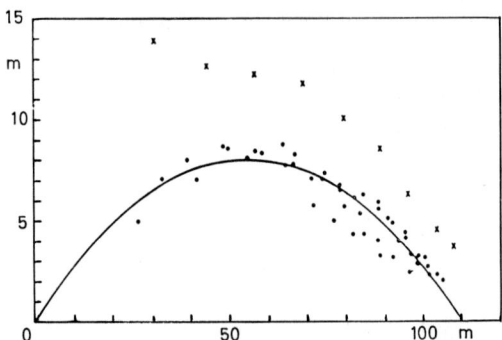

Fig. 2.2. – Increases in global root-mean-square 500 mb height differences $E_{j+1,k+1} - E_{jk}$ plotted against average height differences $(E_{j+1,k+1} + E_{jk})/2$, in m, for each one-day segment of each thin curve in fig. 1 (large dots) and increases $E_{0,k+1} - E_{0k}$ plotted against average differences $(E_{0,k+1} + E_{0k})/2$ for each one-day segment of left curve in fig. 2.1 (crosses). Parabola of best fit to large dots is shown. (From [6].)

(left curve in fig. 2.1 and crosses in fig. 2.2) does not exhibit this behavior. There is a large apparent increase of error between days zero and one (fig. 2.1) and no indication of a maximum in error growth during the 10-day period.

The different growth of real forecast errors and identical-twin comparisons is not surprising, since the ECMWF model, although one of the most accurate models presently available, is still a relatively crude representation of the continuous atmosphere. This difference between a model and the real atmosphere is an « external » source of errors which must be included if we want to represent real forecast error growth. LEITH [13-15] has, therefore, modeled the actual forecast error variance V growth (at short times) by the equation

$$\text{(2.2)} \qquad \frac{dV}{dt} = \alpha V + S,$$

where α is the growth rate of the variance (twice as that of the r.m.s. error) and S is the error growth due to external sources. Assuming an initial value of the error V_0 characteristic of the atmospheric analysis (initial data), he obtains the following equation for the *perceived error variance* (difference between the forecast and an independent analysis):

$$\text{(2.3)} \qquad V_p(t) = 2V_0 + (V_0 + S/\alpha)(\exp[\alpha t] - 1).$$

We can derive a parameterization of the actual variance error growth which includes both saturation effects, as suggested by LORENZ, and external sources of error growth, as well as the effect of analysis errors on the perceived error variance, as suggested by LEITH:

$$\text{(2.4)} \qquad V_p(t) = 2V_0 + V(t),$$

$$\text{(2.5)} \qquad \frac{dV}{dt} = \alpha(V + S)(V^\infty - V).$$

The solution of (2.5) is

$$\text{(2.6)} \qquad V(t) = V_\infty \frac{\mu}{1+\mu} - S \frac{1}{1+\mu},$$

where $\mu = c \exp[\alpha(V_\infty + S)t]$.

We have tested the validity of this formula using the same 100-day data set as in [6], but with error variances rather than r.m.s. errors. Figures 2.3a) and b) show the excellent fit of (2.6) to the observed ECMWF forecast error variances in the winter and summer of 1981, respectively. On the other hand, no good fit was possible when the external error growth S was assumed to be zero.

We conclude this section by pointing out that there are several reasons why the usefulness of the concept of doubling time for real forecast error growth may be limited, as the widely varying estimates of its value perhaps indicate.

First, the uncertainties present in atmospheric analysis are large enough to make difficult detection of errors after short times, before finite-amplitude effects become important. The large perceived error growth on the first day in fig. 2.1 may be an indication of the effect of analysis errors (term $2V_0$ of eq. (2.3)), although it is small in our parameterization (fig. 2.3).

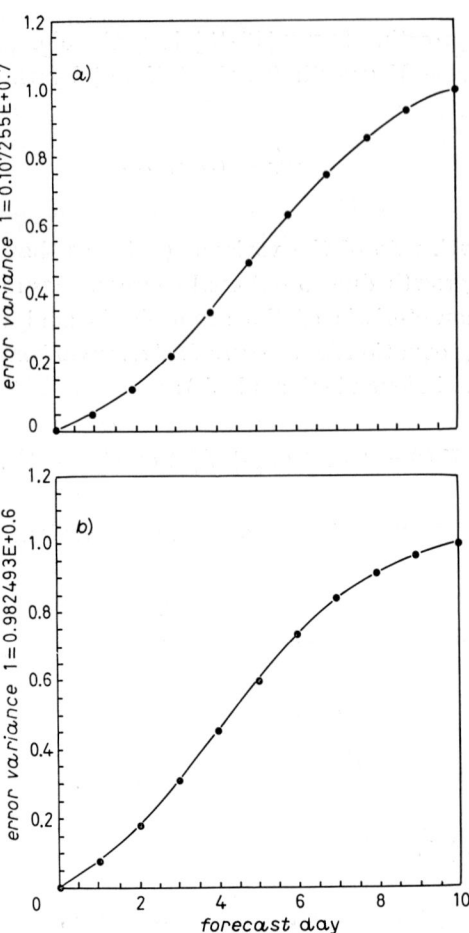

Fig. 2.3. – Global 500 mb height forecast error variance from the ECMWF operational model as a function of forecast length. The dots are the actual error variances and the curve is the best fit according to formula (2.6). a) 100 days during winter of 1981, b) 100 days during summer of 1981. (From [16]).

Second, a sizable portion of the errors in the initial conditions may project on the model's fast normal modes (inertia-gravity waves) and be dispersed in a day or so. This can lead to an actual *decrease* in the r.m.s. error during the first day, as observed in several predictability studies (*e.g.*, [9]). During the process of nonlinear normal-mode initialization, as practiced by operational centers such as the National Meteorological Center (NMC) and ECMWF, the time tendencies of the model's fast modes are forced to remain small. However, this and other characteristics of the analysis/initialization methods in NWP may introduce *systematic* errors which are difficult to estimate, and whose consequences are not well known.

Finally, it is important to remember that the error growth varies widely between different individual forecasts [7]. Forecast error growth can occur in sudden bursts, during which there is a serious breakdown in forecast skill, as shown by fig. 2.4 (from [3]). The nonidentical-twin study of Hoffman and

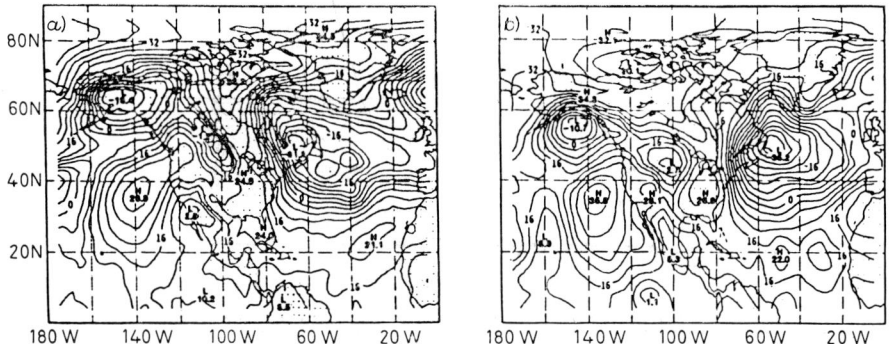

Fig. 2.4A. – a) Eight-day sea level pressure forecast for 0000 GMT 6 February 1979 with the 2.5° × 3° GLAS fourth-order model from initial conditions provided by the FGGE assimilation at 0000 GMT 29 January 1979. Sea level pressure minus 1000 is in millibar. b) The FGGE sea level pressure analysis at 0000 GMT 6 February 1979. (From [3].)

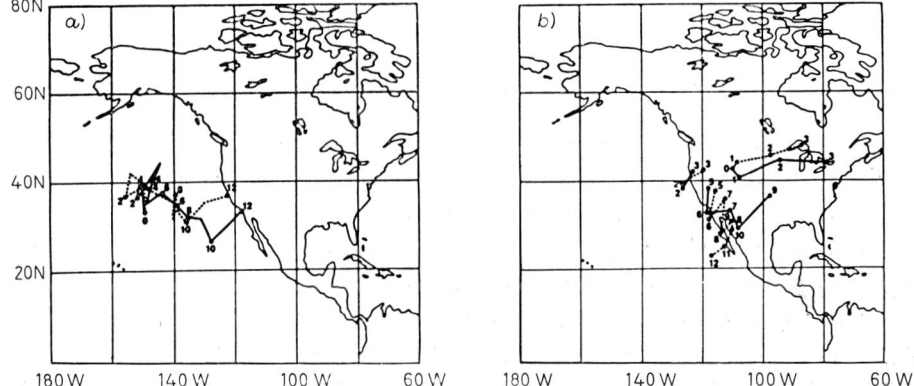

Fig. 2.4B. – a) Comparison between observed (solid line) and predicted (dotted line 500 mb) positions of a Pacific anticylcone during the 0000 GMT 29 February 1979 forecast. The numbers indicate the number of elapsed days in the forecast. b) As in a) except for three cyclones over North America. (From [3].)

Kalnay [17] shows that these short periods of explosive growth can occur at very different times for individual forecasts (fig. 2.5). In fig. 2.5 we see that the resulting ensemble average error grows almost linearly with time, even though the individual forecast errors clearly do not. The linear growth rate apparent in the real forecast errors of L after the first day (fig. 2.1 and 2.2)

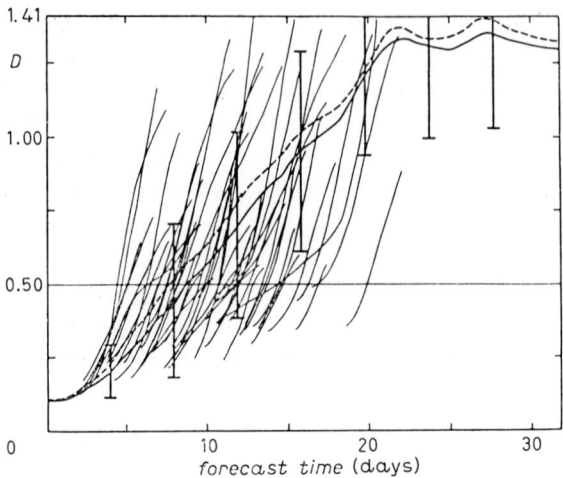

Fig. 2.5. – Time evolution of the scaled r.m.s. error D for 50 individual dynamical forecasts. Shown are selected portions of the evolution of D for each case (light curves), $\langle D^2 \rangle^{\frac{1}{2}}$ (dashed curve) and $\langle D \rangle$ (heavy curve). Error bars are calculated as $(\langle D^2 \rangle - \langle D \rangle^2)^{\frac{1}{2}}$. (From [17].)

may, therefore, be due not only to external sources of error (eq. (2.5)) but also to the ensemble averaging process.

3. – Predictability of time averages: internal dynamics *vs.* boundary forcing.

We have seen that the limit of dynamic predictability of individual weather events is not more than two weeks. Furthermore, since this limit depends on the dominant atmospheric instabilities, it is also a function of season, latitude and geographical location. Nevertheless, there are good reasons to believe that *time means may be predictable well after weather has ceased to be predictable*.

The aim of short-range weather forecasting is to accurately predict the changes of weather, and, therefore, the skill of the forecasts is usually compared with persistence, *i.e.* a « no change » forecast. In long-range weather prediction the goal is to forecast anomalies, *i.e.* the extent by which monthly means, for example, differ from their expected climatological values.

Anomalies in the monthly or seasonal means may be due to *internal* variability, associated with the nonlinear dynamics of the atmosphere, and to *external* variability, due to anomalies in the boundary forcing of the atmosphere. Anomalies of dynamical origin might be expected to have a limited period of deterministic predictability. At longer time scales, they can be a source of unpredictable noise. Anomalies due to changes in the external forcing, such as sea surface temperature, may be predictable for longer periods if the boundary forcing anomalies themselves are predictable, or if they change slowly.

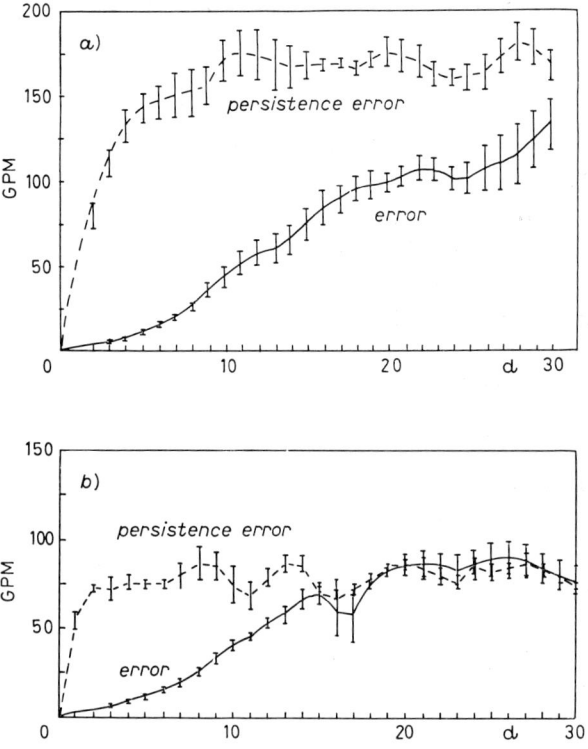

Fig. 3.1. – a) Root-mean-square error, averaged for six pairs of control and perturbation runs and averaged for latitude belt $(40 \div 60)°$ N for 500 mb geopotential height (GPM), for wave numbers $0 \div 4$. Dashed line is the persistence error averaged for the three control runs. Vertical bars denote the standard deviation of the error values. b) As in a) except for wave numbers $5 \div 12$. (From [18].)

Because of the dependence of dynamic predictability on the spatial scale (sect. **2**), planetary (long) waves have a longer theoretical limit of predictability than synoptic (weather) scales (fig. 3.1a), b), from [18]). This suggests that even anomalies due to internal dynamics may be partially predictable beyond two weeks. SHUKLA [18] performed several 60-day numerical integrations with the GLAS climate GCM. In one set of experiments he used slightly perturbed initial conditions and computed the corresponding variance of monthly means. He then compared this variance with that obtained using initial conditions corresponding to different years. He found that monthly means retained some predictability (i.e. the variance of monthly means from the perturbed forecasts was smaller than that of runs corresponding to different years) for 1 to 1.5 months, even though no anomalous boundary forcings were included. After two months, however, no further dynamical predicability in the monthly means was found. SEIDMAN [19] also performed identical-twin experiments using a Goddard Institute for Space Studies (GISS) 3-level model and found that regional space averaging and 5-day time averaging increased the predictability time (defined by a correlation of 0.5 between observed and forecasted surface temperature fields over the U.S.) from 11 days to 19 days.

LEITH [20] has suggested a method to determine the percentage of variance of monthly or seasonal means which can be ascribed to changes in the boundary conditions. This percentage, denoted *potential predictability*, is obtained by subtracting from the observed variance of monthly means that which can be expected from *natural variability*, or *climate noise*. The natural variability or climate noise is due to the fact that dynamic day-to-day weather variations will produce sampling fluctuations in the monthly means even if there are no changes in the external forcings. Therefore, it represents the percentage of observed variance which is not predictable by dynamic methods (in the absence of anomalous forcings) beyond the first couple of weeks of determinisitic predictability.

Leith's method has been applied by MADDEN [21, 22] and by MADDEN and SHEA [23]. In order to determine the natural variability, it is necessary to know the power spectral density of weather fluctuations. MADDEN [21] estimated it by assuming that it can be modeled by a first-order autoregressive (Markov) process, and, therefore, that it becomes white noise (constant spectrum) at low frequencies. He estimated the value of the spectrum at low frequency as equal to that of periods of 96 days or longer. SHUKLA [24] pointed out that this may over-estimate natural variability, and hence underestimate potential predictability, because time scales of ~ 90 days already include the effect of changing boundary conditions. MADDEN [25], however, indicated that reasonably moderate modifications of the assumptions would only change the estimation of natural variability by about 20%.

Maps of potential predictability of seasonal means of surface air temperature over the U.S. (from [26]) are shown in fig. 3.2. It can be seen that the percentage of variance due to boundary forcing depends on both season and geographical location. It is maximum near the East and West coasts, pointing out the possible importance of sea surface temperature anomalies, or the preferred position of planetary waves. A comparison with hindsight and actual skill in the prediction of monthly means will be presented in sect. **5**.

REINHOLD and PIERREHUMBERT [27] performed long integrations with a low-order truncated model which included both planetary-scale and synoptic-scale baroclinic waves. They found that the presence of the unstable baroclinic waves resulted in the existence of « weather regimes ». The model atmosphere resided in each of several regimes for aperiodic lengths of time. This almost intransitivity, if applicable to the real atmosphere, represents another source of natural variability not associated with boundary forcing. It has the potential of increasing predictability, if the residence time is predictable, or of reducing it, by increasing natural variability beyond that of a single weather regime.

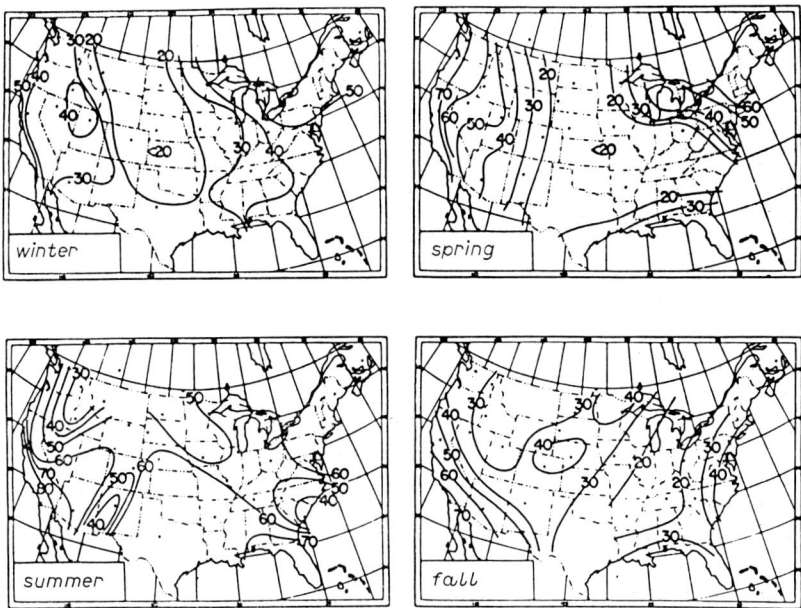

Fig. 3.2. – Distribution of potential predictability as a percent of variance. (From [26].)

It is interesting to note that VAN DEN DOOL [28] has recently suggested that most of the long-range forecasting skill is due to persistence of the anomalies, and that the enhanced skill during winter, and to some extent summer, is due to the fact that during these extreme seasons the mean flow of the atmosphere is most stationary, and, therefore, the response to boundary forcing is most persistent.

The existence of a long-lasting anomaly in the boundary conditions will not necessarily result in an anomalous circulation, as many sensitivity experiments have shown. It is important to know under what circumstances boundary anomalies will be effective in influencing the atmospheric circulation, and also whether observational and model errors will allow a faithful reproduction of these effects. SHUKLA [29] discussed in some detail the physical mechanisms by which anomalous bondary forcings can influence the atmospheric circulation and hence provide for potential long-range predictability. He summarized several sensitivity experiments performed with the GLAS, Geophysical Fluid Dynamics Laboratory (GFDL) and GISS climate models, and for more detail the reader is directed to the references therein.

The anomalous forcings that SHUKLA considered were sea surface temperatures, soil moisture, surface albedo, snow cover and sea ice. Briefly, « the effectiveness of a boundary anomaly on the circulation depends on the existence of a favorable dynamical environment in which the surface forcing can be transformed into a three-dimensional heat source, and on the ability of this influence to propagate away from this source». For example, a 1 K anomaly in the tropical SST is more effective than in colder mid-latitudes because it produces a much larger change in saturation vapor pressure, and hence of latent heating, due to the nonlinearity of the Clausius-Clapeyron equation. Similarly, a 1 K anomaly in areas of large-scale convergence of mass and moisture, such as the ascending branches of the Hadley or Walker cells, will also be more effective than in the descending branches, where precipitation is inhibited by the circulation. Furthermore, the influence of a stationary SST anomaly within an easterly regime will not easily propagate into mid-latitudes because of the barrier effect of critical surfaces separating easterlies from westerlies. In a similar way, anomalous soil moisture will have negligible influence during the winter, when there is little evaporation. During the summer, on the other hand, it can affect the circulation in two ways: modifying the evaporation rate over continents and changing the heat capacity of the soil, and, therefore, the intensity and position of heat lows. A positive feedback between increased subtropical albedo and consequent desertification was studied by CHARNEY et al. [30], and relationships between snow cover and the atmospheric circulation have been found both observationally and numerically.

Having reviewed the factors which can affect the potential success of long-range forecasting (LRF), we turn now to its methods. There are three basic approaches to LRF. The first one, and perhaps the most successful so far, is essentially statistical modeling, which will be discussed in sect. **5** and **6**. The second one is the use of *simplified dynamical models*, which intend to represent only the main regional budgets in the thermodynamic and vorticity equations, taking into account anomalous forcing. Two such examples which have been used in LRF are ADEM [31] and the Republic of China's anomaly model [32]. Other simplified modes have been developed by WEBSTER [33] and OPSTEEGH and VAN DEN DOOL [34], and active research in this area is presently going on. Chao's model attempts to determine atmospheric anomalies by requiring that the atmospheric vorticity equation be in balance given slowly varying ocean and land surface temperature anomalies. MIYAKODA and CHAO [32] present 14 cases of monthly prediction with Chao's model of which *only* 4 show more skill than persistence. One of them, however, corresponding to January 1977, exhibited a remarkable score of 72% correlation with observed anomalies in the Northern Hemisphere (fig. 3.3a) and c), reproduced from [32]). In this case persistence also had a high skill of 59% in the anomaly correlation.

The most expensive, but in principle the most promising method, is to use *comprehensive general circulation models*, as detailed as those used for short-range weather prediction, and with the most accurate parameterizations of subgrid-scale physics possible. This approach is beginning to become feasible because of two relatively recent developments. One is the availability of new very fast vector processing computers like the Cray-1 and Cyber-205 array processors. The GLAS Fourth Order GCM, for example, can now be integrated

Fig. 3.3. – Comparison of the monthly mean predicted anomalies at 500 mb geopotential height: the observation (left), the prognostic of free mode by the N48L9-E4 GFDL GCM (middle) and the prognostic of forced model by the one-layer anomaly model (right). The contour interval is 30 m, and the negative regions are shaded. (From [32].)

on the Cyber-205 for 20 days in the same time it took the Amdahl-V6 computer to integrate it for one day. The second factor, equally important, is that satellite-based remote-sensing systems, and new algorithms for the processing of their observations, are starting to be able to provide *global* observations of weather and climate anomaly fields. For example, a temperature retrieval algorithm has been developed by SUSSKIND *et al.* [35], based on the inversion of the radiative transfer equation using simultaneously several infra-red and microwave frequencies observed by the operational satellite TIROS-N HIRS/MSU sensors. This retrieval system produces, not only the atmospheric temperature soundings for which the sensors were designed, but, as a by-product, unexpectedly accurate global fields of sea surface temperature (SST), snow and ice cover, and cloud heights and amounts.

An important remaining difficulty of the GCM approach is the problem of *climate drift*: present models have become more and more accurate in the short run, but their climatology is still different from the observed one. As a result, in the long run, these models drift towards their own climatology and away from the real climatology. Further improvements in the models (resolution, orographic forcing and especially parameterization of subgrid-scale physical processes) should not only improve deterministic predictability but, through the reduction of systematic errors which result in climate drift, produce more realistic responses to boundary forcing anomalies. WALLACE *et al.* [36], for example, have shown how the use of topographic forcing enhanced in regions with rugged orography resulted in a significant reduction of the ECMWF model systematic errors.

Two pioneering efforts in long-range forecasting with general circulation models are those of Miyakoda *et al.* [37] (reviewed also in [32]) and of the British Meteorological Office ([38] and BMO: GILCHRIST, presented at the 1983 LRF Princeton Meeting). Miyakoda's experiment was very successful in predicting the January 1977 monthly average 500 mb anomaly (fig. 3.3a) and b)) and, to a certain extent, the 10-day average evolution during that month. The result was sensitive to the model's numerical scheme and physical approximations, as well as to the initial conditions. GILCHRIST showed several examples of skillful monthly predictions, using a global BMO GCM. Both of these results were obtained using *climatological* boundary forcing, indicating that in these cases the extended predictability was dynamical in origin.

We have seen that the skill in short-range forecasting varies strongly from case to case, from region to region, and that LRF skill may depend on other factors such as season, the state of the atmosphere and the type of boundary anomalies. It is very desirable to develop a system that can provide an *a priori* estimation of the extent to which a particular forecast is more skillful than persistence or climatology. A *stochastic-dynamic* prediction method was developed for this purpose by EPSTEIN [39] including forecast equations for the probability distribution of the atmospheric variables. Because of the large number of equations and of closure problems the implementation of Epstein's method is unfeasible except for the simplest models. Therefore, LEITH [40] proposed instead the use of an ensemble forecast method such as Monte Carlo forecasting. In this method a small number ($N \sim 8$) of forecasts are performed from initial states selected by random perturbations of the best estimate of the initial conditions.

HOFFMAN and KALNAY [17] formulated the lagged averaged forecast (LAF) method, which has all the advantages of Monte Carlo forecasting and can be attained at virtually no cost. It makes use of not only the latest operational forecast, but also of forecasts for the same verification time started one or more days earlier than the latest one. The LAF method, in which the forecasts

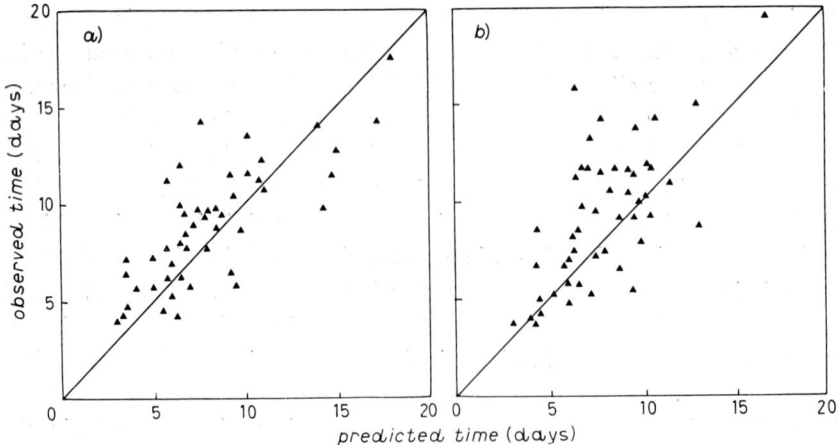

Fig. 3.4. – Scatter plots of predicted *vs.* observed time for the scaled r.m.s. forecast error to reach a value of 0.5 for the a) lagged average forecast scheme ($r = 0.79$) and b) the Monte Carlo forecast scheme ($r = 0.68$). (From [17].)

are weighted by a simple regression scheme based on a parameterization of the forecast error growth, was tested using low-order « nature » and « forecast » models. The most important result of this study was that the spread of members of the ensemble forecast was a good *a priori* estimator of the skill of the ensemble forecast (fig. 3.4, from [17]).

The LAF method seems particularly appropriate for long-range forecasting, in combination with a comprehensive GCM and with observed boundary conditions. In this context, the longer predictability expected for time averages will naturally give rise to regression weights different from zero for periods beyond the limit of deterministic predictability. The LAF method should provide more accurate forecasts than a single dynamical forecast because the unpredictable components will be filtered out by the ensemble averaging, and it may also generate meaningful estimates of the confidence limits for individual anomaly forecasts.

4. - Predictability of long-lived atmospheric phenomena.

As discussed in sect. **2**, the existence of atmospheric instabilities precludes any possibility of periodic behavior. Nevertheless, there are some observed quasi-periodicities and long-lived phenomena, which, if properly observed and/or properly predicted, can contribute under certain situations to extended predictability.

Atmospheric blocking is an example of a rather long-lasting phenomenon: It is frequently observed that a center of high pressure develops at high latitudes, becomes stationary for about a week or longer and « blocks » the normal passage of cyclones, producing a split of the westerly flow [41]. There is not yet a universally accepted theory to explain blocking (*e.g.*, [42-46]).

Ten years ago, atmospheric blocking was considered to be difficult to reproduce in a GCM. Today, with the advent of better models, prediction of blocking is more successful, and cases of exceptional extended predictability are usually associated with blocking [3, 37, 47]. It is not surprising that blocking should be highly predictable because it is a rather quiescent phenomenon, and, therefore, errors due to local instabilities and to numerical truncation can be expected to grow slowly.

The existence of *atmospheric teleconnections* is also associated with time scales beyond a week. Strong correlations between time means of sea level pressure (SLP) or 500 mb anomalies have been observed at widely separated points of the globe [48-51]. WALLACE and GUTZLER [52] performed a systematic study of teleconnection patterns in winter monthly mean anomalies using 15-year analysis of the Northern Hemisphere (NH). They found five teleconnection patterns in the 500 mb winter anomalies. These patterns (Pacific-North American, East Atlantic, West Atlantic, West Pacific and Eurasian) appear as centers of simultaneous positive and negative correlations (fig. 4.1).

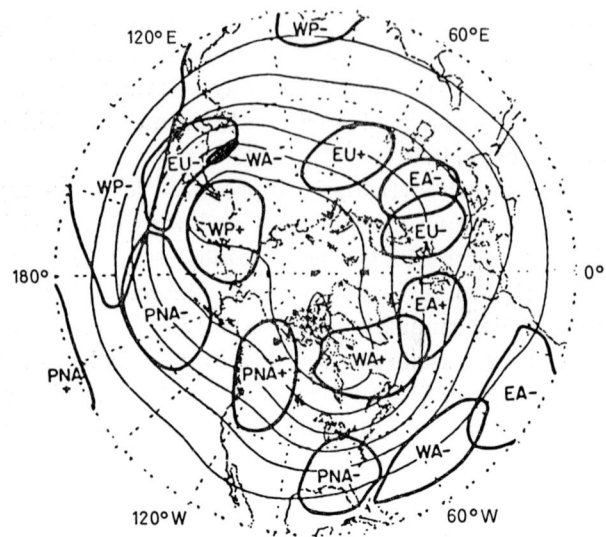

Fig. 4.1. – ± 0.6 isopleths of correlation coefficient between each of five pattern indices (Pacific-North American, West Atlantic, East Atlantic, Eurasia and West Pacific) and local 500 mb heights (heavy lines). On the background is the wintertime mean 500 mb height contours (light lines). (From [52].)

BLACKMON et al. [53, 54] did a study similar to Wallace and Gutzler's, but on 500 mb time-filtered fields separated into three time scales: periods longer than 30 days, intermediate scales with periods between 10 and 30 days and short time scales with periods of 2.5 to 6 days. They found that the longer time scales had teleconnection patterns characterized by North-South elongated dipoles geographically located at the exit of the main winter jets (fig. 4.2a)) and with no clear direction of propagation.

Intermediate time scales exhibit teleconnection centers showing an eastward and poleward propagation, and, although their origin is loosely associated with the jet entrances, they do not have a fixed geographical location (fig. 4.2b)). The short-time-scale correlation fields show very zonal trains of Rossby waves. Lagged correlation maps of the short-time-scale fields indicate an eastward propagation with speeds similar to the mean 700 mb geostrophic wind (fig. 4.3). The lagged correlation maps of intermediate time scales show little phase propagation (which is consistent with their time window) but again eastward and poleward energy propagation (fig. 4.4).

The importance of teleconnections, especially in the intermediate and longer time scales, is that they may be interpreted as trains of Rossby waves which are the atmospheric response to anomalous forcing [44, 55]. Just like the extratropical-cyclone and polar-front models provide guidance to subjective weather forecasting, the knowledge of the location, phase and direction of

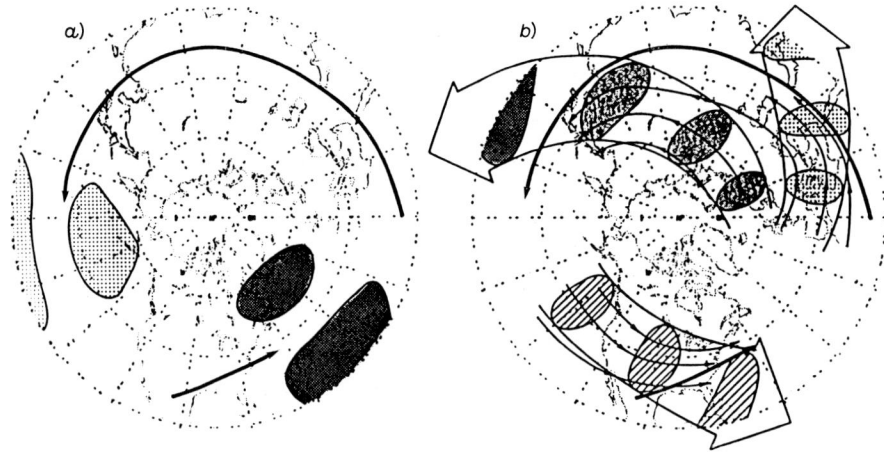

Fig. 4.2. – Schematic model of teleconnections: Heavy arrows indicate axis of climatological mean wintertime jetstreams and lighter lines with arrows indicate a few selected jetstream level geopotential height contours. Shading indicates centers of action of the teleconnection patterns. a) Geographically fixed dipole patterns in jet exit regions, dominant in the long-time-scale variability, b) mobile patterns in the jet entrance regions, dominant in the intermediate-time-scale variability. (From [53].)

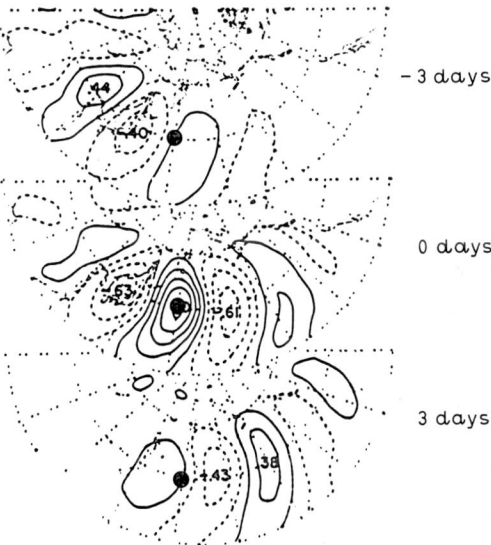

Fig. 4.3. – Short-time-scale (2.5 to 6 days) lagged correlation patterns for 500 mb heights for the base point 40° N, 160° E (indicated by dot) for lags relative to the time series at the base point of −3 days, 0 days and +3 days. Contours 0.1, 0.3, ... solid, −0.1, −0.3, ... dashed. (Adapted from [54].)

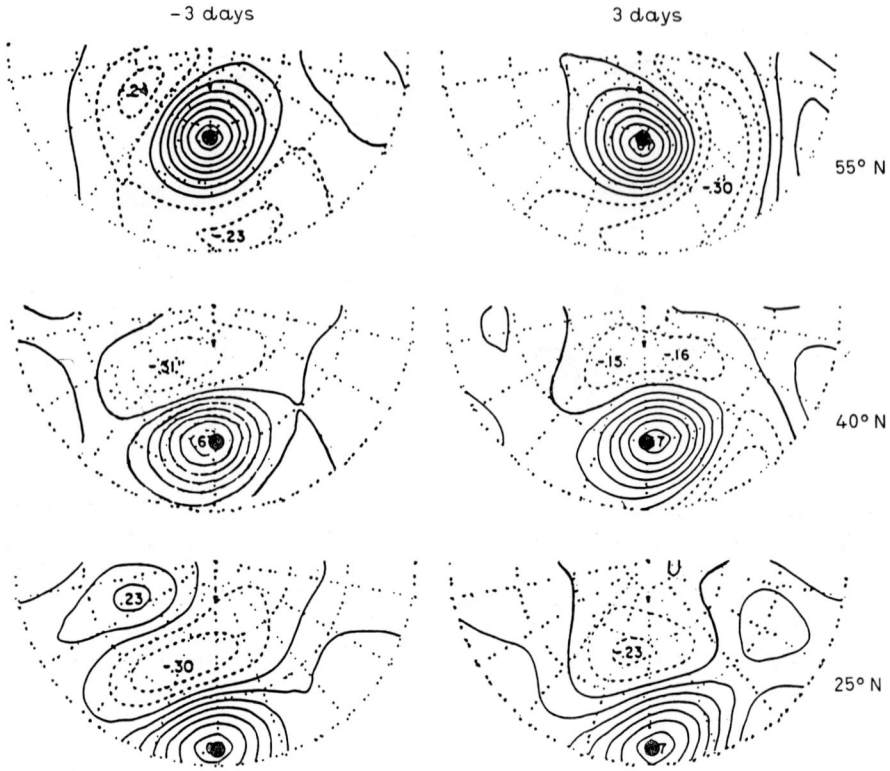

Fig. 4.4. – Composite lagged correlation patterns for intermediate time scales (10 to 30 days) for lags of -3 days and $+3$ days. Base points (dots) at 55° N, 40° N and 25° N. (Adapted from [54].)

propagation of atmospheric teleconnections is an important tool in the preparation of monthly and seasonal forecasts (sect. **5**).

The strongest observed climate oscillation signal, which is closely related to the subject of teleconnections, and which has dominated the field of climate research in the last few years, is that of *El Niño-Southern Oscillation* (ENSO). Several recent papers have reviewed the atmospheric an oceanic phenomena associated with ENSO episodes (*e.g.*, [56-60]).

The Southern Oscillation is a manifestation of variations in the intensity and position of the Walker cell, *i.e.* the East-West Pacific tropical circulation with upward motion in the Indonesian region (associated with strong precipitation and latent-heat release), eastward flow at upper levels, sinking motion in the dry South Pacific anticyclone near South America and westward flow at low levels. When the maximum-precipitation region shifts eastwards towards the dateline, the intensity of both the thermal low in Indonesia and the southeast Pacific high diminishes. This results in the correlation pattern observed by WALKER in the 1920's and 30's and denoted « the Southern Oscillation ».

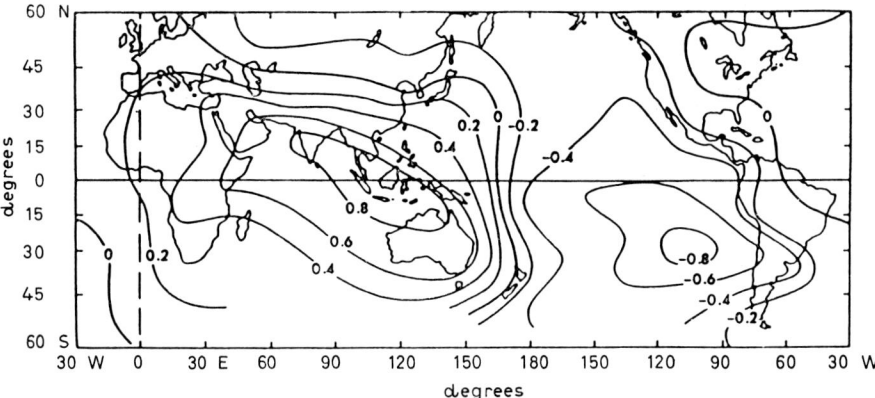

Fig. 4.5. – The correlation of monthly mean surface pressure with that of Djakarta. The correlations are large and negative in the South Pacific High Pressure Zone and are large and positive in the Australian-Indonesian Low Pressure Zone. The Southern Oscillation is not a standing oscillation, so that correlations do not have a maximum at zero lag. (From [60].)

It is a remarkable example of teleconnection of planetary scale (fig. 4.5). BJERKNES [48] pointed out the relationship between this atmospheric oscillation and the warming of the equatorial Pacific SST (usually referred to as « El Niño », because it is associated with an enhancement of the seasonal warming of SST that takes place near Christmas off the coast of Peru).

Although the complex ocean-atmosphere interactions that give rise to this phenomenon are not yet thoroughly explained, a brief description of a typical « warm » event is as follows (fig. 4.6): around October-November before the onset of El Niño, for reasons not yet understood, the upward branch of the Walker circulation shifts eastward from Indonesia towards the dateline. With this shift, both the Indonesian low-pressure zone and the eastern South Pacific anticyclone weaken. Trade winds in the western Pacific diminish and this results in a weakening of the strong East-West SST gradient in the equatorial Pacific (fig. 4.7). Warm water moves eastward [61], and maximum warming is observed off the Coast of Peru (« El Niño »), where normally coastal and equatorial upwelling forced by the South Pacific anticyclone and the trade winds maintain the minimum SST's.

After this preliminary period, starting around January, the warm anomaly intensifies and moves westward (*) [58] until, by October, the whole equatorial

---

(*) The westward propagation and intensification of the anomaly may be an instability of the coupled ocean-atmosphere system [60]: A warm equatorial SST anomaly produces upward motion, latent heating and surface convergence. This results in weakened easterlies (trades) to the west of the anomaly, where solar heating becomes more effective in warming the water. At the same time, the weakened easterlies result in warm-water propagation to the east, so that the original anomaly also intensifies.

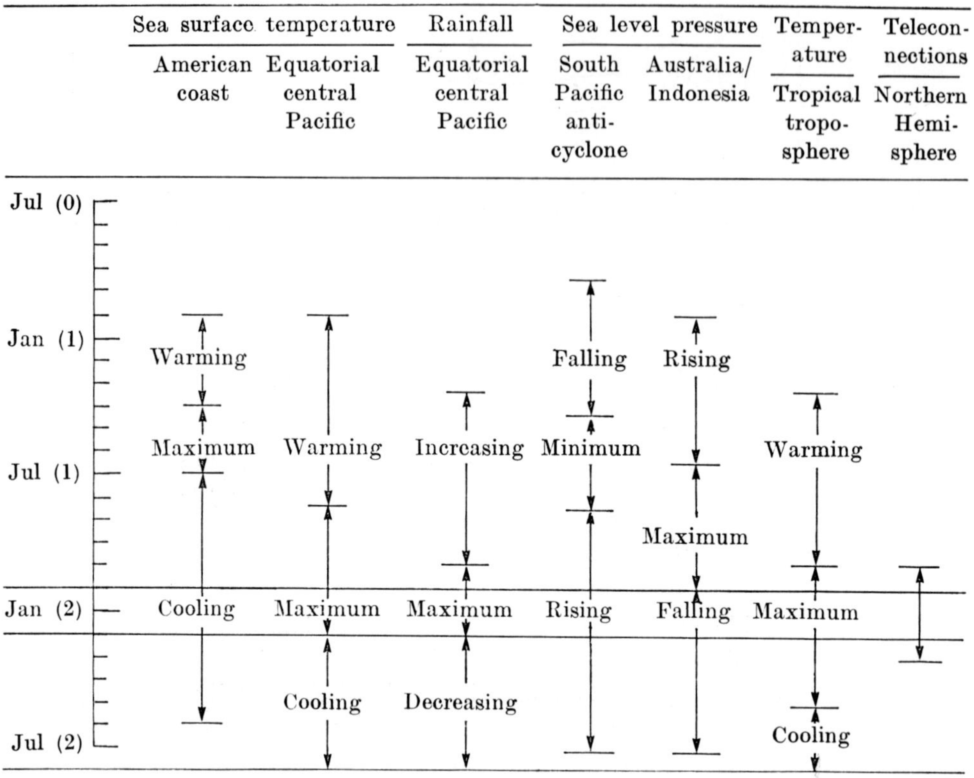

Fig. 4.6. – Schematic representations of the changes in selected parameters during the evolution of a typical warm episode. Dates in the left-hand margin indicate a continuous time sequence covering the periods preceeding, during and shortly after the second Northern Hemisphere winter during which the dynamical conditions favor the influence of tropical anomalies upon the Northern Hemisphere planetary waves. The calendar year (1) refers to an El Niño (*e.g.* 1957, 1972, 1976). (From [59].)

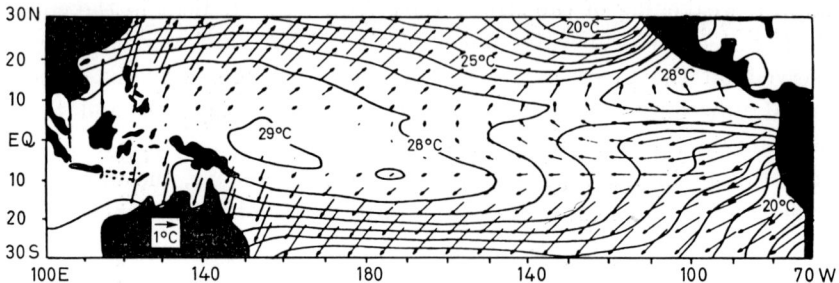

Fig. 4.7. – Mean sea surface temperatures (solid lines) and the amplitude (length of arrow) and phase (direction of arrow) of the annual cycle of sea surface temperature in the tropical Pacific Ocean. Maximum sea surface temperature is in January for an arrow that points downward and in April for an arrow that points to the left. (From [60].)

Pacific is very warm (fig. 4.8). The SST anomalies reach their maximum around the second NH winter, and then slowly decay for the next half year or so.

It is during this second NH winter that the SST anomalies, which result in maximum absolute temperature and latent heating near the dateline, may force teleconnections of anomalous atmospheric circulation in the Northern Hemisphere [59]. Figure 4.9 shows schematically the Rossby waves that might be forced by the SST anomaly, forming a Pacific-North American (PNA) teleconnection pattern. Recent experiments by SHUKLA and WALLACE [62] have confirmed this relationship between equatorial SST heating and the PNA pattern.

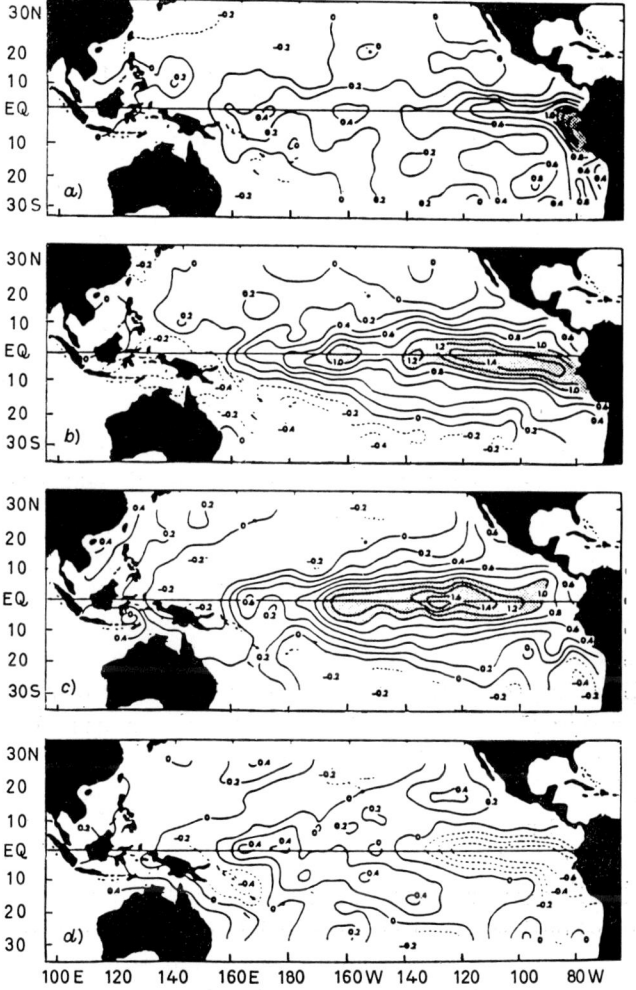

Fig. 4.8. – Sea surface temperature anomalies in °C during a typical ENSO event for a) March, April and May after the onset, b) the following August, September and October, c) the following December, January and February and d) May, June and July, more than a year after the onset. (From [58].)

Fig. 4.9. – Teleconnections to the extratropics during ENSO (schematic). (Adapted from [59].)

Unfortunately, even in this long sequence of events, which is qualitatively predictable once the precursors are established, it is not easy to produce skillful LRF. The scale of the anomalies and their propagation is so large that small differences in their evolution result in markedly different regional weather patterns in different episodes (fig. 4.10). More detailed GCM studies, using the observed atmospheric circulation and SST's, will be necessary in order to explore predictability of weather associated with ENSO events.

The most recent ENSO episode, in 1982-1983, was unusual both in strength and in evolution. Warm Pacific anomalies were observed in early 1982, but not near Peru. Even after October 1982, when the pattern was more similar to that of a normal mature El Niño event, the anomalous heating pattern was much more intense and further East than in other events. This episode was associated with intense droughts in Australia and Indonesia and very strong precipitation in Southern Peru and Central Pacific. Other anomalies such as the droughts in Southern Africa and NE Brasil and record floods in SE Brasil and NE Argentina may have also been related to this event. But, although the observed winter of 1983 teleconnections had some crude similarity to fig. 4.9, the differences were large enough to make this winter not particularly predictable (sect. **5**).

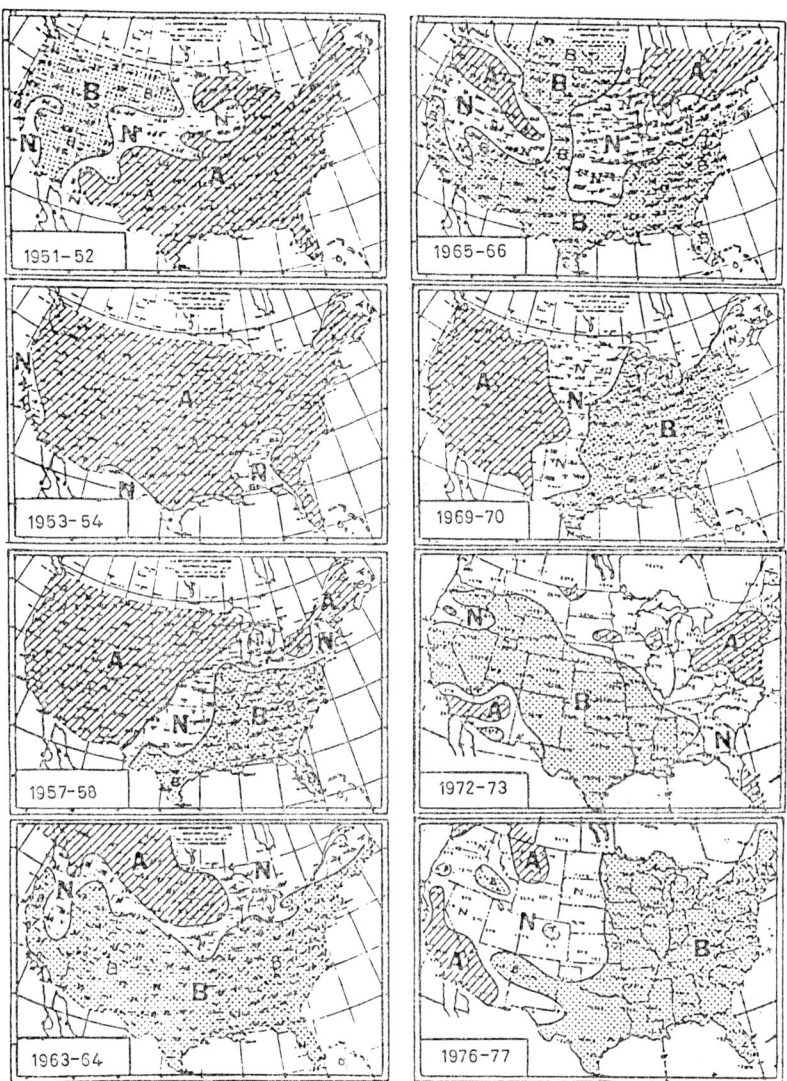

Fig. 4.10. – Surface temperature departures for the winters following the eight post World War II eastern equatorial Pacific warm episodes. A indicates above normal, B below normal and N normal. (From the *Special Climate Diagnostics Bulletin*: *A Major Warm Episode in the Eastern Equatorial Pacific Ocean*, Climate Analysis Center, NMC/NWS/NOAA, Washington, D.C. 20233, November 1982.)

Finally, before closing this section, we should mention that there has been a large number of scientific publications on the controversial problem of variations in solar radiation and their relationship to weather and climate quasi-periodicities. This subject was thoroughly reviewed by PITTOCK [63, 64]. He points out that there are two main difficulties with the majority of the papers on the subject: there are no clear physical mechanisms to support the claimed relationships, and, most important, the finite length of the observational records implies that there are not enough independent data to ensure the statistical significance of the results and to accept the hypothesis. As PITTOCK points out, the situation is also often ambiguous, because there are also not enough data to reject the hypothesis.

In one of the most promising studies, MITCHELL et al. [65] have found an apparently significant relationship between the quasi-periodic droughts in the U.S. west of the Mississippi River and the phase of the double sunspot cycle (about 22 years), and perhaps of the lunar nodal cycle (18.6 years [66]).

Even if this quasi-periodic relationship is confirmed, the explained variance is, so far, too small to make it a useful long-range forecasting tool, but further research may change this situation. It should be mentioned that the expectations raised by early observations of a short-term Sun-weather relationship [63] were not borne out by more independent data [64].

5. – Current status of long-range forecasting.

As an example of the current status of LRF, we will discuss the operational products and methodology of the U.S. Climate Analysis Center (NOAA/CAC). Many of the methods utilized were developed by NAMIAS (*e.g.*, [50, 67]) and the present experience is, to a large extent, based on his many years of work.

Table 5.I summarizes the forecast products, the lead time between the forecast preparation and the beginning of the verification period and the basic methods utilized. We will first discuss the long-range products and their verification skill.

TABLE 5.I. – *Summary of CAC long-range forecast products and methods utilized in their preparation.*

Climate Analysis Center Forecast Products		
	90-Day Outlook	30-Day Outlook
Lead time	$(4 \div 5)$ days	$(2 \div 3)$ days
Frequency	monthly	twice monthly
Products	temperature and precipitation, in 3 classes (normal 30/40/30)	
Methods	empirical (statistical and case analysis)	empirical, kinematic and dynamic
Inputs	upper air and surface condition histories antecedent, external conditions (sea surface temperatures, snow-ice lines, soil moisture content, etc.) on limited basis	

The NOAA/CAC forecasts of monthly or seasonal means are made in three classes: above, normal and below average, defined from climatological observations as occurring, on the average, 30%, 40% and 30% of the cases, respectively. Figure 5.1 shows an example of monthly prediction for June 1983 in this recently

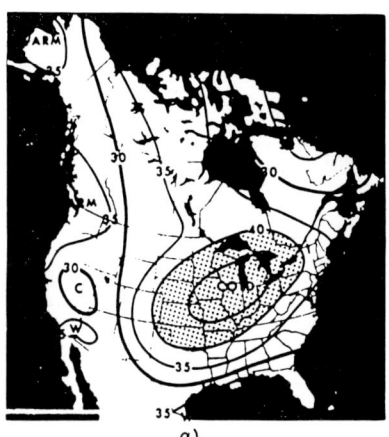 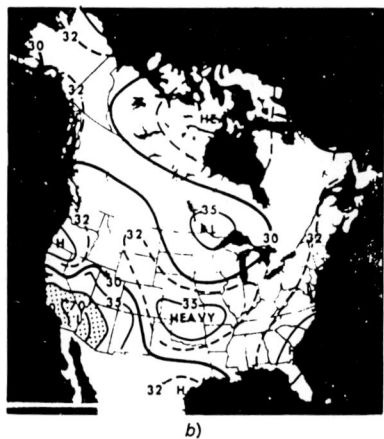

Fig. 5.1. – Example of monthly and seasonal weather outlook for June 1983: *a*) temperature probabilities, *b*) precipitation probabilities (U.S. Department of Commerce, National Oceanic and Atmospheric Administration, National Weather Service).

introduced probability format. The normal class is supposed to have always 40% chance of occurring. The remaining 60% must be shared by the preferred class (which appears in the map) and the least likely class. For example, the 45% COLD line must be interpreted as indicating 45% probability of colder than average, 40% normal and 15% above average. A line of 30% indicates the normal or indifferent situation. This presentation is somewhat more complicated than a simple « above », « normal », « below », but provides the user an indication of the confidence that the forecaster has placed on the forecast. More important, it allows the forecaster to give different confidence limits to forecasts where experience or theory indicates high or low potential skill.

The results of a recent *preliminary, unofficial* evaluation of the monthly and seasonal prediction skills are presented in table 5.II. The skill is defined as

$$S = \frac{C-E}{T-E} \times 100,$$

where C is the number of stations where the 3-class forecast was correct, E is the number that could be expected just from chance, and T is the total number of stations. It represents the percentage of skill above climatology. Several points should be noticed: *a*) The monthly skill, as could be expected, is greater than the seasonal skill. *b*) The temperature forecast skill is greater than the

TABLE 5.II. – *Preliminary estimation of the monthly and seasonal CAC forecast skill scores.*

Climate Analysis Center Forecast Skill Scores				
$S = ((\# \text{ correct} - \# \text{ expected})/(\text{total } \# - \# \text{ expected})) \times 100$				
Scores for three-class forecasts				
Monthly means (7÷8 years)	Summer	Fall	Winter	Spring
temperature	10.3	10.9	18.3	9.0
precipitation	4.1	1.6	10.7	−2.3
Seasonal means (22÷23 years)				
temperature	7.8	7.6	14.7	3.2
precipitation	4.4	2.5	6.6	3.2
Scores for two-class forecasts				
Seasonal				
temperature	annual: 20 (60% correct), winter 30 (65% correct)			
precipitation	annual: 10 (55% correct)			

precipitation skill, which is negligible except in winter. *c*) Winter skill is greater than that of other seasons, and winter temperature skill is clearly useful.

All these results are in agreement with what may be expected from the characteristic time and space scales of the atmospheric circulation, the corresponding signal-to-noise ratios and the fact that precipitation has smaller characteristic scales and is influenced by more parameters than temperature. The 2-class forecasts of « above » or « below » have more skill but are less useful than 3-class forecasts. For comparison, it may be pointed out that the CAC 2-class winter temperature forecast has $S = 30\%$ skill (forecast is correct 65% of the time), whereas the Farmers' Almanac has the clearly insignificant skill of $S = 6\%$ (correct 53% of the time) [68].

It is interesting to compare regional forecast skills for winter and summer with the potential predictability as defined in sect. **3** (fig. 5.2 and 3.2). Although the units are different, the agreement between potential and actual skill is excellent in the winter, with high predictability in the eastern and western thirds of the country, and low predictability in the Midwest. One region of disagreement occurs in the Northeast, where the high potential predictability is not matched by high actual skill, possibly because of the difficulties of snow prediction. The other is on the California coast, where, for reason so far unclear, potential predictability is maximum but actual skill minimum.

In the summer, on the other hand, the forecast skill is lower, much less than expected by potential predictability arguments. The problem of what causes the lack of agreement is an important research topic. One possibility

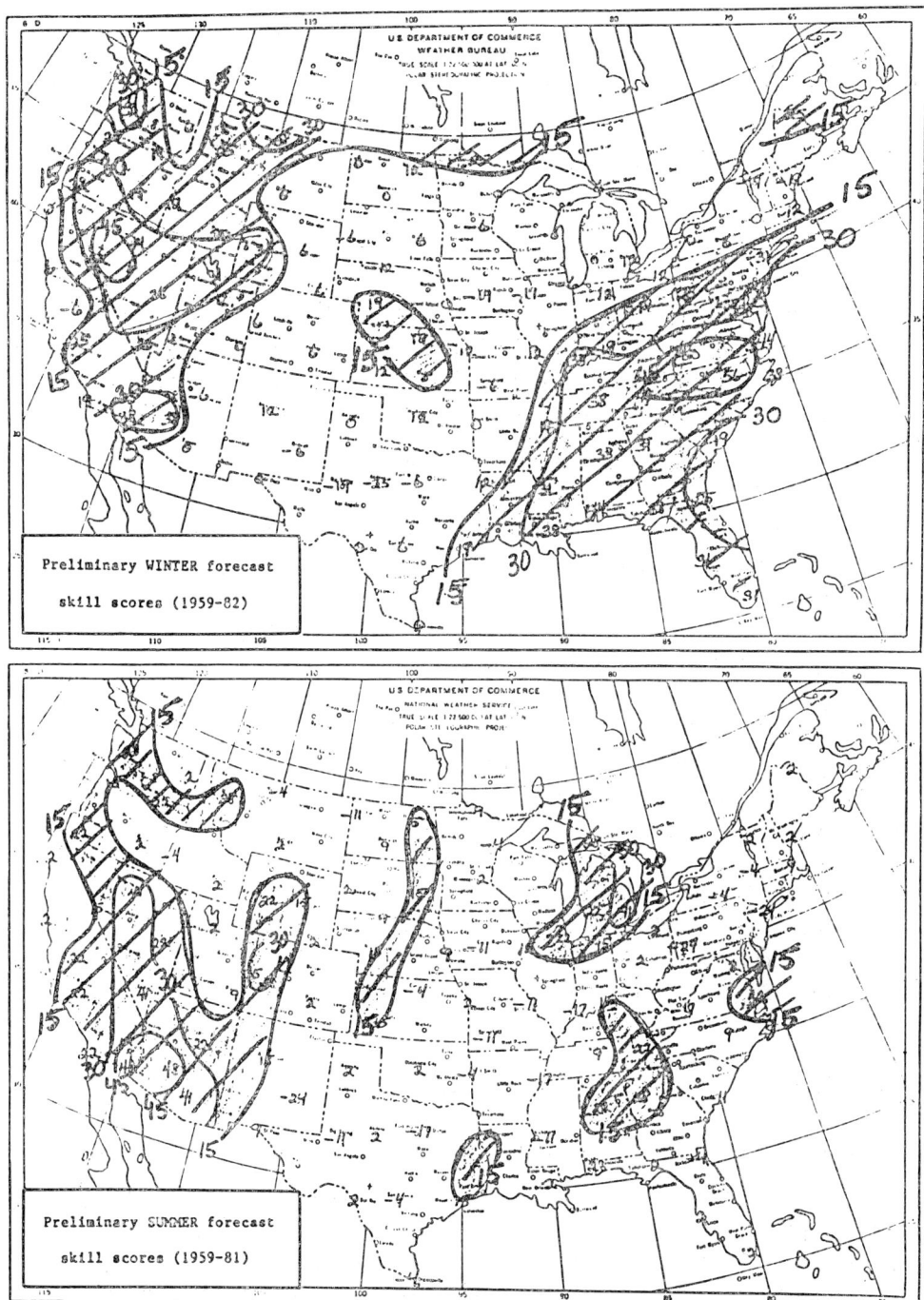

Fig. 5.2. – 3-class CAC seasonal forecast skill scores (preliminary evaluation).

is that, so far, forecast methods may fail to include soil moisture feedbacks properly, and these become important in the summer.

It should be noticed that the winter temperature forecast skill shows quite large regional values. For example, the skill over most of the Southeastern United States is between 30% to 50%, already a level comparable to potential predictability, and high enough to make it very beneficial to certain users such as power utilities.

We will now discuss the present method by which the monthly and seasonal CAC forecasts are prepared. Both forecasts consist of a three-step process:

1) *prognosis* of the 700 mb height anomalies;

2) *specification* from the height anomalies of the expected temperature and precipitation, using statistical methods such as in [69, 70];

3) assignment of *probabilities* based on theoretical predictability, the history of past skill, agreement and strength of different predictors and last, but not least, forecasters agreement.

Presently, the first step of forecasting 700 mb height anomaly maps is performed differently for monthly and seasonal predictions. For the *monthly forecast*, the lead forecaster blends subjectively the results provided by three available indicators: *a*) kinematic/synoptic extrapolation of recent anomaly fields, *b*) statistical forecasts derived from lagged correlation fields and *c*) a five-day averaged NMC forecast centered at the 1st day of the month. Of these, experience indicates that the dynamical forecast is the most skillful. Although the forecast extends only two or three days into the verification period, the five-day averaging process is an effective filter that seems to accurately reveal the low-frequency trend.

The *seasonal forecast* is made independently by four experienced forecasters, based on lagged correlations of height anomalies, the knowledge of present and past SST and snow cover anomalies and the historic data set. No dynamical forecast is used in the seasonal predictions, and lagged correlations are found to be more useful than in the monthly forecasts.

6. – Current research and future outlook.

We expect that progress in LRF will occur in the following areas:

a) *Extended use of global* NWP *models* for monthly predictions. This will require improvements such as *a posteriori* removal of systematic errors or, preferably, their reduction through the improvement of models. Better parameterization of orographic forcing, cumulus parameterization, precipitation, cloud-radiation interaction and appropriate resolution of the stratosphere are areas where significant progress may be expected in the near future. Statistical methods such as lagged-averaged forecasting may help to optimize the utilization of the model's forecast skill. The corrected model output may be also combined with antecedent information for statistical extension of the forecast applicability.

b) Prediction of boundary anomalies. We have so far avoided the discussion of how the boundary anomalies are generated or maintained. The simplest procedure, and the one used so far, is to assume that boundary forcing anomalies observed at the beginning of a numerical forecast are longer lasting than the atmospheric circulations, and, therefore, that to a first approximation the anomalies will either persist at full strength or with some damping. This may underestimate the influence of the boundary anomalies, because during a long forecast new anomalies (SST, snow cover, soil moisture, sea ice, albedo) will be generated as a *result* of the atmospheric circulation. A clear example of this is the El Niño phenomenon discussed in sect. 4. By maintaining only the initial anomalies, the influence of new anomalies is modeled as natural variability rather than as part of potential predictability. Ideally, the GCM should have fully interactive model oceans, soil moisture, etc., but the difficulties in modeling the physical processes involved in the interactions imply that there is a serious danger of *increasing* the climate drift problem. Furthermore, oceans have much longer time scales and much smaller space scales than the atmosphere, and, therefore, for numerical weather prediction special simplified models, including, for example, only the upper ocean, must be developed.

A second significant problem in this respect is that the signal-to-noise ratio for anomaly observations is clearly much smaller than that for the climatological values, and this may be a serious source of error. Further development of satellite sensors and retrieval methods is necessary.

c) Linear statistical prediction. Recent progress in this area is exemplified by [26, 71, 72]. DAVIS [73], following a suggestion of Namias, pointed out that it is essential to take into account seasonality in developing statistical climate prediction methods. Thus progress in this area will result from more emphasis in both regional and seasonal statistics. Also further stratification by circulation regime may result in improved skill under certain circumstances. An example of this is the recent work of Shukla and Paolino [74], who found that high-pressure *and* positive-pressure tendencies at Darwin are associated with wetter than normal monsoons.

d) Analog methods. BARNETT and PREISENDORFER [75] developed a method to order analog candidate years for the preceding seasonal heights and SST's from best to worst. The season following the best analog then constituted their analog prediction. LIVEZEY (unpublished) has recently modified this method to base the forecast on composite analogs instead, and obtained state-of-art surface temperature forecasts. He found particularly large enhancement of skill in the spring, when operational forecasts are generally poor.

e) Simple models. Although the simple dynamical models with realistic behavior are difficult to formulate because of the complexity of processes involved, further work in this area may provide some guidance in the assessment of different possible responses to persistent anomalies. The theoretical guidance provided by such models as [33, 76, 77] is particularly important.

Finally we want to point out that there are other areas in operational LRF which deserve attention beside skill enhancement. Among them, we mention *extension of the lead time*, which is difficult but feasible; *forecast of the variances as well as the means*: little work in this area has been done yet, and ensemble forecasting may be helpful; *forecast of extremes*: it may be possible to forecast more than three classes in certain pre-stratified, pre-identifiable situations; *development* of techniques appropriate to all areas of the world, especially the tropics and the southern hemisphere. This will require acquisition of suitable global data sets.

* * *

We are grateful to M. GHIL for his careful review of the manuscript and many useful suggestions, to J. SHUKLA for providing us with several preprints, to D. GILMAN for supplying the CAC monthly and seasonal forecast skill maps, to M. HALEM for helpful discussions, to L. THOMPSON for her careful preparation of the manuscript and to L. RUMBURG for her help with the illustrations.

REFERENCES

[1] E. KALNAY-RIVAS, A. BAYLISS and J. STORCH: *Beitr. Phys. Atmos.*, **50**, 299 (1977).
[2] W. E. BAKER: *Mon. Weather Rev.*, **111**, 328 (1983).
[3] M. HALEM, E. KALNAY, W. E. BAKER and R. ATLAS: *Bull. Am. Meteorol. Soc.*, **63**, 407 (1982).
[4] WMO: Long-range forecasting research publication series No. 1, *Proceedings of the WMO-CAS/JSC Expert Study Meeting on Long-Range Forecasting*, Princeton, N. Y., 1-4 December 1982.
[5] N. NICHOLLS: *Rev. Geophys. Space Phys.*, **18**, 771 (1980).
[6] E. N. LORENZ: *Tellus*, **34**, 505 (1982).
[7] E. N. LORENZ: *J. Atmos. Sci.*, **20**, 130 (1963).
[8] E. N. LORENZ: *Tellus*, **17**, 321 (1965).
[9] J. G. CHARNEY, R. G. FLEAGLE, H. RIEHL, V. E. LALLY and D. Q. WARK: *Bull. Am. Meteorol. Soc.*, **47**, 200 (1966).
[10] J. SMAGORINSKY: *Bull. Am. Meteorol. Soc.*, **50**, 286 (1969).
[11] E. N. LORENZ: *Tellus*, **21**, 2890 (1969).
[12] G. D. ROBINSON: *Q. J. R. Meteorol. Soc.*, **93**, 409 (1967).
[13] C. E. LEITH: *Nature (London)*, **276**, 352 (1978).
[14] C. E. LEITH: *Annu. Rev. Fluid Mech.*, **10**, 107 (1978).

[15] C. E. LEITH: *Statistical methods for the verification of long and short range forecasts*, in *European Centre for Medium Range Weather Forecasts, Seminar 1981, Problems and Prospects in Long and Medium Range Weather Forecasting* (14-18 September, 1982), p. 313.
[16] A. DALCHER, E. KALNAY and R. N. HOFFMAN: *Application of the lagged averaged forecast method to ECMWF 10 day forecasts*, to be published.
[17] R. N. HOFFMAN and E. KALNAY: *Tellus A*, **35**, 100 (1983).
[18] J. SHUKLA: *J. Atmos. Sci.*, **38**, 2547 (1981).
[19] A. N. SEIDMAN: *Mon. Weather Rev.*, **109**, 1367 (1981).
[20] C. E. LEITH: *J. Appl. Meteorol.*, **12**, 1066 (1973).
[21] R. A. MADDEN: *Mon. Weather Rev.*, **104**, 942 (1976).
[22] R. A. MADDEN: *J. Geophys. Res.*, **86**, 9817 (1981).
[23] R. A. MADDEN and D. J. SHEA: *Mon. Weather Rev.*, **106**, 1695 (1978).
[24] J. SHUKLA: *Mon. Weather Rev.*, **111**, 581 (1983).
[25] R. A. MADDEN: *Mon. Weather Rev.*, **111**, 586 (1983).
[26] T. P. BARNETT: *Mon. Weather Rev.*, **109**, 1021 (1981).
[27] B. B. REINHOLD and R. T. PIERREHUMBERT: *Mon. Weather Rev.*, **110**, 354 (1982).
[28] H. M. VAN DEN DOOL: *Mon. Weather Rev.*, **111**, 539 (1983).
[29] J. SHUKLA: *Predictability of time averages: the influence of the boundary forcing*, NASA Tech. Memo. 85092, NASA/Goddard Space Flight Center, Greenbelt, Md. 20771 (1982).
[30] J. G. CHARNEY, W. J. QUIRK, S. CHOW and J. KORNFIELD: *J. Atmos. Sci.*, **34**, 1366 (1977).
[31] J. ADEM: *Mon. Weather Rev.*, **92**, 91 (1964).
[32] K. MIYAKODA and J.-P. CHAO: *J. Meteorol. Soc. Jpn.*, **60**, 292 (1982).
[33] P. J. WEBSTER: *J Atmos. Sci.*, **38**, 554 (1981).
[34] J. D. OPSTEEGH and H. M. VAN DEN DOOL: *J. Atmos. Sci.*, **36**, 1862 (1979).
[35] J. SUSSKIND, J. ROSENFIELD, D. REUTER and M. T. CHAHINE: *The GLAS physical inversion method for analysis of HIRS2/MSU sounding data*, NASA Tech. Memo. 84936, NASA/Goddard Space Flight Center, Greenbelt, Md. 20771 (1982).
[36] J. M. WALLACE, S. TIBALDI and A. J. SIMMONS: *Q. J. R. Meteorol. Soc.*, **109**, 683 (1983).
[37] K. MIYAKODA, C. T. GORDON, R. CAVERLY, W. F. STERN, J. SIRUTIS and W. BOURKE: *Mon. Weather Rev.*, **111**, 846 (1983).
[38] A. GILCHRIST: *Beitr. Phys. Atmos.*, **50**, 25 (1977).
[39] E. S. EPSTEIN: *Tellus*, **21**, 739 (1969).
[40] C. E. LEITH: *Mon. Weather Rev.*, **102**, 409 (1974).
[41] R. M. DOLE: *Persistent anomalies of the extratropical northern hemisphere wintertime circulation*, in *Large-scale Dynamic Processes in the Atmosphere*, edited by B. J. HOSKINS and R. P. PEARCE (New York, N. Y., 1983), p. 95.
[42] J. G. CHARNEY and J. G. DEVORE: *J. Atmos. Sci.*, **36**, 1205 (1979).
[43] K. K. TUNG and R. S. LINDZEN: *Mon. Weather Rev.*, **107**, 714 (1979).
[44] E. KALNAY-RIVAS and L. MERKINE: *J. Atmos. Sci.*, **38**, 2077 (1981).
[45] B. LEGRAS and M. GHIL: *Blocking and variations in atmospheric predictability*, in *Predictability of Fluid Motions*, edited by G. HOLLOWAY and B. WEST (American Institute of Physics, New York, N.Y., 1983), p. 87.
[46] J. C. MCWILLIAMS: *Dyn. Atmos. Oceans*, **5**, 43 (1980).
[47] L. BENGTSSON: *Tellus*, **33**, 19 (1981).
[48] J. BJERKNES: *Mon. Weather Rev.*, **97**, 163 (1969).
[49] J. BJERKNES: *J. Phys. Oceanogr.*, **2**, 212 (1972).
[50] J. NAMIAS: *Mon. Weather Rev.*, **97**, 173 (1969).
[51] J. NAMIAS: *J. Phys. Oceanogr.*, **6**, 130 (1976).

[52] J. M. WALLACE and D. S. GUTZLER: *Mon. Weather Rev.*, **109**, 784 (1981).
[53] M. L. BLACKMON, Y.-H. LEE and J. M. WALLACE: *Horizontal structure of 500 mb height fluctuations with long, intermediate and short time scales*, Contribution No. 659, Department of Atmospheric Sciences, University of Washington (1983).
[54] M. L. BLACKMON, Y.-H. LEE, J. M. WALLACE and H.-H. HSU: *Time variation of 500 mb height fluctuations with long, intermediate and short time scales as deduced from lag-correlation statistics*, Contribution No. 660, Department of Atmospheric Sciences, University of Washington (1983).
[55] B. J. HOSKINS and D. J. KAROLY: *J. Atmos. Sci.*, **38**, 1179 (1981).
[56] P. R. JULIAN and R. M. CHERVIN: *Mon. Weather Rev.*, **106**, 1433 (1978).
[57] K. E. TRENBERTH: *Q. J. R. Meteorol. Soc.*, **102**, 639 (1976).
[58] E. M. RASMUSSON and T. H. CARPENTER: *Mon. Weather Rev.*, **110**, 354 (1982).
[59] J. D. HOREL and J. M. WALLACE: *Mon. Weather Rev.*, **109**, 813 (1981).
[60] S. G. H. PHILANDER: *Nature (London)*, **302**, 295 (1983).
[61] K. WYRTKI: *J. Phys. Oceanogr.*, **5**, 572 (1975).
[62] J. SHUKLA and J. M. WALLACE: *J. Atmos. Sci.*, **40**, 1613 (1983).
[63] A. B. PITTOCK: *Rev. Geophys. Space Phys.*, **16**, 400 (1978).
[64] A. B. PITTOCK: *Q. J. R. Meteorol. Soc.*, **109**, 23 (1983).
[65] J. M. MITCHELL, C. W. STOCKTON and D. M. MEKO: *Evidence of a 22-year rythm of drought in western U.S. related to the Hale solar cycle since the 17th century*, in *Solar Terrestrial Influences on Weather and Climate* (Dordrecht, 1979), p. 125.
[66] R. G. CURRIE: *J. Geophys. Res.*, **86**, 11055 (1981).
[67] J. NAMIAS: *Long range weather and climate predictions*, in *Geophysical Predictions*, National Research Council Studies in Geophysics (Washington, D.C., 1978), p. 103.
[68] J. E. WALSH and D. ALLEN: *Testing the Farmer's Almanac-Weatherwise* (1981), p. 212.
[69] W. H. KLEIN: *Regional and seasonal differences in specifying monthly mean surface temperature from the 700 mb height field*, in *Proceedings of the VII Annual Climate Diagnostics Workshop, Boulder, Co., 18-22 October 1982*, Department of Commerce (1982), p. 499.
[70] W. H. KLEIN: *Mon. Weather Rev.*, **111**, 674 (1983).
[71] T. P. BARNETT and K. HASSELMANN: *Rev. Geophys. Space Phys.*, **17**, 949 (1979).
[72] R. P. HARNACK and A. J. BROCCOLI: *J. Phys. Oceanogr.*, **9**, 1232 (1979).
[73] R. E. DAVIS: *J. Phys. Oceanogr.*, **8**, 233 (1978).
[74] J. SHUKLA and E. A. PAOLINO: *Mon. Weather Rev.*, **111**, 1830 (1983).
[75] T. P. BARNETT and R. W. PREISENDORFER: *J. Atmos. Sci.*, **35**, 1771 (1978).
[76] A. J. SIMMONS, J. M. WALLACE and G. W. BRANSTATOR: *J. Atmos. Sci.*, **40**, 1363 (1983).
[77] J. D. OPSTEEGH and H. M. VAN DEN DOOL: *J. Atmos. Sci.*, **37**, 2169 (1980).

Theoretical Climate Dynamics: an Introduction.

M. GHIL

Courant Institute of Mathematical Sciences, New York University - New York, NY 10012

1. – Introduction.

Theoretical climate dynamics is a relatively new member of the family of geophysical sciences. Descriptive, static climatology goes back, of course, at least to the ancient Greeks, who realized the importance of the Sun's mean zenith angle in determining the climate of a given latitude belt, as well as that of land-sea distribution in determining the regional, zonally asymmetric characteristics of climate. The general human perception of climate change is also preserved in numerous written records throughout history, starting with Noah's flood.

Only recently did the possibility of global climate monitoring present itself to the geophysical community, through ground-based observational networks and space-borne instrumentation. This increase in quantitative, detailed knowledge of the Earth's current climate was supplemented by the development of elaborate, geochemical and micropaleontological methods for sounding the planet's climatic past.

Observational information about present, spatial detail, and about past, temporal detail, was accompanied in the 1960s by an increase of computing power used in the processing of climatic data, as well as in the modeling and simulation of the seasonally varying general circulation. The knowledge thus accumulated led to an increase of insight which was distilled in simple models trying to analyze the basic ingredients of climatic mechanisms and processes.

In these lectures, I describe a few simple models and try to convey the flavor of the new, theoretical climate dynamics. As in every area of the exact sciences, the fundamental ideas suggested by simple models have to be tested by further observations and detailed simulations of the phenomena under study. I hope that this description of preliminary, theoretical results will stimulate the comparisons and verifications required, and help to bring therewith the needed corrections to and further development of the theory.

Section 2 describes the radiation budget of the Earth and its distribution with latitude. Global energy-balance models based on this budget are introduced in sect. 3. These energy-balance models exhibit multiple equilibria, and the stability of model equilibria and their dependence on parameters is analyzed. Section 4 carries out the same analysis for latitude-dependent models.

Section 5 introduces observational evidence about Quaternary glaciation cycles. Physical mechanisms which influence the glaciations are continental ice-sheet dynamics, the visco-elastic rebound of the Earth's upper strata under the ice load and insolation changes. They are described and modeled in sect. 6. Section 7 presents free and forced behavior of a nonlinear climatic oscillator based on the coupling between radiation balance, a highly simplified hydrologic cycle, ice-sheet dynamics and isostatic rebound.

The amplitude of this model's stable, self-sustained oscillations is comparable to the variations in global temperature and ice volume suggested by Quaternary proxy data. Its period is roughly 10 ka, with 1 ka = 1000 years. When subject to small, quasi-periodic insolation forcing, whose dominant periods are near 19 ka, 23 ka and 41 ka, the model's forced oscillations exhibit nonlinear resonance, with locking of the response frequency to multiples of a forcing frequency. Combination tones in the response among forcing frequencies include prominent peaks near 100 ka, the dominant periodicity of Quaternary glaciations, and near 10 ka. The sharp peaks of the response are superimposed on a continuous spectral density, whose appearance is related to a transition from quasi-periodic to aperiodic model behavior.

Implications of this complicated spectrum for the predictability and accurate simulation of climate on long time scales are discussed in sect. 8. Concluding remarks follow in sect. 9.

2. – Radiation budget of the Earth.

The major characteristics of a physico-chemical system, such as the climatic system, are given by its energy budget. The climatic system's energy budget is dominated by the short-wave radiation, R_i, coming in from the Sun, and the long-wave radiation, R_0, escaping back to space. The approximate balance between R_i and R_0 defines the mean temperature of the system. The distribution of radiative energy within the system, in height, latitude and longitude, determines to a large extent the distribution of climatic variables, such as temperature, throughout the system.

I consider in this section first the global radiation balance, then its variation with latitude. The discussion is based on [1].

2'1. *Global radiation balance.* – Figure 1 shows the annually and globally averaged radiation budget for the Earth-atmosphere system. The budget components in the figure are expressed as fractions of the solar radiative flux normally incident at the top of the atmosphere, along a straight line connecting the Earth and the Sun. This flux value, the so-called solar constant, S.C., is currently believed to be 1370 Wm^{-2}, to within ± 2 Wm^{-2}. The quantity $Q_0 = \frac{1}{4}$ S.C. is the value of the incoming solar radiative flux averaged over the year and over the surface of the Earth; the factor $\frac{1}{4}$ results from the Earth's sphericity.

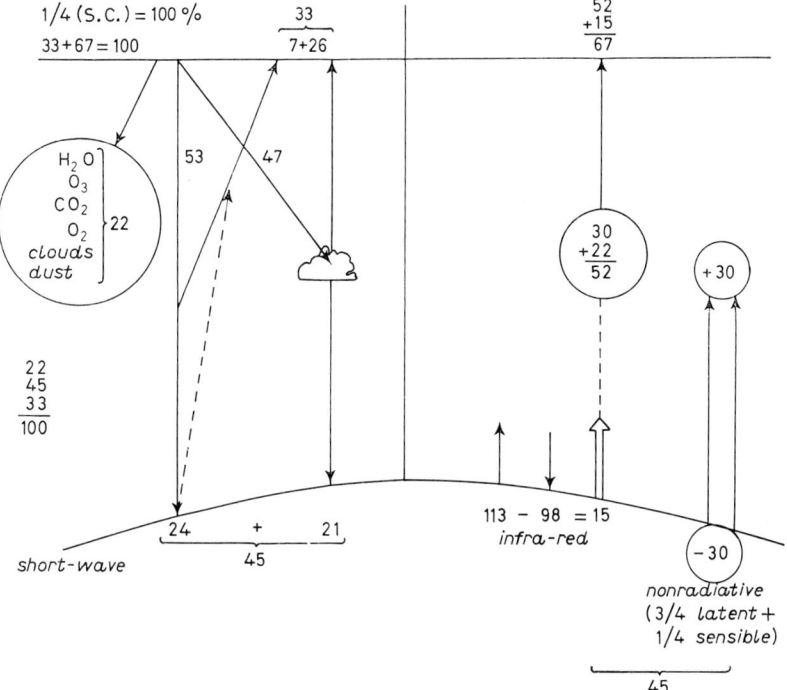

Fig. 1. – Globally and annually averaged radiation budget of the Earth.

The radiation coming in from the Sun is dominated by short-wave components due to the Sun's very high temperature. This radiation is either absorbed by the atmosphere (22%), or transmitted to the ground (45%), or reflected back to space (33%). The short-wave radiation balance appears on the left of fig. 1.

A particularly important role in both the absorption and reflection of solar radiation is played by clouds: they cover on the average 50% of the Earth's surface. The average reflectivity of the Earth-atmosphere system, 0.33, is called the planetary *albedo*. It includes the contribution of clouds. The radiation reflected at the Earth's surface also contributes to the total short-wave radiation returned to space at the top of the atmosphere. The surface albedo, *i.e.* the amount of radiation reflected as a fraction of the radiation received $(0.45\,Q_0)$ at the surface, is on the average considerably less than 0.33; it depends locally on the nature of the surface.

The short-wave radiation absorbed by the atmosphere throughout its depth (22%) and by the lithosphere and hydrosphere at the surface (45%) heats them. They cool off by emitting long-wave, infra-red (IR) radiation. The net IR emission at the surface is 15% and the emission by the atmosphere out to space is 52%; together they total 67%. Thus the radiative flux of the system to outer space is made up of 33% short-wave and 67% long-wave radiation. The IR radiation balance appears at center right of fig. 1.

To equilibrate the radiative transactions at the ground and throughout the atmosphere, we have to include the nonradiative fluxes on the right of fig. 1. Sensible (7%) and latent heat (23%) flow from the surface into the atmosphere. This 30%, together with 15% emitted by the surface as IR radiation, balances the 45% received by it in short-wave flux. Together with the 22% solar radiation absorbed by the atmosphere, this 30% makes up the 52% the atmosphere radiates to space.

The numerical values given in fig. 1 are constantly being revised due to new data provided by meteorological satellites. The purpose of the present discussion is merely to understand qualitatively the different radiative and nonradiative fluxes involved in the Earth-atmosphere's global energy budget and not to provide the most accurate or latest numbers.

An important consequence of this discussion is that, within the accuracy of the available data and calculations, the climatic system is close to radiative equilibrium, when considering global and annually averaged quantities. This equilibrium determines global temperature, which makes the Earth a relatively pleasant place to live, on the average. A very simple model for this mean temperature will be dealt with in the following section.

Local radiative imbalances make certain zones more pleasant to live in than others. This is our concern in the remainder of the present section.

2˙2. *Local imbalances and meridional fluxes.* – It was already clear to the ancient Greeks that the inclination (κλισις) of the Sun, which changes with latitude, was the most important factor in determining the climate (κλιμα) of a zone, or latitude belt. Figure 2 shows the latitudinal distribution of annually and zonally averaged radiation, absorbed and emitted, for the Earth-atmosphere system.

The absorbed flux, R_i, which averages $0.67 Q_0$, falls off sharply from the equator towards the poles, with the mean annual zenith angle. The emitted flux, R_o, which has approximately the same average as R_i, originates mostly within the atmosphere ($0.52 Q_0$ on the average), rather than at the surface ($0.15 Q_0$). Free-air temperature, at mid-troposphere say, is observed to be a much weaker function of latitude than the Sun's zenith angle. Hence the outgoing radiation R_0, emitted according to the Stefan-Boltzmann law, is much more constant from equator to pole than the absorbed one. This results in a local excess of mean annual, radiative energy in the tropics (33° N to 37° S,

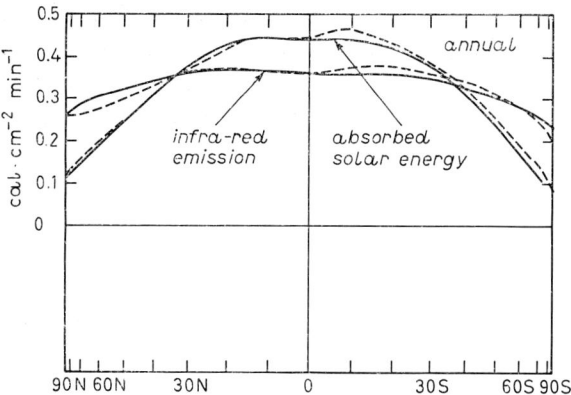

Fig. 2. – Zonally and annually averaged radiation budget of the Earth: ——— conventional data [2], – – – satellite data [3].

approximately) and in a deficit at higher latitudes. As a consequence, the excess has to be carried off from low to high latitudes, by the atmosphere and by the oceans.

The average annual energy transport required by local energy-balance considerations is a horizontal flux from one latitude belt to the next, F, which is taken positive northward. If we assume an interannually stationary climate, the local excess of energy is the divergence of this flux,

$$(2.1) \qquad R_i(\varphi) - R_0(\varphi) = \frac{1}{a \cos \varphi} \frac{\partial F}{\partial \varphi},$$

where φ is latitude and a the radius of the Earth.

Numerical integration, according to eq. (2.1), of either curve in fig. 2 yields the flux values shown in fig. 3. The level of current observational uncertainty is reflected by the difference between the solid curve and the dashed curve in both figures; the two curves are based on data sets obtained from different observing systems about twenty years apart [2, 3]. It becomes clear, therefore, that the problem of climatic change, i.e. of the difference on various time scales between the corresponding averages of the two sides of eq. (2.1), is difficult: climatic change involves the small difference of two large, poorly observed and poorly understood quantities.

The energy transport in fig. 3, assuming stationarity, achieves the temperature distribution which resulted in the curve for R_0 in fig. 2. In fact, both the latitudinal distributions of R_i and R_0 depend on properties internal to the system: cloud distribution, surface albedo, temperature. To understand better these interdependencies and the climatic features they determine, let us turn now to the formulation and analysis of some models.

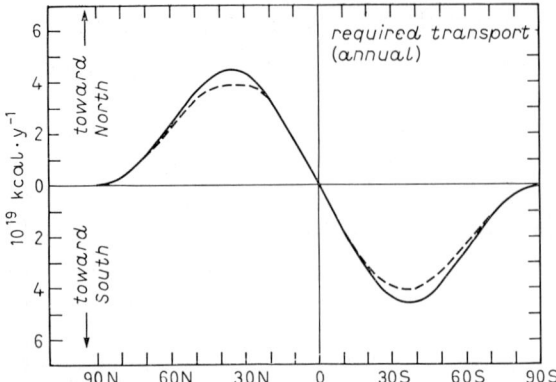

Fig. 3. – Meridional heat fluxes balancing the radiation budget of the previous figure: ⎯⎯⎯ conventional data [2], – – – satellite data [3].

3. – Global energy-balance models.

The variable perceived most widely as defining climate is temperature, T. It is also most important in determining the components of radiation balance. I start, therefore, with the simplest model of climate, one for the annually averaged temperature of the Earth-atmosphere system, $\bar{T}$.

3'1. *A model for global temperature.* – The equation governing the model is

$$(3.1) \qquad c\frac{d\bar{T}}{dt} = Q_0\{1 - \alpha(\bar{T})\} - \sigma g(\bar{T})\bar{T}^4 \,.$$

It expresses the approximate radiation balance between absorbed radiation

$$(3.2a) \qquad R_i = Q_0\{1 - \alpha(\bar{T})\}$$

and emitted radiation

$$(3.2b) \qquad R_o = \sigma g(\bar{T})\bar{T}^4 \,.$$

Q_0 is the global mean solar radiative input, α the planetary albedo, σ the Stefan-Boltzmann constant and g the grayness of the system, *i.e.* its deviation from black-body radiative emission, $\sigma \bar{T}^4$; α and g are functions of $\bar{T}$. Any slight imbalance between R_i and R_o leads to a change in the temperature of the system, at the rate $d\bar{T}/dt$, t being time; c is the heat capacity of the system, whose heat storage is $c\bar{T}$. This type of model derives from the work of Budyko [4] and Sellers [5], and has been investigated extensively since.

In principle, the dependence of α on $\bar{T}$ should express the change in both cloud and surface albedo with $\bar{T}$. But the dependence of cloud albedo on $\bar{T}$ is still not well understood, not even to the extent of knowing whether, or under which circumstances, it increases or decreases with $\bar{T}$. On the other hand, surface albedo varies most strongly with the presence or absence of snow and ice, on land or sea. Hence we take α to decrease linearly with $\bar{T}$, as ice cover decreases over part of the Earth, and to be constant for all $\bar{T}$ for which the Earth would be either entirely ice covered, $\bar{T} < T_l$, or ice free, $\bar{T} > T_u$. This is the *ice-albedo feedback* of [4, 5].

The dependence of emissivity σg on $\bar{T}$ has to express the *greenhouse effect*, *i.e.* the process by which outgoing IR radiation is partly screened by various atmospheric gaseous absorbers and by clouds. I choose

(3.2c) $$g(\bar{T}) = 1 - m \, \text{tgh} \, (\bar{T}/T_0)^6 \, ,$$

with $m = 0.5$, for 50 percent cloud cover, and $T_0^{-6} = 1.9 \cdot 10^{-15} \, \text{K}^{-6}$ (cf. [5]). This expression indicates that, as temperature increases, g decreases, so that the greenhouse effect becomes stronger. The shapes of $\alpha(\bar{T})$ and $g(\bar{T})$ are reflected in the graphs of R_i and R_0 in fig. 4.

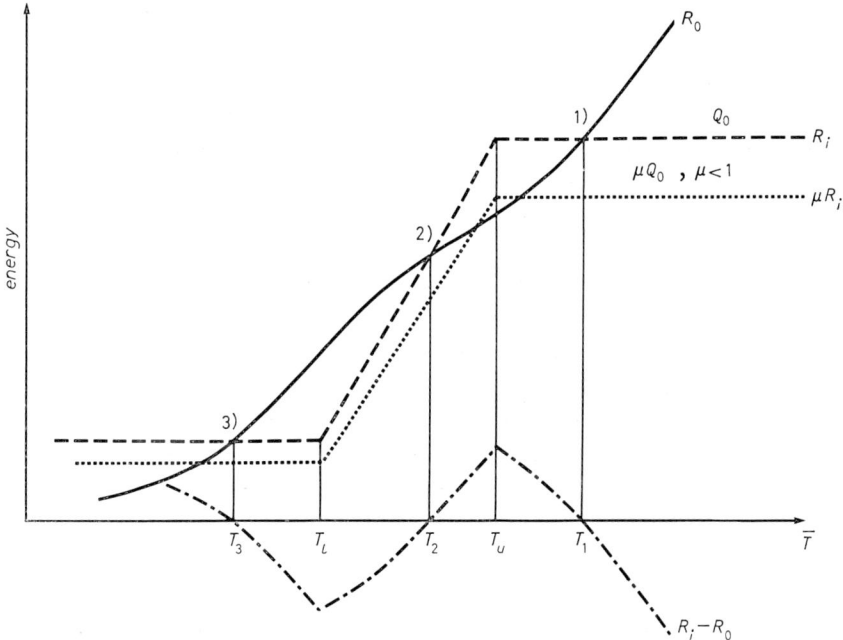

Fig. 4. – Outgoing radiation R_0 (solid) and absorbed incoming radiation R_i (dashed) for a global energy-balance model. The model has three stationary solutions $\bar{T} = T_1, T_2, T_3$.

3'2. Stationary solutions and stability to perturbations. – The intersections of $R_i(\bar{T})$ and $R_0(\bar{T})$, or those of $R_i - R_0$ with the $\bar{T}$-axis, determine the steady states, or *equilibria*, of model (3.1). We notice in fig. 4 that, for α and g as we have taken them, there are three stationary solutions of eq. (3.1): T_1, T_2 and T_3. Hence, even as simple a model as the present one shows the possible existence of more than one climate, if we are willing to interpret the equilibria of the model as steady-state climates of the Earth; T_1 would represent the present climate, and T_2, T_3 would be colder climates, possibly ice ages.

In the real world, small deviations from a given temperature regime always appear, due to a multitude of mechanisms not included in the model. It is important, therefore, to investigate the stability of the model's solutions to small perturbations. Assume, for instance, that at time $t = 0$, $\bar{T}(0) = T_1 + \theta_0$. We are interested in knowing how the deviation $\theta(t)$ of $\bar{T}$ from T_1, which is initially equal to θ_0, evolves in time.

If we assume that $\theta(0) = \theta_0$ is suitably small, it suffices at first to consider a linearized equation for $\theta(t)$, obtained by expanding (3.1) in $\bar{T} = T_1 + \theta$. If we use $\mathrm{d}T_1/\mathrm{d}t = 0 = R_i(T_1) - R_o(T_1)$ and neglect the terms of higher order in θ, such an expansion yields

(3.3a) $$\frac{\mathrm{d}\theta}{\mathrm{d}t} = \lambda_1 \theta,$$

where

(3.3b) $$\lambda_1 = -(Q_0 \alpha_1 + \sigma g_1)/c,$$

(3.3c, d) $$\alpha_1 = \left.\frac{\mathrm{d}\alpha}{\mathrm{d}\bar{T}}\right|_{\bar{T}=T_1}, \quad g_1 = 4g(T_1)T_1^3 + \left.\frac{\mathrm{d}g}{\mathrm{d}\bar{T}}\right|_{\bar{T}=T_1} \cdot T_1^4.$$

This is a linear ordinary differential equation (ODE) for the deviation $\theta(t)$ of $\bar{T}(t)$ from the equilibrium value T_1.

The solution of eq. (3.3) is

(3.4) $$\theta(t) = \theta_0 \exp[\lambda_1 t].$$

Hence θ will grow exponentially if $\lambda_1 > 0$ and decay to zero with time if $\lambda_1 < 0$. Thus T_1 is linearly *stable* if $\lambda_1 < 0$, and *unstable* if $\lambda_1 > 0$. A similar analysis holds for T_2 and T_3.

In fig. 4, an equilibrium T_j, $j = 1, 2, 3$, can be seen to be stable if the corresponding slope of $R_i(\bar{T}) - R_o(\bar{T})$ is negative at $\bar{T} = T_j$, and unstable if it is positive. It follows that T_1 and T_3 are stable, while T_2 is unstable. Indeed, near T_1, we have $\mathrm{d}\bar{T}/\mathrm{d}t < 0$ when $\bar{T} > T_1$, causing a decrease of $\bar{T}$ with time, while $\mathrm{d}\bar{T}/\mathrm{d}t > 0$ when $\bar{T} < T$, causing an increase. In other words, the energy balance (3.1) tries to restore $\bar{T}$ to its equilibrium value T_1; the same situation occurs for T_3. Near T_2, quite the opposite happens: once $\bar{T} > T_2$ it will increase further, while $\bar{T}(0) < T_2$ will lead to further decrease of $\bar{T}$, away from T_2.

This stability discussion suggests that it is reasonable to identify T_1 with the present climate; T_2, however, is not a good candidate for a colder climate, such as an ice age, since it is unstable and could never have persisted for any length of time. The equilibrium T_3 corresponds to a completely ice-covered Earth, or a « deep freeze ». While the model shows it to be stable, it was never observed in the paleoclimatological record of the past; we hope it will not occur in the future either.

3˙3. Structural stability. – It remains to be seen how the existence of one or more solutions for eq. (3.1), and their stability, can be affected by changes in the model's parameters. The most important among these is the value of the solar input, Q_0. It could change as a result of variation in the Sun's energy output, in its distance from the Earth, or, equivalently, in the atmosphere's optical properties. Within our simple model, either one of these changes can be expressed by replacing Q_0 in (3.1), (3.2a) with μQ_0, and taking $\mu \neq 1$. Such a change will leave R_0 as it is in fig. 4, but the graph for μR_i, $\mu \neq 1$, will be either above or below the curve representing R_i.

In fig. 4, I have drawn the situation for a certain value of μ less than 1, *i.e.* solar input smaller than that corresponding to present-day conditions (dotted line). Clearly, T_1 and T_2 in this new situation lie closer to T_u, while T_3 lies more to the left of its value for $\mu = 1$. If μ were to decrease further, a situation would obtain in which the graph of μR_i just touches that of R_0, at $T_2 = T_u = T_1$. Let the corresponding value of μ be μ_c. For $\mu < \mu_c$, the solutions T_1 and T_2 of eq. (3.1) disappear altogether, and T_3 is the only solution left.

If we recall now that T_1 represents the present climate within the model, it follows that, while the solar input μQ_0 decreases, the Earth's climate cools off slowly at first, as T_1 moves to the left. Then, as μ crosses the value μ_c, the temperature of the system would have to decrease dramatically to that of a completely ice-covered Earth, T_3. It is this type of result that attracted attention to the original models of Budyko and of Sellers, as well as the possibility of human activities reducing μ in the future, more than natural events had done in the past.

In the opposite situation, of μ increasing, the curve μR_i moves up. Then T_1 moves to the right, while T_2 and T_3 approach T_l. As μ increases through and beyond a critical value μ_d, T_2 and T_3 coalesce, then disappear, while T_1 keeps increasing. This situation is, therefore, less dramatic, although even a gradual increase of T_1 might eventually lead to a climate much less desirable than the present one.

It should be noticed that, for any value of μ, the analysis of the stability for the model's equilibria can be carried out as it was for $\mu = 1$. Indeed, eqs. (3.3) become

$$(3.5a) \qquad \frac{d\theta}{dt} = \lambda_j(\mu)\theta \,,$$

$$(3.5b) \qquad \lambda_j(\mu) = -\{\mu Q_0 \alpha_j(\mu) + \sigma g_j(\mu)\}/c \,,$$

$$(3.5c) \qquad \alpha_j(\mu) = \frac{d\alpha}{d\bar{T}}\bigg|_{\bar{T}=T_j(\mu)} \,,$$

$$(3.5d) \qquad g_j(\mu) = 4g(T_j(\mu))T_j^3(\mu) + \frac{dg}{d\bar{T}}\bigg|_{\bar{T}=T_j(\mu)} \cdot T_j^4(\mu) \,,$$

where $T_j(\mu)$, $j = 1, 2, 3$, is the position of solution T_j as μ increases or decreases. Clearly, each T_j is a continuous function of μ near $\mu = 1$, and, therefore, $\alpha_j(\mu)$ and $g_j(\mu)$ are continuous, so that λ_j depends continuously on μ near $\mu = 1$.

In particular, the sign of λ_j will not change near $\mu = 1$. The sign of λ_j is, in fact, the same as that of the slope of the curve $R_i(\overline{T}; \mu) - R_0(\overline{T}; \mu)$ in fig. 4, whether the curve moves with μ or not. This suggests, and careful evaluation of eqs. (3.5b)-(3.5d) confirms, that the sign of $\lambda_j(\mu)$ stays the same until the solution $T_j(\mu)$ coalesces with another solution, or until it disappears. Thus T_1 is stable to small perturbations θ_0 in $\overline{T}$ for $\mu > \mu_c$, T_3 is stable for $\mu < \mu_d$, and T_2 is unstable for $\mu_c < \mu < \mu_d$. To summarize, T_1 and T_3 are stable, and T_2 is unstable, over the entire μ interval over which they exist.

We have just considered the way in which the number of solutions and their *internal* stability to small perturbations in initial conditions changes when one of the model's parameters is changed. It is customary to call these properties the model's *structural* or external stability.

4. – Latitude-dependent models for surface temperature.

We have seen already in subsect. 2'2 that the radiation balance of the Earth-atmosphere system changes with latitude φ (fig. 2). This latitudinal dependence of $R_i - R_0$ gives rise to and is maintained by zonally averaged heat fluxes, $F(\varphi)$ (fig. 3). In sect. 3, a spatially zero-dimensional (0-D) model was formulated for the globally averaged radiation balance, corresponding to fig. 1. In a similar way, we wish to formulate now a model for the latitudinally dependent energy balance of fig. 2 and 3. The results obtained with the 0-D model (3.1) will guide us in the study of the following, spatially one-dimensional (1-D) model.

4'1. *Horizontal heat transport.* – To formulate such a model, it is necessary to consider horizontal heat transfers, in the same way in which the discussion of fig. 1 concerned itself with vertical heat transfers. There are three kinds of heat transfer: radiative, conductive and convective. In *radiative* transfer, energy passes between the medium's molecules, by electromagnetic radiation due to photon emission. Radiative transfer is important in the vertical distribution of temperature, which will not be discussed here. It is negligible in horizontal heat transport.

In *conduction*, energy passes from one molecule to another, by thermal agitation. It is transported, therefore, from parts of the system with higher temperature to those where temperature is lower, without the medium's motion contributing to the transport. The corresponding heat flux F_c can be taken in many cases as being proportional to the local temperature difference, or gradient,

(4.1) $$F_c = -k_c \nabla T.$$

In eq. (4.1) ∇ is the gradient operator, and k_c is the so-called conduction, or diffusion coefficient; depending on the system, k_c can be constant, a function of position, of T itself or of other quantities. In heat transfer, (4.1) is often called Fourier's law. In the diffusion of trace constituents, a similar formula is also valid, and called Fick's law.

Conduction is the main form of heat transport in solid media, such as the lithosphere. In a solid, in the absence of internal sources of energy, temperature will change in time only due to the convergence or divergence of the conductive heat flux at a given location,

(4.2) $$c \frac{\partial T}{\partial t} = -\nabla \cdot F_c = \nabla \cdot k_c \nabla T;$$

here $\nabla \cdot$ is the divergence operator.

In fluid systems, the medium often is in motion, and thermal energy of the molecules is carried along with this motion. This type of heat transfer is called in fluid dynamics in general *convective*. In meteorology and oceanography, the term convection is mostly reserved for transfer by vertical, small-scale motions of the fluid, while large-scale, horizontal motions are said to *advect* heat. The advective heat flux, F_a, modifies the local temperature T of the fluid according to the equation

(4.3) $$c \frac{\partial T}{\partial t} = v \cdot \nabla T,$$

v being the velocity of the fluid which carries internal energy at the temperature T with it.

It is clear from (4.3) that atmospheric and oceanic dynamics play an important role in the climatic system's local energy balance and temperature distribution T via the velocity field v. To determine v at the same time as T is a considerably more complicated task than computing T alone. Also, more is known about T for climates different from the present one than about v.

When considering only the largest, planetary scales of the temperature field, $O(10^4 \text{ km})$, and time scales longer than months and years, it is reasonable to attempt to eliminate the velocity field from our considerations, by using the so-called eddy diffusive approximation

(4.4) $$-\nabla \cdot F_a \simeq \nabla \cdot k_e \nabla T.$$

In eq. (4.4), k_e is an *eddy diffusion* coefficient. This type of approximation is used in many areas of fluid dynamics, with k_e being usually much larger than k_c. It often gives acceptable qualitative results when the time and space scales of T in which one is interested are much larger than the spatial extent and life span of the fluid motions which have the highest velocities. This is the case when studying planetary-scale, long-term climate change, given the typical scales of atmospheric and oceanic eddies, $O((10^2 \div 10^3)\,\text{km})$ and $O(\text{weeks} \div \text{months})$, respectively. Notice that, like k_c, k_e can be a function of position, T itself and, furthermore, of ∇T. In the sequel, the zonally averaged, meridional heat fluxes will be expressed by combining eqs. (4.1)-(4.4) in

$$-\nabla \cdot F = \nabla \cdot k \nabla T \, ; \tag{4.5}$$

here $F = F_c + F_a$, $k = k_c + k_e$, and k is taken to be a function of latitude φ alone, $k = k(\varphi)$.

4˙2. *Model formulation.* – The energy balance of a zonal slice of the climatic system, located at latitude φ and extending to the top of the atmosphere and down to a prescribed depth in the oceans and the continents, is governed by the equation

$$C(\varphi) \frac{\partial T}{\partial t} = \frac{1}{\cos \varphi} \frac{\partial}{\partial \varphi} [k(\varphi) \cos \varphi] \frac{\partial T}{\partial \varphi} + \tag{4.6}$$

$$+ \mu Q_0 S(\varphi) \{1 - \alpha(\varphi, T)\} - \sigma g(\varphi, T) T^4 \, .$$

This equation summarizes our discussion of vertical—radiative—and horizontal —advective—heat fluxes. Here T is the sea-level temperature, $T = T(\varphi, t)$.

In comparing (4.6) with (3.1), we notice that R_i and R_o have become functions of φ and T, rather than of the single, global variable $\overline{T}$. Model (3.1) was spatially zero-dimensional and hence governed by a nonlinear ODE. Model (4.6) is one-dimensional, the spatial dimension taken into account is latitude, and it is governed by a partial differential equation (PDE). The ratio $\partial T/\partial t$ is the rate of change of T in time at location φ, while $\partial T/\partial \varphi$ is the rate of change of T in latitude at time t; they are related to each other by (4.6).

The zonal slice at φ exchanges $R_i - R_o$ energy with outer space and $\nabla \cdot F$ energy with adjacent slices; as a result its temperature changes at the rate $\partial T/\partial t$, inversely proportional to its heat capacity $C = C(\varphi)$. The function $S(\varphi)$ gives the distribution of solar radiation incident at the top of the atmosphere with latitude; its average is one, so that Q_0 is the same constant as in eq. (3.1). The first term on the right expresses horizontal flux divergence $\nabla \cdot F$ in the meridional direction, according to (4.5).

Before proceeding with an analysis of solutions to (4.6), it is interesting to compare this model with other 1-D energy-balance models (EBMs). The original models of Budyko [4] and of Sellers [5] were one-dimensional and we shall set them up side by side in table I.

TABLE I. – *Comparison of Budyko's and of Sellers' models.*

Heat flux	BUDYKO	SELLERS
$R_i = Q(1-\alpha(T))$ absorbed solar radiation, as a function of ice-albedo feedback	step function albedo $\alpha = \begin{cases} \alpha_M, & T < T_s, \\ \alpha_m, & T \geqslant T_s, \end{cases}$ $\alpha_M > \alpha_m,$ $T_l \leqslant T_s \leqslant T_u$	ramp function albedo $\alpha = \begin{cases} \alpha_M, & T < T_l \\ \alpha_M - \dfrac{T-T_l}{T_u-T_l}(\alpha_M-\alpha_m) \\ \quad T_l \leqslant T < T_u, \\ \alpha_m, & T_u \leqslant T \end{cases}$
R_o outgoing IR radiation	linear, empirical $A + BT$	Stefan-Boltzmann law with greenhouse effect $\sigma T^4\{1 - m\,\mathrm{tgh}(T^6/T_0^6)\}$
$\nabla \cdot F$ horizontal flux divergence	Newtonian cooling $\varkappa\{T(\varphi) - \bar{T}\}$	eddy-diffusive $\nabla \cdot \{k(\varphi)\nabla T(\varphi)\}$

Here

(4.7) $$\bar{T}(t) = \int_0^{\pi/2} T(\varphi, t) \cos\varphi\,\mathrm{d}\varphi\,,$$

α_M and α_m are albedo values assigned to ice-covered and ice-free surfaces, respectively, and T_s is the assumed temperature of the « ice margin »; notice that the Sellers formulation allows for a partially ice-free zone, $T_l < T < T_u$, rather than requiring a sharp ice margin. The constants A, B and $\varkappa$ in Budyko's formulation of IR flux and horizontal heat flux, like the other constants in table I, are determined by matching the computed fluxes to current climatic data. The explicit dependence of α and g in (4.6) on latitude φ is rather weak and was introduced also to better match the available data.

It became immediately clear [6] that, in spite of its empirical simplicity, Budyko's Newtonian cooling formulation led to certain counter-intuitive results, being compatible with infinitely many positions of the presumed ice margin at once. In the context of a simple, zonally averaged EBM, the ice margin $\varphi_s(t)$ of a solution is defined by

(4.8a) $$T(\varphi_s, t) = T_s\,.$$

If temperature is a monotone function of latitude from pole to equator, and

(4.8b) $$\partial T/\partial \varphi \neq 0 \qquad \text{at } \varphi = \varphi_s,$$

the ice margin thus defined is unique, in each hemisphere, for such a solution.

The indeterminacy of ice margin position in Budyko's model comes from the discontinuous character of its solutions. It can be removed by adding diffusion, which renders solutions continuous. The 1-D EBM formulation which became most widely used is thus the one in heavy outline in the table, which combines Budyko's step function albedo and linear IR flux with Sellers' eddy-diffusive heat flux, taking, moreover, a spatially constant diffusivity [7]. Such a formulation can easily be solved analytically, by expanding $T(\varphi, t)$ in Legendre polynomials with argument $y = \sin \varphi$ and with time-dependent coefficients, and matching the two linear problems for the ice-free and the ice-covered portion of the Earth across the « ice margin ». The simplicity of North's formulation [7], along with the striking results of Budyko and Sellers, led to the great popularity of EBMs in the recent literature.

I pursue here the slightly more complicated model (4.6), with ramp function albedo and greenhouse effect according to SELLERS. The main reason for doing so is to illustrate how careful numerical methods, combined with mathematically informed qualitative reasoning, can give the same detailed information about solution behavior as analytical methods. Such a combination will prove very fruitful for the subsequent material in these lectures, for which no analytic methods are available.

The analysis of the 1-D model (4.6) will follow the steps for the 0-D model (3.1). Results will be compared at the end of this section with those of other 1-D formulations.

4`3. *Stationary solutions.* – Equation (4.6) is to be solved subject to boundary conditions of zero heat flux at pole and equator

$$(4.9a, b) \qquad \frac{\partial T}{\partial \varphi} = 0 \qquad \text{at } \varphi = 0, \pi/2.$$

The condition at the pole is natural, while that at the equator is equivalent to assuming the symmetry of the two hemispheres; such a symmetry assumption is reasonable for the simplicity of the model under consideration. The appropriate initial condition is

$$(4.9c) \qquad T(\varphi, 0) = \mathring{T}(\varphi).$$

We start by considering present solar input conditions, $\mu = 1$. One expects that, if (3.1) is a reasonable first approximation to (4.6), eq. (4.6) will also have steady-state solutions. This turns out to be so, and can be proven to be the case under rather general assumptions on the functions k, α and g in (4.6) [8].

Computation of stationary solutions, using numerical methods of high, controlled accuracy [9], yields the three equilibria $T_1(\varphi)$, $T_2(\varphi)$ and $T_3(\varphi)$ in fig. 5. The model solution $T_1(\varphi)$ closely matches the data for the present climate (open circles in the figure). The « deep freeze » $T_3(\varphi)$ has a mean temperature

$\bar{T}_3$ about 100 K below that of the present, $\bar{T}_1$, and a much smaller pole-to-equator temperature difference

(4.10) $$\Delta T(t) = T(0, t) - T(\pi/2, t) \, .$$

We notice that both $T_1(\varphi)$ and $T_2(\varphi)$ straddle T_l, the perennial snow line temperature, as well as T_u, the snow absence temperature. It should be remembered that (4.6) is still a model for annually averaged temperature; hence it is customary to take $T_l = -10\,°\text{C} \simeq 263$ K and $T_u = 10\,°\text{C} \simeq 283$ K.

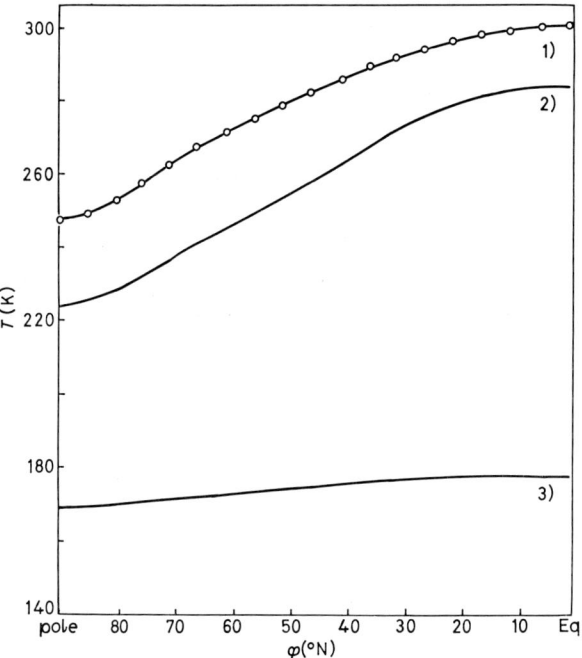

Fig. 5. – Latitude dependence of the three equilibrium solutions of a zonally averaged climate model.

In model (3.1), one had $T_3 < T_l < T_2 < T_u < T_1$. This is, in fact, true of the mean equilibrium temperatures here, i.e.

(4.11a) $$\bar{T}_3 < T_l < \bar{T}_2 < T_u < \bar{T}_1 \, ;$$

actually $T_3(\varphi)$ lies entirely below T_l, $T_3(\varphi) < T_l$. From the results for (3.1) one might have been tempted to guess that also

(4.11b) $$T_l < T_2(\varphi) < T_u < T_1(\varphi)$$

holds for all latitudes φ. This is, however, not the case.

This elementary discussion illustrates both the usefulness and dangers of simple models. Equation (3.1) led us to conjecture that (4.6) would have three equilibria, which proved correct. It also suggested the chain of inequalities (4.11a), which does hold for (4.6). However, model (4.6), in accordance with reality, does not satisfy (4.11b); the latter would have been an unrealistic interpretation of the results for model (3.1).

4'4. *Internal stability.* – We are interested now in the stability to small perturbations of the equilibrium solutions to eq. (4.6). Assume that, for some reason, at $t = 0$ the temperature in a latitude band would be slightly different from the « present climate », i.e.

(4.12) $$T(\varphi, 0) = T_1(\varphi) + \varepsilon \theta_0(\varphi),$$

for ε small and $\theta_0(\varphi)$ arbitrary. We would like to know how the temperature difference $\theta(\varphi, t; \varepsilon) = T(\varphi, t; \varepsilon) - T_1(\varphi)$, initially equal to $\theta(\varphi, 0; \varepsilon) = \varepsilon \theta_0(\varphi)$, evolves in time.

A linear PDE for $\theta(\varphi, t)$ can be derived from (4.6) in a way similar to that in which the linear ODE (3.3) was derived from (3.1). Substitution of $T(\varphi, t; \varepsilon) = T_1(\varphi) + \theta(\varphi, t; \varepsilon)$ into (4.6) and differentiation with respect to ε yields, at $\varepsilon = 0$,

(4.13) $$\theta_t = - L_1 \theta,$$

with $(\ )_t = \partial(\ )/\partial t$. Here L_1 is a linear operator obtained from the right-hand side of (4.6),

(4.14a) $$L_1 \theta = - \frac{1}{r(x)} \{p(x) \theta_x\}_x q(x) \theta,$$

with x normalized co-latitude, $x = (\pi/2 - \varphi)/(\pi/2)$, so that $x = 0$ at the pole and $x = 1$ at the equator, and $(\ )_x = \partial(\ )/\partial x$. The functions r, p and q are given by

(4.14b) $$r(x) = C(x) \sin \frac{\pi x}{2},$$

(4.14c) $$p(x) = \frac{4}{\pi^2} \sin \frac{\pi x}{2} k(x),$$

(4.14d) $$q(x) = \{\mu Q_0 S(x) h(x, T_1) - d(x, T_1(x))\}/C(x)$$

and

(4.14e) $$d(x, T) = c(x) \frac{\partial}{\partial T} \sigma g(T) T^4,$$

where $c(x)$ contains the explicit x-dependence of $g(x, T)$, while $h(x, T_1)$ is a piecewise constant function equal to zero when $\alpha(x, T_1(x))$ is constant and equal to the T-slope of α when T_1 is along the ramp portion

(4.14f) $$T_l < T_1(x) < T_u .$$

All this, although lengthy, is quite straightforward and rather general. Similar linearizations L_j, $j = 2, 3$, are obtained about $T_2(x)$ and $T_3(x)$. For convenience, we drop the subscript $j = 1$ from L_j in eq. (4.13).

The role of the coefficient λ_1 in (3.3) will now be played by the eigenvalues $\lambda^{(k)}$ of L, defined by

(4.15a) $$L\psi_{(k)}(x) = \lambda^{(k)} \psi_{(k)}(x) .$$

It is again possible to show [8, 9], under rather general assumptions on the form of $k(x)$, $\alpha(x, T)$ and $g(x, T)$, that the eigenvalues $\lambda^{(k)}$ are all discrete, real and tend to infinity,

(4.15b) $$\lambda^{(k)} \to +\infty \qquad \text{as } k \to \infty .$$

It follows that solutions to (4.13) can be expanded in the eigenfunctions $\psi_{(k)}(x)$ of L,

(4.16) $$\theta(x, t) = \sum_0^\infty a_k \exp[-\lambda^{(k)} t] \psi_{(k)}(x) ,$$

which is the 1-D form of (3.4). If $k \equiv \text{const}$, $g(T)$ is chosen so that gT^4 may be a linear function of T, and α is piecewise constant, then $\psi_{(k)}(y)$, with $y = \sin \varphi$, are Legendre polynomials [7].

From (4.15), (4.16) it follows that at most a finite number of $\lambda^{(k)}$ will be negative and that a stationary solution $T_j(x)$ will be stable if and only if no $\lambda_j^{(k)}$ is negative. Neutral stability corresponds to a zero eigenvalue, $\lambda^{(0)} = 0$, strict (or asymptotic) stability to $\lambda^{(0)} > 0$.

At this point, the linear stability problem for stationary solutions of (4.6) has been reduced to determining the sign of $\lambda_j^{(0)}$ for each solution $T_j(x)$. It turns out that this sign is easily determined by a single numerical integration of the linear ODE

(4.17a) $$L_j w(x) = 0 ,$$

with initial conditions

(4.17b, c) $$w_x(0) = 0 , \quad w(0) = 1 .$$

If $w(x)$ is nonnegative for $0 \leqslant x \leqslant 1$, then $\lambda_j^{(0)}$ has the same sign as $w_x(1)$, so that $T_j(x)$ is stable according to whether w_x at $x=1$ is positive or not. This result is a simple extension of comparison and oscillation theorems for Sturm-Liouville problems.

For (4.6), the numerical integration of (4.17) shows, as expected from (3.3), that the « present climate » $T_1(x)$ and the « deep freeze » $T_3(x)$ are stable, while $T_2(x)$ is unstable; in fact, for the latter exactly one eigenvalue, $\lambda_2^{(0)}$, is negative.

4'5. *Structural stability.* – With the results for $\mu = 1$ in hand, we can turn to the study of stationary solutions and their internal stability for all values of the insolation parameter μ. Figure 6 shows the average temperature $\bar{T}$ of stationary solutions as a function of μ. For the 0-D model (3.1), such a graph would contain all the information required, since $\bar{T}$ determines its solution completely.

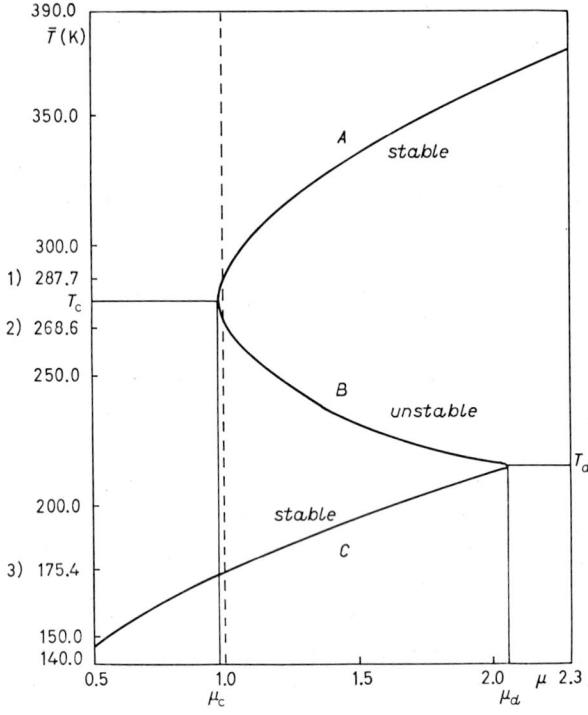

Fig. 6. – Dependence of solutions on the insolation parameter $\mu = \Delta Q/Q_0$.

For a 1-D model, infinitely many numbers $a_k(\mu)$ are required to determine a stationary solution (cf. (4.16)); one can choose, for instance, $\bar{T}(\mu) = a_1(\mu)$, $\Delta T(\mu) = a_2(\mu)$, etc. Still, fig. 6 is very informative: for $\mu = 1$ one can see the three solutions, represented by $\bar{T}_1, \bar{T}_2, \bar{T}_3$.

As μ decreases, climate branch A, which passes through the point ($\mu = 1$, $\overline{T} = \overline{T}_1$), and branch B, passing through ($1, \overline{T}_2$), approach each other and merge, at a point ($\mu_c, \overline{T}_c$). For $\mu < \mu_c$, only the « deep freeze » branch C exists. For $\mu > 1$, branches C and B approach each other, merging at ($\mu_d, \overline{T}_d$). Beyond $\mu = \mu_d$, only the « warm climate » branch A exists.

The dependence of solution number on μ is thus quite analogous to (3.1), as discussed in connection with fig. 4. From fig. 4 and eqs. (3.3), (3.4), we were also able to infer, using the T-slope of $R_i(\overline{T}; \mu) - R_0(\overline{T}; \mu)$, that all solutions of (3.1) corresponding to branches A and C were stable, while branch B in the 0-D case is unstable. There is no simple counterpart to this geometric argument in the general 1-D case; for a partial result see [10].

We have seen in the 0-D case that $\lambda_j(\mu)$, $j = 1, 2, 3$, were continuous functions of μ near $\mu = 1$ (cf. eqs. (3.5)). This continuity provided an alternative argument for the stability persistence of $T_j(\mu)$. It is this argument which can be extended in all generality to the 1-D case.

In fact, we noticed that for $T_2(x; \mu)$, $\mu = 1$, exactly one eigenvalue, $\lambda_2^{(0)}$, was negative. It turns out that $\lambda_j^{(k)}(\mu)$ are all continuous functions of μ, for each k. The function $\lambda_j^{(0)}(\mu)$ changes sign where two branches, A and B or B and C, coalesce, i.e. at $\mu = \mu_c$ and $\mu = \mu_d$, respectively, while all other $\lambda_j^{(k)}(\mu)$, $k \geqslant 1$, stay positive for all j and all μ. This mathematical analysis confirms the physical intuition, embodied in fig. 4, that a stationary climate branch $T(x; \mu)$ is stable if $\overline{T}$ is an increasing function of insolation μ along the branch, and unstable if $\overline{T}(\mu)$ is decreasing.

4˙6. *Concluding remarks on energy-balance models.* – The independent analyses of 1-D EBMs by [6-9] and other investigators gave the following picture of these models:

a) Multiple stationary solutions exist for fixed solar input; among these solutions, one is close to the present climate and one represents a completely ice-covered Earth.

b) The « present climate » and the « deep freeze » are stable to internal perturbations.

c) All stationary solutions depend continuously on the insolation parameter μ, with various solution branches merging at certain critical values of μ.

In fact, NORTH [11] showed that the diffusively modified Budyko model with constant k (outlined in table I) and the original Budyko model with Newtonian cooling [4] are equivalent if exactly two even Legendre polynomials are retained in the expansion of the problem (cf. (4.16)). In a two-mode approximation, various other forms of 1-D EBMs are also equivalent.

Unfortunately, the number of solution branches changes in North's [7, 11] model if an additional Legendre polynomial is retained in the expansion: the three-mode model has five solution branches, rather than three, as we found here. The diffusive Budyko model does actually have five solutions, for all truncations higher than three, as shown already by HELD and SUAREZ [6].

The two additional solutions, related to the so-called « small ice cap » instability, are an artifact of the step function formulation of albedo: they disappear when a ramp function is introduced, as in [5] and here. Furthermore, the two-mode step function formulation leads to a « present climate » branch which is entirely ice free, like the 0-D model (3.1), in contradiction to current climatic data [12, fig. 6 and 8].

In spite of such technical differences, the fundamental agreement among EBM investigators led to verification of these results with more detailed models. Clearly, the « present climate » branch $T_1(x; \mu)$ alone is of interest in studying climatic change on the interannual and decadal time scale. Thus the most important feature of EBMs which had to be verified by general circulation models (GCMs) is the dependence of solutions on solar input near present conditions.

A great deal of effort went into « tuning » EBMs, on the one hand, and GCMs, on the other, to yield the same « sensitivity » of climate to insolation variations, $d\bar{T}/d\mu$ at $\mu = 1$, $T = T_1(x)$. Agreement today is satisfactory, at approximately 1 K for one-percent change in solar input. This sensitivity determines the quasi-equilibrium response of such a model to slow changes in insolation.

Aside from the derivative of $\bar{T}(\mu)$ at $\mu = 1$, the shape of the curve $\bar{T}_1(\mu)$ for $\mu \geqslant \mu_c$ is of interest. This shape (cf. fig. 6) is parabolic, with vertical slope at the limit point $\mu = \mu_c$, $\bar{T} = \bar{T}_1(\mu_c) = \bar{T}_2(\mu_c)$. A parabolic dependence, with slope increasing as μ decreases, was obtained in the GCM experiments of Wetherald and Manabe [13, fig. 5]. This nontuned qualitative result offers rather striking confirmation to the simple story that EBMs are telling about climate.

5. – Quaternary glaciations.

The previous sections dealt with the climatic system's radiation balance, which led to the formulation of equilibrium models. Slow changes of these equilibria due to external forcing, internal fluctuations about an equilibrium and transitions from one possible stable equilibrium to another are discussed in [14, Chapt. 10] and [15].

5`1. *Climatic variability.* – Climatic records exist on various time scales, from instrumental records on the time scale of months to hundreds of years, to geological proxy records on the time scale of thousands to millions of years. These records indicate that climate varies on all time scales in an irregular fashion. It is difficult to imagine that a model's stable equilibrium, whether slowly shifting or randomly perturbed, can explain all this variability.

A summary of climatic variability on all time scales appears in [16]. The most striking feature is the presence of sharp peaks superimposed on a continuous background. The relative power in the peaks is poorly known; it depends, of course, on the climatic variable whose power spectrum is plotted, which is left undefined in [16, fig. 1]. Furthermore, phenomena of small spatial extent will contribute mostly to the high-frequency end of the spectrum, while large spatial scales play an increasing role towards the low-frequency end.

Many phenomena are believed to contribute to changes in climate. Anomalies in atmospheric flow patterns affect climate on the time scale of months and seasons. On the time scale of tens of millions of years, plate tectonics and continental drift play an important role. Variations in the chemical composition of the atmosphere and oceans are essential on the time scale of billions of years.

The appropriate definition of the climatic system itself depends on the phenomena one is interested in, which determine the components of the system active on the corresponding time scale. No single model could encompass all temporal and spatial scales, include all the components, mechanisms and processes, and thus explain all the climatic phenomena at once.

My goal in the remaining sections will be much more modest. I shall concentrate on the phenomena of Quaternary glaciation cycles, whose time scale ranges from thousands to millions of years. The components of the climatic system active on these time scales—atmosphere, ocean, continental ice sheets, the Earth's upper strata—and their nonlinear interactions are described and modeled.

These interactions lead in a natural way to the formulation of a model which exhibits stable, self-sustained, periodic oscillations. Changes in the orbital parameters of the Earth on the Quaternary time scales provide small changes in insolation. These quasi-periodic changes in the system's forcing will produce forced oscillations of a quasi-periodic or aperiodic character. Their power spectra show the peaks which fall within the range under discussion in [16, fig. 1], as well as the continuous background.

5˙2. *Paleoclimatic evidence of glaciations.* – Changes in the extent of Alpine glaciers were known to Swiss mountaineers for many generations [17, Part I]. Huge erratic boulders lying in the low, currently unglaciated valleys and parallel striations of permanently exposed, flat rock surfaces were easily associated with like boulders carried on the surface of glaciers flowing in the higher valleys and with the aspect of present-day glacier beds temporarily exposed by unusual summer melting or deep cracks. The boulders and striations spoke clearly to the unbiased eye of times when the glaciers extended further down into the valleys and even filled them to the crest of their dividing ranges.

Until the early Nineteenth Century, however, the nascent science of geology had another explanation for these and related observations: the Biblical flood. It was only with difficulty that a number of courageous scientists, led by L. AGASSIZ, faced the facts and gradually convinced their colleagues of the past extent of glaciations and their alternation in time with warmer periods, such as the one in which we live now.

Eventually, the geology of the last two million years of the Earth's history, the Quaternary period, became intimately linked with the study of the changing extent of continental ice sheets. The Laurentide ice sheet at various times covered most of Canada and New England, and much of the Middle West, Great Plains and Rockies. The Fenno-Scandian ice sheet extended into Eastern and Central Europe, without quite linking up with the smaller Alpine ice cap, mentioned before. In the Southern Hemisphere, ice sheets developed over parts of Australia and New Zealand, and extended out from the Andes. In between its extreme advances, the ice retreated to areas comparable to those it occupies today: Antarctica, Greenland, the Canadian Archipelago and small mountain glaciers.

On the whole, the Quaternary in its entirety has to be considered an ice age when compared with the mean temperature of much longer, completely ice-free periods of the Earth's history, such as the Mesozoic era, the age of dinosaurs, which lasted for 160 million years. Other ice ages like the Quaternary occurred in the Earth's past, 2.3 billion and 1 billion years ago and 700, 450 and 250 million years ago. All the episodes for which the presence of ice is recorded and temporally resolved in the geological record seem to show higher climatic variability than the entirely ice-free episodes. The total duration of the ice-free episodes in the geological past seems to be much longer than that of episodes where ice of variable extent was present [18].

Plate tectonics, continental drift and orographic changes probably play a role in creating geographic situations in which land ice can appear and develop. We shall not concern ourselves, however, with the time scales of hundreds of millions of years on which these phenomena are important. The presence of ice in the system will be considered as given hereafter and the variations in its extent will be studied, along with the variations in temperature with which it interacts.

5`3. *Geochemical proxy data.* – Variations in continental ice extent leave clear traces in the geological record. Besides the previously mentioned erratic boulders and striations, classical stratigraphic methods record glacial till, the coarse, unstratified debris left behind when glaciers melt, alternating with fine, stratified, interglacial deposits. The uplift of the Earth's crust as ice sheets melt is recorded in shifting strand lines along the Baltic Sea and the East Coast of North America. The pollen of temperate-climate plants alternates in stratigraphic sequences with that deposited during cold climates.

But the most important, detailed evidence of glaciations started to accumulate with the advent of geochemical, radiometric methods in the 1950s. Long piston cores raised from the bottom of the sea by oceanographic research vessels contain fossils of micro-organisms living near the bottom, called benthic foraminifera. The calcium carbonate in their shells contains oxygen whose *isotopic* abundance *ratio* $R = {}^{18}O/{}^{16}O$ reflects approximately the per mil difference $\delta^{18}O$ between the water in which they were deposited, R_{sample}, and that of current ocean water, R_{std},

$$(5.1) \qquad \delta = \{R_{sample}/R_{std} - 1\} \times 1000 \;.$$

The standard for water is called Standard Mean Ocean Water (SMOW). The normal abundance of the heavy ^{18}O isotope is only 0.2% approximately and it varies little, hence the use of the factor 1000 in formula (5.1) for $\delta^{18}O$. Currently available mass spectrometers determine $\delta^{18}O$ in calcium carbonate with a nominal precision of $\pm\,0.05$ promil [19].

As part of the hydrologic cycle, water molecules which evaporate from the surface of the oceans contain preferentially the light isotope ^{16}O. If these molecules precipitate into high-latitude ice sheets and are fixed there, the ^{16}O cannot return to the ocean as it does during warm episodes, with runoff and river water. Hence the ocean becomes impoverished in the light, abundant isotope ^{16}O when large ice sheets are present, so that $\delta^{18}O$ of ocean water, and of benthic foraminiferal shells deposited in it, is higher during glacial episodes.

The mixing time of the world ocean is of the order of 1000 years, and bottom waters never have an entirely uniform $\delta^{18}O$. Moreover, the isotopic ratio of the shells differs from one species to another, is influenced by the temperature of the ambient water, as well as by its $\delta^{18}O$, and is not quite in thermochemical equilibrium with the water. Still, the $\delta^{18}O$ of the microscopic shells in the deep-sea cores appears to be a relatively reliable proxy indicator of global ice volume on the time scale of thousands of years and is positively correlated with the ice volume.

5˙4. *The phenomenology of glaciation cycles.* – What is the evidence from the deep-sea cores as to climatic change during the Quaternary? Figure 7 shows the $\delta^{18}O$ record of a deep-sea core from the western equatorial part of the Pacific Ocean.

The scale on top is simply the depth in the core. To translate this into a time scale, absolute dates for points along the core are necessary. Other isotopic methods, in particular the study of the potassium-argon ratio in lava flows, helped establish in the 1960s *polarity reversals* of the Earth's magnetic field [20]. The current polarity is called normal, the opposite one is reversed. The first reversal can be dated now to 730 thousand years ago, to within 20 thousand years.

Two additional, short episodes of normal polarity have occurred during the

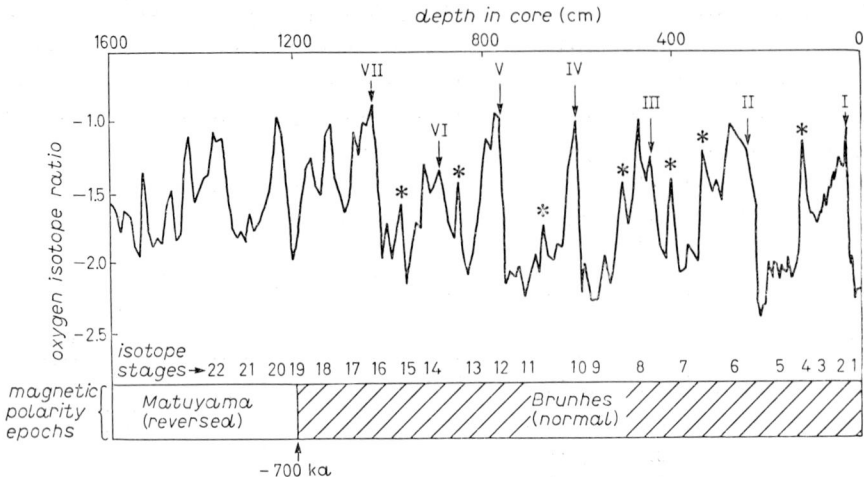

Fig. 7. – Oxygen isotope record of deep-sea core V28-238 (after [17]).

Quaternary. They are called the Jaramillo and Olduvai normal events. The beginning of the Olduvai event, 1.8 million years ago, is now agreed upon as the beginning of the Quaternary period. Our unit of time henceforth will be 1 ka = 1000 years, so that the Quaternary is 1800 ka long.

Aside from the Brunhes/Matuyama polarity reversal, believed at that time to have occurred 700 ka ago, no absolute dates were available for the analysis of the core in fig. 7 [21]. So a uniform sedimentation rate is assumed in the time scale of the figure. The isotopic stages marked along this scale represent episodes of algebraically higher (even numbers) or lower (odd numbers) $\delta^{18}O$ ratios.

The record in fig. 7 shows an irregular evolution of $\delta^{18}O$, and hence ice volume, over its entire length. Prominent are sharp drops in ice volume approximately every 100 ka, called *terminations* [22]. The terminations are indicated by arrows in the figure and numbered backwards in time by Roman numerals. The growth of ice volume in between terminations seems more gradual in certain records, particularly in continental stratigraphic sequences [23], leading to the idea of a roughly sawtoothlike shape of *glaciation cycles*.

The « terminations » are not equally spaced, nor do the segments of record between terminations look alike: irregular variability on time scales shorter than 100 ka is evident in fig. 7. A few sharp spikes as high as the terminations, and many smaller ones, occur repeatedly, but aperiodically. The most prominent spikes are marked by stars in the figure. Long plateaus with very little high-frequency variability appear in the $\delta^{18}O$ curve during some of the major cycles, but not others. These plateaus are mostly near the mean value of the record $\delta^{18}O \simeq -1.5$ promil.

Figure 8 shows the power spectrum of a combined $\delta^{18}O$ record taken from two cores in the southern part of the Indian Ocean, one of which goes back 450 ka [24]. The largest peak is at 106 ka, with smaller peaks at 43 ka, 24 ka and 19 ka. These are superimposed on a red-noise-like continuous background (stippled in the figure).

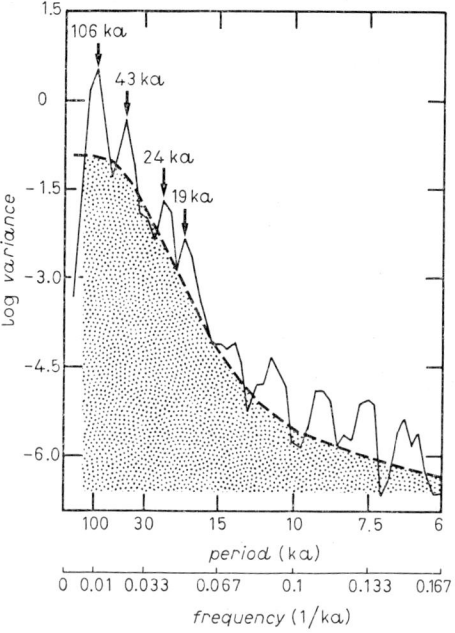

Fig. 8. – Power spectrum of a patched $\delta^{18}O$ record based on deep-sea cores RC11-120 and E49-18 (after [17]).

Other deep-sea cores and other proxy indicators, both isotopic and micropaleontological, all seem to show irregular variations like fig. 7, with a major near-periodicity close to 100 ka, terminations, spikes and plateaus. Their power spectra show peaks near 100 ka, 40 ka and 20 ka, as well as occasional additional peaks. These are superimposed on a continuous background which decreases from low frequencies towards the higher ones.

6. – Physical mechanisms of glaciations.

The amplitude of glaciation cycles, as inferred from the proxy data, is of a few degrees in temperature and of a factor of two or three in global ice volume. Such variations of temperature and ice volume cannot be accounted for by the quasi-equilibrium response of energy-balance models to small changes in the external radiation budget [12]. A large portion of these variations must be due to the interaction of mechanisms internal to the climatic system.

The ice-albedo feedback central to sect. **3** and **4** is but the simplest way that the presence of ice in the system acts on its temperature evolution. In subsect. **6**`1 and **6**`2, the mass balance and visco-plastic flow of ice sheets, and the response of the Earth's lithosphere and mantle to the ice load, will be introduced and their interaction with the radiation balance will be described. Subsections **6**`3 and **6**`4 deal with the description and modeling of insolation changes external to the system.

6`1. *Internal mechanisms: model formulation.* – The model to be discussed in the remainder of these lectures is a climatic oscillator [25-28]. It is governed by three nonlinear, coupled equations, which describe

the evolution of the global temperature of the ocean-atmosphere system owing to changes in the radiation balance,

the evolution of continental ice sheets owing to changes in precipitation budget and to motions of the underlying bedrock,

the visco-elastic response of the lithosphere and upper mantle to ice load variations.

A systematic derivation of these equations appears in [14, Chapt. 11], where an elementary discussion of ice-sheet dynamics and geodynamics is also given.

The surface features participating in the radiation balance are illustrated in fig. 9. They are bare land, open ocean, sea ice and ice sheets. The simplified geometry of the ice sheets, the underlying lithosphere and upper mantle is shown in the meridional cross-section of fig. 10. Conditions corresponding to

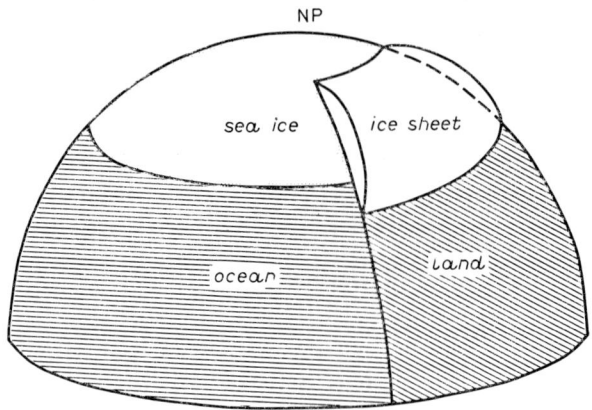

Fig. 9. – Surface features of radiation balance (after [25]).

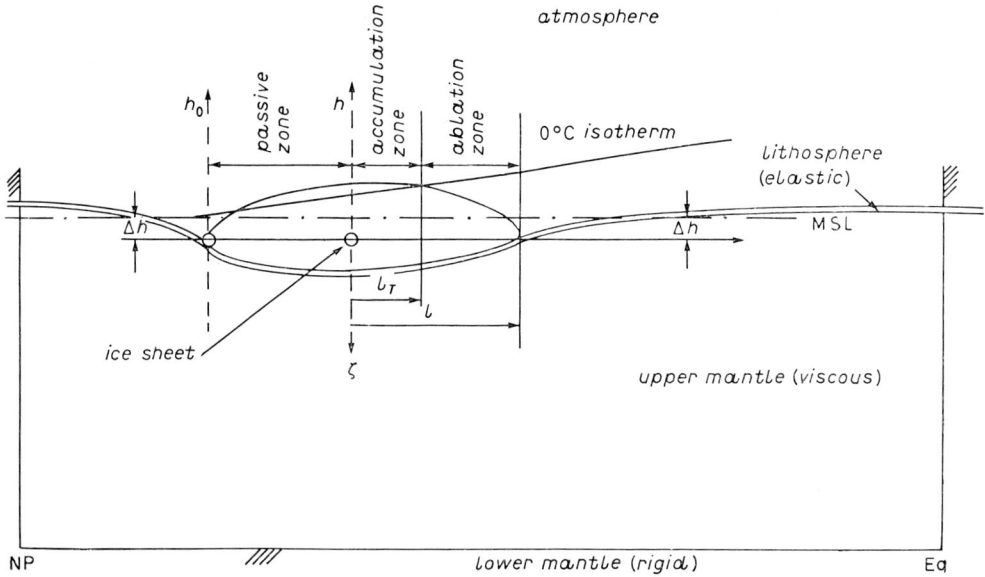

Fig. 10. – Meridional cross-section through the Earth's upper strata (after [28]).

the Northern Hemisphere are considered, and symmetry with respect to the equator is assumed for the sake of simplicity.

The model equations are

$$(6.1a) \quad \frac{dT}{dt} = Q\{1 - [\gamma(\alpha_0 + \alpha_1 L^* l) + (1-\gamma)\alpha_{oc}(T)]\} - \varkappa(T - T_\varkappa),$$

$$(6.1b) \quad \frac{dl}{dt} = \left(1 + (2/3\sqrt{L^*})(\zeta/\sqrt{l})\right)^{-1}\{(3/2)(c_T/c_L)(1/\sqrt{l})[(1 + \varepsilon(T))l_T(T, l) - l] - (2/3)\sqrt{l}(A\zeta l^{-2} + Bl^{-3/2})\},$$

$$(6.1c) \quad \frac{d\zeta}{dt} = A\zeta l^{-2} + Bl^{-3/2}.$$

The notation is as follows:

T globally averaged temperature of the ocean-atmosphere-cryosphere system;

l nondimensional extent of the ice sheet, scaled by L^*;

ζ nondimensional deflection of the local bedrock under the ice sheet, scaled by $\sqrt{\lambda L^*}$;

L^* the scaling factor for the total latitudinal extent $2L$ of the zonally symmetric ice sheet, $L = L^* l$,

$$(6.2a) \quad L^* = \varepsilon_M(\varepsilon_M + 1)/s^2(\varepsilon_M + 2)^2;$$

λ a coefficient with the dimensions of length defined in terms of the density ϱ_i of the ice and of its yield stress;

s the slope of the 0 °C isotherm: its intersection with the upper surface of the ice is taken as the snowline,

(6.2b)
$$s = \sigma \cdot 10^{-3} \, \mathrm{m}^{-1/2} \, ;$$

$\varepsilon = a/a'$ temperature-dependent ratio between the snow accumulation rate a and the ablation rate a'; its minimum value is ε_m, the maximum value is ε_M, with ε increasing linearly with T from T_m to T_M (solid line in fig. 11);

$d(T, l, \zeta)/dt$ rate of change of the model variables (T, l, ζ);

t nondimensional time, scaled by c_T;

c_T the heat capacity of the ocean-atmosphere-cryosphere system;

c_L time constant governing the plastic flow of the ice sheet, given in terms of the prescribed model parameter $c = c_L/c_T$, i.e. $c_L = c c_T$;

Q solar radiation incident at the top of the atmosphere;

γ ratio of the surface of the continents to the total surface of the Earth;

$\alpha_0 + \alpha_1 L^* l$ albedo of continents, with dependence on the ice-sheet extent;

α_oc temperature-dependent albedo of the ocean, determined by the extent of sea ice; its maximum value is α_M, the minimum is α_m, with α decreasing linearly from $T_{\alpha,l}$ to $T_{\alpha,u}$ (dash-dotted line in fig. 11);

$\varkappa(T - T_\varkappa)$ outgoing infra-red radiation;

$l - l_T$ (nondimensional) width of the ablation zone, between the southern edge of the ice sheet and the snowline;

l_T (nondimensional) extent of the accumulation zone, in the southern half of the ice sheet, North of the snowline; the northern half of the ice sheet is assumed to be in passive equilibrium with the southern half, yielding

(6.2c) $l_T = (L^* s^2)^{-1} \{ (2 s^2 L^* l + s h_0 + 1/4)^{1/2} - (s^2 L^* l + s h_0 + 1/2) \} \, ;$

h_0 height of the 0 °C isotherm above the northern edge of the ice sheet,

(6.2d)
$$h_0 = \beta(T - T_{00}) + \Delta h \, ;$$

β a proportionality constant,

(6.2e)
$$\beta = 2s(\alpha_\mathrm{M} - \alpha_0)/(T_{00} - T_{\alpha,l}) \alpha_1 \, ;$$

Δh the global net sinking of the ice sheet under mean sea level (MSL).

Also A and B,

(6.2f)
$$A = -2D(1-q) c_T/L_*^2 ,$$

(6.2g)
$$B = 2Dq c_T/L_*^2 ,$$

are constants which determine the viscous response of the upper mantle to the ice load, where D is a viscosity coefficient, and q is the ratio of the density of the ice sheet ϱ_i to that of the upper mantle, ϱ_a.

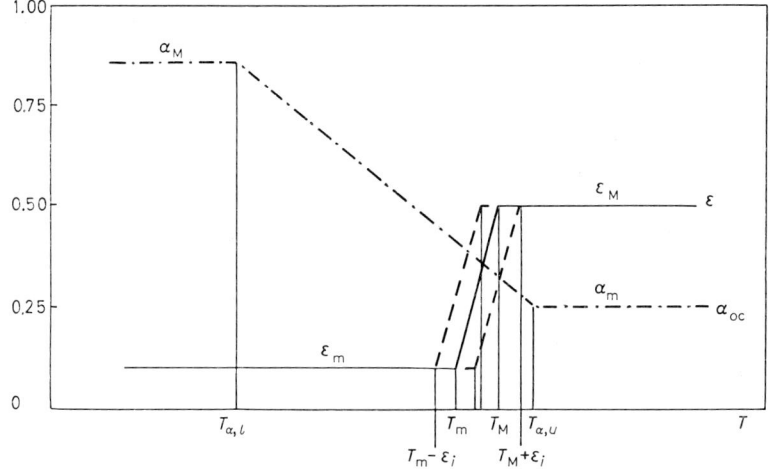

Fig. 11. – The oceanic albedo $\alpha_{oc}(T)$ and the accumulation-to-ablation rate $\varepsilon(T; t)$ as a function of global, mean annual temperature T (from [28]).

6'2. Internal mechanisms: discussion. – The first equation, (6.1a), expresses the *radiation balance* at the Earth's surface between absorbed solar radiation, $Q(1-\alpha)$, and emitted terrestrial radiation, $\varkappa(T - T_\varkappa)$. We saw in subsect. 4'6 that linearizing eq. (3.2b) does not modify qualitatively the results of EBMs. The same is true for the coupled model (6.1), and a linear form of R_0 is retained for convenience.

The global albedo, $\alpha = \alpha(T, l)$, is computed in terms of the amount of sea ice and the meridional extent of the ice sheet (fig. 9). Sea ice has huge seasonal changes of areal extent. Hence it can be safely assumed to respond instantaneously to temperature changes on the time scale of the ocean's overturning time, $c_T = 1$ ka, which is our time unit.

Continental ice sheets take a long time, however, to change their extent. Hence the need to compute their evolution explicitly on the time scale of interest. Equation (6.1b) expresses the *mass balance* of the continental ice sheet in terms of the accumulation and ablation of snow masses. The accumulation and ablation zones are separated by the snowline, which is attached to the 0 °C isotherm (fig. 10). The northern half of the ice sheet is in passive equilibrium with the southern half, due to ice calving from its northern edge.

The geometry and the dynamics of the zonally symmetric ice sheet are determined using a *plastic flow* assumption [29]. The three-dimensional dynamics and thermodynamics of continental ice sheets are considerably more complicated and rather incompletely understood. Our simplifying assumption is both reasonable [30, fig. 5 and 6] and sufficient for a zero-dimensional model, the purpose of which is exploration rather than simulation. The value of c_L, the ice sheet's flow constant, is taken to be $O(1\ \text{ka})$, as suggested by wet friction at the base of the sheet, due to basal melting and consequent sliding.

The accumulation-to-ablation ratio, $\varepsilon = \varepsilon(T)$ (solid line in fig. 11), reflects the increase in activity of the *hydrological cycle* with temperature: warmer winters have more snowfall. The relevance of this simple observation to the ice age problem was probably first pointed out by SIMPSON [31]. Evidence for net accumulation rates increasing with temperature on geological time scales is reviewed in [28]. It centers on stratigraphic and radiometric studies of ice cores in Greenland and the Antarctica, and on detailed numerical simulations of atmospheric circulation during the last glacial maximum and the interglacial present.

The mechanism which couples eqs. (6.1a), (6.1b) for ζ constant can be written in simplified short-hand as

(6.3a, b) $$\dot{T} = -\alpha, \qquad \alpha = m,$$

(6.3c, d) $$\dot{m} = p, \qquad p = T.$$

Here equality signs stand for rough proportionality, m is ice mass and p is precipitation, while T and α are temperature and albedo, as before. Equations (6.3b), (6.3d) are the short-hand representation of the ice-albedo and *precipitation-temperature feedbacks*, respectively.

Eliminating α and p yields

(6.4a) $$\dot{T} = -m,$$

(6.4b) $$\dot{m} = T.$$

Equations (6.4) represent the standard form of a linear oscillator. It was shown, in fact, that, for certain reasonable parameter values, eqs. (6.1a), (6.1b), with ζ as a prescribed function of l, produce self-sustained oscillations with an amplitude of $O(10\ \text{K})$ in T and a period $O(10\ \text{ka})$ [25, 27]. These oscillations, although nonlinear, can be understood at the most basic level in terms of the two opposing feedbacks of (6.4).

The third model equation, (6.1c), represents the zero-dimensional *viscous response* of the mantle [26, 32] to the ice-sheet load. The *elasticity* of the lithosphere leads to a difference Δh between the assumed point of contact of lithosphere and cryosphere, on the one hand, and global mean sea level (MSL), on the other (fig. 10). This difference is taken to be

(6.5) $$\Delta h = k(h + \zeta),$$

where k is an elasticity constant with a value $O(10^{-2})$ [32, 33] and $h + \zeta$ is the total thickness of the ice sheet; the height h is given by the geometry of the ice sheet in terms of the width l, $h = \sqrt{L^* l}$.

The mechanical part of the system, governed by eq. (6.1c), corresponds now to a mass, the ice sheet, on a spring, the lithosphere, with a dashpot viscous damping, given by the upper mantle. This mechanical oscillator, however, is strongly overdamped, i.e. *isostatic adjustment* of bedrock to ice load takes place.

The effect of isostatic rebound on the mass balance of the ice sheet can be expressed, in the simplifying short-hand notation of eqs. (6.3), (6.4), by

$$\dot{p} = -m. \tag{6.6a}$$

In words, the net accumulation of snow on the ice sheet, which is p, is reduced as the total mass m increases and the sheet sinks into the bedrock. As a result of this sinking, the snowline moves poleward, reducing the accumulation area A and hence the total accumulation.

Coupling the load-accumulation feedback of eq. (6.6a) with the precipitation-temperature feedback of eq. (6.3c),

$$\dot{m} = p, \tag{6.6b}$$

can lead to an oscillatory mechanism, independent of the one studied in [25, 27]. Whether a self-sustained oscillation actually occurs in eqs. (6.1b), (6.1c), keeping T fixed, depends on the parameter values used, in particular on the viscous relaxation time of the mantle.

Postglacial-uplift data and various geodynamical models used to explain these data yield multiple relaxation times [14, Chapt. 11; 32]. In these lectures, I shall use a single relaxation time $O(1 \text{ ka})$ (cf. eqs. (6.2f), (6.2g)) for which the oscillator denoted by (6.6) is damped. A careful analysis, along the lines of [26, 27], of this oscillator, given in its complete, nonlinear form by eqs. (6.1b), (6.1c) with T fixed, will probably give parameter ranges for which the oscillation is self-sustained. The coupling of oscillators (6.4) and (6.6) when both are self-sustained would be an interesting theoretical undertaking, which might explain certain features of Quaternary glaciations.

GHIL and LETREUT [26] showed that, in the absence of any forcing, and for values of the density ratio $q = \varrho_i/\varrho_a$ near the usually accepted $q = 1/3$, and values of the nondimensional viscosity coefficient L_*^2/Dc_T corresponding to the mantle's shorter relaxation times, system (6.1) with lithospheric elasticity parameter $k = 0$ and with snowline parameter $\sigma = 0.3$ exhibits natural oscillations very similar to those of [25, 27]. Such an oscillation is illustrated in fig. 12.

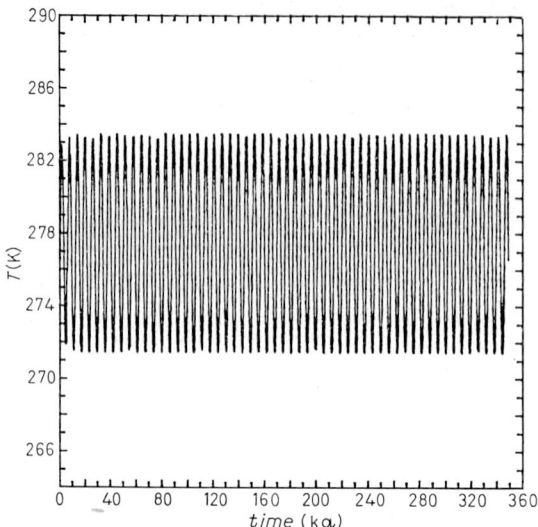

Fig. 12. – Self-sustained oscillation of the climate model (6.1), $\sigma = 0.3$, $k = 0$ (from [28]).

The parameter values used in the figure and kept constant hereafter are

$$(6.7)\quad\begin{cases}\lambda = 14\text{ m}, & \gamma = 0.3, \\ \alpha_0 = \alpha_m = 0.25, & \alpha_1 = 4.1\cdot 10^{-7}\text{ m}^{-1}, & \alpha_M = 0.85, \\ T_{\alpha,l} = 217\text{ K}, & \mathring{T}_m = 273\text{ K}, & T_{\alpha,u} = \mathring{T}_M = 283\text{ K}, \\ \varepsilon_m = 0.1, & \varepsilon_M = 0.5, & \varkappa = 1.74\text{ Wm}^{-2}\text{ K}^{-1}, \\ T_\varkappa = 154\text{ K}, & T_{00} = 283\text{ K}, & c = 0.7, \\ D = 40\text{ km}^2/\text{y}, & q = 0.33.\end{cases}$$

The oscillation of fig. 12 exhibits the amplitude implied by paleoclimatic proxy data, but does not have the multiple periodicities of fig. 7 and 8. To account for the peaks and continuous density of geologic power spectra, it appears worthwhile to turn now to the small, but complex forcing provided by slow changes in insolation.

6˙3. *Orbital changes.* – Since the mid Nineteenth Century and the beginnings of ice age theory (cf. subsect 5˙2), many hypotheses have been advanced about their causes. The global scale of the phenomenon has prompted a search for causes on the scale of the planet or even larger.

At the same time, the first half of the Nineteenth Century saw the triumph and popularization of celestial mechanics in the work of Lagrange, Laplace and LeVerrier. It is not suprising, therefore, that speculation about causes for ice ages turned to changes in insolation due to the changing distance between Earth and Sun. Indeed, the most important changes in the climatic system's external conditions, on the time scale of $(10 \div 1000)$ ka, are probably those related to the Earth's *orbital elements*.

In the absence of any other celestial bodies, the Earth would describe an elliptic orbit with the Sun as one of the foci: the shape and position in space of this ellipse, called the ecliptic, would be fixed. Furthermore, if the Earth were a perfectly spherical, homogeneous body, its axis of rotation would have a fixed direction in space. In particular, it would make a constant angle with the normal to the plane of the ecliptic, which in turn is fixed in space in the presence of Sun and Earth only.

The gravitational attraction of the other planets in the solar system perturbs the Earth's elliptic motion around the Sun. The masses of the planets are small compared to that of the Sun and hence these perturbations are small. It is practical, therefore, to describe the motion of the Earth as an ellipse whose orbital elements vary slowly in time.

The choice of elements is somewhat arbitrary, and depends on the problem under consideration. When dealing with the particular problem of insolation changes it is customary to choose *eccentricity e* in order to describe changes in the *shape* of the ellipse. In defining the ellipse's *position* in space, one uses the fact that the direction of the axis of rotation with respect to the fixed stars is mostly affected by the gravitational pull of Sun and Moon on the Earth's equatorial bulge. In fact, the Earth's axis of rotation describes a circular cone in space, under the influence of the Sun and Moon, with negligible interference from the other planets. As a result, the angle of tilt of the axis of rotation on the ecliptic, or *obliquity*, reflects the effect of the planets on the position of the ecliptic in space, independently of the motion of the axis itself. The obliquity is chosen, therefore, as the element describing the slow change in the position of the ecliptic.

In order to describe the slowly varying position of the Earth's elliptic orbit within the plane of the ecliptic, it is customary to use the *precession* angle $\tilde{\omega}$ of the perihelion—the point on the ellipse closest to the Sun—with respect to a point of intersection of the ecliptic with a plane of reference. In fact, the total precession is due to a combination of the luni-solar effect above on the Earth's mass distribution, on the one hand, and of the planetary effect on its total mass, on the other.

The position of the Earth itself on its elliptic orbit, within a year, is not of interest, since it varies too rapidly on the time scale under investigation. The other six orbital elements, such as the semi-major axis of the ecliptic, necessary in order to describe the position and velocity of the Earth in space and that of its axis of rotation, vary on even slower time scales than eccentricity, obliquity and precession.

As a result of the interactions between the nine planets and the Sun, the orbital elements of all planets, in particular those of the Earth, change in a *quasi-periodic* manner. A quasi-periodic function $F(t)$ can by definition be represented as

(6.8) $$F(t) = F_n(\omega_1 t, \omega_2 t, \ldots, \omega_n t),$$

with F_n being τ_j-periodic of period $\tau_j = 2\pi/\omega_j$ in each one of its n arguments; the n periods are rationally independent, *i.e.* they have no common, longer period. When n is larger than two or three, a quasi-periodic function can actually look rather irregular.

In fact, the behavior of quasi-periodic solutions to nonlinear differential equations is very difficult to compute accurately. Even the *three-body problem*, in which the motion of the Earth around the Sun is perturbed by a single planet, Jupiter, cannot be solved either in closed form or with prescribed accuracy. A large body of modern mathematics has been created in the effort to solve this problem and to settle the stability question for the planetary system, which already preoccupied NEWTON.

Various methods have been devised to compute approximate solutions to the planetary N-body problem. Most of these methods are based on the smallness of the masses of the planets compared to that of the Sun, on the smallness of the planetary eccentricities and on the smallness of the mutual angles of inclination of the planets' ecliptics. They use perturbations of the orbits about the hypothetical quasi-periodic phase-space flow of the planetary system represented by the independent revolutions of each planet around the Sun, in the absence of mutual attraction between the planets.

Perturbation methods fall into two broad categories [34]. Time-dependent methods consider, in fact, perturbations of the initial-value problem for the system of ODEs governing planetary motion. The most widespread one is *variations of constants*, where the « constants » are the orbital elements of the Earth, say, which would be constant in the absence of the other planets. The slow variation of these elements is the result of the small gravitational attraction of the other planets. The effect of the Earth on the other planets is neglected altogether in this method.

Variation of constants is entirely appropriate for the computation of the short-term orbital changes called *ephemerides* and used in classical navigation. Its validity, however, is limited in time in the same way in which a truncated power series expansion of purely periodic motion fails to capture the periodicity, no matter how high the truncation.

Time-independent perturbation methods attempt to follow the deformation of the entire phase-space torus on which the unperturbed, quasi-periodic flow lies, as the small parameters increase from zero. This is similar to the study of a stationary solution as a function of the parameter μ in sect. **3** and **4**; the zero-dimensional « torus » of such a solution is represented simply as a point on the curve in fig. 6.

The perturbation study of higher-dimensional tori is more difficult than the above-mentioned study, or that of Hopf bifurcation of stationary solutions into one-dimensional tori, *i.e.* limit cycles [14, Chapt. 12; 27]. This added difficulty is due to the appearance of *resonances* between the frequencies, as the small parameters change. Classical methods for this study, such as *perturbation of co-ordinates*, arbitrarily neglect certain terms in the perturbation expansion. Even so, they lead to asymptotic series for the elements which were shown already by POINCARÉ in the 1880s to be nonconvergent.

It is only in the 1950s and 1960s that KOLMOGOROV, ARNOLD and MOSER were able to show that the problem of resonances can be avoided under certain technical conditions. This provided an abstract proof that the motion of the planetary system is actually quasi-periodic with nonzero probability. So far, the methods of KAM theory have not produced realistic computations of planetary orbits.

I shall only quote, therefore, the results of calculations based on perturbation of co-ordinates. For the duration of the Quaternary, the *mean* periodicities of the Earth's orbital elements obtained by these calculations are probably rather accurate, although the periodicities themselves are slowly changing due to the neglected terms. As a result of these and other inaccuracies, the phases of the elements are likely to be in error by as much as $180°$ beyond 1000 ka ago.

It should be noted that these estimates are based on intercomparisons of different methods. No observational verification exists until now of long-term astronomical computations. In fact, accurately computing the effect of orbital changes on the climate would be important for celestial mechanics. If paleoclimatological proxy data could be combined with a few absolute dates to determine a geological time scale with known error bars, this would allow then the first observational verification of long-term orbital changes. Such a verification would turn celestial mechanics, which has become a branch of pure mathematics, back into a branch of physics, permitting a serious evaluation of nongravitational effects in the planetary system, such as tides, gaseous and electromagnetic drag, radiation pressure and relativistic effects [34, 35].

6˙4. *Insolation changes and their climatic effect.* – Tables II, III and IV (after [36]) give the first ten terms each in an asymptotic series expansion of trigonometric type for the obliquity, the precessional parameter $e \sin \tilde{\omega}$ and the eccentricity e, respectively. The terms are ordered by decreasing amplitude (second column), with the period of the term in the last column.

The series used here, as in many perturbation methods of celestial mechanics, are so-called *d'Alembert series* $\sum_j \mathscr{C}_j \cos \mathscr{D}_j$. Any argument $\mathscr{D}_j$ is a linear combination with integer coefficients $j_1, j_2, ..., j_n$ of «fast-periodic» elements, such as angular precession $\tilde{\omega}$ or annual position λ on its ecliptic of each of the planets involved, $j = (j_1, j_2, ..., j_n)$. The coefficients $\mathscr{C}_j$ are monomials in the perturbing masses, the eccentricities and the mutual inclinations (tg I) of the eclipties of the planets involved, with the masses (m') expressed in solar units. The sum of the powers of the constant masses is the *order* of the term, the sum of the powers of the very slowly varying eccentricities and inclinations its *degree*. The amplitudes of the terms in tables II-IV and the periods are thus not constant in time, but they vary slowly on Quaternary time scales.

In table II the largest term has a period of 41 ka, and the next term is

Table II. – *Series expansion of obliquity variations (after [36])*.

Term	Obliquity (")	Period (y)
1	− 2 462.22	41 000
2	− 857.32	39 730
3	− 629.32	53 615
4	− 414.28	40 521
5	− 311.76	28 910
6	308.94	41 843
7	− 162.55	29 678
8	− 116.11	40 190
9	101.12	42 354
10	− 67.69	30 365

TABLE III. – *Series expansion of the precessional parameter $e \sin \tilde{\omega}$* (after [36]).

Term	Amplitude	Period (y)
1	0.018 608	23 716
2	0.016 275	22 428
3	−0.013 007	18 976
4	0.009 888	19 155
5	−0.003 367	19 261
6	0.003 331	23 293
7	−0.002 354	18 873
8	0.001 400	16 907
9	0.001 007	28 818
10	0.000 857	19 445

TABLE IV. – *Series expansion of the eccentricity e (after* [36]).

Term	Amplitude	Period (y)
1	0.001 102 94	412 885
2	− 0.008 732 96	94 945
3	− 0.007 492 55	123 297
4	0.006 723 94	99 590
5	0.005 812 29	131 248
6	− 0.004 700 66	2 305 441
7	− 0.002 544 64	102 535
8	0.002 314 85	1 306 618
9	− 0.002 219 55	136 412
10	0.002 018 68	603 630

three times smaller. Hence the Earth's obliquity has a clear 41 ka periodicity over the Quaternary.

Angular precession $\tilde{\omega}$ (not shown) has a similar dominant period, due to the luni-solar effect, of 22 ka. In insolation calculations, however, it is the precessional parameter $e \sin \tilde{\omega}$ which plays the important role. Table III shows the first four terms to be dominant, with the fifth term three times smaller than the fourth. The first two and the next two terms can be grouped into a periodicity of 23 ka and one of 19 ka, respectively. The «splitting» of the 22 ka angular period into these two precessional periodicities is the result of multiplication by the eccentricity and of the so-called *addition theorem*, which allows the product of two trigonometric functions to be replaced by a sum of two trigonometric functions whose arguments are the sum and difference of the two original arguments.

Table IV shows that the d'Alembert series of the eccentricity has very slowly decreasing amplitudes. The first term has a period of 413 ka, too long to be detected with statistical confidence in Quaternary records. The next four terms all have periods between 95 ka and 131 ka, and similar amplitudes.

How do orbital changes affect the climatic system? The first thing to notice is that their direct effect is very small. Among the three insolation-affecting parameters above, eccentricity is the only one which actually modifies the globally and annually averaged amount of insolation Q received at the top of the atmosphere. This change, however, is proportional to $(\Delta e)^2/2$, Δe being the change in eccentricity, which over the Quaternary has not exceeded 0.04. Thus ΔQ is only about one part in a thousand over the same period. As noticed in subsect. 4˙6, the quasi-equilibrium response of a mean annual, zonally or globally averaged energy-balance model to such an insolation change would not exceed a fraction of one degree.

Obliquity and precession have no net global and annual effect at all. They only change the contrast between seasons. For obliquity this change of contrast is the same in both hemispheres, for precession the contrast is largest in one hemisphere when it is smallest in the other.

The precessional effect $e \sin \tilde{\omega}$ is the same at all latitudes within one hemisphere, while the obliquity effect increases with latitude. As a result, insolation changes due to obliquity might dominate at higher latitudes, while precessional changes might dominate at lower latitudes. Changes of insolation due to either effect averaged over a half-year and a hemisphere are larger than the direct effect of Δe, but still quite small.

The quest for a mechanism which might translate minute yearly and small seasonal insolation changes into large changes in global temperature and ice volume has mostly pointed in the direction of the hydrological cycle and the annually averaged net accumulation rate of snow. Larger snow accumulation in winter or less ablation in summer would both increase the ice sheets' mass over one year, while less snowy winters and warmer or less cloudy summers would both act in the opposite direction.

Considerable controversy has existed over just which season and which latitudes, or zones, would be most decisive in the net annual ice balance. ADHÉMAR and CROLL in the Nineteenth Century thought that high-latitude winter was crucial, while KÖPPEN early in this century believed it was mid-latitude summer [17, Chap. 8]. More recent arguments, emphasizing precessional effects, even brought in low latitudes, where the latter effects dominate [17, Chap. 12].

In the absence of a convincing answer to the quandary of critical latitude and season, I shall merely let the accumulation-to-ablation rate $\varepsilon(T)$ vary in time, according to the orbital periodicities. Within the degree of simplicity and abstraction of the model to be studied here, this seems to be a reasonable assumption.

More precisely, the extreme values ε_m and ε_M in fig. 11 are kept fixed. The only thing allowed to vary is the mean annual temperature T_m at which ε starts to increase with T, and the temperature T_M at which it reaches its maximum value. Thus, instead of being constant in time, as in fig. 12, T_m and T_M vary quasi-periodically,

$$(6.9a) \qquad T_m(t) = \overset{\circ}{T}_m \left\{ 1 + \sum_1^5 \varepsilon_j \sin\left(\frac{2\pi}{\tau_j} t + \varphi_j\right) \right\},$$

$$(6.9b) \qquad T_M(t) = \overset{\circ}{T}_M \left\{ 1 + \sum_1^5 \varepsilon_j \sin\left(\frac{2\pi}{\tau_j} t + \varphi_j\right) \right\},$$

with periods τ_j and phases φ_j. The phases φ_j will at first be taken equal to each other, $\varphi_j = 0$, $1 \leqslant j \leqslant 5$.

The variation of ε with time is shown by the two dashed lines in fig. 11: the ramp portion of $\varepsilon(T)$ can move parallel to itself in time. Hence ε becomes a function of time t both via the temperature variable $T(t)$ and via the prescribed effective hydrologic temperatures $T_m(t)$ and $T_M(t)$,

(6.9c) $$\varepsilon = \varepsilon(T(t); T_m(t), T_M(t));$$

it reduces in the autonomous case illustrated in fig. 12 to

(6.9d) $$\varepsilon(T) = \varepsilon(T(t); \mathring{T}_m, \mathring{T}_M).$$

I shall consider, in order of increasing periodicity, the following approximate values for the orbital periods:

(6.9e) $\quad \tau_1 = 19 \text{ ka}, \quad \tau_2 = 23 \text{ ka}, \quad \tau_3 = 41 \text{ ka}, \quad \tau_4 = 100 \text{ ka}, \quad \tau_5 = 400 \text{ ka}.$

The first two periodicites, τ_1 and τ_2, correspond to the precessional effect, $e \sin \tilde{\omega}$, with the precessional angular frequency $1/22$ ka modulated by an eccentricity frequency near $1/100$ ka. The third period, τ_3, is associated with the obliquity, while τ_4 and τ_5 are approximate, simplified periodicites related to eccentricity changes.

The insolation Q will also be allowed to vary according to

(6.9f) $$Q = Q_0 \left\{ 1 + \delta_4 \sin \frac{2\pi}{\tau_4} + \delta_5 \sin \left(\frac{2\pi}{\tau_5} + \psi_5 \right) \right\}.$$

In this model, changes in $\varepsilon(T)$ and Q with time represent the net effect of orbital variations, integrated over latitude and season. These changes do not appear in the model equations as additional terms on the right-hand side, but rather as variations in the coefficients. In this sense, orbital forcing is different from periodic or quasi-periodic forcing in many mechanical and electronic oscillators. It could be called multiplicative, or *parametric*, rather than additive *forcing*.

7. – The forced behavior of a climatic oscillator.

To summarize, the model studied in this section is governed by three coupled ODEs, repeated here for convenience in more compact notation:

(7.1a) $$\dot{T} = P(T, l, t; \boldsymbol{\delta}),$$

(7.1b) $$\dot{l} = R(T, l, \zeta, t; \mu, k, s, \boldsymbol{\eta}'),$$

(7.1c) $$\dot{\zeta} = S(l, \zeta; D, q).$$

Notice that all three variables, global temperature T, ice extent l and bedrock deflection ζ are coupled in the ice-sheet evolution equation (7.1b), which represents the tie between the two oscillation mechanisms (6.4) and (6.6).

The parameters $\boldsymbol{\delta} = (\delta_4, \delta_5)$ and $\boldsymbol{\eta}' = (\varepsilon_1, \varepsilon_2, \varepsilon_3, \varepsilon_4, \varepsilon_5)$ represent the intensity of the insolation and of the hydrologic forcing, respectively. We know that $\delta_4 + \delta_5 \leqslant 0.001$ and that $\boldsymbol{\eta}'$ is also small.

The parameter values (6.7) and $Q_0 = 362.2$ Wm^{-2} used in fig. 12 will be kept fixed in the sequel. The values of k and σ and the nonzero values of the forcing parameters $\delta_4, \delta_5, \varepsilon_1, ..., \varepsilon_5$ will be indicated for each figure.

This section follows [28] and will study the forced oscillations of the model governed by eqs. (7.1), as model parameters k and σ, and the forcing, change. Again, this study is undertaken as an illustration of the mathematical methods which I believe are useful in this type of problem. Some of the results, however, are probably of direct physical relevance to an explanation of the paleoclimatological spectra introduced in sect. **5**.

7˙1. *Free oscillations.* – A nonzero elasticity parameter, $k \neq 0$, tends to damp the model's oscillations seen in fig. 12. For $\sigma = 0.3$ and $k = 0.03$, the self-sustained oscillations decay to a stable steady state $Q'_s = (T_s, l_s, \zeta_s) \simeq$
$\simeq (281$ K, $0.79, 0.44)$.

The first two eigenvalues of the system's linearization about Q'_s are complex conjugate. The third eigenvalue is real and negative, so that Q'_s is attractive in the ζ-direction, for all parameter values considered here. The sign of the real part of the complex conjugate pair becomes negative as k increases from zero, turning the focus Q'_s into a stable one.

The effect of the slope parameter σ is similar to k. When σ decreases from $\sigma = 0.3$, say $\sigma = 0.28$ and $k = 0$, model oscillations increase out of the physically realistic bounds. An increase in σ stabilizes the model. Hence larger k and smaller σ act in opposite directions.

The period τ_0 of the self-sustained oscillation depends on k and σ, $\tau_0 = \tau_0(k, \sigma)$. The value corresponding to fig. 12 is $\tau_0 \simeq 6.25$ ka, or $f_0 = 1/\tau_0 \simeq$
$\simeq 0.16$ ka^{-1}. For $k = 0.03$ and $\sigma = 0.28$, $f_0 \simeq 0.18$ ka^{-1}. For other values of the relaxation time parameter $c = c_L/c_T$, $c > 0.7$, much longer periods obtain [27]. As we shall see in the sequel, the behavior of the forced oscillations does not depend very much on the value of τ_0. Only values of $f_0 = O(10^{-1}$ ka$^{-1})$ will be considered here.

7˙2. *Nonlinear resonance.* – To discuss the response of our climate model to external forcing, it is convenient to introduce the frequencies $f_j = 1/\tau_j$, $j = 1, 2, 3, 4, 5$. Notice that $f_0 > f_1 > f_2 > ... > f_5$.

Response to insolation forcing with $\delta_4 = 0.001$ in eq. (6.9f) depends dramatically on the presence of self-sustained oscillations (fig. 13b)) or their absence (fig. 13a)). For $(\sigma, k) = (0.3, 0.03)$, the model's quasi-equilibrium response has the small amplitude which could be inferred from its insolation sensitivity of 1 K for $\Delta Q/Q_0 = 1$ percent. When $(\sigma, k) = (0.3, 0)$ as in fig. 12, the model responds resonantly, with the amplitude of its self-sustained oscillation at f_0 modulated by that of the forcing at f_4.

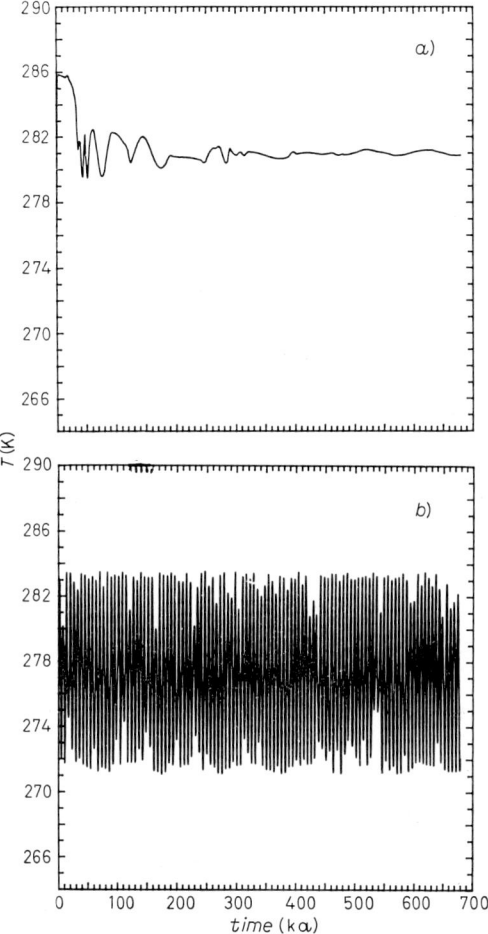

Fig. 13. – Model response to eccentricity forcing: a) quasi-equilibrium response, $\sigma = 0.3$, $k = 0.03$, $\delta_4 = 0.001$; b) resonant response, $\sigma = 0.3$, $k = 0$, $\delta_4 = 0.001$ [28].

It is well known that a nonlinear oscillator like (7.1) can respond resonantly when the forcing frequency f_4 is rather different from its own frequency f_0. To understand better how this nonlinear resonance occurs for such a large «detuning» as $f_0 - f_4$ or even $f_0 - f_5$, the resonance mechanism needs to be investigated in further detail.

7'3. *Entrainment and combination tones.* – As a first step, we verify that changes in the hydrologic cycle, eqs. (6.9a), (6.9b), also lead to resonant response, even at constant Q. Figure 14 illustrates a series of model responses to obliquity forcing, $\varepsilon_3 \neq 0$.

For all the solutions given in fig. 14, $(\sigma, k) = (0.3, 0)$, as in fig. 12. Figure 14a) shows the temperature evolution $T(t)$ for $\varepsilon_3 = 0.0025$. The modulation of the free oscillations with amplitude $O(10 \text{ K})$ is similar to that in fig. 13b). Figure 14b) gives the corresponding spectral density $\hat{T}(f)$, in logarithmic units. The nonlinear eigenfrequency f_0 dominates the spectrum, followed by the forcing frequency f_3. Additional frequencies visible are a harmonic of the forcing, $2f_3$, and the *combination tones* $f_0 + f_3$, $f_0 - f_3$ and $f_0 - 2f_3$. Harmonics and combination tones are also a characteristic of nonlinear system response to additive or multiplicative periodic forcing.

In fig. 14c), $\varepsilon_3 = 0.0035$ and both amplitude and character of the temperature oscillations appear rather similar to fig. 14a). A close inspection of the spectrum $\hat{T}(f)$ in fig. 14d), shows, however, a remarkable difference: the peak at $f_0 \simeq 0.16 \text{ ka}^{-1}$ has been replaced by a peak at $f_0 \simeq 0.17 \text{ ka}^{-1}$, which corresponds to $7f_3$. Smaller peaks at $2f_3$ and $6f_3$ are also visible.

Between the two values of ε_3 in fig. 14b) and in fig. 14d), *entrainment* has taken place—the frequency of the free oscillation has locked onto the frequency of the forcing or, more precisely, onto a harmonic of the forcing. This phenomenon, also called *frequency locking*, is common in nonlinear oscillations. Its simplest form was observed first by HUYGENS, who noticed that two clocks, which had slightly different pendulum periods when far apart, would synchronize when brought close together.

In the case observed by HUYGENS, as well as in the entrainment of most living organisms to the diurnal light cycle, *phase locking* occurs along with frequency locking. In many other oscillators, however, frequency locking does not necessarily imply synchronization of phases. The complexity of the climatic spectra discussed in sect. **5** and the presence of the continuous background seem to preclude simple phase locking between response and forcing for the system at hand. I shall return to this question in the next section.

For $\varepsilon_3 = 0.009$, fig. 14e) shows oscillations with an amplitude noticeably smaller than in fig. 14a) and c), and with a dominant period of 41 ka. These oscillations, however, are not perfectly periodic: a slight irregularity is clearly visible. Even slight aperiodicity of the temperature evolution is associated with a *continuous* component of its *spectral density*, which increases with decreasing frequency [14, sect. **10**'4; 37]. Figure 14f) shows, besides the now dominant peak at f_3, a continuous background which looks like that produced by random *red noise* [14, eq. (10.54); 15]. Upon this background one can still distinguish, as in [37, fig. 5d)-f)], some harmonics of the fundamental, namely $2f_3$, $4f_3$, $5f_3$ and $6f_3$.

We have herewith an example of deteministic transition to aperiodic, or

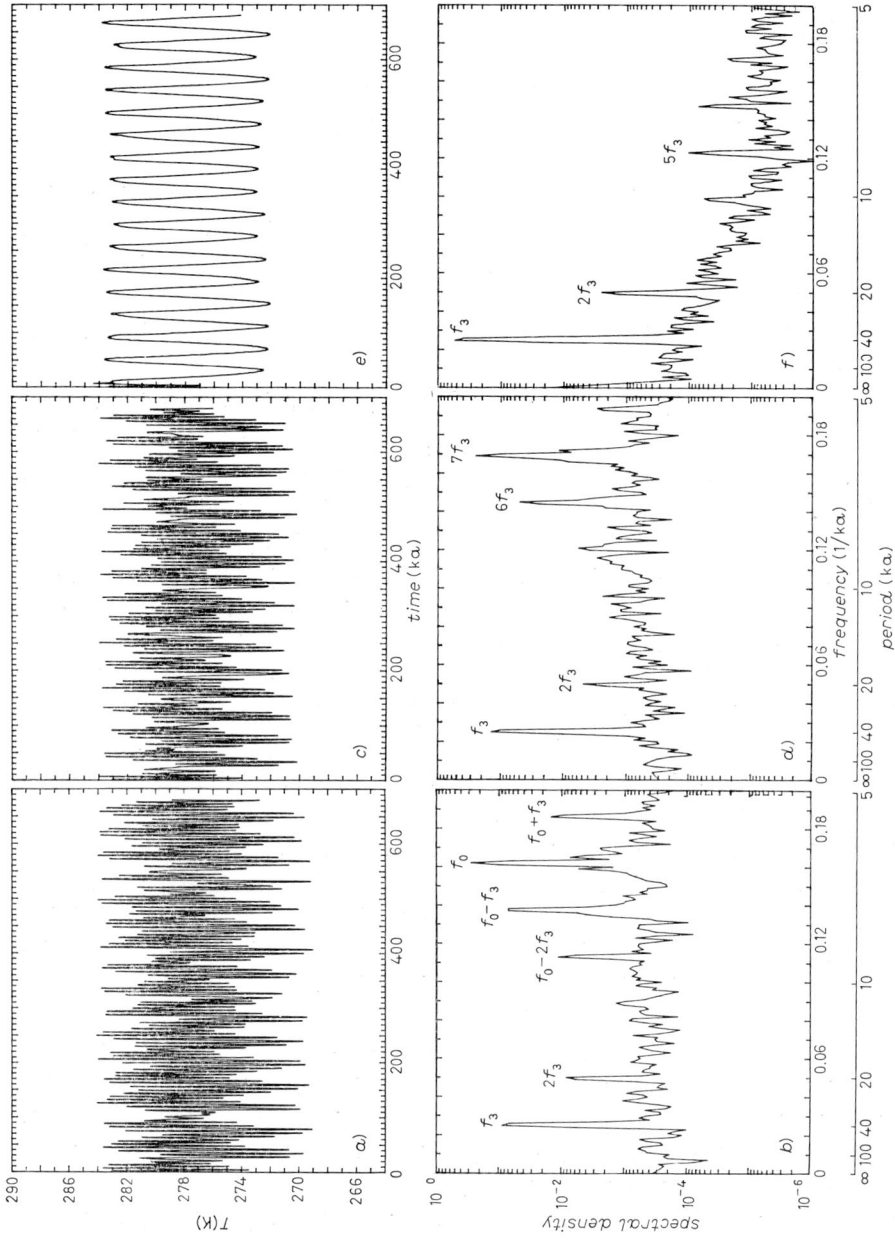

Fig. 14. – Model response to obliquity forcing, for $\sigma = 0.3$, $k = 0$; a, c, e) temperature evolution: a) $\varepsilon_3 = 0.0025$, c) $\varepsilon_3 = 0.0035$, e) $\varepsilon_3 = 0.009$; b, d, f) power spectrum of temperature: b) $\varepsilon_3 = 0.0025$, d) $\varepsilon_3 = 0.0035$, f) $\varepsilon_3 = 0.009$ [28].

chaotic, behavior. The mechanism of transition is loss of entrainment, which has also been observed experimentally in fluid systems [38], as well as numerically and analytically in various model equations [39]. Notice in fig. 14f) that the peak $7f_3$ to which f_0 was entrained is about to disappear into the background noise.

Similar results, with entrainment producing a large number of sharp peaks in the spectrum, and its loss leading to a continuous, red-noise–type background, were obtained for other forcing frequencies. Aperiodic model behavior is observed for forcing amplitudes as small as $\varepsilon_j = 0.001$ or $\delta_4 = 0.001$. In fact, aside from a red-noise–type continuous background, decreasing with frequency to the right of the forcing peak which dominates the spectrum, one obtains also « band-limited chaos »: a high, almost flat plateau in spectral density near f_0, left behind by a « cock's comb » of combinations tones $f_0 \pm kf_5$, $1 \leqslant k \leqslant 3$, as it is first entrained by harmonics of f_5, then detrained as in fig. 14f).

7˙4. *Multiple forcing: more combination tones.* – We have seen the effects of single-frequency forcing in insolation (fig. 13), as well as in the hydrologic cycle (fig. 14), at the eccentricity frequency f_4 and the obliquity frequency f_3, respectively. It is natural to inquire next about the effects of multiple-frequency forcing, in particular at the two precessional frequencies, f_1 and f_2.

A series of numerical experiments with $(\sigma, k) = (0.3, 0)$ and $\varepsilon_1 = \varepsilon_2 = 0.001$, 0.003 and 0.005 [28, fig. 8a)-f), not shown here] give results very similar to those in fig. 14: At $\varepsilon_1 = \varepsilon_2 = 0.001$ the f_0 peak is most prominent, with smaller peaks at f_1, f_2 and $f_0 \pm (f_1 - f_2)$. At $\varepsilon_1 = \varepsilon_2 = 0.003$, entrainment of f_0 to a fractional harmonic of the forcing, $(15/4) f_2$, obtains. At $\varepsilon_1 = \varepsilon_2 = 0.005$, detrainment has occurred, the high-frequency peaks at $f \geqslant 0.1$ ka^{-1} have disappeared, and the peaks at f_1 and f_2 are dominant, superimposed on low-frequency, red-noise–type background.

Figure 15 presents the results for $(\sigma, k) = (0.28, 0.05)$. Nonzero lithospheric elasticity has the effect [28, fig. 7; $\sigma = 0.3$, $k = 0.03$, $\varepsilon_3 = 0.005$, not shown here] of leading to a slower growth and more rapid decay of ice sheets, giving a sawtooth appearance to the ice extent and bedrock evolution curves, while the temperature evolution is still roughly sinusoidal.

The model's self-sustained oscillation has a frequency of $f_0 \simeq 0.155$ ka^{-1} for the parameter values in the figure. At $\varepsilon_1 = \varepsilon_2 = 0.005$, this oscillation has been entrained already. The ice sheet evolution $l(t)$ in fig. 15a) shows an irregular behavior with intermittent, large-amplitude fluctuations every 100 ka, approximately. The behavior for $k = 0$, $\sigma = 0.3$, discussed before, is also modulated at roughly 100 ka, but less intermittent.

Figure 15b) gives the power spectrum of the ice volume, $\hat{V}(f)$, for the solution in fig. 15a), where V is proportional to $l(\zeta + l^{\frac{1}{2}})$, and the spectrum is normalized. The most prominent peak is at $f_1 - f_2$. With our choice of approximate values for f_1 and f_2, this peak has a period of 109 ka.

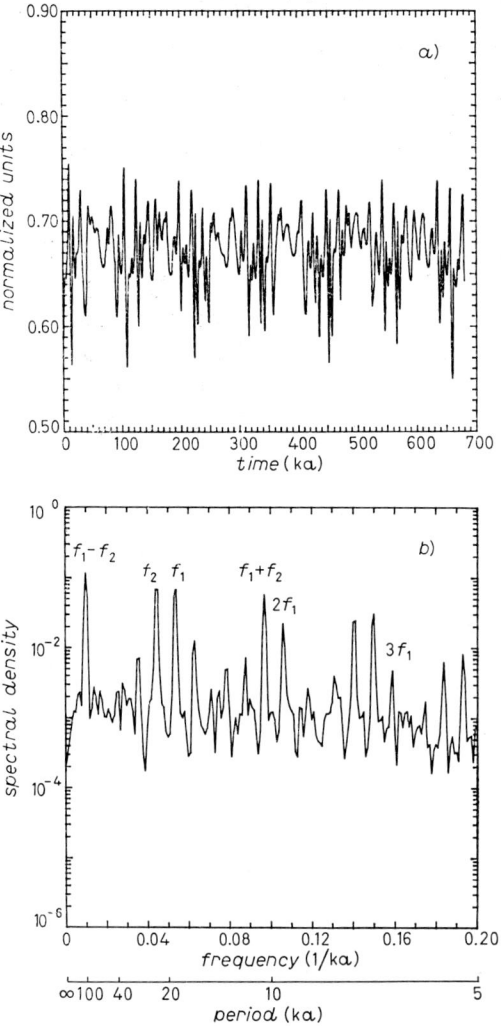

Fig. 15. – Model response to precessional forcing: a) ice extent evolution, b) power spectrum of ice volume; $\sigma = 0.28$, $k = 0.05$, $\varepsilon_1 = \varepsilon_2 = 0.005$ [28].

Recall from the discussion of insolation changes that the precessional effect $e \sin \tilde{\omega}$ had two apparent periods, of approximately 19 ka and 23 ka, due to the modulation of angular precession $\tilde{\omega}$ by the eccentricity e in the range 95 ka to 130 ka. The climatic system can thus resynthesize and focus, by its combination tones, a response to the original, diffuse frequency range of eccentricity, hidden in the precessional effect.

After the peaks at f_1 and f_2, the next largest peak in fig. 15b) is at $f_1 + f_2$, which corresponds to a period of roughly 10 ka. This peak is flanked by two peaks at $2f_1$ and $2f_2$. Peaks between 9.5 ka and 11.5 ka appear in various marine and continental records with high sedimentation rates. The stratigraphic, radiometric and micropaleontological analysis of such records will lead in the near future to more refined Quaternary time scales and more reliable proxy records of glaciations. Careful harmonic analysis will then permit the high-resolution discrimination between peaks suggested by the theoretical spectra presented here.

7˙5. *Sharp peaks, aperiodicity and terminations.* – We have examined the effect of insolation forcing at each eccentricity frequency, as well as that of mean annual hydrologic forcing at the obliquity frequency and at the two precessional frequencies. The effects of insolation changes and of forced changes in the hydrologic cycle are essentially independent of each other and can be conceptually superposed. It is interesting, however, to see the combined effect of all three frequencies which affect the hydrologic cycle exclusively, f_1, f_2 and f_3.

For the sake of definiteness, I choose the values of ε_1, ε_2 and ε_3 which correspond approximately to July insolation at 65° N. That is, the ratio is $\varepsilon_1 : \varepsilon_2 : \varepsilon_3 = 3 : 4 : 2$, with the amplitude taken close to those previously considered, *i.e.* $\varepsilon_1 = 3 \cdot 10^{-3}$, etc. This is, of course, arbitrary, since the present model avoids the changes of forcing with latitude and season, by integrating their net effect over the globe and the year. The choice is only made for comparison purposes with models which might include more details of insolation distribution in space and time. Figure 16a) shows the evolution of bedrock deflection. Terminations are clearly visible at roughly 100 ka, 200 ka, 330 ka, 520 ka and 640 ka. The slow growth of ice sheets does not, according to paleoclimatic records, occur monotonically, but rather in successive rises and smaller falls (cf. fig. 7). This feature of small oscillations superimposed on a large trend is well captured by the model solution in fig. 16a).

Occasionally, one or two spikes appear in the solution between a termination and the subsequent slow growth, for instance at about 220 ka in fig. 16a). Such a spike is apparent in the Pacific core V28-238, as well as in the combined record of cores RC11-120 and E49-18 from the Indian Ocean, during isotopic stage 7 (fig. 7 and 8). A similar spike is now believed to have occurred during the last deglaciation, where one distinguishes between termination I*A* and I*B*. The exact position, amplitude and duration of this spike is still a matter of debate, but its existence is gaining wide acceptance.

The lack of exact periodicity in the terminations, as well as the irregularity of smaller peaks and plateaus within each major cycle, suggests again the presence of a complex spectrum. Figure 16b) shows the Fourier spectrum of the total ice volume for the solution at hand. It is dominated by the peaks at f_1, f_2, $f_1 - f_2$ and $f_1 + f_2$. A number of combination tones lying between $f_1 - f_2$ and f_2 are as prominent as the peak at f_3.

The striking feature of fig. 16b) is the presence, as in fig. 14f), of the continuous background, decreasing with frequency (stippled in the figure). This

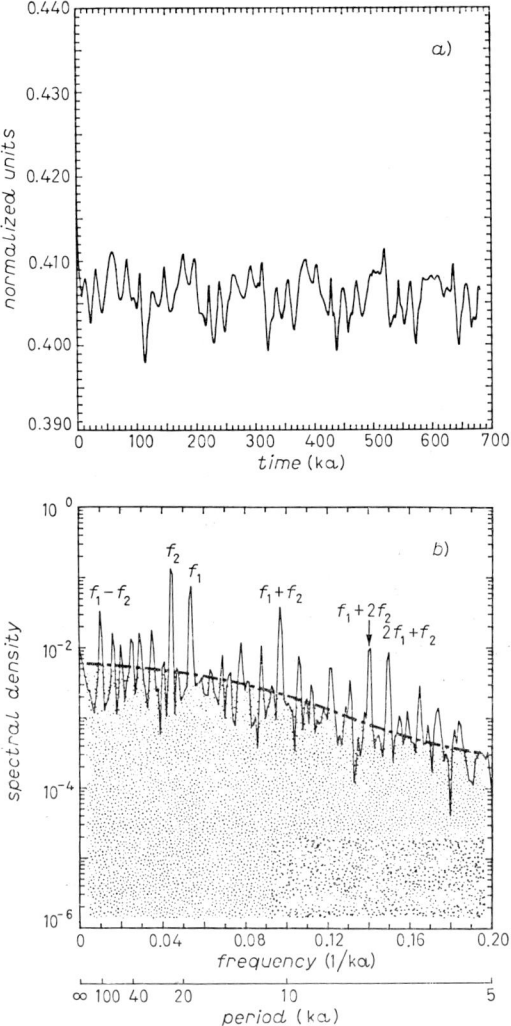

Fig. 16. – Model response with irregular terminations and sharp peaks on a continuous background: a) bedrock evolution, b) power spectrum of ice volume; $\sigma = 0.28$, $k = 0.05$, $\varepsilon_1 : \varepsilon_2 : \varepsilon_3 = 3:4:2 \cdot 10^{-3}$ [28].

background is produced by the deterministic nonlinear interactions in the model, rather than by any stochastic forcing. At least part of the continuous spectral density observable in paleoclimatic records is thus probably due to internal variability on the time scale of glaciation cycles, rather than to mere passive filtering of higher-frequency weather noise by the slow components of the climatic system.

The temperature spectrum (not shown) has the same frequency-dependent continuous background. It shows strong peaks at f_1, f_2 and $f_1 + f_2$, but only a very weak peak at $f_1 - f_2$. On the other hand, two peaks at f_3 and at $f_3 - f_1 + f_2$ are prominent. The latter corresponds to a period of roughly 66 ka.

Model solutions for $T(t)$ and $V(t)$ were also translated into simulated isotopic records in ice cores and deep-sea cores by H. LETREUT and J. JOUZEL (personal communication, 1983). A simple, robust model of fractionation processes taking place during the hydrologic cycle [40] is used for this purpose (compare subsect. 5'3). The power spectra of the simulated $\delta^{18}O$ records also show sharp peaks at $f_2 + f_3 \simeq 1/15$ ka and $f_1 + f_3 \simeq 1/13$ ka.

The relative power in each peak changes from one simulated record to another, but their frequencies are the same. Some of these and other combination tones are found, along with a few of their harmonics, in four carefully dated and analyzed rapid-sedimentation cores from the Indian Ocean (P. PESTIAUX and J.-C. DUPLESSY: pre-publication manuscript, 1984). These results and related ones suggest in particular that the spectral peak at $\tau \simeq 2.5$ ka [16, fig. 1], which appears, for instance, in ice cores from Greenland and the Antarctica, might be the fourth harmonic of the $f_1 + f_2$ combination tone emphasized in this section, with $\tau = 10.4$ ka$/4 = 2.6$ ka.

Nonlinear resonance, harmonics, combination tones and deterministic aperiodicity thus appear to explain many characteristics of paleoclimatic time series and of their Fourier spectra. An important feature of these results is that they are insensitive to the exact nature of the system's self-sustained oscillations, their cause and period. These oscillations only act as an amplifier of the forcing and lead to a partial transfer of the system's internal variability from the high frequencies, $O(10^{-1}$ ka$^{-1})$, at which most physical mechanisms involved are active, to the low frequencies, $O(10^{-2}$ ka$^{-1})$, which are also observed in paleoclimatic records. It is time to turn now to the consequences of quasi-periodicity and aperiodicity for the predictability of climate on the time scales under discussion.

8. – Periodicity and predictability.

The search for periodicity in geophysical time series is obviously driven by the desire to understand, as well as to predict. Constant behavior is the most predictable; it is without surprises, but seldom encountered in Nature. The next most predictable type of behavior is purely periodic.

8'1. *Phase errors and frequency errors.* – For periodic behavior it becomes important to ascertain the phase of the phenomenon. An error of phase determination at one time, however, will merely result in a similar error at any other time, past or future. Not so far an error in period: even the smallest error in period will result, after a sufficient number of periods, in a completely erroneous forecast, or hindcast, as the case may be.

For a quasi-periodic function with two or more rationally independent periods, things become rapidly more complicated. Errors in either period will result in complete loss of predictive skill after a small multiple of the shortest period involved. The following discussion will be restricted to the type of error which is both less severe and more inevitable—phase error.

Phase errors can arise in the forcing, as well as in the internal oscillation of the system. The initial values $P_0 = (T_0, l_0, \zeta_0)$ used for the model variables so far were equal to

(8.1a) $\qquad P_1 = (T_1, l_1, \zeta_1) = (276 \text{ K}, 0.65, 0.42)$,

i.e. $P_0 = P_1$. In order to study the effect of the internal oscillations' phase on the predictability of model variables, the points $P_0 = P_j$, $j = 2, 3, 4$, were also used as initial data, where

(8.1b) $\qquad \begin{cases} P_2 = (280 \text{ K}, 0.65, \zeta_1), \\ P_3 = (272 \text{ K}, 0.70, \zeta_1), \\ P_4 = (284 \text{ K}, 0.80, \zeta_1). \end{cases}$

The points P_1, P_2, P_3 and P_4 are approximately at a quarter phase from each other along the limit cycle of the model. The phase of the forcing is varied by taking $\varphi_j = k\pi/2$, $k = 0, 1, 2, 3$, in eqs. (6.9a), (6.9b). So far $\varphi_j = 0$ in all solutions discussed.

Figure 17 illustrates the differences between the solution in fig. 14, with $P_0 = P_1$ and $\varphi_3 = 0$, and those obtained for $\varphi_3 = \pi/2$ (fig. 17a)) and for $P_0 = P_2$ (fig. 17b)), respectively. The two frequencies in the solution are f_0 and f_3. They are actually not rationally independent, but the corresponding common period, close to 250 ka, is sufficiently long for the very small background noise to make the resulting solution appear quasi-periodic.

Comparison between fig. 17a) and 14a) shows a clear phase shift of the longer, modulating period τ_3, as expected. But the high-frequency oscillations within each long period are not at all alike for two corresponding periods, say $t_0 \leqslant t \leqslant$ $\leqslant t_0 + \tau_3$ in fig. 14a) and $t_0 - \tau_3/4 \leqslant t \leqslant t_0 + 3\tau_3/4$ in fig. 17a). For instance, the « quiet episode » near 100 ka when $\varphi_3 = 0$ has no « shifted » counterpart in fig. 17a), nor does the quiet episode near 300 ka when $\varphi_3 = \pi/2$ have its nearby counterpart in fig. 14a).

This observation is confirmed by fig. 17b), in which $P_0 = P_2$, while $\varphi_3 = 0$ again, as in fig. 14a). The initial difference between the temperature values does not disappear with time, given the identical forcing, but continues to produce differences between the two figures for the entire duration of the display. The « missing spikes » in fig. 17b) just before 200 ka and just after 250 ka have

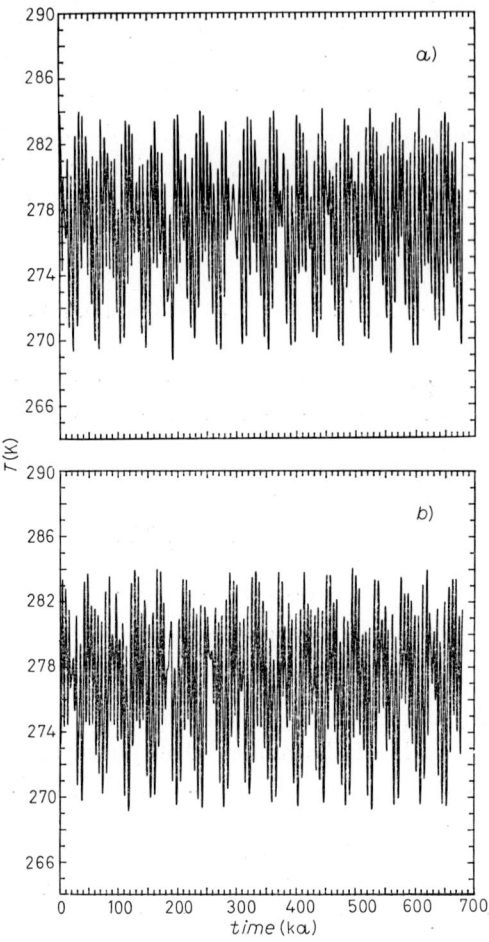

Fig. 17. – Effect of phase errors on a quasi-periodic model solution, for $\sigma = 0.3$, $k = 0$, $\varepsilon_3 = 0.0025$: a) error in the forcing, $\varphi_3 = \pi/2$; b) error in the initial data, $P_0 = P_2$ [28].

no counterpart in fig. 14a), while the solution in fig. 14a) is more quiescent near 100 ka than in fig. 17b).

Figure 18 shows the actual difference between the temperature evolution in fig. 17a), $T(t; \varphi_3 = \pi/2)$, and that in fig. 14a), $T(t; \varphi_3 = 0)$. The peak-to-peak amplitude of the difference is twice that of either solution. This maximum is reached within 100 ka, *i.e.* roughly twice the forcing period. The difference exhibits the aspect of « beats », with the two frequencies present producing quiet intervals, like that between 250 ka and 400 ka, and agitated episodes, like that near 100 ka or 550 ka. Similar results obtain by taking the difference between the solution in fig. 17b) and that in fig. 14a).

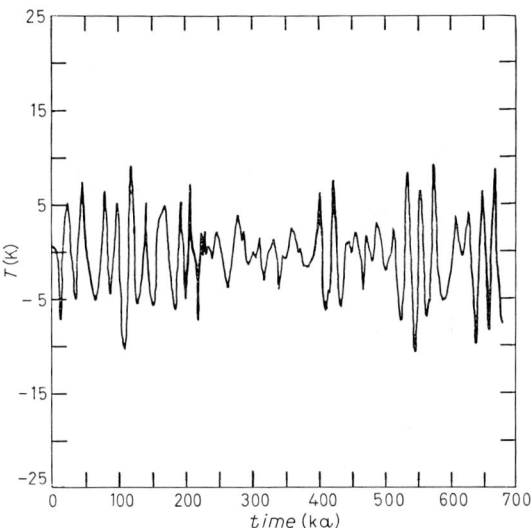

Fig. 18. – Difference between two solutions distinguished by the phase of the forcing: $T(\varphi_3 = 0) - T(\varphi_3 = \pi/2)$, $\sigma = 0.3$, $k = 0$, $\varepsilon_3 = 0.0025$ (courtesy of H. LeTreut).

The corresponding temperature spectra for the three solutions above (not shown) are practically indistinguishable from each other. All the peaks which are marked in fig. 14b) appear in the identical location and with like magnitude for the solutions in fig. 17a), b). Thus spectral information characterizes solution behavior, independently of phase information.

8'2. *Measures of predictability*. – Various measures of predictive skill can be used to make the difficulties illustrated above more precise [41]. The most natural one in the present context is the lagged correlation of solutions, $R(s)$:

$$(8.2) \qquad R(s) = \int_0^\tau T(t)\, T(t+s)\, \mathrm{d}t \bigg/ \int_0^\tau T^2(t)\, \mathrm{d}t\,.$$

In principle, $R(s)$ depends on the length of the averaging interval, τ, and on its initial point, $t = 0$. If τ is sufficiently long, and the solutions take on all possible values under the given conditions, both dependences are very weak.

For a constant solution, $R \equiv 1$. For a periodic or quasi-periodic solution, $R(s)$ is itself a periodic or quasi-periodic function. For random white noise, the predictive skill will be a delta-function concentrated at the origin, $R(s) = \delta(s)$. Hence $R(s)$ takes on the correct behavior in the two extreme cases, of perfect predictability and of complete lack of predictability. Its behavior in the intermediate case of quasi-periodicity is also suggestive.

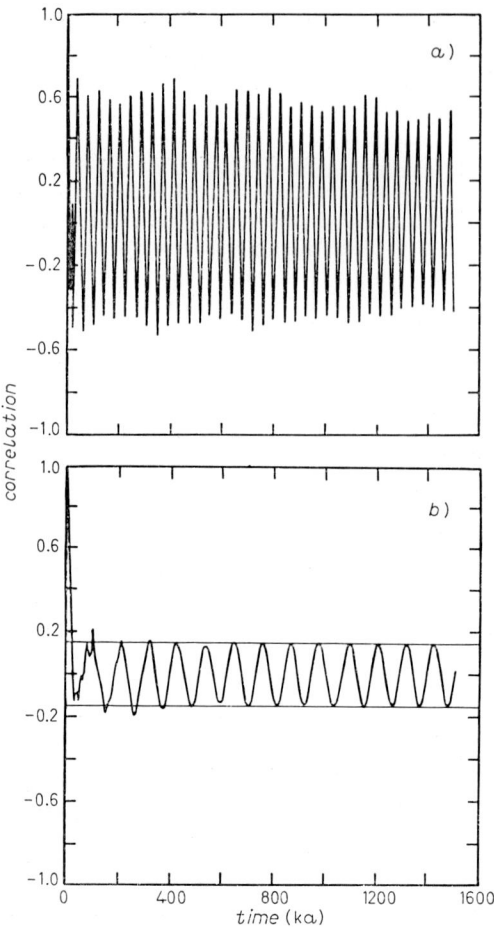

Fig. 19. – Lagged correlations for a solution a) with large quasi-periodic component, $\sigma = 0.3$, $k = 0$, $\varepsilon_3 = 0.0035$, and b) with large aperiodic component, $\sigma = 0.3$, $k = 0.03$, $\delta_4 = 0.001$ (courtesy of H. LeTreut).

Figure 19 shows the behavior of the lagged correlation function for solutions containing a quasi-periodic, as well as an aperiodic spectral component. We saw that for a spectral-density distribution which is perfectly flat, i.e. white noise, $R(s) = \delta(s)$. For a spectral density decreasing with frequency, like red noise, $R(s)$ will be an exponential function. Similar results would obtain if the initial conditions were random and the averaging in (8.2) were with respect to these initial data, rather than with respect to time [14, eqs. (10.29), (10.32); 37, appendix B].

In all solutions of (7.1) considered so far, there is a small component of continuous spectral density, due to numerical noise in the computations. This component leads to the initial, rapid decay of the correlation function in fig. 19a)

with lag time. Such a decay of predictability with time will be present in any numerical integration, and the more complicated the model, the stronger this purely numerical decay will be.

The decay in fig. 19b) is much larger. After the rapid, small decay caused by the numerical component of the spectral density, the exact model's true aperiodicity takes over. The correlation time of this portion of the continuous spectral density is longer, and hence the larger portion of the predictability decay occurs more slowly, over a time $O(10^2 \text{ka})$. Finally, the pure-line part of the spectrum leaves a periodic correlation function of amplitude $O(0.1)$ at large lag times [14, Chapt. 12; 41].

These results about the damped periodicity of correlations are again model-independent. They can be obtained from the power spectrum of the data themselves by the Wiener-Khintchine formula. Hence, detailed simulation of the Earth's paleoclimatic history is not possible.

On the other hand, much can be learned about climatic mechanisms, and about orbital changes, by studying paleoclimatic records in the spectral domain. The distribution of power between the continuous background and the peaks, as well as the position of the latter and their power relative to each other, give tell-tale indications about the climate's internal workings and the external changes which affect it.

9. – Concluding remarks.

In these lectures I have covered two areas of theoretical climate dynamics: the study of the Earth's radiation balance by energy-balance models and the study of Quaternary glaciation cycles by forced oscillator models. The purpose of the lectures was to give a relatively self-contained introduction to the field, rather than an exhaustive review or survey.

Energy-balance models (EBMs) have provided the first theoretical insights into the previously descriptive field of climatic change. Their literature, quite extensive by now, is thoroughly reviewed in [12]. A rigorous theory of climate model sensitivity is currently being developed [42].

Radiative-convective models (RCMs) use greater vertical detail in studying the Earth's energy budget than EBMs, with less horizontal details [43]. General circulation models (GCMs) combine radiative and fluid-dynamical processes, along with vertical and horizontal resolution, at the cost of great computational expense, difficulty of analysis and a closer adjustment to present-day conditions of all model processes [44].

EBMs, GCMs and RCMs are all used in attempting to assess the impact of natural and anthropogenic changes on the environment. Among these are the radiative effects of massive volcanic eruptions and of CO_2 increases [45, 46]. Each type of model has its own advantages, and confidence can be derived from the different models' answers to the extent that they agree [47, 48].

All models mentioned so far are quasi-equilibrium models: their stationary states, or the statistics of their aperiodic solutions, shift slowly in response to small changes of external parameters. Paleoclimatological proxy data of Quaternary glaciations suggest a temporal evolution of climate on the time scale of millenia with higher variability and more complexity than explained by quasi-equilibrium models. The harmonic analysis of the proxy records shows a line spectrum with multiple, relatively narrow peaks near the periods 100 ka, 40 ka, 20 ka and 10 ka. These are superimposed on a continuous spectral density which increases with decreasing frequency.

Oscillatory models with a natural frequency $O(10^{-1} \, \text{ka}^{-1})$ seem to explain much of the total variability of global temperature and ice volume on the Quaternary time scale. The self-sustained oscillations can result from various interactions of climatic components with relaxation times $O((1 \div 10) \, \text{ka})$.

One such interaction was emphasized here, between the ice-albedo feedback of EBMs and the precipitation-temperature feedback [26-28]. Equally important is the coupling between precipitation-temperature feedback and the load-accumulation feedback of ice-sheet models (ISMs) with bedrock deflection [32, 33, 49-51].

Orbital changes provide a small, quasi-periodic forcing in the absorbed energy and in the hydrologic cycle of the oscillatory model discussed here. This forcing results in a transfer of variability from the relatively high frequencies of the climatic mechanisms to the low frequencies documented by the records, $O((10^{-2} \div 10^{-1}) \, \text{ka}^{-1})$. It also synchronizes the variability at the observed peaks, overcoming in part the effect of random perturbations [52, 53]. Similar entrainment by periodic forcing of the system's natural frequency, as well as loss of entrainment, appears in a simple sea-ice/ocean model [54].

Modeling the internal variability of the climatic system on geologic time scales, and its interaction with external forcing, is enriching the tool kit of climate dynamics with an understanding of oscillatory models having periodic and aperiodic behavior. The application of this new understanding to shorter time scales, from weeks to months [55], is an important direction of development, which will bring theoretical climate dynamics in contact with dynamic meteorology and extended-range weather prediction [44].

* * *

I am grateful to all my associates over the years, E. BATTIFARANO, K. BHATTACHARYA, M. BUYS, C. CRAFOORD, E. KÄLLÉN, H. LeTREUT and J. TAVANTZIS. Our joint work has brought whatever understanding of climate dynamics is reflected in these pages. Conversations and correspondence with many charter members of our young discipline have also been invaluable. The most frequent contacts were with G. E. BIRCHFIELD, T. GAL-CHEN, C. NICOLIS, G. R. NORTH, B. SALTZMAN, S. H. SCHNEIDER and A. SUTERA. The writing of these notes was supported in part by the National Science Foundation under grant ATM-8214754, while the lectures themselves were supported by the Italian Physical Society. The typing of the manuscript was done by C. ENGLE at the Courant Institute, and figures were drafted by L. RUMBURG at NASA's Laboratory for Atmospheric Sciences, Goddard Space Flight Center.

REFERENCES

[1] M. GHIL: in *Climatic Variations and Variability: Facts and Theories*, edited by A. BERGER (Dordrecht, 1981), p. 461.
[2] J. LONDON and T. SASAMORI: in *Man's Impact on the Climate*, edited by W. H. MATTHEWS, W. W. KELLOG and G. D. ROBINSON (Cambridge, Mass., 1971), p. 141.
[3] T. H. VONDER HAAR and V. E. SUOMI: *J. Atmos. Sci.*, **28**, 305 (1971).
[4] M. I. BUDYKO: *Tellus*, **21**, 611 (1969).
[5] W. D. SELLERS: *J. Appl. Meteorol.*, **8**, 392 (1969).
[6] I. M. HELD and M. J. SUAREZ: *Tellus*, **26**, 613 (1974).
[7] G. R. NORTH: *J. Atmos. Sci.*, **32**, 2033 (1975).
[8] M. GHIL: *Steady-State Solutions of a Diffusive Energy-Balance Climate Model and Their Stability* (New York, N. Y., 1975).
[9] M. GHIL: *J. Atmos. Sci.*, **33**, 3 (1976).
[10] R. F. CAHALAN and G. R. NORTH: *J. Atmos. Sci.*, **36**, 1205 (1979).
[11] G. R. NORTH: *J. Atmos. Sci.*, **32**, 2033 (1975).
[12] G. R. NORTH, R. F. CAHALAN and J. A. COAKLEY jr.: *Rev. Geophys. Space Phys.*, **19**, 91 (1981).
[13] R. T. WETHERALD and S. MANABE: *J. Atmos. Sci.*, **32**, 2044 (1975).
[14] M. GHIL and W. S. CHILDRESS: *Topics in Geophysical Fluid Dynamics: Atmospheric Dynamics, Dynamo Theory and Climate Dynamics* (New York, N. Y., 1985), in press.
[15] R. BENZI and A. SUTERA: this volume, p. 403.
[16] J. M. MITCHELL, jr.: *Quat. Res.* (*N. Y.*), **6**, 481 (1976).
[17] J. IMBRIE and K. P. IMBRIE: *Ice Ages, Solving the Mystery* (Short Hills, N. J., 1979).
[18] T. J. CROWLEY: *Rev. Geophys. Space Phys.*, **21**, 828 (1983).
[19] J.-C. DUPLESSY: in *Climatic Change*, edited by J. GRIBBIN (Cambridge, 1978), p. 46.
[20] W. S. CHILDRESS: this volume, p. 200.
[21] N. J. SHACKLETON and N. D. OPDYKE: *Quat. Res.* (*N. Y.*), **3**, 39 (1973).
[22] W. S. BROECKER and J. VAN DONK: *Rev. Geophys. Space Phys.*, **8**, 169 (1970).
[23] G. J. KUKLA: *Geol. Mijnbouw*, **48**, 307 (1969).
[24] J. D. HAYS, J. IMBRIE and N. J. SHACKLETON: *Science*, **194**, 1121 (1976).
[25] E. KÄLLÉN, C. CRAFOORD and M. GHIL: *J. Atmos. Sci.*, **36**, 2292 (1979).
[26] M. GHIL and L. LETREUT: *J. Geophys. Res. C*, **86**, 5262 (1981).
[27] M. GHIL and J. TAVANTZIS: *SIAM J. Appl. Math.*, **43**, 1019 (1983).
[28] H. LETREUT and M. GHIL: *J. Geophys. Res. C*, **88**, 5167, 8623 (1983).
[29] J. WEERTMAN: *Nature* (*London*), **261**, 17 (1976).
[30] W. S. B. PATTERSON: in *Dynamics of Snow and Ice Masses*, edited by S. C. COLBECK (New York, N. Y., 1980), p. 1.
[31] G. SIMPSON: *Nature* (*London*), **141**, 591 (1938).
[32] R. PELTIER: *Adv. Geophys.*, **24**, 1 (1983).

[33] G. E. Birchfield, J. Weertman and A. T. Lunde: *Quat. Res. (N. Y.)*, **15**, 126 (1981).
[34] M. Buys and M. Ghil: in *Milankovitch and Climate: Understanding the Response to Orbital Forcing*, edited by A. Berger, J. Hays, J. Imbrie, G. Kukla and B. Saltzman (Dordrecht, 1984), p. 55.
[35] R. D. Vicente: in *Long-Time Prediction in Dynamics*, edited by C. W. Horton jr., L. E. Reichl and V. G. Szebehely (New York, N. Y., 1983), p. 235.
[36] A. Berger: *J. Atmos. Sci.*, **35**, 2362 (1978).
[37] K. Bhattacharya, M. Ghil and I. L. Vulis: *J. Atmos. Sci.*, **39**, 1747 (1982).
[38] A. Libchaber: this volume, p. 17.
[39] D. Ruelle: this volume, p. 3.
[40] J. Jouzel and L. Merlivat: *J. Geophys. Res. D*, **89** (1984), in press.
[41] E. N. Lorenz: in *Proceedings of the III Conference on Probability and Statistics in Meteorology* (Boston, Mass., 1973), p. 1.
[42] T. L. Bell: this volume, p. 424.
[43] V. Ramanathan and J. A. Coakley: *Rev. Geophys. Space Phys.*, **16**, 465 (1978).
[44] E. Kalnay and R. Livezey: this volume, p. 311.
[45] R. E. Dickinson: in *Carbon Dioxide Review 1982*, edited by W. C. Clark (Oxford, 1982), p. 103.
[46] M. E. Schlesinger: *Int. J. Environ. Stud.*, **20**, 103 (1983).
[47] S. H. Schneider and R. E. Dickinson: *Rev. Geophys. Space Phys.*, **12**, 447 (1974).
[48] U. S. Committee for the Global Atmospheric Research Program: *Understanding Climatic Change, A Program for Action* (Washington, D.C., 1975).
[49] V. Ya. Sergin: *J. Geophys. Res. C*, **84**, 3191 (1979).
[50] J. Oerlemans: *Climatic Change*, **4**, 353 (1982).
[51] D. Pollard: *J. Geophys. Res. C*, **88**, 7705 (1983).
[52] B. Saltzman, A. Sutera and A. Evenson: *J. Atmos. Sci.*, **38**, 494 (1981).
[53] C. Nicolis: in *Milankovitch and Climate: Understanding the Response to Orbital Forcing*, edited by A. Berger, J. Hays, J. Imbrie, G. Kukla and B. Saltzman (Dordrecht, 1984), p. 637.
[54] P. Lundberg and L. Rahm: *Dyn. Atmos. Ocean*, **8**, 59 (1984).
[55] B. Legras and M. Ghil: *J. Méc. Théor. Appl.*, special issue (1983), p. 45; in *Predictability of Fluid Motions*, edited by G. Holloway and B. J. West (New York, N. Y., 1984), p. 87.

The Mechanism of Stochastic Resonance in Climate Theory.

R. BENZI

Centro Scientifico IBM - Roma, Italia

A. SUTERA

The Centre for the Environment and Man, Inc.
275 Windsor Street - Hartford, CT 06120

1. – Introduction.

The Earth's climate is determined by complex interactions of the physical processes governing the bio-cryo-ocean-atmosphere system and the upper-mantle dynamics. The latter are mainly fixed by the amount of emerged land and its geographical position which in the last 10^6 years have been practically unchanged. It follows that the climate changes are fully described by the dynamics and thermodynamics governing the bio-cryo-ocean-atmosphere. Recent developments in acquiring and interpreting climate records indicate that a dominant feature of the late Quaternary is characterized by seven major glaciations that occurred with an apparent periodicity of about 10^5 y (fig. 1a), b)). Interest in this problem has generated a number of possible explanations. Among the theories we might distinguish two groups. One group invokes factors external to the climate system as the cause of the glaciations. Other theories are based upon internal elements believed to have response times on the order of the observed main periodicity. However, in our opinion, a definitive answer to these questions is not available yet and, although any of these theories has its own appeal at the present, we like to think of them as being only indicative of the real physical process governing the system.

The absence of a definitive answer to the cause of the climate's changes resides in the difficulty of extracting from the laws of the dynamics and thermodynamics the slow components of the system that ultimately govern the climate's evolution. On the other hand, at the present an *ab initio* derivation of such slow physics appears impractical because of the complexity of the system and the long time scales. In this lecture, we will attempt to describe the glacial cycles following a strategy that, in the last decade or so, has become popular in climate theory. This approach consists of identifying, among the many compo-

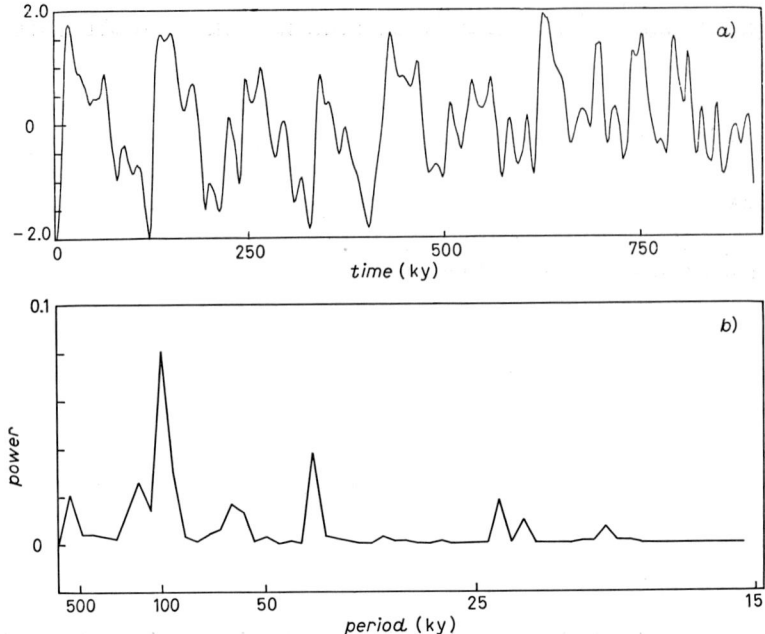

Fig. 1. – a) $\delta(^{18}O/^{16}O)$ vs. time from J. D. Hays, J. Imbrie and N. J. Shackleton: Science, **194**, 1121 (1976); b) power spectrum of the signal.

nents of the system, a few climate variables thought, on intuitive grounds, to be the carriers of the slow physics and parametrize the physical processes in the system in terms of these variables. Pioneering work along this line was attempted by Budyko [1] and independently by Sellers [2]. In their work, see [3] for more details, the climate variables were reduced to one, namely a vertically integrated temperature T. In this case, the other variables are diagnostically related and derivable from T. Models considering only T as a climate variable are known as Budyko-Sellers models or alternatively energy budget models, and in this lecture we refer to them as such.

In what follows we will describe an almost trivial Budyko-Sellers model, and we will show that in the framework of our model two climatic features may be explained: a) the apparent 10^5 y periodicity of the climate records and b) a variance of T consistent with climate observations, namely a few degrees. Part of the material found here is based upon already published work of a few authors, viz. [4-8].

2. – Formulation of the model.

The thermal regime of the Earth is determined by the balance between the amount of incoming and outgoing radiation. A globally averaged (per unit of area) climate system can be expressed using the law of energy conservation (see [3])

$$c \frac{\mathrm{d}}{\mathrm{d}t} T = R_{\text{in}} - R_{\text{out}}, \tag{2.1}$$

where R_{in} and R_{out} are the incoming and the outgoing radiation at the top of the atmosphere. If the climate variable T is interpreted as a vertically averaged temperature, its value is representative of an average surface temperature of an ocean on a spherical planet subject to radiative heating. In this case, on the left-hand side of eq. (2.1), c can be interpreted as a constant thermal inertia. If a mixed reservoir of fixed depth d and area coverage a is considered:

$$c = \varrho_w c_w d a, \tag{2.2}$$

where c_w and ϱ_w are, respectively, the specific heat and density of the ocean layer.

To close eq. (2.1), we must express R_{in} and R_{out} as functions of T. For simplicity, we assume that the Earth emits radiation as a slightly modified black-body radiator. Hence, R_{out} can be expressed by the Stephan-Boltzmann law. Although infra-red radiation, for the Earth's system, is better correlated with a mass-averaged temperature of the atmosphere rather than with a surface temperature, we assume that

$$R_{\text{out}} = \varepsilon \sigma T^4, \tag{2.3}$$

where σ is the Stephan-Boltzmann constant and ε is the Earth's emissivity. ε can be expressed as a function of T and, following SELLERS [2], we assume

$$\varepsilon = 1 - m \operatorname{tgh} c_1 T^6, \tag{2.4}$$

where m and c_1 are empirical constants.

The reduced radiation expressed by eq. (2.3) accounts for the Earth's atmosphere's chemical composition which absorbs in the infra-red band, trapping in the system part of the radiation that otherwise would escape into space. Hence, eq. (2.4) models the greenhouse effect.

Most of the available energy in the Earth is provided by the Sun. If we denote by I_0 the amount of radiation reaching the top of the atmosphere, the incoming radiation can be expressed as

$$R_{\text{in}} = \frac{I_0}{4}(1-\alpha) = Q(1-\alpha), \tag{2.5}$$

where α is the Earth's reflectivity or albedo and the factor $\frac{1}{4}$ derives from the global average.

A relevant difficulty with the derivation of a zero-dimensional energy balance model resides in the parametrization of α in terms of T. In the work of Sellers [2], α is considered a function of a latitudinal dependent temperature

$$(2.6) \qquad \alpha_{\rm S} = \begin{cases} \alpha_1, & \text{if } T < T_{\rm s}, \\ \alpha_2, & \text{if } T > T_{\rm s}. \end{cases}$$

In [1], α is only a function of the latitude

$$(2.7) \qquad \alpha_{\rm B} = \alpha(x, x_{\rm s}),$$

where $x_{\rm s}$ is the sine of the latitude where ice is present. Both eqs. (2.6) and (2.7) are empirically derived. The scarce knowledge of the dependence of the planetary albedo on the surface temperature has rendered it very difficult to translate eqs. (2.6) or (2.7) for a globally averaged temperature. In the present work we proceed as follows. GHIL [9], studying a one-dimensional model where (2.6) was used, found three stationary solutions corresponding to the following globally averaged temperatures:

$$(2.8) \qquad T_{\rm ic} = 175.43 \text{ K}, \qquad T_{\rm in} = 267.44 \text{ K}, \qquad T_{\rm p} = 287.76 \text{ K}.$$

The very cold temperature $T_{\rm ic}$ represents a state in which the ice line extends nearly to the equator. CRAFOORD and KALLEN [10] showed that, considering a zero-dimensional model and an albedo expressed as

$$(2.9) \qquad \alpha_{\rm KC} = \begin{cases} \text{const}, & \text{if } T \leqslant T_i, \\ a + bT, & \text{if } T_i \leqslant T \leqslant T_n, \\ \text{const}, & \text{if } T \geqslant T_n, \end{cases}$$

they obtained the same equilibria as in [9]. BHATTACHARYA, GHIL and VULIS [11], hereafter BGV, employing a different parametrization of the albedo, showed that five equilibria were obtained at global temperatures

$$(2.10) \qquad \begin{cases} T_{\rm ic} = 175.17 \text{ K}, \quad T_{\rm in} = 241.96 \text{ K}, \quad T_{\rm g} = 272.94 \text{ K}, \\ T_{\rm u} = 281.67 \text{ K}, \quad T_{\rm p} = 286.61 \text{ K}. \end{cases}$$

The most natural choice, in replacing a discontinuous function as eq. (2.9) with a continuous one, is the hyperbolic tangent (as in [12]). In our zero-dimensional model we use the same functional dependence. We choose the following function

$$\alpha(T) = \frac{d_1}{2} - d_2 \tgh d_3(T - T_2) - d_4 \tgh d_5(T - T_4),$$

where the constants d_i, $i = 1, ..., 5$, T_2 and T_4 will be determined in such a way that our model has approximately the same equilibria as in BGV.

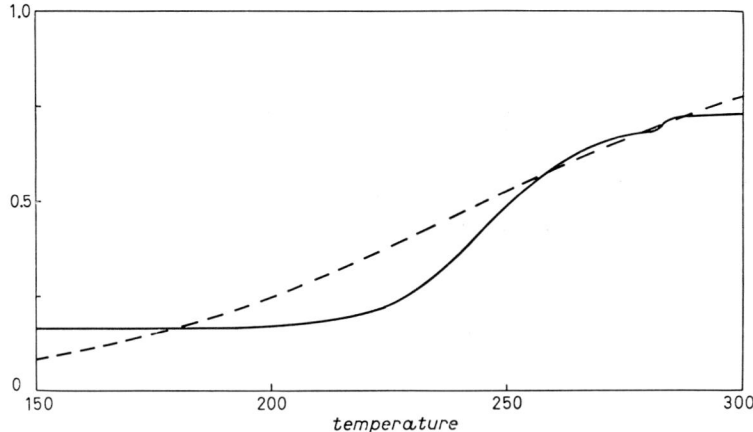

Fig. 2. – R_{in} (solid line), R_{out}/Q (dashed line). The values of the parameter are in table I.

In fig. 2 we have plotted R_{in} and R_{out}/Q vs. T for selected values of the parameters (see table I). Before considering the mathematical properties of our model, let us briefly discuss the physical causes that have produced the steady states T_g and T_u. In BGV, in parametrizing the albedo in middle latitude, the effect of the intensified cloud coverage on it was taken into account. This produced a « kink » in the albedo around $T = 280$ K. Whether this effect is powerful enough to produce global equlibria such as T_g is an open question, though, recently, some favourable observational evidence has emerged (GHIL:

TABLE I. – *Values of constants.*

$a =$	1	$d_4 =$	0.54
$c_w =$	$4.186 \cdot 10^3$ J kg^{-1} K^{-1}	$d_5 =$	$0.03 \times \pi/2$
$c_1 =$	$1.415 \cdot 10^{-15}$	$m =$	0.54
$d =$	70 m	$T_2 =$	282.15 K
$d_1 =$	0.55	$T_4 =$	245 K
$d_2 =$	0.03/2	$\sigma =$	$5.67 \cdot 10^{-8}$ W m^{-2} K^{-4}
$d_3 =$	$0.5 \times \pi/2$	$\varrho_w =$	10^3 kg m^{-3}

private communication; see also [13]). In what follows, the stationary solution T_g plays a central role. In fact, T_g will be considered to be the climatic state of the Earth in its glacial phase. It is somewhat disturbing to develop a theory based upon an unsubstantiated characteristic of the system; however, as we shall see, the existence of T_g is a necessity if eq. (2.1) is to describe climate changes. If the existence of this steady state is not deducible from first principles or supported by observational evidence, it should be assumed, and this would be the predictive element of our theory.

3. – Some mathematical properties of Budyko-Sellers models.

At this stage, it is useful to review some mathematical properties of eq. (2.1). It is easily seen that eq. (2.1) can be expressed as

$$(3.1) \qquad \frac{d}{dt}T = -\frac{d}{dT}V,$$

where

$$(3.2) \qquad V = -\int \frac{1}{c}\{R_{in} - R_{out}\}\,dT.$$

It follows that V is the Liapunov function for our system. Thus any steady state *will be* asymptotically stable (unstable) if

$$(3.3) \qquad \left.\frac{d^2}{dT^2}V\right|_{T_i} < 0 \quad (>0).$$

This implies that, unless other sources of instability are hypothesized for any initial condition, the solution of eq. (3.1) will approach one of the stable (unstable) steady states for $t \to +\infty$ ($t \to -\infty$). While property (3.1) is trivially deduced for our zero-dimensional model, it is interesting to notice that it is preserved also in one-dimensional models. In fact, as we will review, NORTH et al. [14], extending a result of Ghil [9], demonstrated that, if c is a constant, one-dimensional Budyko-Sellers models may be written as

$$(3.4) \qquad c\frac{\partial T}{\partial t} = -\frac{\delta}{\delta T}\mathscr{V}(T, T_x).$$

Here, $\mathscr{V}$ is a functional rather than a function and $\delta/\delta T$ denotes a variational derivative.

To demonstrate eq. (3.4), we write the general form of Budyko-Sellers models

$$(3.5) \qquad c\frac{\partial}{\partial t}T = R_{\text{in}} - R_{\text{out}} + \frac{\partial}{\partial x}F$$

(where x denotes the sine of the latitude). F (the flux of energy) is generally parametrized as

$$(3.6) \qquad F = K(x)\frac{\partial}{\partial x}T\,.$$

Now eq. (3.5) can be derived as the first variation of the functional

$$(3.7) \qquad \mathscr{V} = \int \mathrm{d}x\, [\tfrac{1}{2}K(x)(T_x)^2 + A(T) - B(T)]\,,$$

where

$$(3.8) \qquad A(T) = -\int R_{\text{in}}\,\mathrm{d}T\,,$$

$$(3.9) \qquad B(T) = \int R_{\text{out}}\,\mathrm{d}T\,.$$

In fact,

$$(3.10a) \qquad \delta V = \frac{\delta V}{\delta T}\delta T + \frac{\delta V}{\delta T_x}\delta T_x\,.$$

Calculating the variations, we find

$$(3.10b) \qquad \delta V = \int \mathrm{d}x\left[-\frac{\partial}{\partial x}\left(K(x)\frac{\partial}{\partial x}T\right) - R_{\text{in}} + R_{\text{out}}\right]\delta T\,,$$

where we have used $\delta(T_x) = (\delta T)_x$, and we have integrated once by parts the following terms

$$\int \mathrm{d}x\,[K(x)\,T_x]\,\delta T_x\,.$$

Equation (3.10b) is zero for any arbitrary variation if the integrand is zero and eq. (3.5) follows. Now, as in the zero-dimensional case, the functional

$$(3.11) \qquad \mathscr{V} = \frac{1}{c}\int \mathrm{d}x\left[\frac{1}{2}K(x)(T_x)^2 + A(T) - B(T)\right]$$

is the Liapurow function of eq. (3.5) and hence the previous comments about stationary solutions hold.

4. – Stochastic perturbations.

In principle, eq. (2.1) describes the instantaneous evolution of the climate variable T. However, a characteristic feature of climate records is their pronounced variability. As illustrated in fig. 3, the spectral density of these records

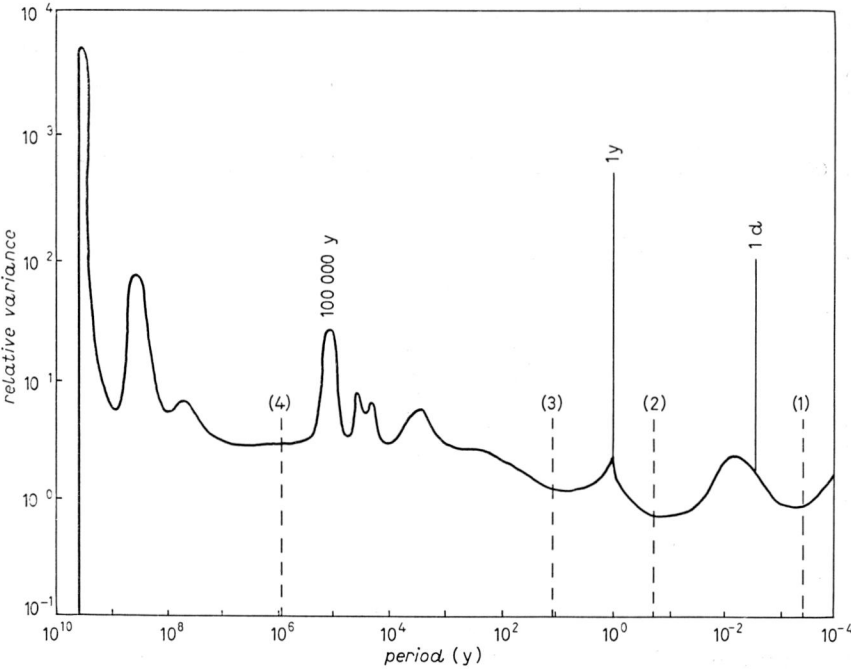

Fig. 3. – Schematic power spectrum of climatic variability from B. SALTMAN: *Adv. Geophys.*, **25**, 173 (1983).

encompasses all the frequencies from a few hours up to 10^6 years and more. In modelling such phenomena, a useful distinction consists of separating high-frequency processes from those evolving at slower rates. On the other hand, the r.h.s. of eq. (2.1) is better related to a time-averaged temperature rather than to an instantaneous value. This implies that in deriving eq. (2.1) we have conceptually averaged the fundamental law over time intervals long enough that very fast phenomena have been averaged out. However, these high frequencies may modify the system, so we should parametrize their effect.

HASSELMAN [15] introduced the idea that a proper representation of the effect of high-frequency variabilities is a stochastic process and, further, he assumed that this process has a white spectrum. If we take into account Hasselman's approach, eq. (2.1) becomes

(4.1)
$$c \frac{\mathrm{d}}{\mathrm{d}t} T = R_{\mathrm{in}} - R_{\mathrm{out}} + f(t),$$

where

(4.2) $$Ef(t) = 0,$$

(4.3) $$Ef^2(t) = \varepsilon_n \delta(t),$$

where E is the mathematical expectation operator and ε_n is a constant measuring the noise level. The idealization of eqs. (4.1)-(4.3) recasts the climate problem into the realm of the theory of diffusion processes. As we shall see, this new outlook on the climate system has very interesting consequences.

Before studying the effect of $f(t)$ on our climate equation, we shall review a few facts about stochastic equations (see [16] for more detail). First we assume that $f(t)$ is the differential of the Wiener process W,

(4.4) $$E[dW] = 0,$$

(4.5) $$E[dW]^2 = \varepsilon_n dt.$$

In this case, the correct form of the Langevin equation is

(4.6) $$dT = \frac{1}{c}\{R_{in} - R_{out}\} dt + \varepsilon_n^{\frac{1}{2}} dW.$$

By generalizing the Liouville theorem on the conservation of volume, we can derive the equation for the joint-probability density of T, $P(T, t)$. It is a fundamental solution of

(4.7) $$\frac{\partial}{\partial t} P = \frac{\varepsilon}{2} \frac{\partial^2}{\partial T^2} P - \frac{\partial}{\partial T}\left\{\frac{1}{c}(R_{in} - R_{out})\right\} P$$

with natural boundary conditions.

Let us introduce the random variable

(4.8) $$\tau(T, T(0)) = \{\inf t | T(0) \in D, T(\tau) \notin D\},$$

where D is any domain on which T is defined. If $R_{in} - R_{out}$ does not depend explicitly on t, denoting M_n the n-th moment of $P(\tau)$, we find that M_n is a solution of the Dirichlet problem

(4.9) $$\frac{\varepsilon}{2} \frac{d^2}{dT^2} M_n + \frac{1}{c}\{R_{in} - R_{out}\} \frac{d}{dT} M_n = -nM_{n-1}, \qquad M_n = 0 \text{ on } \partial D$$

where $M_0 = 1$ and ∂D denotes the boundary of D. In the case in which the drift of eq. (4.6) depends on t also, then M is a fundamental solution of

$$(4.10) \qquad -nM_{n-1} + \frac{\partial}{\partial t} M_n = \frac{\varepsilon}{2} \frac{d^2}{dT^2} M_n + \frac{1}{c}\{R_{in} - R_{out}\} \frac{d}{dT} M_n .$$

Equations (4.9) and (4.10) are consequences of the Ito formula (see [16]). This is enough to keep the text self-contained and we shall assume that the reader is already familiar with the notation and derivation of the equations stated in this section.

5. – Effects of stochastic perturbations.

In this section we study the effects of stochastic perturbations on our climate model:

$$(5.1) \qquad dT = \frac{1}{c} \left\{ Q \left[\frac{d_1}{2} - d_2 \operatorname{tgh} d_3(T - T_2) - d_4 \operatorname{tgh} d_5(T - T_4) \right] - \right.$$
$$\left. - (1 - m \operatorname{tgh} c_1 T^6) \sigma T^4 \right\} dt + \varepsilon_n^{\frac{1}{2}} dW .$$

Let us study the behaviour of eq. (5.1) around one of its stable equilibria, say T_p. Around T_p and for short times, eq. (5.1) can be written as

$$(5.2) \qquad dT = \frac{1}{\lambda_p}(T - T_p) dt + \varepsilon_n^{\frac{1}{2}} dW ,$$

where

$$(5.3) \qquad \lambda_p^{-1} = \frac{1}{c} \frac{d}{dT} F(T) \bigg|_{T=T_p} .$$

For the parameter values selected in the previous section, $\lambda_p \simeq \frac{1}{8}$ y (see table II). We notice that eq. (5.2) is an Ornstein-Uhlenbeck process (see [16]) and we know that

$$(5.4) \qquad \begin{cases} E(T - T_p) = 0 , \\ E(T - T_p)^2 = \frac{1}{2} \varepsilon_n \lambda_p . \end{cases}$$

From observations $E(T - T_p)^2$ can be estimated [17] on the order of

$$(5.5) \qquad E(T - T_p)^2 \simeq 0.4 K^2 .$$

Combining eqs. (5.4) and (5.5), we get

$$(5.6) \qquad \varepsilon_n \simeq 0.1 K^2 \, \text{y}^{-1} .$$

For longer times, eq. (5.2) does not hold, since the noise introduces a new time scale, namely the exit time. For our model, we can estimate the exit times τ_{ic}, τ_g^-, τ_g^+ and τ_p, where

τ_{ic} = the exit time from the attraction domain of T_{ic},

τ_g^- = the exit time from the attraction domain of T_g to that of T_{ic},

τ_g^+ = the exit time from the attraction domain of T_g to that of T_{ic} and finally

τ_p = the exit time from the attraction domain of T_p.

TABLE II. – Equilibria, decay times, barrier heights and exit times for $Q = Q_0 = 340$ W m^{-2}, $Q^+ = (1 + 0.0015)Q_0$ and $Q^- = (1 - 0.0015)Q_0$.

Equilibria (K)	T_{ic}	T_{in}	T_g	T_u	T_p
Q_0	178.7	257.3	276.9	282.1	286.8
Q^-	178.6	257.5	276.4	282.2	286.5
Q^+	178.8	251.1	277.3	282.0	287.1

Decay time (y)	λ_{ic}	λ_{in}	λ_g	λ_u	λ_p
Q_0	-6.7	6.6	-9.5	2.7	-7.0
Q^-	-6.8	6.9	-9.7	2.7	-7.0
Q^+	-6.7	6.4	-9.3	2.7	-6.9

Barrier height (K^2/y)	ΔV_1	ΔV_2	ΔV_3	ΔV_4
Q_0	287.0	8.3	1.0	1.0
Q^-	288.4	7.5	1.3	0.8
Q^+	285.7	9.1	0.8	1.2

Exit time (y)	$E\tau_{ic}$	$E\tau_g^-$	$E\tau_g^+$	$E\tau_p$
Q_0	10^{30}	10^{30}	94 000	80 000
Q^-	10^{30}	10^{30}	18 595 000	6 000
Q^+	10^{30}	10^{30}	6 000	13 626 000

The general formula for the exit time for a system of gradient type (as ours) is Kramer's formula [18]

$$(5.7) \qquad E\tau(x_s, x_{in}) \leqslant \frac{\pi}{|V''(x_s) V''(x_{in})|^{\frac{1}{2}}} \exp\left[\frac{2\Delta V}{\varepsilon_n}\right],$$

where

(5.8) $$V = -\int F(x)\,dx, \qquad \Delta V = \left| \int_{x_{st}}^{x_{in}} F(x)\,dx \right|$$

and x_s (x_{in}) denote the stable (unstable) steady solutions. Our case is slightly more complicated since we have three stationary stable solutions T_{ic}, T_g and T_p. In particular, τ_g^-, as calculated through eq. (5.7), will be a rough lower bound. In fact, we will calculate it as if T_p were absent. It would not be difficult to derive the proper formula, but in our case, as we shall see, such rigour would not be needed.

By using eq. (5.7) we obtain the results in table II. In fig. 4 we have also plotted the potential function for eq. (5.1). As we see, τ_{ic} and τ_g^- are practically

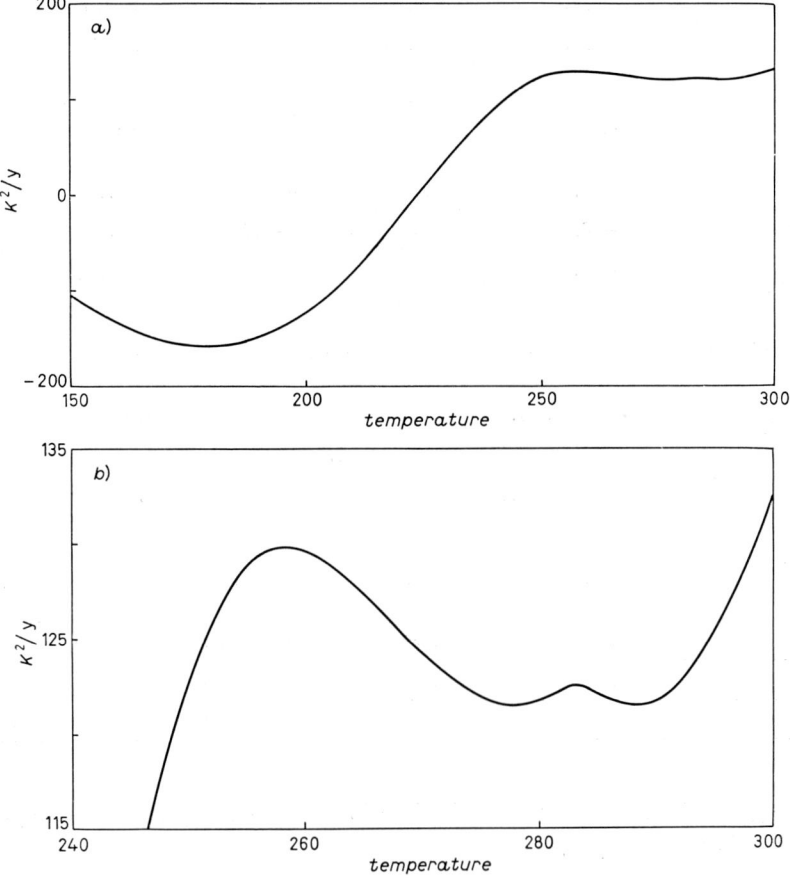

Fig. 4. – a) Potential for eq. (5.1), b) potential for eq. (5.1) in a temperature range $(240 \div 300)$ K.

infinite, so that we must not worry about the fact that the calculation actually concerns a lower bound. We notice also that τ_g^+ and τ_p are both around 10^5 y, i.e. the same time scale on which climate changes occur. However, as we shall see next, the power spectrum of our model would not show any apparent periodicities, in contrast to the observational evidence. Let us consider the power spectrum of a process defined by the stochastic differential equation

$$(5.9) \qquad \mathrm{d}x = -\frac{\mathrm{d}}{\mathrm{d}x} V \, \mathrm{d}t + \varepsilon_n^{\frac{1}{2}} \mathrm{d}W \, ,$$

where V is a double-well potential. The power spectrum is defined as the Fourier transform of the correlation function

$$\mathscr{C}(\delta) = \frac{\langle x(\delta) x(0) \rangle}{\langle x \rangle^2} \, .$$

It is easy to show that

$$\mathscr{C}(\delta) = \exp\left[-\delta/\tau\right] + O(\varepsilon_n) \, ,$$

where τ is the first exit time. If $\mathscr{C}$ is a periodic function, we will observe a spectral line in the power spectrum of x. However, for eq. (5.9) it can be demonstrated that τ is Poisson distributed. It follows that $\mathscr{C}$ is not a periodic function. To calculate the probability distribution of τ we can use the moments M_n. We will demonstrate in the appendix that

$$M_n \simeq n! \, (M_1(x))^n$$

from which our assertion follows.

6. – The orbital forcing.

In the previous sections, we have assumed that the geometry of the Earth's orbit is constant. GHIL [3] has described its changes and the effects of them on the radiation balance of the Earth. Since our model is zero-dimensional, we shall consider only that orbital parameter that directly influences the annual radiation budget, namely the eccentricity $r(t)$ (or mean Earth-Sun distance). Such a parameter has a secular variation with periodicities in the range of (90 000, 125 000) years all with approximately the same amplitude. In what follows, we shall assume that $r(t)$ is a pure periodic function with period 100 000 y and amplitude $3 \cdot 10^{-3} Q$. It is customary, in climate studies, to consider the sensitivity of the present climate to orbital changes. For our model, the periodic forcing, assuming that no stochastic forcing is acting on the system, only changes the equilibria a few tenths of a degree (see table II), in agreement with other calculations (see [14]). However, as we already mentioned, stochastic perturbations can drive any initial condition outside the attraction domain of stable solutions, while the periodic forcing may favour exits at this frequency. To illustrate qualitatively this mechanism, let us choose the initial condition at $T = T_p$. We know that eventually with an exponentially distributed probability the solution will jump to the neighbourhood of T_g. Since the mean exit time depends exponentially on the potential barrier height, it is important to calculate its behaviour. Setting

$$|V(T_{ic}) - V(T_{in})| = \Delta V_1,$$

$$|V(T_{in}) - V(T_g)| = \Delta V_2,$$

$$|V(T_g) - V(T_u)| = \Delta V_3,$$

$$|V(T_u) - V(T_p)| = \Delta V_4,$$

we have calculated

$$\Delta V_i \qquad \text{for } Q^{\mp} = Q_0(1 \mp 3\cdot 10^{-3}).$$

The results are summarized in table II. Inspection of this table shows that ΔV_4 has changes of about 50%. So, if we start with initial conditions, say at T_p, after 50 000 y the expected exit time has decreased to about a few thousand years (see fig. 5). It is conceivable that in this position the solution might jump easily across the barrier. If this happens, the solution will fluctuate around T_g and any escape will be improbable because the barrier height is very large. After 50 000 y, however, the barrier height has decreased again

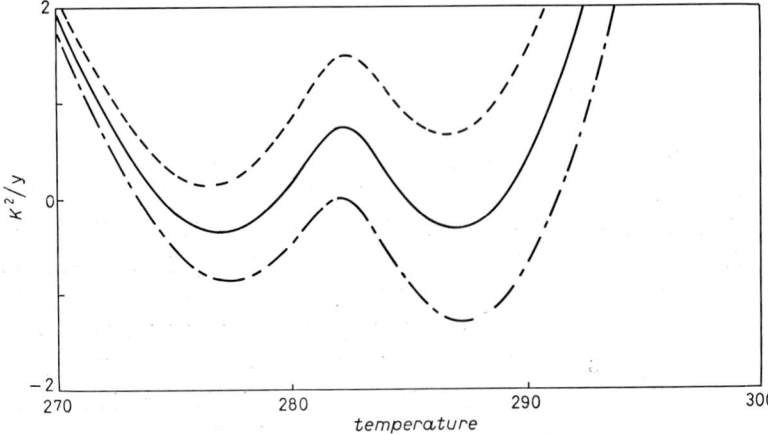

Fig. 5. – Potentials in a temperature range (270 ÷ 300) K, solid line $Q = Q_0$, dashed line $Q = Q_1$, $Q = Q_2$, values of Q_0, Q_1 and Q_2 as in table II.

and the solution will be ready to jump back towards T_p. We have portrayed a picture in which almost periodic solutions are possible. If a mechanism such as this operates, then our model would show a power spectrum with a marked peak at 100 000 y, with a variance of order $25 K^2$. We call this phenomenon stochastic resonance. Finally, we remark that, for a noise amplitude of the magnitude considered here, the barrier and the exit time from T_g to T_{1c} are not significantly decreased. In this case the expected exit time is still very long compared to the time scales of interest. Henceforth, we will neglect the existence of T_{ic} and T_{in} and we will concentrate on studying eq. (5.1) around T_g, T_u and T_p. In fig. 5 we have plotted the potential in this temperature range.

7. – The mechanisms of stochastic resonance.

As we have seen, eq. (5.1) has three stationary solutions in a temperature range of about 10 K around the present average temperature. To study the behaviour of our model in a relatively painless way, we will approximate eq. (5.1) in this range. Let us start by considering

$$B = \frac{1}{T_p - T_g} \int_{T_g}^{T_p} R_{out} \, dT .$$

In this case, eq. (5.1) can be written as

(7.1) $$c \frac{d}{dt} T = Q(1 - \alpha(T)) - B .$$

At the equilibria

(7.2) $$\frac{Q(1 - \alpha(T))}{B} = 1 .$$

Away from the equilibria we assume that

(7.3) $$F(\tau) = \frac{Q(1 - \alpha(T))}{B} = 1 + \beta \left(1 - \frac{T}{T_g}\right)\left(1 - \frac{T}{T_u}\right)\left(1 - \frac{T}{T_p}\right).$$

Now, assuming that

$$Q = Q_0(1 + A \cos \omega t) ,$$

where $2\pi/\omega = 10^5$ y, and using eq. (7.3), we have

(7.4) $$\frac{d}{dt} T =$$
$$= \frac{B}{c} \left\{ (1 + A \cos \omega t) \beta \left(1 - \frac{T}{T_g}\right)\left(1 - \frac{T}{T_u}\right)\left(1 - \frac{T}{T_p}\right) + A \cos \omega t \right\} = g(T) .$$

If we add stochastic perturbations, we obtain

(7.5)
$$dT = g(T)\,dt + \varepsilon_n^{\frac{1}{2}}\,dW.$$

Now β can be calculated by considering

$$\left.\frac{d}{dT}g(T)\right|_{T=T_p} = \frac{1}{\lambda_p} < 0$$

and using λ_p from table I. Setting

$$T = T_u + \Delta T\xi \qquad (\Delta T = T_p - T_u = T_u - T_g)$$

and expanding eq. (7.5) in the Taylor series of $\Delta T/T_u$, we finally obtain

(7.6)
$$d\xi = \left\{\frac{1}{2\lambda_p}\xi(\xi^2 - 1) + \frac{AB}{\Delta Tc}\cos\omega t\right\}dt + \frac{\varepsilon_n^{\frac{1}{2}}}{\Delta T}\,dW.$$

To understand the effect of the periodic forcing, we study eq. (7.6) at $t = 0$ and $t = \pi/\omega$. We obtain the two equations

(7.7a)
$$dx' = \left\{\frac{1}{2\lambda_p}x'(x'^2 - 1) + \frac{AB}{\Delta Tc}\right\}dt + \frac{\varepsilon_n^{\frac{1}{2}}}{\Delta T}\,dW,$$

(7.7b)
$$dx'' = \left\{\frac{1}{2\lambda_p}x''(x''^2 - 1) - \frac{AB}{\Delta Tc}\right\}dt + \frac{\varepsilon_n^{\frac{1}{2}}}{\Delta T}\,dW.$$

We denote by x'_i ($i = 1, 2, 3$) and x''_i ($i = 1, 2, 3$) the stationary solutions of eqs. (7.7a) and (7.7b), respectively ($x'_1 < x'_2 < x'_3$ and $x''_1 < x''_2 < x''_3$). The mean exit times can be calculated by using eq. (5.7)

(7.8a)
$$\begin{cases} E\tau(x'_1, x'_3) = \frac{\pi|\lambda_p|}{\sqrt{2}}\exp\left[\frac{\Delta T^2}{4\varepsilon_n|\lambda_p|}\right]\left(1 + \frac{8AB|\lambda_p|}{\Delta Tc}\right), \\ E\tau(x''_1, x''_3) = \frac{\pi|\lambda_p|}{\sqrt{2}}\exp\left[\frac{\Delta T^2}{4\varepsilon_n|\lambda_p|}\right]\left(1 - \frac{8AB|\lambda_p|}{\Delta Tc}\right). \end{cases}$$

It follows

$$E\tau(x'_1, x'_3) \ll E\tau(x''_1, x''_3).$$

Now if

$$E\tau(x'_1, x'_3) \gg \frac{\pi}{\omega} \quad \text{and} \quad E\tau(x''_1, x''_3) \ll \frac{\pi}{\omega},$$

then we expect that the mean exit time for eq. (7.6), $E\tau$, is bounded by $E\tau(x_1', x_3')$ and $E\tau(x_1'', x_3'')$ and

$$E\tau \simeq \frac{\pi}{\omega}.$$

The inequalities are satisfied if

$$\varepsilon_{n1} \simeq \Delta T^2 \frac{1 - 8AB|\lambda_p|/\Delta Tc}{4|\lambda_p|\ln\left(\sqrt{2}/\omega|\lambda_p|\right)},$$

$$\varepsilon_{n2} \simeq \Delta T^2 \frac{1 + 8AB|\lambda_p|/\Delta Tc}{4|\lambda_p|\ln\left(\sqrt{2}/\omega|\lambda_p|\right)}.$$

For the numerical values here used

$$\varepsilon_{n1} \simeq 0.06, \quad \varepsilon_{n2} \simeq 0.14.$$

As we have seen, our estimates are based on heuristic arguments. In order to establish our results more firmly, we have studied eq. (7.6) numerically. Integrating eq. (7.6) for a total time equivalent to 20 cycles of the forcing, consider the function $\Theta(\varepsilon_n)$, defined as the ratio between the variance of the exit time and its mean value. In the absence of periodic forcing, we know that $\Theta(\varepsilon_n) \simeq 1$. In fig. 6, $\Theta(\varepsilon_n)$ is plotted *vs.* selected values of ε_n. As we can

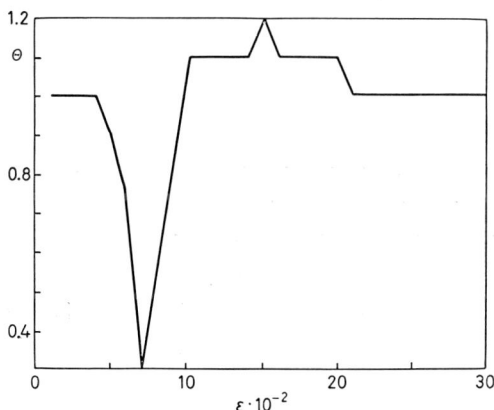

Fig. 6. – $\Theta(\varepsilon_n)$ *vs.* ε_n.

see, Θ decreases dramatically if ε_n is within the range $[\varepsilon_{n1}, \varepsilon_{n2}]$. Outside this range, ε_n is either too small or too large to have periodic exits. In fig. 7a)-c) we show the plot of ξ vs. time for three cases $\varepsilon_n^{(1)}$, $\varepsilon_n^{(2)}$ and $\varepsilon_n^{(3)}$, where $\varepsilon_n^{(1)}$, $\varepsilon_n^{(2)}$, $\varepsilon_n^{(3)}$ have been chosen in such a way that they are either at the left ($\varepsilon_n^{(1)}$), or within ($\varepsilon_n^{(2)}$) or at the right ($\varepsilon_n^{(3)}$) of the range where stochastic resonance occurs.

For x near $\partial\Omega$, $\varrho(x)$ satisfies the differential equation [19]

$$\frac{1}{2}\varepsilon^2 \frac{d^2}{d\xi^2}\varrho(\xi) + f'(P_2)\frac{d}{d\xi}\varrho(\xi) = 0, \quad \xi = \text{distance from } x \text{ to } \partial\Omega.$$

Let

$$Z = \xi/\varepsilon;$$

then

$$\frac{1}{2}\frac{d^2}{dZ^2}\varrho(Z) + f(P_2)Z\frac{d}{dZ}\varrho(Z) = 0$$

with $\varrho(0) = 0$ and $\varrho(\infty) = 1$,

$$\frac{d\varrho}{dZ} = C\int_0^Z \exp[-f'(P_2)]Z^2\,dZ,$$

$$C^{-1} = \int_0^\infty \exp[-f'(P_2)]Z^2\,dZ = \frac{1}{2}\sqrt{\frac{\pi}{f'(P_2)}} = \frac{1}{2}\sqrt{\frac{\pi}{|\Phi''(P_2)|}}.$$

The equation for the moments is

$$\frac{1}{2}\varepsilon^2 \frac{d^2}{dx^2}M_n + f'(x)\frac{d}{dx}M_n = -nM_{n-1}(x).$$

Therefore,

(A.5) $$M_n(x) = +nC_0\varrho(x)M_{n-1}(P_1),$$

where

$$M_0(x) = 1,$$

and

$$M_1(x) = C_0\varrho(x),$$

(A.6) $$M_n(x) = nC_0\varrho(x)(n-1)C_0\varrho(P_1)(n-2)C_0\varrho(P_1)\ldots =$$
$$= n!\,C_0^n\varrho(P_1)^{n-1}\varrho(x) = n!(M_1(x))^n\left[\frac{\varrho(P_1)}{\varrho(x)}\right]^{n-1}.$$

REFERENCES

[1] M. I. Budyko: *Tellus*, **21**, 611 (1969).
[2] W. D. Sellers: *J. Appl. Meteorol.*, **8**, 392 (1969).
[3] M. Ghil: this volume, p. 347.
[4] R. Benzi, G. Parisi, A. Sutera and A. Vulpiani: *Tellus*, **34**, 10 (1982).

[5] C. NICOLIS and G. NICOLIS: *Tellus*, **33**, 225 (1981).
[6] C. NICOLIS: *Tellus*, **34**, 1 (1982).
[7] A. SUTERA: *J. Atmos. Sci.*, **37**, 245 (1980).
[8] A. SUTERA: *Q. J. R. Meteorol. Soc.*, **107**, 137 (1981).
[9] M. GHIL: *J. Atmos. Sci.*, **33**, 3 (1976).
[10] C. CRAFOORD and E. KALLEN: *J. Atmos. Sci.*, **35**, 1123 (1979).
[11] K. BHATTACHARYA, M. GHIL and I. L. VULIS: *J. Atmos. Sci.*, **39**, 1747 (1982).
[12] D. H. GRIFFEL and P. G. DRAZIN: *J. Atmos. Sci.*, **38**, 2327 (1982).
[13] I. SIMMONDS and C. CHIDZEY: *J. Atmos. Sci.*, **39**, 2144 (1982).
[14] G. R. NORTH, L. HOWARD, D. POLLARD and B. WIELICKI: *J. Atmos. Sci.*, **36**, 255 (1979).
[15] K. HASSELMAN: *Tellus*, **28**, 473 (1976).
[16] G. JONA-LASINIO: this volume, p. 29.
[17] J. E. KUTZBACH and R. A. BRYSON: *J. Atmos. Sci.*, **31**, 1958 (1974).
[18] N. WAX: *Selected Topics in Noise and Stochastic Processes* (New York, N. Y., 1954).
[19] Z. SCHUSS: *Theory and Applications of Stochastic Differential Equations* (New York, N. Y., 1980).

Climatic Sensitivity and Fluctuation-Dissipation Relations.

THOMAS L. BELL

Goddard Laboratory for Atmospheric Sciences, NASA/Goddard Space Flight Center
Greenbelt, MD 20771

1. – Introduction.

Weather varies from one day to the next and its changes cannot be predicted very far into the future [1]. The *climate* of an area refers to *probabilities* of meeting with weather of different kinds there. LEITH [2] has suggested a quantitative way to talk about climate and climatic change by introducing the concept of an ensemble. He suggests constructing an ensemble of planets, each identical to the Earth in all respects believed to be important in determining the climate, but differing in the initial conditions for the atmospheres. Averages over the ensemble are assumed to give the same results as long-time averages over a simple planet. In this framework climate refers to the statistics of the ensemble.

In order to make our discussion more concrete, let us concentrate on a single variable describing the atmosphere, the globally averaged surface air temperature $T_0(t)$, which depends on time t. We shall represent ensemble averages by angular brackets $\langle T_0 \rangle$. For a stationary climate, the time-averaged temperature $\bar{T}_0$,

$$(1.1) \qquad \bar{T}_0 \equiv \frac{1}{Y} \int_0^Y dt\, T_0(t),$$

will approach the ensemble average $\langle T_0 \rangle$ for long averaging times Y. The advantage of introducing an ensemble average is that we may discuss climatic change in terms of the time dependence of the nonstationary ensemble-averaged temperature $\langle T_0(t) \rangle$. It is much more difficult to discuss climatic change when climate is defined in terms of time-averaged quantities.

We can write the equations for T_0 schematically as

$$(1.2) \qquad \frac{dT_0}{dt} = I_0(T_0, x_\alpha) + f_0,$$

where I_0 depends, in principle at least, on T_0 itself and all of the other variables x_α of the atmosphere, oceans, etc. that affect the temperature T_0 and cause it to vary in time. The variables x_α in turn have equations of the form

$$(1.3) \qquad \frac{\mathrm{d}x_\alpha}{\mathrm{d}t} = I_\alpha(T_0, x_\alpha) + f_\alpha ,$$

where again we have represented the interactions of the various variables with each other schematically.

In both eqs. (1.2) and (1.3) we have separated off a «forcing» term f that we are interested in changing to see how the system reacts. The forcing f_0 may, for instance, represent solar heating of the atmosphere, and a change Δf_0 might represent the increased heating due to an increase in the luminosity of the Sun.

Climatic sensitivity is represented by the proportionality constant M relating a climatic change to a small change in forcing:

$$(1.4) \qquad \Delta \langle T_0 \rangle = M \, \Delta f_0 .$$

We assume that climatic change is linear in small changes in the forcing, which is plausible for a stable climate. We can, of course, consider the sensitivity of T_0 to changes in forcing f_α of other variables, and write

$$(1.5) \qquad \Delta \langle T_0 \rangle = M_{0\alpha} \Delta f_\alpha .$$

Climate sensitivity $M_{\alpha\beta}$ is thus a matrix relating shifts in climatic means $\Delta \langle x_\alpha \rangle$ to changes in forcing Δf_β.

The importance of knowing the sensitivity matrix **M** to climate research is clear. There is the direct benefit that what we know about changes in the forcing Δf can be translated into its consequences for climate. The classic example of such relationships is the prediction of how much warming of the planet we can expect due to the extra opacity of the atmosphere to infra-red radiation caused by increasing carbon-dioxide concentration.

Another possible benefit lies in the interesting possibility that, if we knew the sensitivity matrix of the atmosphere, we could replace the atmosphere in coupled-ocean-atmosphere models by a «black box» constructed from the sensitivity matrix, so that we could run the model using time steps with a size characteristic of the ocean rather than having to use the much shorter time steps required to integrate properly the full atmospheric equations. Need for some such scheme is exemplified by the discussion of the coupled-ocean-atmosphere problem by DICKINSON [3].

But the sensitivity matrix **M** is not easy to obtain. Some information about it can occasionally be obtained when «natural experiments» such as volcanic eruptions occur that affect the radiative equilibrium of the atmosphere. But the climatic changes that result are not easily distinguished from the natural variability of time averages of atmospheric variables due to day-to-day weather changes. This problem is discussed by LEITH [2].

Another approach to learning about **M** is through computer modeling of the climate system. One tries to construct as good a model as possible, incorporating as many of the physical processes determining atmospheric behavior as possible, and integrates the model on a computer. But models of the atmosphere require a lot of simplification of the equations for the climate system, and it is a difficult problem to determine whether the simplifications used are adequate. To test the model, one tries to compare its behavior with the atmosphere's. But quantitative information about the climatic sensitivity of the atmosphere is rather scanty, and tests of the model tend to be limited to reproducing atmospheric behavior on time scales of a week, by comparing weather forecasts to actual weather developments, and to trying to make the model climate agree with the present climate.

Large models of the atmosphere strain the capacity of the most powerful computers, and this makes experimenting with the models difficult, especially for the long runs required to establish climatic means. Determining the sensitivity of the model to a change in forcing Δf_α requires a separate run for each α.

HALL et al. [4] have proposed using the adjoint method of sensitivity analysis in order to alleviate the computational task. In this method, a set of modified (adjoint) model equations linearized about a solution of the equations is integrated backwards in time. Certain results from functional analysis [5] allow the solution of the backward adjoint problem to be used in obtaining the sensitivity of the model to many different changes in forcing Δf_α from a single computer integration. The method will be difficult to apply to large general-circulation models of the atmosphere, but is worth further investigation.

Finally, LEITH [6] has suggested a method of determining **M** from observations of natural fluctuations of the atmosphere about the mean. He points out that the average manner in which fluctuations of the atmosphere relax back to the mean should give us some information about the dynamical responsiveness of the atmosphere, and suggests using the *fluctuation-dissipation relation* (FDR) to express this relationship quantitatively.

2. – The fluctuation-dissipation relation.

2`1. *Statement of the relation.* – Leith's paper [6] on the FDR is beautifully written and makes excellent reading, and we shall not attempt here to explore the FDR with as much rigor or in as much detail as he has done, but will only try to review the elements necessary for an understanding of the method.

Let us suppose for simplicity that we need consider only one variable of the atmosphere, T_0. In order to state the FDR, we will first need to define two functions related to the evolution of $T_0(t)$. The first is the *lagged autocorrelation* of the temperature,

$$(2.1) \qquad C(\tau) = \langle T_0'(t+\tau) T_0'(t) \rangle / \langle (T_0')^2 \rangle ,$$

where we have assumed that the statistics are stationary in time, so that the autocorrelation is a function of lag τ alone and not of t. The primes indicate variations from the climatic mean, $T_0' = T_0 - \langle T_0 \rangle$.

The second function needed is the *average response* of T_0 to an infinitesimal heat pulse

$$(2.2) \qquad \Delta f_0(t) = \varepsilon \delta(t - t_0)$$

occurring at t_0. To compute this response, consider the perturbed equation obtained by adding Δf_0 to eq. (1.2),

$$(2.3) \qquad \frac{d\tilde{T}_0}{dt} = I_0 + f_0 + \varepsilon \delta(t - t_0) ,$$

where $\tilde{T}_0$ is the solution to the perturbed equation. The solution may be written in terms of the response function $g(t; t_0)$ as

$$(2.4) \qquad \tilde{T}_0(t) = T_0(t) + \varepsilon g(t; t_0) .$$

The average response to a perturbation is obtained from the ensemble mean

$$(2.5) \qquad g(\tau) = \langle g(t; t_0) \rangle , \qquad \tau = t - t_0,$$

where the stationarity assumption has been used again.

The FDR may be stated now for our simplified, one-variable model:

$$(2.6) \qquad g(\tau) = C(\tau) \theta(\tau) ,$$

where $\theta(\tau)$ is the Heaviside step function. In words, the mean response function is identical to the lagged autocorrelation function for the system. Notice that $g(\tau)$ vanishes for negative τ because of causality.

We may express the response of the system to any small perturbation $\Delta f_0(t)$ as

$$(2.7) \qquad \Delta T_0(t) = \int_{-\infty}^{t} dt' \, \Delta f_0(t') g(t; t')$$

and the response to a constant Δf_0 as

(2.8) $$\Delta T_0(t) = \Delta f_0 \int_{-\infty}^{t} dt'\, g(t; t')\,.$$

It follows that the climatic change due to Δf_0 is

(2.9) $$\Delta \langle T_0 \rangle = \Delta f_0 \int_{0}^{\infty} d\tau\, g(\tau)\,,$$

and from eq. (1.4) we identify the sensitivity of this elementary model as

(2.10) $$M = \int_{0}^{\infty} g(\tau)\, d\tau\,.$$

If the FDR, eq. (2.6), is valid, one can obtain the climate sensitivity directly from the autocorrelation function:

(2.11) $$M = \int_{0}^{\infty} C(\tau)\, d\tau\,.$$

The climate sensitivity of a perturbed system obeying the FDR can be obtained from the correlation time of the undisturbed system. This is what makes the FDR so interesting.

2`2. *Markov processes and the* FDR. – An example of a system for which the FDR works, and which helps in analyzing some of the issues that need to be considered in using it, is a multivariate first-order Markov process. Its variables x_α satisfy the stochastic equations

(2.12) $$\dot{x}_\alpha = -\sum_\beta \Lambda_{\alpha\beta} x_\beta + f_\alpha\,,$$

where $\mathbf{\Lambda}$ is a matrix of coefficients and $f_\alpha(t)$ is a zero-mean white-noise forcing function with covariance

(2.13) $$\langle f_\alpha(t) f_\beta(t') \rangle = F_{\alpha\beta}\, \delta(t - t')\,,$$

$\mathbf{F}$ being a symmetric positive-definite matrix. The solution to eq. (2.12) may be obtained by using an integrating factor, which yields

(2.14) $$\mathbf{x}(t) = \int_{-\infty}^{t} dt'\, \exp[-\mathbf{\Lambda}(t - t')]\, \mathbf{f}(t')\,.$$

From (2.14) we immediately identify the response function

(2.15) $$\mathbf{g}(t; t') = \exp[-\mathbf{\Lambda}(t-t')]\theta(t-t').$$

Since (2.12) is linear in $\mathbf{x}$, this response function is independent of any particular solution. Therefore, the mean response function for the system is

(2.16) $$\mathbf{g}(\tau) = \exp[-\mathbf{\Lambda}\tau]\theta(\tau).$$

With a little more algebra one can obtain the lagged covariance matrix of the system

(2.17) $$U_{\alpha\beta}(\tau) \equiv \langle x_\alpha(t+\tau)x_\beta(t)\rangle,$$

which may be written

(2.18) $$\mathbf{U}(\tau) = \exp[-\mathbf{\Lambda}\tau]\mathbf{\Gamma}\theta(\tau) + \mathbf{\Gamma}\exp[\mathbf{\Lambda}^{\mathrm{T}}\tau]\theta(-\tau),$$

where $\mathbf{\Gamma}$ is the zero-lag covariance $U_{\alpha\beta}(0)$ and $\mathbf{\Lambda}^{\mathrm{T}}$ is the matrix transpose of $\mathbf{\Lambda}$. Comparing (2.16) and (2.18), we find the relation

(2.19) $$\mathbf{g}(\tau) = \mathbf{U}(\tau)\mathbf{U}^{-1}(0)\theta(\tau),$$

which is a multivariate generalization of the single-variable expression (2.6). This is the form in which LEITH [6] has expressed the FDR.

We have so far dealt with models assumed to have stationary statistics, but the atmosphere has statistics that vary, for example, with the diurnal and annual cycles. The assumption of stationary statistics can be relaxed in eq. (2.12) by allowing $\Lambda_{\alpha\beta}$ and $F_{\alpha\beta}$ to be time dependent, and a form of the FDR very similar to eq. (2.19) derived appropriate to a system with nonstationary statistics,

(2.20) $$\langle \mathbf{g}(t+\tau; t)\rangle = \mathbf{U}(\tau; t)\mathbf{U}^{-1}(0; t)\theta(\tau),$$

where the definition of $\mathbf{U}(\tau; t)$ is as in eq. (2.17). Nonstationary versions of the FDR are used in constructing statistical theories of turbulence [7].

3. – Will the FDR work?

The fluctuation-dissipation theorem, which states that the FDR, eq. (2.19), holds for a system, is known to be valid for a wide class of dynamical systems studied in statistical mechanics: namely, systems that 1) satisfy the Liouville equation, which states that an ensemble of systems moves as an incompressible gas in phase space, 2) have quadratic constants of the motion, such as energy, and 3) are in thermal equilibrium [8]. The atmosphere is unfortunately not one of these systems. It does, however, have many characteristics of systems that do obey the FDR, and so, as LEITH [6] argues, the theorem may provide serviceable estimates of the climatic sensitivity. One must keep in mind that, even if the FDR is off by a factor of 2, it may still provide useful information where none is otherwise available.

Some of the arguments that the FDR may be applicable to the atmosphere are:

1) The statistics of the atmosphere on large spatial scales and for time scales small compared with a year are probably not too far from Gaussian, if for no other reason than that the central-limit theorem implies that large-scale averages of independent smaller-scale phenomena will tend to be normally distributed. Very little has been done to test systematically the statistics of the large scales [9]. Some model results will be presented later in this section. (The Gaussianity of geopotential height on small scales has been studied by WHITE [10]. Deviations from normality are small but detectable in some regions of the northern hemisphere.)

2) The time-dependent behavior of the atmosphere on large scales can be represented by multivariate Markov processes. This is perhaps more conjecture than the result of observation, since very little research has been done yet to test it systematically. Results of a model study described later in this section are encouraging.

3) Closure models for some statistical properties of turbulence in the atmosphere con be constructed that satisfy a FDR and seem to be able to do a creditable job of representing the statistics [11].

There are indications that the FDR may serve less well for small scales. There is, first, evidence from turbulence closure models [11, 12] that on small scales the correlation times may over-estimate response times by more than a factor of 2. There is also reason to believe that from a practical point of view the FDR will be difficult to use for obtaining small-scale sensitivities because of our inability to collect enough of the necessary data. This will be discussed in more detail later.

The best evidence for the usefulness of the FDR may come from tests of the method using atmospheric models. We present a few results of such tests here, some of which are not yet completed.

With regard to how well the statistics of a model are fitted by a Markov process, we consider first a version of the 2-level Held-Suarez general-circulation model [13]. The model used here has no oceans, no moisture transport and mean annual solar heating (*i.e.* no diurnal or seasonal cycle), in order to reduce the number of variables and time scales involved in carrying out the test.

The amplitudes of large-scale temperature fluctuations are extracted from the data generated by the model, using

$$\bar{\theta}_l \equiv (2\pi)^{-1} \int d\Omega \, P_l(\sin \lambda) \bar{\theta}(\hat{r}) \,, \tag{3.1}$$

where $\bar{\theta}(\hat{r})$ is the vertically averaged potential temperature of the atmosphere at point $\hat{r}$ on the sphere, P_l is a Legendre polynomial, λ is latitude, and integration is over the surface of the sphere, $d\Omega$ being the surface element of the unit sphere. The variable $\bar{\theta}_0$ represents globally averaged temperature of the model, and $\bar{\theta}_2$ represents roughly the equator-to-pole temperature difference averaged over the two hemispheres.

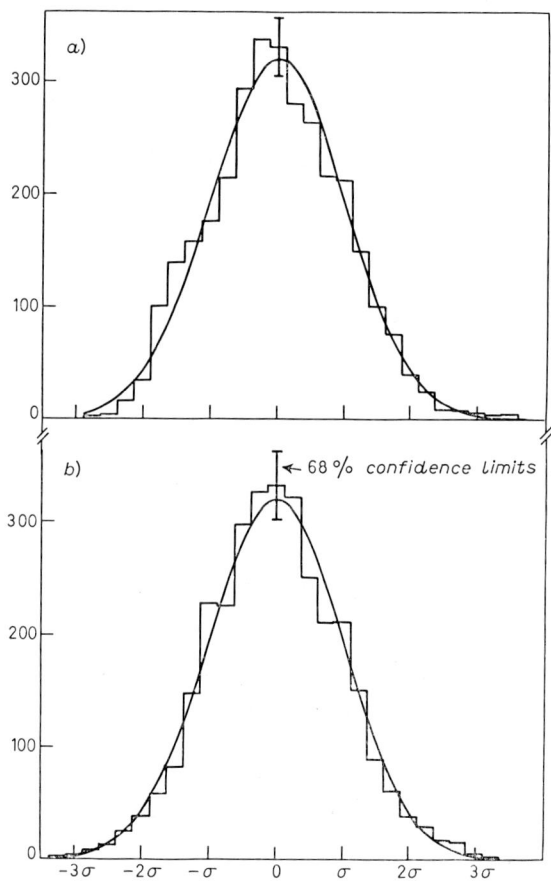

Fig. 1. – Histograms of two of the Legendre amplitudes defined in eq. (3.1) of vertically averaged potential temperature for the Held-Suarez model described in the text: Gaussian curves are fitted to the histograms. *a)* P_2, 3200 days, mean $= -13.79$, $\sigma = 0.68$; *b)* P_0, 3200 days, mean $= 53.168$, $\sigma = 0.074$.

Figure 1 shows histograms of these two variables from a 3200-day run of the model. Gaussian curves are fitted to the histograms. The variables are well approximated by Gaussian variables. The tendency to have more days at large positive deviations of $\bar{\theta}_2$ than at large negative deviations is probably a reflection of the increased stability of the atmosphere when the meridional gradient of temperature is reduced.

Figure 2 shows an attempt to represent the lagged correlation statistics of the variables $\bar{\theta}_2$ using a Markov model. The solid curve shows the autocorrelation function of the variable from data, which is, of course, only imperfectly known because it is estimated from a finite-length time series. Two Markov

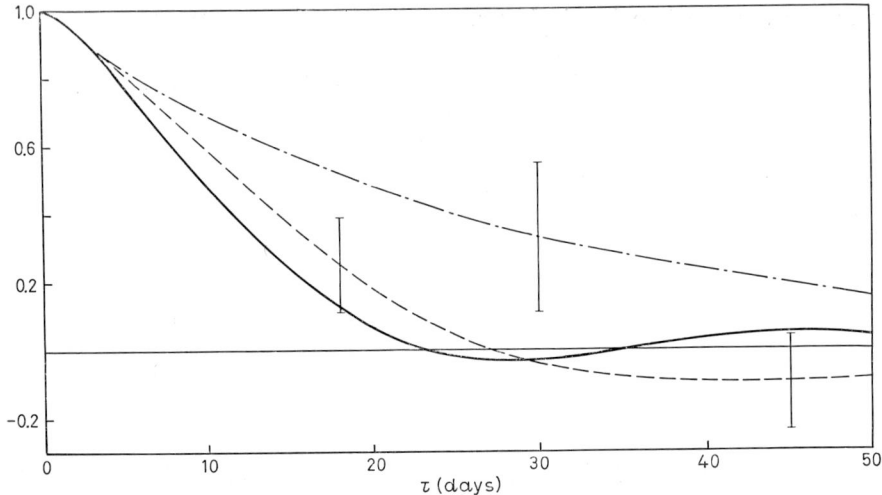

Fig. 2. – Lagged autocorrelation of the amplitude $\bar{\theta}_2$ defined in eq. (3.1) (solid curve) and autocorrelations for a first-order Markov process (2.12) fitted to the data using 2 variables $\bar{\theta}_2$ and $\bar{\theta}_4$ (dash-dotted curve) and using 7 variables $\bar{\theta}_2$, $\bar{\theta}_4$, $\bar{\theta}_6$, $\hat{\theta}_0$, $\hat{\theta}_2$, $\hat{\theta}_4$, $\hat{\theta}_6$ (dashed curve), where $\hat{\theta}$ is the vertical gradient of potential temperature in the Held-Suarez model. Error bars show 68% confidence limits.

models of the form shown in eq. (2.12) are tried, one in which only 2 variables $\bar{\theta}_2$ and $\bar{\theta}_4$ are included, and another in which 7 variables $\bar{\theta}_2$, $\bar{\theta}_4$, $\bar{\theta}_6$, $\hat{\theta}_0$, $\hat{\theta}_2$, $\hat{\theta}_4$, $\hat{\theta}_6$ are used, where $\hat{\theta}_l$ represents the Legendre amplitude of the vertical gradient of temperature. The variable $\hat{\theta}_0$ was left out of the Markov models because its fluctuations are small and so tied to the fluctuations of $\bar{\theta}_2$ that it was not useful to include it.

The 68% confidence limits are shown on the model fits. The sampling error size for the correlation function of $\bar{\theta}_2$ is similar to the 68% confidence limits of the 7-variable model fits; deviations of the correlation function from 0 for lags τ beyond 20 days are not statistically significant. The 7-variable fit, while not perfect, is probably good to 25% or so. Cross-correlations, not shown here, are similarly well fitted.

Note how much error in our knowledge of the true behavior of the correlations is present because of sampling error, even for a relatively long run of the model. When the FDR is used to estimate the response characteristics of the model from the correlation functions, the estimates will suffer the same level of uncertainty. A test of the validity of the FDR for this model is under way but not yet completed.

We shall touch briefly on a test of the FDR for a much simpler model where the FDR proves to function quite well as a predictor of the model's sensitivity even though the model was studied in a regime far from where the fluctuation-dissipation theorem can be proved. Details may be found elsewhere [14].

The model was constructed using the barotropic vorticity equation for incompressible flow on a nonrotating plane. The equation governing the vorticity $\zeta = (\nabla \times v) \cdot \hat{e}_z$, the vertical component of the curl of the velocity, is

$$(3.2) \qquad \frac{\partial \zeta}{\partial t} + (v \cdot \nabla) \zeta = F - D ,$$

where the forcing and dissipation terms F and D will be specified in a moment. If a Fourier mode expansion of vorticity is introduced,

$$(3.3) \qquad \zeta(x, t) = \sum_k \tilde{\zeta}(k, t) \exp [ik \cdot x] , \qquad k = \frac{2\pi}{L} (n_x, n_y) ,$$

after assuming spatial periodicity, the equations of motion of the amplitudes $\tilde{\zeta}(k, t)$ can be obtained using (3.2). These equations are truncated, so that only terms involving Fourier modes with $k^2 = |k|^2 = 1, 2, 4$ and 5 (with $L = 2\pi$) are allowed to be nonzero. This results in 20 coupled nonlinear equations. The forcing term F and the dissipation term D on the right-hand side of (3.2) have the form

$$(3.4) \qquad \frac{d}{dt} \tilde{\zeta}(k) = \text{nonlinear terms} + F(k) - \nu(k) \zeta(k)$$

for the Fourier amplitude equations. The study described here sets $F(k) = 0$ for all k except $F(0, 1) = 1$. The « viscosities » were given values

$$(3.5a) \qquad \nu(k^2 = 1) = \nu(k^2 = 5) = 1 ,$$

$$(3.5b) \qquad \nu(k^2 = 2) = \nu(k^2 = 4) = -0.6 .$$

The modes with $k^2 = 2, 4$ were made artificially unstable by the choice (3.5b) in order to make the model behave turbulently. This is discussed in more detail in ref. [14]. The resulting model is turbulent, but because of the forcing and dissipation terms the FDR cannot be proved for it.

The response functions and the correlation functions for two of the model variables are plotted in fig. 3. They should be equal if eq. (2.6) were satisfied. The response function is determined from an ensemble average and sampling error estimates (due to the finite size of the ensemble) are shown as 68% confidence limit error bars. The uncertainty diverges nearly exponentially with τ because of the turbulent behavior of the model. It is clear that, while the FDR is not exact, it is satisfied to within 25%.

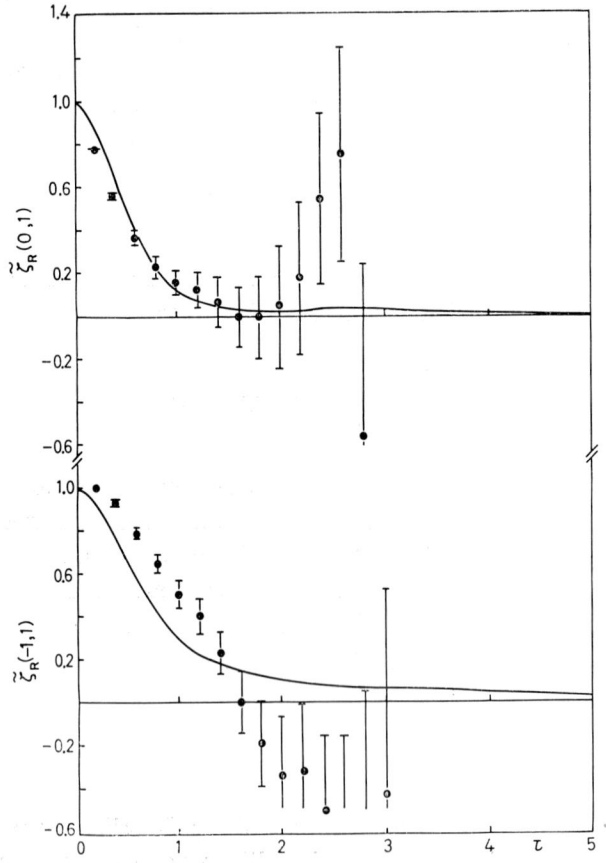

Fig. 3. – Lagged autocorrelation functions $C(\tau)$ (smooth curves) and response functions $g(\tau)$ (dots) for 2 of the 20 variables for the model described in eqs. (3.2)-(3.6). These functions are graphed for the real part of $\tilde{\zeta}(0, 1)$, defined in eq. (3.3), in the lower half of the figure, and for the real part of $\tilde{\zeta}(-1, 1)$ in the upper half. Error bars on the response function are 68% confidence limits determined from sampling errors due to the finite size of the ensemble used in estimating the ensemble averages in eq. (2.5). For further details see ref. [14].

The tests just described explored the validity of eq. (2.6) at each lag τ. To test whether the climatic sensitivity of the model is accurately predicted by eq. (2.11), the strength of $F(0, 1)$ in eq. (3.4) was increased from 1.0 to 1.5. A change

$$\Delta \langle \tilde{\zeta}(0, 1) \rangle = 0.28 \pm 0.05$$

was observed in the climatic mean of variable $\tilde{\zeta}(0, 1)$. The uncertainty in the change is due to sampling error from the finite length of the computer run used to obtain the new climatic mean of the model.

From eq. (1.4), this result corresponds to a value of the climatic sensitivity of the model

(3.6) $$M = 0.56 \pm 0.10 \, .$$

The value of the integral of $C(\tau)$ for variable $\tilde{\zeta}(0, 1)$ estimated from fig. 3 is

(3.7) $$\int_0^\infty C(\tau) \, d\tau = 0.65 \pm 0.05 \, .$$

Estimate (3.7) of the sensitivity M of the model from the FDR using eq. (2.11) agrees with the climatic sensitivity (3.6) actually observed for the variable to within the confidence limits of the estimates.

Similar tests [14] for three other variables of the model showed no statistically significant departures from equality (2.11) predicted by the FDR. The integrated form of the FDR, eq. (2.11), thus seems to be obeyed quite well by this model, even though the dissipation terms in the equations of motion drive it well away from the regime where a fluctuation-dissipation theorem can be proved.

4. – Sampling problems.

Even if we were sure the FDR is exact for the atmosphere, there are practical problems with using it due to the unavailability of data needed to obtain the covariance matrices in (2.19). There is also the obvious problem that the number of variables that may in principle be used to describe fully the atmosphere is enormous. General-circulation models of the atmosphere can integrate equations for $10^5 \div 10^6$ variables. One must find some means of limiting *a priori* the variables that need to be considered.

LEITH [6] suggests using empirical orthogonal functions (EOF's), the eigenfunctions of the covariance matrix

(4.1) $$\mathbf{U}(0) \mathbf{\psi}^{(k)} = \lambda_k \mathbf{\psi}^{(k)} \, ,$$

where $\psi^{(k)}$ is the k-th eigenfunction and λ_k is the corresponding eigenvalue, ordered so that $\lambda_1 \geqslant \lambda_2 \geqslant ... \geqslant \lambda_k \geqslant \lambda_{k+1} \geqslant ...$. EOF's are orthonormal:

(4.2) $$\psi^{(k)} \cdot \psi^{(l)} = \delta_{lm}.$$

They can be used as a basis to represent the original set of variables x as

(4.3) $$x(t) = \sum_k \xi_k(t) \psi^{(k)},$$

(4.4) $$\xi_k = \psi^{(k)} \cdot x.$$

The variance of ξ_k is λ_k.

EOF's have the useful property that a large portion of the total variance of the system $\sum_\alpha U_{\alpha\alpha}$ can be expressed in terms of a small subset of the eigenfunctions, because of the relation

(4.5) $$\sum_\alpha U_{\alpha\alpha} = \sum_k \lambda_k$$

and the empirical fact that the eigenvalues λ_k tend to diminish rapidly with increasing k. It is not uncommon to find that over 90% of the variance can be represented by the first dozen EOF's in expansion (4.3). Moreover, the EOF's in this subset tend to describe large-scale variations and have the longest time scales. They, therefore, tend to have large sensitivities, if the FDR is any guide (see eq. (2.11)).

The FDR, eq. (2.19), has a particularly simple form when expressed in terms of EOF amplitudes:

(4.6) $$\tilde{g}_{kl}(\tau) = \tilde{U}_{kl}(\tau)/\lambda_l,$$

$\tau \geqslant 0$, where the tildes denote quantities expressed in the EOF basis. As LEITH [15] points out, determining the response of EOF variable ξ_k to a perturbation of variable ξ_l using the FDR (4.6) requires knowledge of the covariance of those two variables alone. In contrast, using the FDR expressed in terms of the original variables is considerably more complicated, since it requires knowing the full covariance matrix of all of the variables in order to invert the 0-lag matrix in eq. (2.19). Model tests of the validity of the FDR would be much easier to carry out in this basis, since the tests could focus on a subset of EOF's and escape the cumbersome collection of statistics for thousands of variables.

However, one does not know the EOF's of the atmosphere exactly, owing both to the lack of data for all variables and to sampling errors due to the short time histories of atmospheric data. It may be that EOF amplitudes describing large-scale variability of the atmosphere may escape these problems. Further investigation is needed here.

The usefulness of EOF's in any practical study of climate sensitivity may depend on how well the forcing $\Delta \boldsymbol{f}$ can be expressed in terms of the $\boldsymbol{\psi}^{(k)}$. If the forcing of a single variable x_α is Δf_α, representing some extra heating near a grid point, perhaps, the EOF representation of this forcing would be

$$(4.7) \qquad \Delta \tilde{f}_k = \psi_\alpha^{(k)} \Delta f_\alpha \, ;$$

that is, the forcing amplitude in the EOF representation, $\Delta \tilde{f}_k$, is proportional to the α-th component of EOF $\boldsymbol{\psi}^{(k)}$. But it is an empirical fact that the components of EOF's tend to be about $1/\sqrt{p}$ in size, for a p-variable system, *independent of* k, and so a localized perturbation of variable x_α will tend to appear in the EOF representation evenly spread over all EOF's. The EOF representation is, therefore, more appropriate to studies of sensitivities to large-scale influences, since they are likely to be represented by just a few EOF's.

One can also argue that sensitivities to small-scale disturbances may be difficult to obtain from the FDR because of sampling problems due to insufficient amounts of data for determining the covariance matrices needed. As a simple model designed to illustrate the kinds of problems that may appear, let us suppose that the true lagged covariance matrix of a system may be written

$$(4.8) \qquad \tilde{U}_{jk}(\tau) = \lambda_j \delta_{jk} \exp\left[-|\tau|/\tau_j\right],$$

where we have expressed the statistics for the EOF amplitudes rather than for the original variables x_α. The zero-lag covariance matrix is necessarily diagonal, by definition of EOF's, and we assume that it remains diagonal for all lags τ. Different correlation times τ_j are allowed for different EOF's. It is observed in analyses of atmospheric models that eigenvalues λ_k decrease with k almost exponentially, and much faster than time scales τ_k.

Suppose now that we have a data set covering a time period $0 \leqslant t \leqslant T$ and we try to estimate the covariance function (4.8) from the data. For $j \neq k$, the estimates $\tilde{\mathbf{U}}^{(e)}$ of $\tilde{\mathbf{U}}$ will differ from zero by an amount

$$(4.9) \qquad \tilde{U}_{jk}^{(e)}(\tau) = 0 \pm (\lambda_j \lambda_k)^{\frac{1}{2}} [2 \tau_j \tau_k / ((\tau_j + \tau_k) T)]^{\frac{1}{2}}$$

for $j \neq k$ and large lag τ, a result that can be found in standard textbooks on statistics [16].

The estimate of the sensitivity matrix $\tilde{M}_{jk}$ is obtained by integrating expression (2.19) over τ, as in eq. (2.11). However, we cannot integrate very far in τ, because our data allow us to estimate covariances only for $\tau < T$, in principle, and usefully only for $\tau \ll T$. We must decide somehow where to stop the integration over τ, but wherever we stop we shall have accumulated sampling errors in the integral dictated by the size of errors in (4.9). A more sophisticated approach might be to fit the data to a Markov process and estimate $\tilde{\mathbf{M}}$ from the Markov fit. This reduces the sampling errors somewhat over the brute-force approach just described, but, of course, does not eliminate them. A laborious but straightforward calculation gives an estimate of the errors in what we obtain for $\tilde{M}_{jk}$ of order

$$(4.10) \qquad \tilde{M}_{jk}^{(e)} = \tau_j \delta_{jk} \pm (\lambda_j/\lambda_k)^{\frac{1}{2}} [\tau_j \tau_k (\tau_j + \tau_k)/T]^{\frac{1}{2}}.$$

Sampling errors generate off-diagonal elements in the sensitivity matrix!

Suppose now that we wanted the sensitivity of EOF $\boldsymbol{\psi}^{(j)}$ to a perturbation $\Delta \tilde{f}_k$, $j \ll k$, for which $\lambda_j \gg \lambda_k$ and $\tau_j \gg \tau_k$. Result (4.10) becomes approximately

$$(4.11) \qquad \tilde{M}_{jk}^{(e)} \approx \pm \tau_j [(\lambda_j/\lambda_k)(\tau_k/T)]^{\frac{1}{2}}.$$

But, as mentioned earlier, the ratio τ_k/λ_k increases rapidly with k for the atmosphere. Consequently, if we try to estimate the sensitivity of EOF $\boldsymbol{\psi}^{(j)}$ to a *localized perturbation*, for which $\Delta \tilde{f}_j$ tends to be of the same order of magnitude as $\Delta \tilde{f}_k$ (as we mentioned at the beginning of this section), the sampling noise from the contribution $\tilde{M}_{jk} \Delta \tilde{f}_k$ will tend to swamp the *true* contribution $\tilde{M}_{jj} \Delta \tilde{f}_j = \tau_j \Delta \tilde{f}_j$ by a factor $[(\lambda_j/\lambda_k)(\tau_k/T)]^{\frac{1}{2}}$ unless T is enormous. This argument is by no means rigorous, but it explains sampling problems encountered in trying to obtain statistically stable results for the Held-Suarez model study.

5. – Additional remarks.

The FDR may give us access to the climate sensitivity of the atmosphere without having to construct elaborate climate models, although confidence in the accuracy of the FDR will probably require testing how well it works with models.

LEITH [15] has suggested two areas where the FDR may also be useful for studying and improving models of the atmosphere. The first is more in the realm of weather forecasting than in climate modeling. Suppose, as is likely, that the climatic mean of a forecast model differs from the true climate. Every time atmospheric data are used to initialize the forecast model, part of the time evolution of the models will consist in a drift from the climatic mean of the atmospheric data to the climatic mean preferred by the model. This climatic drift will generate a systematic bias in the model forecast. The bias could be removed if we knew what it was. The mean response function $\mathbf{g}(\tau)$ represents exactly the information we need to remove the bias, and, if the FDR is valid, can be obtained from the covariance statistics of the model.

The second area where the FDR can help is in suggesting tests of climate models to probe how well the model's climatic sensitivity agrees with the real atmosphere's. Even if the FDR were to prove an inaccurate gauge of climatic sensitivity (and there is no reason as yet to suppose this), the FDR suggests that the failure of a model to generate values of $\mathbf{U}(\tau)\mathbf{U}^{-1}(0)$ that agree with values derived from atmospheric data is good cause for concern about the ability of the model to estimate climatic sensitivity well.

6. – Conclusion.

We have discussed how the fluctuation-dissipation relation might be useful to the study of climate dynamics and reviewed some of the reasons for believing it might be an accurate guide to climatic sensitivity of the atmosphere, at least on large scales. Much hard work remains to be done on solving the sampling problems associated with making actual estimates of climate sensitivity from real data, but is justified by the unique contribution such estimates would represent to research on how and why the climate changes.

* * *

I wish to thank C. E. LEITH for his very useful comments during the lectures, M. GHIL for many suggested improvements of the manuscript and all the organizers of the course and the Società Italiana di Fisica for assembling a very stimulating group of students and lecturers.

REFERENCES

[1] E. N. LORENZ: this volume, p. 243.
[2] C. E. LEITH: J. Appl. Meteorol., **12**, 1066 (1973).
[3] R. E. DICKINSON: J. Atmos. Sci., **38**, 2112 (1981).
[4] M. C. G. HALL, D. G. CACUCI and M. E. SCHLESINGER: J. Atmos. Sci., **39**, 2038 (1982).
[5] D. G. CACUCI: J. Math. Phys. (N.Y.), **22**, 2794 (1981).
[6] C. E. LEITH: J. Atmos. Sci., **32**, 2022 (1975).
[7] S. A. ORSZAG: J. Fluid Mech., **41**, 363 (1970); G. F. CARNEVALE and J. S. FREDERIKSEN: J. Fluid Mech., **131**, 289 (1983).
[8] See, for example, R. H. KRAICHNAN: Phys. Rev., **113**, 1181 (1959), and references within.
[9] Normality of large-scale rainfall statistics over India is found by B. PARTHASARATHY and D. A. MOOLEY: Mon. Weather Rev., **106**, 771 (1978). Here the central-limit theorem is almost certainly playing a role.
[10] G. H. WHITE: Mon. Weather Rev., **108**, 1446 (1980).

[11] J. R. HERRING and R. H. KRAICHNAN: in *Lecture Notes in Physics*, Vol. **12**: *Statistical Models and Turbulence*, edited by M. ROSENBLATT and C. VAN ATTA (Berlin, 1972), p. 148.
[12] J. R. HERRING: *J. Atmos. Sci.*, **34**, 1731 (1977).
[13] I. M. HELD and M. J. SUAREZ: *J. Atmos. Sci.*, **35**, 206 (1978).
[14] T. L. BELL: *J. Atmos. Sci.*, **37**, 1700 (1980).
[15] C. E. LEITH: personal communication.
[16] See, for example, G. M. JENKINS and D. G. WATTS: *Spectral Analysis and Its Applications* (San Francisco, Cal., 1968), p. 336.

INDICE ANALITICO

Ablation zone
374
Accumulation-ablation rate
376, 377
Accumulation zone
374
Action functional
31
Advective accelerations
305
Ageostrophic motion
160, 161
Albedo
349, 351, 374, 406
Alpha effect
214
Amplification of vorticity gradients
58
— rate of small-scale disturbances, 51
Amplitude amplification
165
Analog methods
343
Analogues
250, 254, 260, 261
Angular-momentum mixing
179
Annulus
162
Anomalies, persistence of
325
Anticascade
179, 180
Anticipated vorticity
268
Astronomical computations
381

Asymptotic infra-red freedom
81
Attractors
5, 72
Available potential energy
147
Average, time, space, ensemble
292, 293
— occasional, 293
— response, 427
Averaging scale
292
Axis of rotation of Earth's orbit
379

Balance equation
297
— winds, 297
Baroclinic deformation scale
229
— instability, 150, 230, 307
Barotropic instability
234
— stability equation, 235
— vorticity equation, 156, 433
Batchelor spectrum
104
Benthic foraminifera
369
Beta-model
85
Blocking
303, 329
Boundary layer thickness
48
Broken symmetry
76

Buoyancy lenght scale
  95
Busse diagram
  18

Celestial mechanics
  379
Chandrasekhar number
  19
Channel flow
  182
Chaotic
  5, 45, 49, 72, 121
— behaviour, 3, 69
— flows, 121, 268
— solutions, 79
— streamlines, 73
Characteristic acceleration in two-dimensional turbulence
  66
— exponent, 5, 14, 50, 51, 63, 64, 72, 251, 252
— — spectral distribution, 64
— modes, 292
Chebyshev polynomial expansion
  113
Climate branch
  365
— change, 347
— deep freeze state, 354
— drift, 327
— dynamics, 347, 403
— equilibria, 353, 407
— mechanism, 347
— noise, 324
— present state, 354
— records, 366
— sensitivity, 425, 435
— — to insolation radiation, 366
— slow components, 403
— variability, 366, 420, 424
Clouds
  349, 407
Coherent structure
  50, 51, 68, 72, 83, 268
Combination tones
  348, 388
Complex time singularities
  76
Comprehensive general circulation models
  326

Computational complexity
  112
— cost, 111
Conservative systems
  72
Continental drift
  198
Continuous spectral density
  388
Convection in Earth's mantle
  193, 198
Convective turbulence
  57
Correlation separation function
  297, 434
Cox number
  104

d'Alembert series
  381
Data analysis
  295
Decay of predictability
  399
Deterministic noise
  3
— transition to aperiodic chaotic behaviour, 390
Detrainment
  390
Differential dynamical systems
  4
Diffusion coefficient
  73
Dissipated integral
  56, 59, 267
Dissipation of enstrophy
  66
— range, 75
— rate, 50
— structure, 50
Dissipative modes
  307
Diurnal modulation
  307
Divergence, twisting terms
  308
Double diffusive process
  90
Dynamical systems
  4, 20

Dynamic renormalization group
  126
Dynamo theory
  200

Earth's atmosphere chemical composition
  405
— mantle, 193, 198, 201
— magnetic field, 201
— — secular variation field, 201
— thermal regime, 404
Eccentricity
  379, 415
Ecliptic
  379
Eddy diffusive heat flux
  360
— enstrophy equation, 54, 57
— strain rate, 60
— viscosity, 74, 293
— vorticity, 60
Ekman layer
  180, 183
El Niño southern oscillation
  332
Elliptic orbit
  379
Empirical orthogonal functions
  435
Energy balance models
  348
— — — quasi-equilibrium response, 371
— — — 1-D formulation, 358
— cascade, 55, 89, 173, 226, 227
— cascading range, 59, 141
— spectra, 103
— transport, 176, 351
Enstrophy
  47, 53, 82
— amplification, 55
— budget, 54
— cascading range, 59, 65, 68, 84, 141, 226, 227, 266
— conservation, 53
— equation, 53
Entrainment
  10
— and combination tones, 388
Entropy
  15, 46, 49, 50, 138
Ephemerides
  380

Equilibrium measure
  31
— range, 81
— temperatures, 361
Error growth
  45, 50, 64, 243-265, 282, 288
— — in GCM, 316
— doubling time, 252, 259, 316
— propagation, 67
— variance, 298, 319
Ertel's potential vorticity
  153
Exit time
  31, 34, 413

Farey fractions
  22
Finite differences
  109
— element, 109
Flip bifurcation
  8
Fluctuation-dissipation relation
  426, 429, 433
— — nonstationary version, 429
Fourier-Bessel expansion
  109
— transform, 299
Fractal dimension
  85
— set, 78
Free oscillations in climate models
  386
Frequency errors
  394
— locking, 10, 20, 388
— spectrum, 304
Frictional dissipation
  295
Frontogenesis
  152, 284
Fully developed turbulence
  76, 85, 86

General circulation
  45
Genericity
  4
Geochemical proxy data
  368

Geodynamical models
  377
Geodynamics
  372
Geological record
  368
Geostrophic turbulence
  47, 227
Gibbs' phenomenon
  109
Glaciation cycle
  371, 403
— — sawtoothlike shape, 370
Global climate
  347
— circulation models, 256, 260, 311, 430
— energy balance models, 352
— kinetic-energy cycle, 307
— radiation balance, 348
— temperature, 372
Gortler vortices
  182
Great Red Spot
  170, 229, 234
— — — vertical structure, 235
Greenhouse effect
  353, 405

Hadley overturning
  231
Hausdorff dimension
  85, 86
Heat fluxes
  352
— transfer, radiative, conductive, convective, 356
— transport, 356
Hele-Shaw cell
  193
Helicity
  74
Henon attractor
  11
Homogeneous turbulence
  173
Hopf bifurcation
  9, 26
Hydrological cycle
  348, 376
— forcing, 392
Hydrodynamics turbulence
  3

Hydromagnetic dynamo problem
  205
— equations, 202
— theory, 203
Hyperbolic distribution
  286

Ice ages
  354
— albedo feedback, 353
— cores, 376
— load, 348, 375
— sheet, 348, 372
— volume terminations, 370, 392
Inertial range spectrum
  60, 75, 81, 150, 151, 294
— wave, 173, 174, 176
Infinite-dimensional dynamical systems
  37
— gradient systems, 38
Information loss and production
  50
Infra-red radiation
  374, 405
Insolation
  348
— calculations, 383
— changes, 379
Intermittency
  12, 22, 75, 76, 79, 82, 143, 281, 285
Intermittent scenario
  12, 23
Internal stability of climate equilibria
  356, 362
— waves, 89, 97
Invariant measure
  34, 35
Irreversible thermodynamics
  49
Isostatic adjustment of bedrock to ice load
  377
— rebound, 348
Isotopic abundance ratio
  369

Jet stream
  168, 169
Jovistrophic turbulence
  229, 230

Jupiter's atmosphere
169
— general circulation, 231

KAM theory
381
Karman constant
57, 122
Kelvin-Helmholtz instability
182, 185
Kinematic dynamo problem
204
Kinetic-energy equation
48
Kolmogorov scale
56, 98, 103
Korteweg-de Vries equation
268
Kramer's formula
413

Laboratory experiments in rotating fluids
159, 172
— measurements in stratified flows, 90
Lagged averaged forecast
328
— correlations, 397, 427
Lagrangian tracers
227
— turbulence, 73
Laminar windows
74
Langevin equation
411
Large eddies
181
— eddy simulation, 126
— roller eddies, 184
— scale, 46
— vortical flow structure, 121
Latitude-dependent climate models
348
Legendre transform
86
Linear statistical prediction
343
Lithosphere
350
— elasticity, 377, 390

— visco-elastic response, 372
Local cascade
61
Localized initial errors
257
Logarithmic profile
57, 122
Lognormal distribution
86
Long-range weather prediction
322, 338
Long-wave radiation
348
Lorenz attractor
11
— model in dynamo theory, 207
Low-order finite-difference approximation
115
Lyapounov characteristic exponent, see characteristic exponent
— functional, 268, 279, 408

Magnetic reversal
207, 369
Marine and continental records
392
Markov chain
35
— process, 428, 438
— — fit to GCM statistics, 432
Mass averaged temperature
405
Measure of predictability
397
Meridional heat fluxes
358
Mesoscale
46
Mixing hypothesis
148
Modon
69, 270, 271
Momentum conservation equation
291
Monthly forecast
342
Multifractal
85, 86, 87
Multiple equilibria
348
Multiplicative forcing
390

— ergodic theorem, 14
Multivariate Markov process
    430

Natural variability
    324
Navier-Stokes equations
    47
N-body problem
    380
Newtonian cooling
    359
NMC grid
    300
Noise (deterministic)
    3
Nonlinear normal mode initialization
    320
— processes in shear flow, 117
— resonance, 386
— response to periodic forcing, 388
Nonlocal cascade
    68, 83
Normal modes on the sphere
    301
Numerical studies of transition to turbulence
    117
— turbulence, 122

Obliquity
    379
Ocean atmosphere models
    425
— microstructure, 90
— turbulence, 91
Oceanic albedo
    375
— eddies, 358
Onset of chaos
    118
— of convection, 18
— of turbulence, 3
Orbital changes
    378, 381
— elements, 379
— forcing, 415
— parameters of the Earth, 367
— variations, 385
Ornstein-Uhlenbeck process
    412

Orr-Sommerfeld equation
    118
Oscillation mechanism in climate models
    386
Oxygen heavy and light isotope
    369
Ozmidov scale
    95, 103

Paleoclimatic evidence of glaciations
    367
Parallel shear flows
    118
Parametric forcing in climate models
    385
Pattern analysis
    302
Period-doubling cascade
    13, 24
Phase errors
    394
— locking, 388
— space torus, 380
Piston cores
    4, 369
Pitchfork bifurcation
    8, 24
Planetary system, motion
    381
Poincaré map
    4
Pole wandering
    202
Potential vorticity conservation
    52
— absolute vorticity, 154
— predictability, 324
Prandtl number
    18, 193
Precession angle
    379
— effect, 384
Precipitation-temperature feedback
    376
Predictability
    45, 49, 142, 243-265
— of climate, 348, 395
— of monthly means, 323
— of time averages, 322
— range of, 67
— upper limit, 316

Proxy indicator of global ice volume
369
Pseudospectral transform method
113

Quasi-geostrophic approximation
152
Quasi-periodic insolation forcing
348
— phase-space flow, 379, 380
— scenario, 13, 26
Quaternary glaciation cycles
366
— period, 368
— proxy data, 348

Radiation absorbed, emitted
350, 352, 404, 405
— budget, 347, 351, 415
— outgoing, 353, 404, 405
Radiative heating
405
Radius of deformation
146
Rankine vortex
287
Red noise background
371
Regularity times
142
Renormalization group methods
81, 218
Resonance between frequencies
381
Resonant wave interactions
229
Response function
427, 434
Reversibility
49
Reynolds averaging
54
— equations, 293
— magnetic number, 213
— number, 47, 71, 86, 193
— stress, 182, 186, 187
Rossby number
153, 181, 183
— wave energetics, 234
Rotating fluids
159, 172

Rotation destabilizing effect
181
— stabilizing effect, 173, 181
Rugged integral
59, 267

Saddle node bifurcation
7, 22
Sampling errors
433, 438
— problems, 435
Scale analysis
295
— interactions, 305
Scales of motion
291
Scaling transformations
76
Sea ice
376
— level temperature, 358
— surface temperature anomalies, 324
Seasonal forecast
342
Selective decay hypothesis
269
Self-similarity
77
Self-sustained oscillation in climate models
386
Sensitive dependence on initial conditions
5, 10, 49, 51, 68
— matrix, 425
Sensitivity analysis, adjoint method
426
Shear layers free
182, 190
Shell model
78, 79
Short-wave radiation
348
Simplified dynamical models for the atmosphere
326
Singularities in the Navier-Stokes equations
84, 86, 87
Slantwise convection
160

Smallest eddies life span
    61
Small-scale predictability
    281
— turbulence, 46, 285
Small stochastic perturbations
    29, 410
Smoluchowsky approximation
    38
Snow, accumulation ablation rate
    374
Snowline
    361, 374, 377
Solar constant
    349
Solitary Rossby wave
    226, 234, 235
Soliton
    177
Spatially decaying homogeneous turbulence
    95
Spectral expansion
    109
— flux of energy, 55, 56
— — of enstrophy, 58, 59, 61, 65
— gap, 80, 83, 264
— methods, 107
Spin down
    180
Stability
    138
— neutral, linear, 363
Stabilization problem
    37
Stable invariant measure
    37
Statistical equilibria
    136, 148
— forecast, 342
— properties of turbulence in two and three dimensions, 46
Stewartson layers
    189
Stochastic behaviour
    193
— differential equations, 30, 415, 428
— dynamic prediction, 328
— forcing in climate models, 393
— process, 29, 410
— resonance, 417

Strain rate
    48, 64
Strange attractor
    3, 6, 50, 51, 69
Stratigraphic methods
    368
Structural stability in climate models
    355, 364
Structure functions
    77, 79, 84, 85, 86
Sturm-Liouville problems
    110
Subgrid scale closures
    118
— — enstrophy, 54
— — flux, 63
— — — of momentum, 67
— — parametrization, 63
— — turbulent closures, 122
— — variance, 301
Subsynoptic scales spectrum
    283
Surface albedo
    351
Suspension
    6
Synoptic scales
    291
Systematic errors
    320, 327

Tangent plane
    299
Taylor-Green vortex
    122, 124
Taylor transformation
    303
Teleconnections
    329, 330
Temperature equation
    53
— spectrum, 53
Terrestrial planets
    198
Theory of large deviations
    29
Thermal boundary layer
    193, 194
— convection, 17, 159
— — in rotating fluids, 159
Thermals
    194, 195

Thermodynamic efficiency
  307
Thorpe scale
  99
Three dimensionality of transition
  118
Time-one map
  4
Tollmien-Schlichting instability
  182
Transitional instability
  118, 119
Transition probability
  40
Transport
  73, 174
Truncated spectral dissociation
  294
Turbulence
  45
— buoyancy influenced, 89
— closure models, 430
— decay, 173
— energy production, 182
— front, 174, 176
— grid, 173
— homogeneous, 173
— in rotating fluids, 159, 172
— three-dimensional, 54
— two-dimensional, 65, 179, 191, 226, 266
— — decaying, 180, 267
— — experiments, 180
Turbulent channel flow
  122
— diffusion coefficients, 73
— spot, 124
Turn-over time
  64, 141, 282

Variability, internal, external
  322, 393
Variance spectrum
  294, 305

Variation of constants
  380
Velocity vector
  291
— — gradient tensor, 292
— spectra, measured, 103
Viscosity
  47
Viscous boundary layer
  47
— convective range, 53
— dissipation, 48
— response of the upper mantle, 375
Vortex core
  177
— formation, 179
— stretching, 45, 46, 52
Vortices
  144, 177, 271, 272
— longitudinal, 182
Vorticity dynamics of two- and three-dimensional flow
  46
— equation, 52, 64
— expulsion, 179, 180

Weak turbulence
  17
Weather noise
  393
White noise
  397, 428
— Ovals, 171, 227, 234
Wiener process
  30, 411
Wind tunnels
  173

Zero-dimensional climate models
  356
Zonally symmetric stress
  231
Zonal mean wind
  292

PROCEEDINGS OF THE INTERNATIONAL SCHOOL OF PHYSICS
« ENRICO FERMI »

Course I
Questioni relative alla rivelazione delle particelle elementari, con particolare riguardo alla radiazione cosmica
edited by G. Puppi

Course II
Questioni relative alla rivelazione delle particelle elementari, e alle loro interazioni con particolare riguardo alle particelle artificialmente prodotte ed accelerate
edited by G. Puppi

Course III
Questioni di struttura nucleare e dei processi nucleari alle basse energie
edited by C. Salvetti

Course IV
Proprietà magnetiche della materia
edited by L. Giulotto

Course V
Fisica dello stato solido
edited by F. Fumi

Course VI
Fisica del plasma e applicazioni astrofisiche
edited by G. Righini

Course VII
Teoria della informazione
edited by E. R. Caianiello

Course VIII
Problemi matematici della teoria quantistica delle particelle e dei campi
edited by A. Borsellino

Course IX
Fisica dei pioni
edited by B. Touschek

Course X
Thermodynamics of Irreversible Processes
edited by S. R. de Groot

Course XI
Weak Interactions
edited by L. A. Radicati

Course XII
Solar Radioastronomy
edited by G. Righini

Course XIII
Physics of Plasma: Experiments and Techniques
edited by H. Alfvén

Course XIV
Ergodic Theories
edited by P. Caldirola

Course XV
Nuclear Spectroscopy
edited by G. Racah

Course XVI
Physicomathematical Aspects of Biology
edited by N. Rashevsky

Course XVII
Topics of Radiofrequency Spectroscopy
edited by A. Gozzini

Course XVIII
Physics of Solids (Radiation Damage in Solids)
edited by D. S. Billington

Course XIX
Cosmic Rays, Solar Particles and Space Research
edited by B. Peters

Course XX
Evidence for Gravitational Theories
edited by C. Møller

Course XXI
Liquid Helium
edited by G. Careri

Course XXII
Semiconductors
edited by R. A. Smith

Course XXIII
Nuclear Physics
edited by V. F. Weisskopf

Course XXIV
Space Exploration and the Solar System
edited by B. ROSSI

Course XXV
Advanced Plasma Theory
edited by M. N. ROSENBLUTH

Course XXVI
Selected Topics on Elementary Particle Physics
edited by M. CONVERSI

Course XXVII
Dispersion and Absorption of Sound by Molecular Processes
edited by D. SETTE

Course XXVIII
Star Evolution
edited by L. GRATTON

Course XXIX
Dispersion Relations and Their Connection with Causality
edited by E. P. WIGNER

Course XXX
Radiation Dosimetry
edited by F. W. SPIERS and G. W. REED

Course XXXI
Quantum Electronics and Coherent Light
edited by C. H. TOWNES and P. A. MILES

Course XXXII
Weak Interactions and High-Energy Neutrino Physics
edited by T. D. LEE

Course XXXIII
Strong Interactions
edited by L. W. ALVAREZ

Course XXXIV
The Optical Properties of Solids
edited by J. TAUC

Course XXXV
High-Energy Astrophysics
edited by L. GRATTON

Course XXXVI
Many-Body Description of Nuclear Structure and Reactions
edited by C. BLOCH

Course XXXVII
Theory of Magnetism in Transition Metals
edited by W. MARSHALL

Course XXXVIII
Interaction of High-Energy Particles with Nuclei
edited by T. E. O. ERICSON

Course XXXIX
Plasma Astrophysics
edited by P. A. STURROCK

Course XL
Nuclear Structure and Nuclear Reactions
edited by M. JEAN and R. A. RICCI

Course XLI
Selected Topics in Particle Physics
edited by J. STEINBERGER

Course XLII
Quantum Optics
edited by R. J. GLAUBER

Course XLIII
Processing of Optical Data by Organisms and by Machines
edited by W. REICHARDT

Course XLIV
Molecular Beams and Reaction Kinetics
edited by CH. SCHLIER

Course XLV
Local Quantum Theory
edited by R. JOST

Course XLVI
Physics with Storage Rings
edited by B. TOUSCHEK

Course XLVII
General Relativity and Cosmology
edited by R. K. SACHS

Course XLVIII
Physics of High Energy Density
edited by P. CALDIROLA and H. KNOEPFEL

Course IL
Foundations of Quantum Mechanics
edited by B. D'ESPAGNAT

Course L
Mantle and Core in Planetary Physics
edited by J. COULOMB and M. CAPUTO

Course LI
Critical Phenomena
edited by M. S. GREEN

Course LII
Atomic Structure and Properties of Solids
edited by E. BURSTEIN

Course LIII
Developments and Borderlines of Nuclear Physics
edited by H. MORINAGA

Course LIV
Developments in High-Energy Physics
edited by R. R. GATTO

Course LV
Lattice Dynamics and Intermolecular Forces
edited by S. CALIFANO

Course LVI
Experimental Gravitation
edited by B. BERTOTTI

Course LVII
History of 20th Century Physics
edited by C. WEINER

Course LVIII
Dynamic Aspects of Surface Physics
edited by F. O. GOODMAN

Course LIX
Local Properties at Phase Transitions
edited by K. A. MÜLLER and A. RIGAMONTI

Course LX
C*-Algebras and their Applications to Statistical Mechanics and Quantum Field Theory
edited by D. KASTLER

Course LXI
Atomic Structure and Mechanical Properties of Metals
edited by G. CAGLIOTI

Course LXII
Nuclear Spectroscopy and Nuclear Reactions with Heavy Ions
edited by H. FARAGGI and R. A. RICCI

Course LXIII
New Directions in Physical Acoustics
edited by D. SETTE

Course LXIV
Nonlinear Spectroscopy
edited by N. BLOEMBERGEN

Course LXV
Physics and Astrophysics of Neutron Stars and Black Holes
edited by R. GIACCONI and R. RUFFINI

Course LXVI
Health and Medical Physics
edited by J. BAARLI

Course LXVII
Isolated Gravitating Systems in General Relativity
edited by J. EHLERS

Course LXVIII
Metrology and Fundamental Constants
edited by A. FERRO MILONE, P. GIACOMO and S. LESCHIUTTA

Course LXIX
Elementary Modes of Excitation in Nuclei
edited by A. BOHR and R. A. BROGLIA

Course LXX
Physics of Magnetic Garnets
edited by A. PAOLETTI

Course LXXI
Weak Interactions
edited by M. BALDO CEOLIN

Course LXXII
Problems in the Foundations of Physics
edited by G. TORALDO DI FRANCIA

Course LXXIII
Early Solar System Processes and the Present Solar System
edited by D. LAL

Course LXXIV
Development of High-Power Lasers and their Applications
edited by C. PELLEGRINI

Course LXXV
Intermolecular Spectroscopy and Dynamical Properties of Dense Systems
edited by J. VAN KRANENDONK

Course LXXVI
Medical Physics
edited by J. R. GREENING

Course LXXVII
Nuclear Structure and Heavy-Ion Collisions
edited by R. A. BROGLIA, R. A. RICCI and C. H. DASSO

Course LXXVIII
Physics of the Earth's Interior
edited by A. M. DZIEWONSKI and E. BOSCHI

Course LXXIX
From Nuclei to Particles
edited by A. MOLINARI

Course LXXX
Topics in Ocean Physics
edited by A. R. OSBORNE and P. MALANOTTE RIZZOLI

Course LXXXI
Theory of Fundamental Interactions
edited by G. COSTA and R. R. GATTO

Course LXXXII
Mechanical and Thermal Behaviour of Metallic Materials
edited by G. CAGLIOTI and A. FERRO MILONE

Course LXXXIII
Positrons in Solids
edited by W. BRANDT and A. DUPASQUIER

Course LXXXIV
Data Acquisition in High-Energy Physics
edited by G. BOLOGNA and M. VINCELLI

Course LXXXV
Earhquakes: Observation, Theory and Interpretation
edited by H. KANAMORI and E. BOSCHI

Course LXXXVI
Gamow Cosmology
edited by F. MELCHIORRI and R. RUFFINI

Course LXXXVII
Nuclear Structure and Heavy-Ion Dynamics
edited by L. MORETTO and R. A. RICCI

TIPOGRAFIA COMPOSITORI - BOLOGNA

RAYMOND H. FOGLER LIBRARY
DATE DUE

BOOKS ARE SUBJECT TO